Chemistry and Pharmacology of Drug Discovery

Chemistry and Pharmacology of Drug Discovery

Edited by

Jie Jack Li

WILEY

Published by John Wiley & Sons, Inc., Hoboken, New Jersey.
Published simultaneously in Canada.

For general information on our other products and services or for technical support, please contact our Customer Care Department within the United States at (800) 762-2974, outside the United States at (317) 572-3993 or fax (317) 572-4002.

Wiley also publishes its books in a variety of electronic formats. Some content that appears in print may not be available in electronic formats. For more information about Wiley products, visit our web site at www.wiley.com.

Library of Congress Cataloging-in-Publication Data applied for
Hardback ISBN: 9781394225125

Cover Design: Wiley
Cover Image: Courtesy of Faridoon

Contents

Preface

Our first five installments *Wiley's Drug Synthesis Series*, *Contemporary Drug Synthesis, The Art of Drug Synthesis, Modern Drug Synthesis, Innovative Drug Synthesis,* and *Current Drug Synthesis* were published in 2004, 2007, 2010, 2015, and 2022, respectively. They have been warmly received by the drug discovery community. The current title, *Chemistry and Pharmacology of Drug Discovery*, is our sixth installment of this series.

This book has five sections, reviewing a total of 17 drugs. Section I, "Drugs Treating Infectious Diseases," covers six drugs; Section II, "Oncology Drugs," reviews four drugs; Section III, "CNS Drugs," covers four drugs; Section IV Anti-inflammatory Drugs, reviews only one drug; and Section V, "Miscellaneous Drugs," covers two additional drugs.

Each chapter is divided into seven sections as before:

1. Background
2. Pharmacology
3. Structure–activity relationship
4. Pharmacokinetics and drug metabolism
5. Efficacy and safety
6. Syntheses
7. Summary
8. References

I am very much indebted to all contributing authors from both industry and academia. Many of them are veterans and well-known experts in medicinal chemistry. Some of them discovered the drugs that they reviewed. As a consequence, their work tremendously elevated the quality of this book as a teaching tool.

Meanwhile, I welcome your critique and suggestions so we can make this *Wiley's Drug Synthesis Series* even more useful to the drug discovery/development community.

Jie Jack Li
Ann Arbor, Michigan
February 1, 2024

Contributing Authors

Prof. Timothy A. Cernak
Department of Chemistry
University of Michigan
500 S State St.
Ann Arbor, MI 48109, USA

Dr. Dao-Qian Chen
STA, 90 Delin Road
Pudong New District, Shanghai 200131,
P. R. China

Prof. Ke Ding
State Key Laboratory of Chemical
Biology
Shanghai Institute of Organic Chemistry,
Chinese Academy of Sciences
345 Fenglin Road
Shanghai 200032, P. R. China

Dr. Faridoon
Genhouse Bio
Floor 4, Building No.8, No.1 Xinze
Road, SIP, Suzhou 215000, P. R. China

Prof. Timothy J. Hagen
Department of Chemistry and
Biochemistry
Northern Illinois University
Faraday Hall
DeKalb, IL 60115, USA

Charles L. Lail III
Department of Chemistry and
Biochemistry
Northern Illinois University
Faraday Hall
DeKalb, IL 60115, USA

Dr. Jie Jack Li
Genhouse Bio
Floor 4, Building No.8, No.1 Xinze
Road, SIP, Suzhou 215000, P. R. China

Dr. Xiang Li
Beijing Kawin Technology
5 Rongjing E. St
BDA, Beijing 100176, P. R. China

Dr. Guanglin Luo
Discovery Chemistry
Bristol-Myers Squibb Co.
3551 Lawrenceville Road
Lawrence Township, NJ 08648, USA

Dr. Daljit Matharu
Medicinal Chemistry
Sanofi
350 Water Street
Cambridge, MA 02141, USA

Andrew Outlaw
Department of Chemistry
University of Michigan
500 S State St
Ann Arbor, MI 48109, USA

Dr. Yan Wang
ChemPartner
280 Utah Avenue, Suite 100
South San Francisco, CA 94080, USA

Dr. Tao Wang
Beijing Kawin Technology
5 Rongjing E. St
BDA, Beijing 100176, P. R. China

Dr. Dexi Yang
QuantX Biosciences
214 Carnegie Center, Suite 108
Princeton, NJ 08540, USA

Dr. Shaohui Yu
NuChem Sciences
2350 Cohen Street, Suite 201
Ville St-Laurent, QC
H4R 2N6 Canada

Yuqi Lavender Zha
Sanegene Bio
Room 301, Building 2, Zone B, Phase III
of BioBAY, No.99 Jingu Road, SIP,
Suzhou 215000, P. R. China

Dr. Guiping Zhang
Genhouse Bio
Floor 4, Building No.8, No.1 Xinze
Road, SIP, Suzhou 215000, P. R. China

Dr. Ji Zhang
HEC Pharm R&D Center
Pharmaceutical Science
Dongguan, Guangdong, P. R. China

Ruheng Zhao
Department of Chemistry
University of Michigan
500 S State St
Ann Arbor, MI 48109, USA

Section I. DRUGS TREATING INFECTIOUS DISEASES

1

Nirmatrelvir (Paxlovid with Ritonavir): A 3-Chymotrypsin-like Protease Inhibitor for Treating SARS-CoV-2 Infection

Jie Jack Li

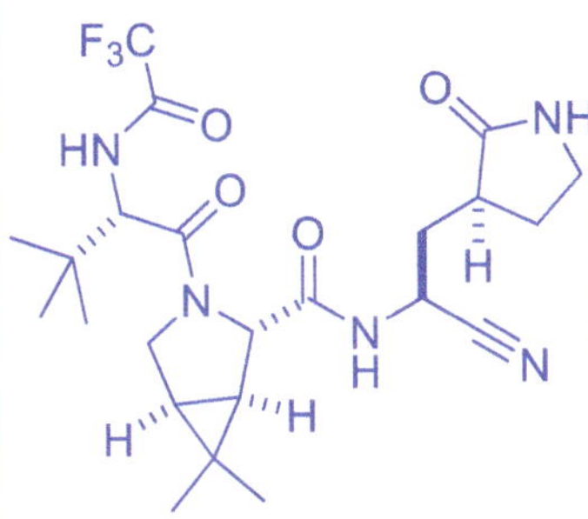

Nirmatrelvir (**1**, Paxlovid with ritonavir)
Pfizer, 2021
SARS-CoV-2 3CL protease inhibitor

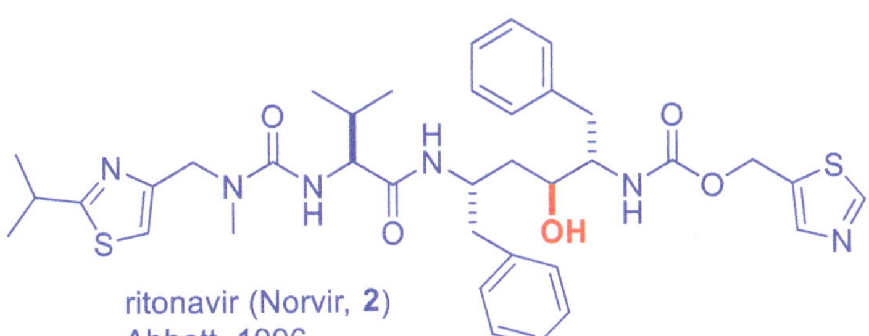

ritonavir (Norvir, **2**)
Abbott, 1996
HIV protease inhibitor/
PK enhancer

Chemistry and Pharmacology of Drug Discovery, First Edition. Edited by Jie Jack Li.

1. Background

The coronavirus disease-2019 (COVID-19) pandemic began in December 2019. Since then on, it has infected over 537 million people and led to more than 6.5 million deaths worldwide.

Some 20 years ago in 2002, severe acute respiratory syndrome (SARS) flared up. In order to discover drugs to treat SARS, Pfizer carried out a fluorescence resonance energy transfer (FRET)-based substrate cleavage assay. PF-00835231 (**3**) was identified as a potent inhibitor of 3-CL^{pro} of recombinant SARS-CoV-1. But since SARS petered out quickly, Pfizer subsequently discontinued the project.[2]

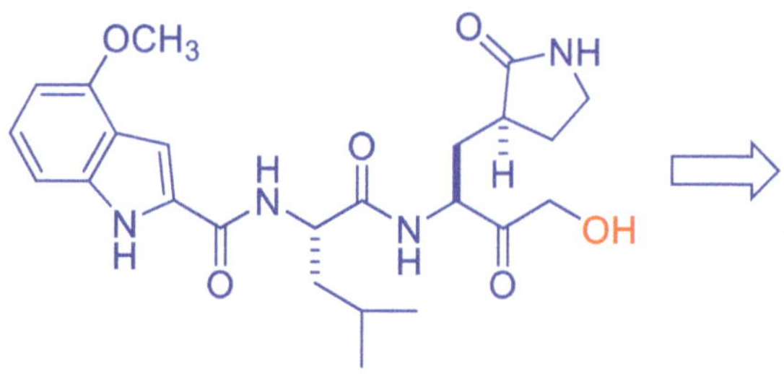

PF-00835231 (**3**)

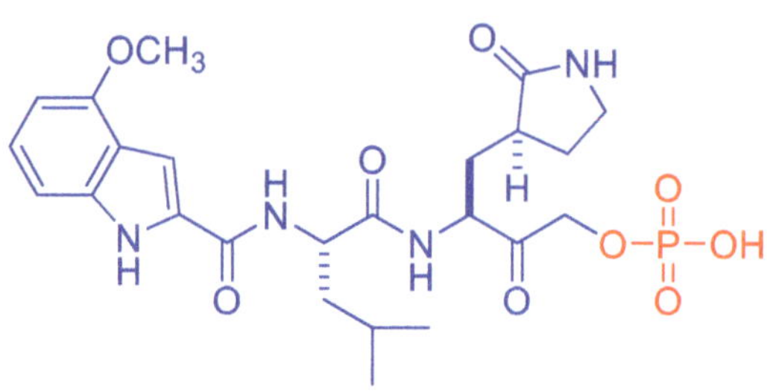

PF-07304814 (**4**)

After the explosion of COVID-19 in 2020, Pfizer prepared PF-00835231 (**3**)'s phosphate prodrug PF-07304814 (**4**) in an effort to boost the solubility. But PF-07304814 (**4**) still lacked oral bioavailability and had to be given intravenously. Later on, Pfizer discontinued clinical trials for PF-07304814 (**4**) when their orally bioavailable 3-CL^{pro} inhibitors became promising. After the discovery of orally bioavailable nirmatrelvir (**1**), its combination drug with ritonavir (**2**), Paxlovid, was approved by the FDA in December 2021.[3]

In November 2022, Shionogi received Japanese government's approval for its oral 3-CL^{pro} inhibitor, ensitrelvir (**5**, Xocova), which is not a peptidomimetic and is orally bioavailable drug by itself without adding a pharmaco-enhancer.[4]

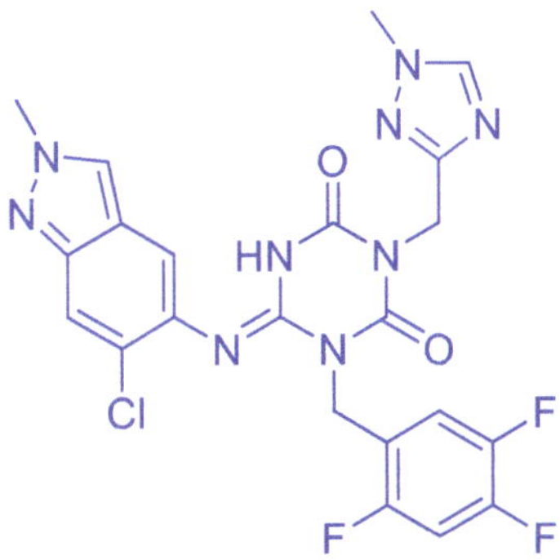

ensitrelvir (**5**, Xocova)
Shionogi, 2022
SARS-Cov-2 3CL^{pro} inhibitor

2. Pharmacology

2.1. The Coronavirus

SARS-CoV-2 is a positive-sense single-stranded RNA (+ssRNA) virus surrounded by an envelope. The virus's genome (Figure 1) consists of 11 *open reading frames* (ORFs) and it has about 30,000 RNA nucleotides in total.

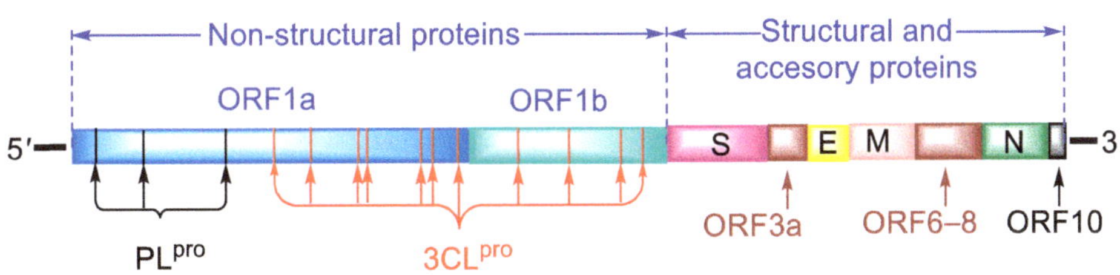

Figure 1. Coronavirus RNA genome

At the left of Figure 1, located at the 5′-end of the genome are the first two open reading frames (ORF1a and ORF1b) that occupy approximately two-thirds of the genome and encode 16 nonstructural proteins. At the right, the other ORFs are located at the 3′-end of the genome and encode four common structural proteins including spike (S), envelop (E), membrane (M), and nucleocapsid (N) proteins. The E and M proteins are responsible for the shape of the virus, while the S protein mediates receptor attachment and viral and host cell membrane fusion. The nucleocapsid (N) protein binds to the viral RNA and forms a ribonucleoprotein that is packaged in the virus envelope (Figure 2).[5]

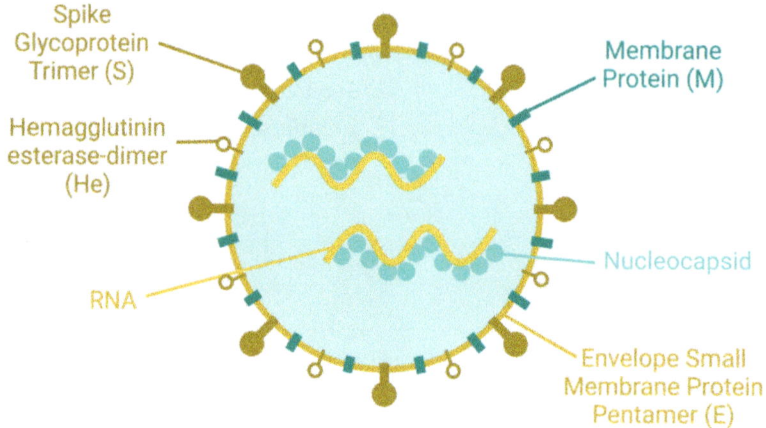

Figure 2. Coronavirus's structure and functions

ORF1a and ORF1b produce polyproteins 1a (pp1a, ~450 kDa) and 1b (pp1b, ~750 kDa), respectively, for which the lengths and amino acid sequences are rather conserved among all known coronaviruses. Among the nonstructural proteins are two very large polyproteins (pp1a and pp1b) that are cleaved by two or three viral proteases.[6]

2.2. The 3CL Protease

Historically, proteases have been tractable drug targets for treating a variety of diseases. Drugs targeting proteases include angiotensin converting enzyme (ACE) inhibitors such as enalapril (**6**, Vasotec) for treating hypertension; neuraminidase inhibitors for treating influenza; dipeptidyl peptidase-4 (DPP-4) inhibitors such as vildagliptin (**7**, Galvus) for treating type II diabetes; HIV protease inhibitors as represented by ritonavir (**2**) for treating HIV/AIDS; and HCV NS3/4A serine protease inhibitors, e.g., boceprevir (**8**, Victrelis) and narlaprevir (**9**, Arlansa), for treating HCV infection. Therefore, 3CL protease is considered as a prominent target for antiviral drugs.

Almost all protease inhibitors are transition-state mimics that are peptidomimetics resulted from truncation and de-peptization of endogenous substrates. Influenza neuraminidase inhibitors are the exceptions. This strategy had paved the road for the discovery of nirmatrelvir (**1**). In fact, some of nirmatrelvir (**1**)'s building blocks were directly "borrowed" from older protease inhibitors such as DPP-4 inhibitor vildagliptin (**7**) and HCV NS3/4A protease inhibitors boceprevir (**8**) and narlaprevir (**9**, *vide infra*).

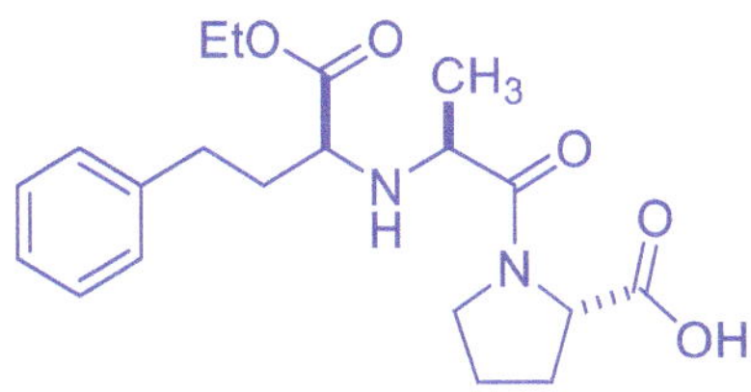

enalapril (**6**, Vasotec)
Merck, 1985
ACE inhibitor

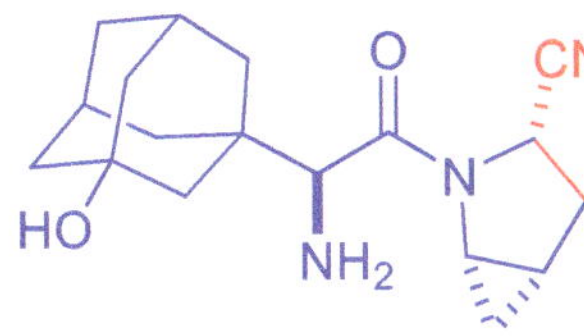

saxagliptin (**7**, Onglyza)
Bristol-Myers Squibb, 2009
DPP-4 inhibitor

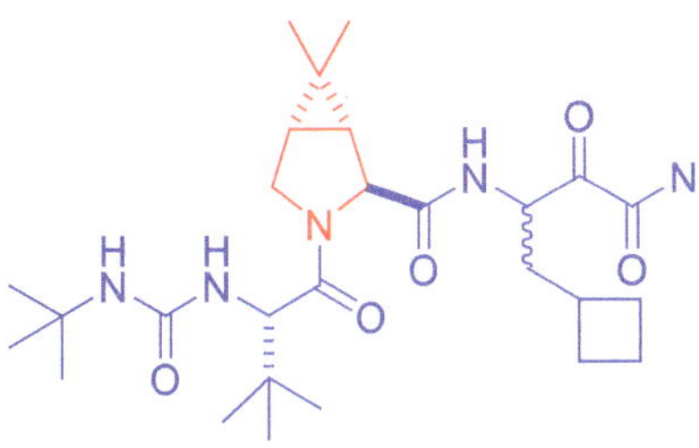

boceprevir (**8**, Victrelis)
Schering-Plough/Merck, 2011
HCV NS3/4A
Serine Protease Inhibitor

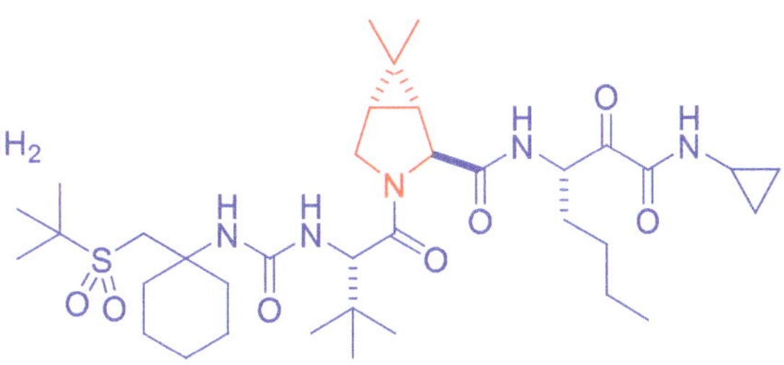

narlaprevir (**9**, Arlansa)
R-Pharm/Merck, 2016, Russia
HCV NS3/4A
Serine Protease Inhibitor

Coronavirus's two cysteine proteases papain-like cysteine protease (PL^{pro}) and 3-CL^{pro} are responsible for cleaving polyproteins. The combined proteolytic actions of 3-CL^{pro} and PL^{pro} produce various shorter, nonstructural proteins vital to viral replication such as RNA-dependent RNA polymerase and helicase that are required in viral life cycle. 3-CL^{pro} itself cleaves two polyproteins (pp1a and pp1b) at 11 different sites (see Figure 1).[7]

Structurally, 3-CL^{pro} is a three-domain cysteine protease. It is a homodimer composed of two protomers that consist of three domains, namely I, II, and III (Figure 3). The homodimer forms due to the interactions between the N-terminus of domain I + I and the C-terminus of domain III. This dimer is reversible and more stable when a substrate is bound. The catalytic dyad Cys145–His41 is located in a cleft between the domains I and II, whereas domain III is just a cluster of helices. The protease is a highly conserved key protease for SARS-CoV-2 replication and no relevant homologous protein with a similar cleavage site to 3CL^{pro} has been identified in humans. Therefore, development of 3CL^{pro} inhibitors offers great promise for treatment of COVID-19.[7]

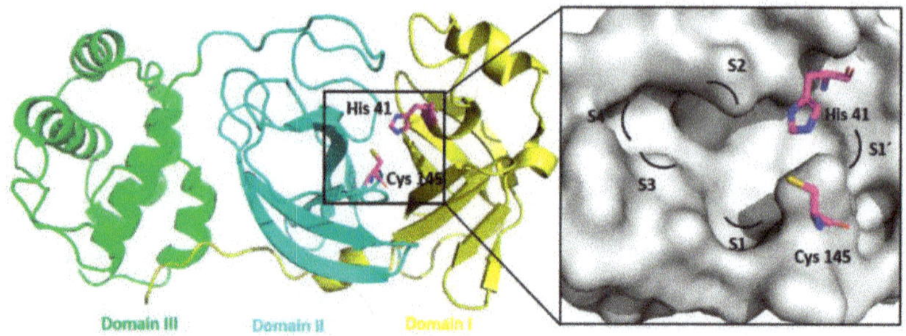

Figure 3. The structure of coronavirus's 3-CL protease, drawn from PDB 6UL7

Unlike the conventional Ser(Cys)–His–Asp(Glu) *triad* found in other chymotrypsin-like enzymes, 3-CL^{pro} of SARS-CoV-2 has a catalytic *dyad* formed by His41 and Cys145 that catalyzes the hydrolysis of the peptide bond at highly specific sites of a polypeptide chain through a common nucleophilic-type reaction. It was suggested that a water molecule might complete the catalytic triad by mediating crucial interactions between His41 and other important conserved residues, such as His164 and Asp187.[8]

A nomenclature carton is shown below to better orient us with regard to the binding pockets of protease and substituents of endogenous ligands or protease inhibitors (Figure 4). In essence, a protease normally has an active catalytic site. It was defined that the first binding pocket on the left of the active site as S_1 and the second on the left as S_2 and so on. The first binding pocket on the right of the catalytic metal is defined as S_1' and the second one on the right as S_2'. For the endogenous ligand in the form as a peptide chain, the fragment occupying the S_1 pocket is defined as P_1 region. Meanwhile, the peptide fragment that is occupying the S_1' pocket is known as the P_1' region. This is known as the Schechter–Berger nomenclature,[9] which will be used throughout this chapter to illustrate drugs' binding and structure–activity relationship (SAR), etc.

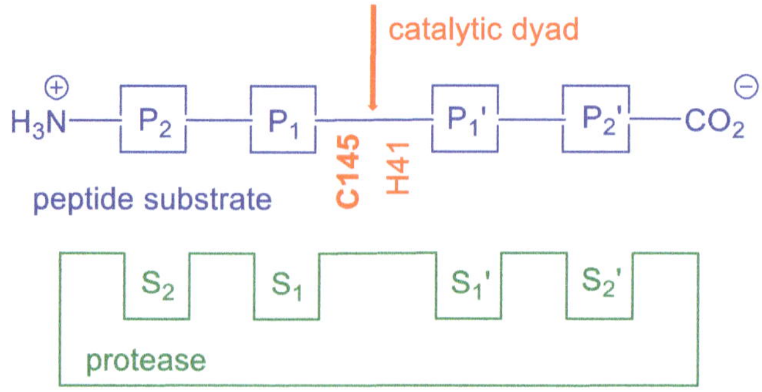

Figure 4. Schechter–Berger nomenclature for protease and its substrate-binding subsites

2.3. The Mechanism of Action of Nirmatrelvir

Nirmatrelvir (**1**) is a *reversible covalent inhibitor* of coronavirus's 3CL^{pro}, eliciting prolonged enzyme inhibition. The recovery of >50% 3CL^{pro} activity after incubation with **1** indicates that inhibition of SARS-CoV-2 3CL^{pro} is reversible.

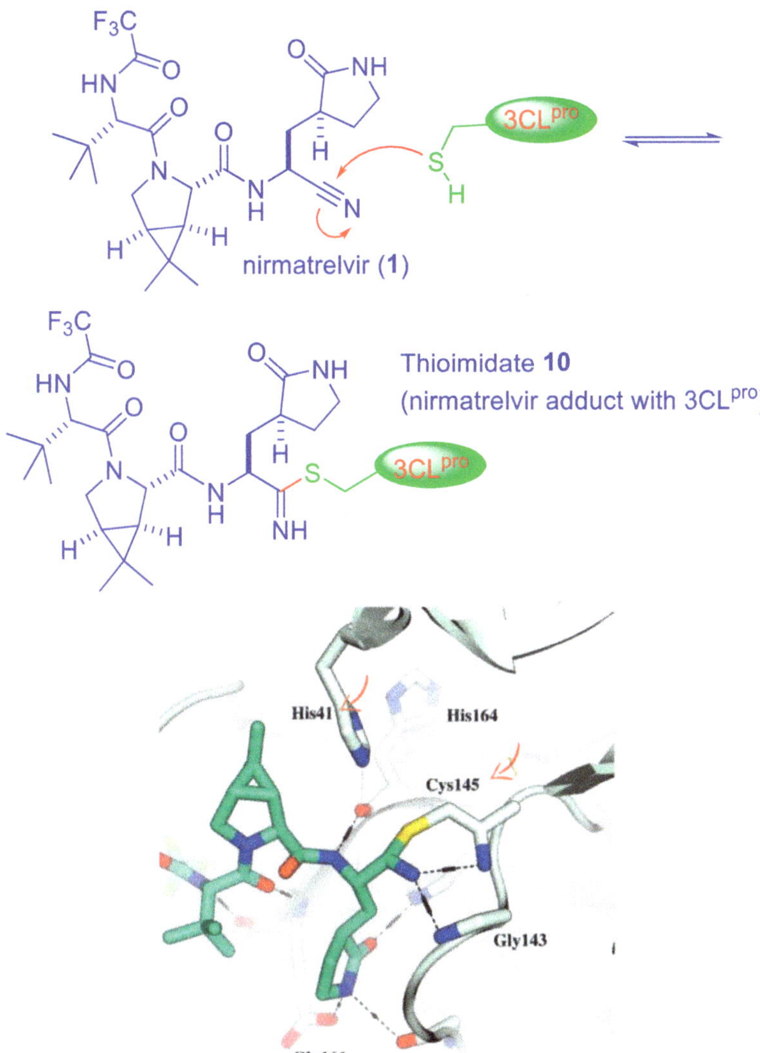

Figure 5. Reversible covalent bond between nirmatrelvir (**1**) and 3CL^{pro}

The substrates for 3CL^{pro}, pp1a and pp1b, share several common features including the omnipresence of a glutamine (Glu) residue at P_1. No known human cysteine protease cleaves after Glu, thus offering potential selectivity for this viral target over the human proteome. Once nirmatrelvir (**1**) properly binds to the 3CL^{pro}, cysteine-145 of the enzyme attacks the nitrile of the inhibitor in a fashion similar to the Pinner reaction to form a reversible S–C covalent bond on thioimidate **10** (Figure 5).[10]

Once 3CL^{pro} is inhibited, the protease is then unable to cleave the polyproteins. As a consequence, the cell fails to produce various shorter, nonstructural proteins vital to viral replication.

3. Structure–Activity Relationship (SAR)

The 3-CL^{pro} sequence between SARS-CoV-1 and SARS-CoV-2 is highly conserved. In fact, they are 100% identical in the catalytic domain that carries out polyprotein cleavage. Consequently, some previously discovered compounds developed over 15 years ago to treat SARS-CoV-1, such as PF-00835231 (**3**), showed high *in vitro* potency against SARS-CoV-2 as well.[2,10]

PF-00835231 (**3**) was a good starting point for lead optimization. But at first, let us examine the origin of the three key fragments on nirmatrelvir (**1**), namely, the pyrrolidone substituent at P_1, the rigid bicyclic dimethylcyclopropylproline at P_2 and the nitrile warhead at P_1'.

3.1. The Pyrrolidone Substituent at P_1

The pyrrolidone substituent in the P_1 pocket has been ubiquitous in many 3-CL^{pro} inhibitors. Why?

Wang and coworkers provided an excellent pedagogical explanation. Coronavirus's 3-CL^{pro} invariably cleaves the Glu residue at the P_1 position of the substrate. In theory, glutamine would be an ideal choice of the P_1 substituent of the protease inhibitors. But glutamine's side chain amide can react with the warhead to form the cyclized product as a piperidinone thus not a good choice after all.[11]

In contrast, the cyclic pyrrolidone provides rigidity over the higher flexibility of the glutamine side chain, likely providing *a reduced loss of conformational entropy* upon binding to the target.[11]

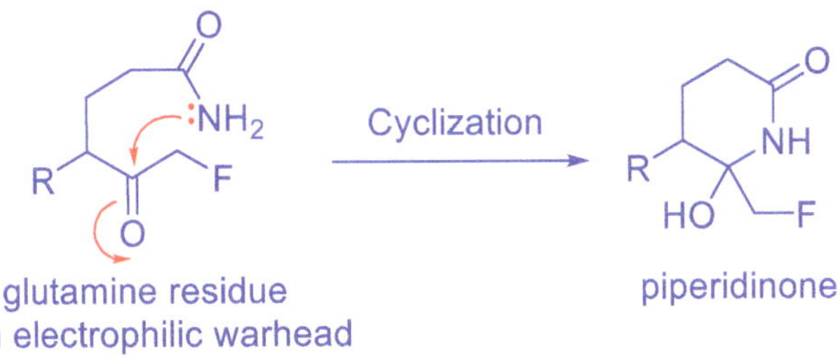

Indeed, Pfizer employed the constrained lactam at the P_1 site as the isostere of the highly conserved P_1 glutamine present in all SARS CoV substrates. The pyrrolidone on PF-00835231 (**3**) is well positioned in the S_1 pocket, making a favorable hydrogen bond to His163. The NH of the lactam is within the hydrogen bond distance to Glu166 and the backbone oxygen of Phe140 in the 3CL^{pro}.[2,10]

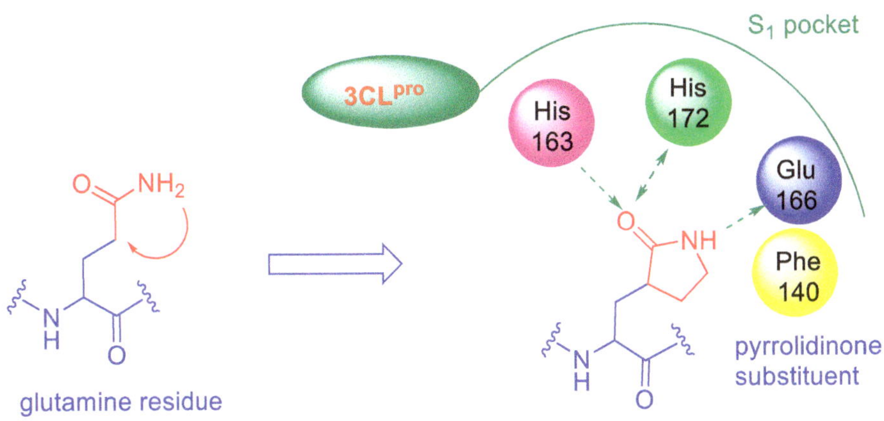

3.2. Dimethylcyclopropylproline at P_2

In order to have good oral bioavailability, a drug has to penetrate into the cell through the cell membrane at first. That requires balanced physiochemical properties of the drug that allow good permeability.

PF-00835231 (**3**) has five hydrogen bond donors, which make the drug too polar to have good passive absorptive permeability for the cell membrane. The obvious tactic would be removal of one or more hydrogen bond donors—the trick is to maintain the peptidomimetics' antiviral activity at the same time. Pfizer eliminated one hydrogen bond donor by switching the warhead from α-hydroxymethyl ketone to benzothiazol-2-yl ketone. Concurrently, they used a fused cyclopropyl bicycle to replace PF-00835231 (**3**)'s leucine portion at the middle of the molecule, the P_2 region, to remove another hydrogen bond donor. The resultant benzothiazol-2-yl ketone **11**, now only has three hydrogen bond donors.

Regrettably, even though benzothiazol-2-yl ketone **11** saw a 49-fold increase in permeability over PF-00835231 (**3**). But its binding was significantly lower than the prototype because **11** lost an important hydrogen bond interaction with glutamine-189 that was present on **3**. The lost potency was made up later on the left side P_3 in the form of trifluoroacetamide (*vide infra*).

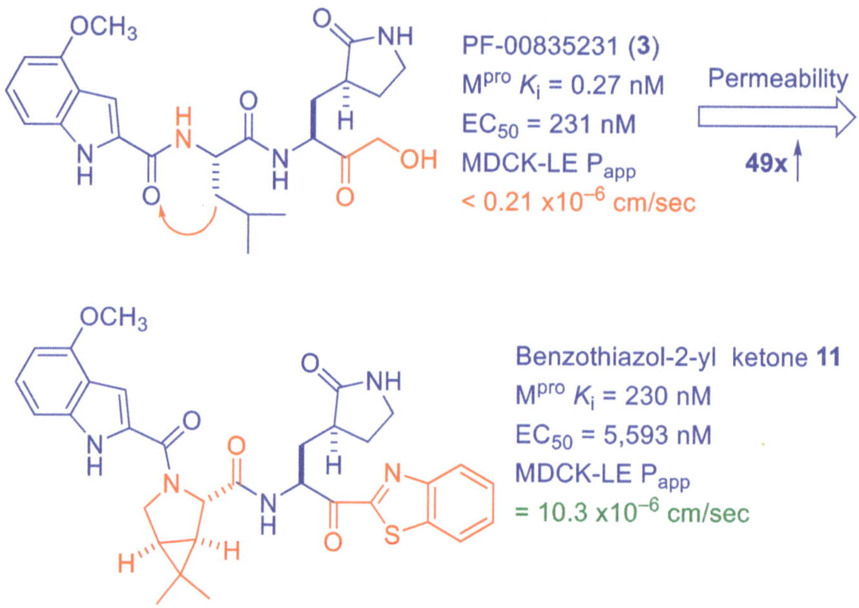

The rigid bicyclic ring was actually an old trick. In the past, dimethylcyclopropylproline (the 6,6-dimethyl-3-azabicyclo[3.1.0]-hexane) substituent was employed in two marketed HCV NS3/4A serine protease inhibitors, e.g., boceprevir (**8**) and narlaprevir (**9**). Last year, Wang[12] and others[13] discovered that boceprevir (**8**) also inhibited SARS-CoV-2's $3CL^{pro}$ as did an older broad-spectrum viral 3CL protease inhibitor GC-376 (**12**), which is an investigational veterinary drug developed for feline infectious peritonitis caused by a coronavirus. It was designed to target the viral 3CL protease and had potent antiviral activity against multiple viruses including MERS, feline infectious peritonitis coronavirus, and norovirus.

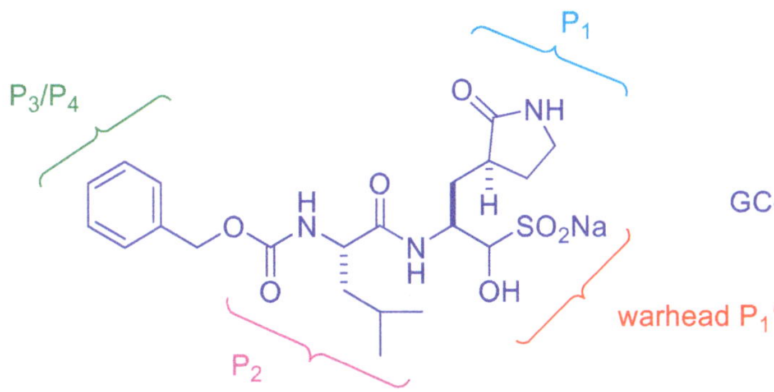

GC-376 (**12**)

3.3. The Nitrile Warhead at P1′

There are numerous reports of reversible cysteine protease inhibitors. Their warheads include aldehydes, thio- or oxymethyl ketones [e.g., PF-00835231 (**3**)], cyclic ketones, amidomethyl ketones, nitriles, or various 1,2-dicarbonyl motifs. The electrophilic carbon of these chemotypes reacts reversibly with the sulfur atom of an active-site cysteine, forming a covalently bound tetrahedral complex.

Rather than keeping the benzothiazol-2-yl ketone as the warhead on **11**, Pfizer chose nitrile instead as the warhead at the end for nirmatrelvir (**1**). Nitrile has several advantages. It is easier to make. It is smaller thus might have better physiochemical properties. Indeed, the nitrile was more soluble than the benzothiazol-2-yl ketone in this series of compounds.

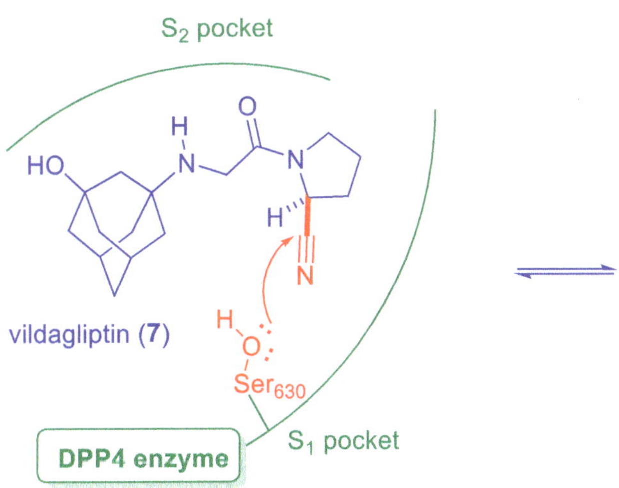

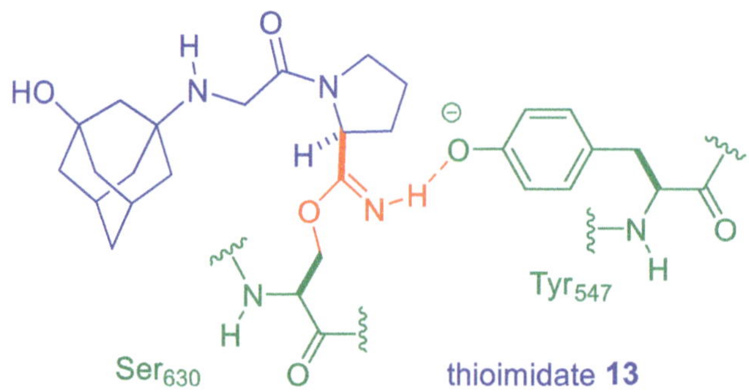

Historically, nitrile warhead has been present in several successful drugs on the market. DPP-4 inhibitors alone boast four nitrile-containing drugs including vildagliptin (7), saxagliptin (Onglyza), alogliptin (Nesina), and anagliptin (Suiny). Take vildagliptin (7) as an example. A reversible Pinner reaction between its nitrile warhead and the serine-630 in the S_1 pocket of the DPP-4 enzyme forms a time-dependent covalent adduct thioimidate **13** with the aid of tyrosine-547 by protonation of the nitrile.[14]

3.4. Lead Optimization

Switching the α-hydroxymethyl ketone warhead on PF-00835231 (**3**) to nitrile **14** resulted in a boost of rat oral bioavailability with a reasonable intrinsic clearance and stability in human liver microsomes (HLMs).

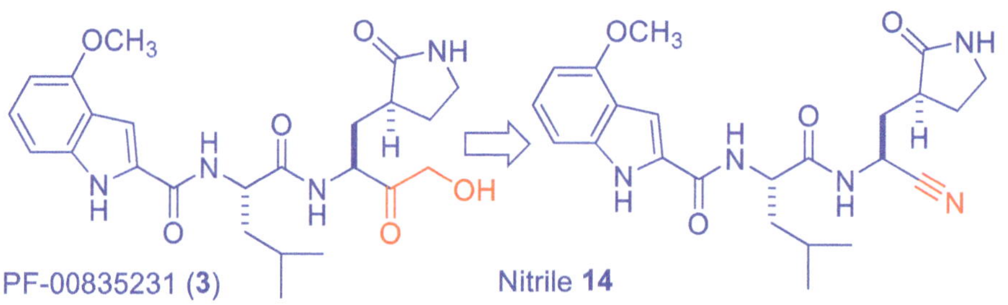

PF-00835231 (**3**)
SARS-CoV-2 3CL^{pro} K_i = 0.27 nM
VeroE6-enACE2 CPE EC_{50} = 231 nM
HLM CL_{int} = 7.47 μL/min/mg
F = 1.4%

Nitrile **14**
SARS-CoV-2 3CL^{pro} K_i = 27.7 nM
VeroE6-enACE2 CPE EC_{50} = 1,364 nM
HLM CL_{int} = 34.4 μL/min/mg
F = 7.6%

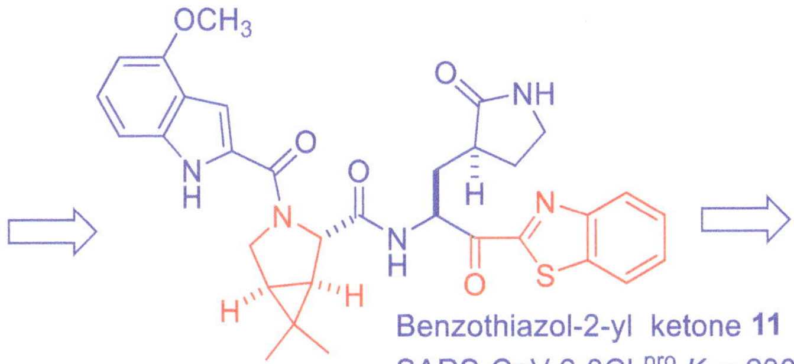

Benzothiazol-2-yl ketone **11**
SARS-CoV-2 $3CL^{pro}$ K_i = 230 nM
VeroE6-enACE2 CPE EC_{50} = 5,593 nM
HLM CL_{int} = 337 μL/min/mg

But nitrile **14** suffered lower affinity toward $3CL^{pro}$ and inferior antiviral efficacy in comparison to **3**. As mentioned before, benzothiazol-2-yl ketone **11** saw 49-fold increase in permeability over **3**. But the former was significantly less potent than the latter because benzothiazol-2-yl ketone **11** could no longer form a productive hydrogen bond with glutamine-189 even though its rigid bicycle at P_2 could effectively occupy the lipophilic S_2 pocket. Compound **11** also suffered from high clearance in HLMs (CL_{int} = 337 μL/min/mg), making it not suitable for further investment.[10]

In order to regain the interaction with glutamine-189, Branched, acyclic P_3 groups were introduced in place of indole. The (*S*)-val-sulfonamide in compound **15** extends underneath Gln-189, productively engaging P_3 pocket residues and achieves improved hydrogen-bonding interactions with the Glu-166 backbone. Sulfonamide **15** regained back much affinity to $3CL^{pro}$ and VeroE6 cell antiviral activity. As an added benefit, it was also much more stable in HLMs in comparison to its precursor **11**. Further optimization of the P_3-capping group led to trifluoroacetamide **16**, which saw significantly improved antiviral efficacy although the biochemical assay showed similar potency to sulfonamide **15**.

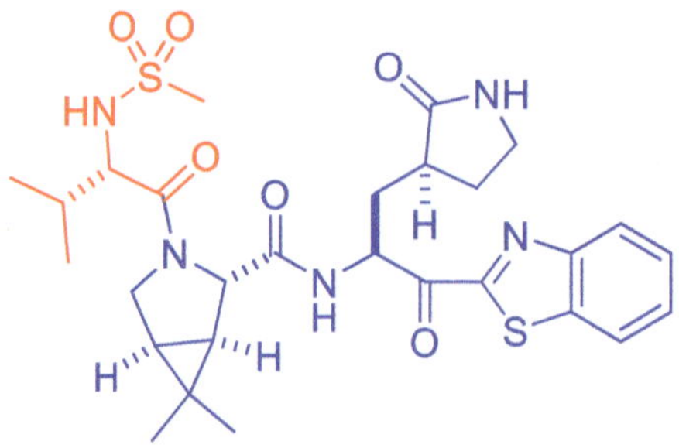

Sulfonamide **15**
SARS-CoV-2 $3CL^{pro}$ K_i = 7.93 nM
VeroE6-enACE2 CPE EC_{50} = 909 nM
HLM CL_{int} = 127 μL/min/mg
F = 10%

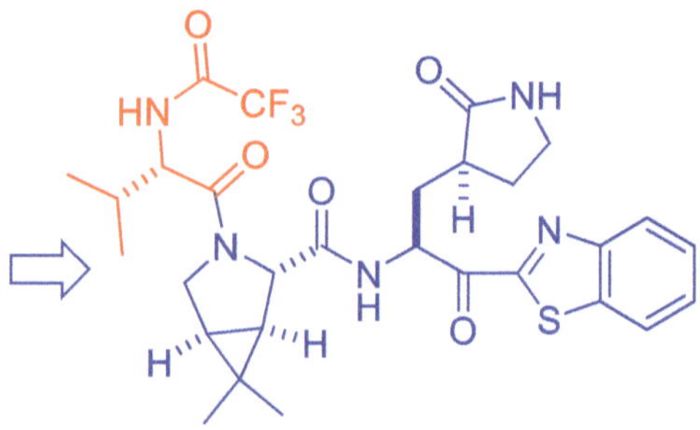

Trifluoroacetamide **16**
SARS-CoV-2 $3CL^{pro}$ K_i = 12.1 nM
VeroE6-enACE2 CPE EC_{50} = 85.3 nM
HLM CL_{int} = 33.3 μL/min/mg
F = 33%

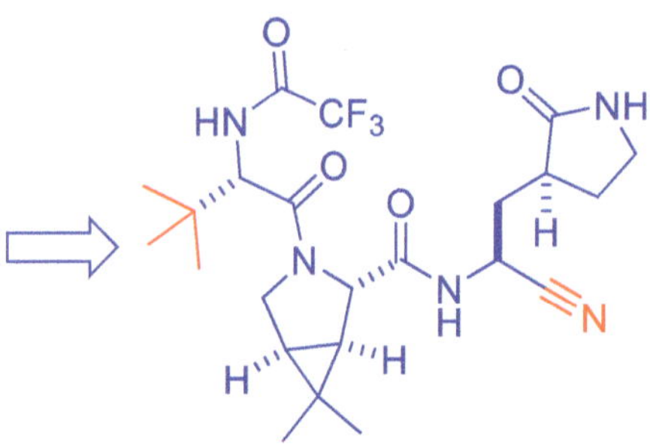

Nirmatrelvir (**1**)
SARS-CoV-2
$3CL^{pro}$ K_i = 3.11 nM
VeroE6-enACE2
CPE EC_{50} = 74.5 nM
HLM CL_{int} = 24.5 μL/min/mg
F = 50%

More excitingly, trifluoroacetamide **16** had increased metabolic stability in HLMs and greatly improved oral pharmacokinetics in both rats and monkeys. The trifluoroacetamide stood out in its ability to permeate the gut barrier in assays in comparison to other bioisosteres. Even through there was some uncertainty with regard to its stability, trifluoroacetamide here proved to be robust. Finally, introduction of the P_1' nitrile and replacing the (*S*)-valine group with the (*S*)-*tert*-leucine group at the P_3 substituent to this scaffold led to the identification of the clinical candidate nirmatrelvir (**1**).[10]

The rest, like they say, is history.

4. Pharmacokinetics and Drug Metabolism

4.1. Pharmacokinetics of Nirmatrelvir

Although nirmatrelvir (**1**) had moderate oral bioavailability in rats (10 mpk, $F = 50\%$). It exhibited poor oral bioavailability in monkeys (10 mpk, $F = 8.5\%$) due to first-pass metabolism along the gastrointestinal tract by cytochrome P450 (CYP) enzymes. This is consistent with a rapid ($t_{1/2}$ = 20.5 min, CL_{int} = 33.8 mL/min/mg), NADPH-dependent metabolic turnover of nirmatrelvir (**1**) in monkey intestinal microsomes. Nirmatrelvir (**1**) exhibited moderate plasma clearance (CL_p) in rats and monkeys, with elimination half-lives ($t_{1/2}$) of 5 h and <1 h, respectively, after intravenous dosing.[10]

4.2. Metabolism of Nirmatrelvir

Nirmatrelvir (**1**) demonstrated moderate clearance, which was significantly inhibited (≥82%) by the selective CYP3A4/5 inhibitor ketoconazole. Moreover, the oxidative metabolic profile of nirmatrelvir (**1**) in HLMs, which includes modifications on the P_2 6,6-dimethyl-3-azabicyclo[3.1.0]hexane (e.g., **17**), the *tert*-butyl group at the P_3 position as represented by **18** and the P_1 pyrrolidinone ring (**19**, as an example), was reproduced in incubations of nirmatrelvir (**1**) with recombinant human CYP3A4.[10]

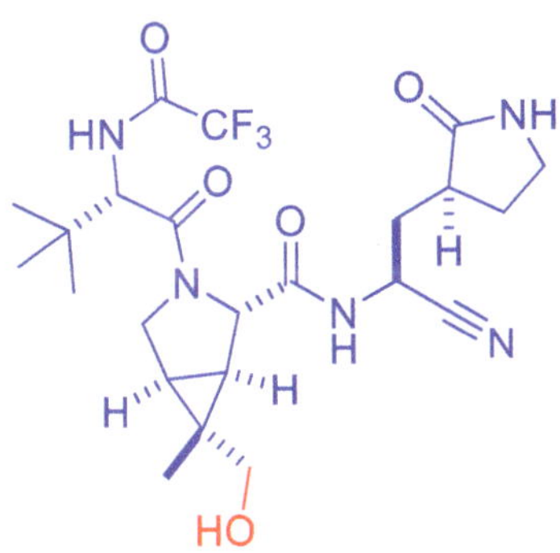

metabolite **17**

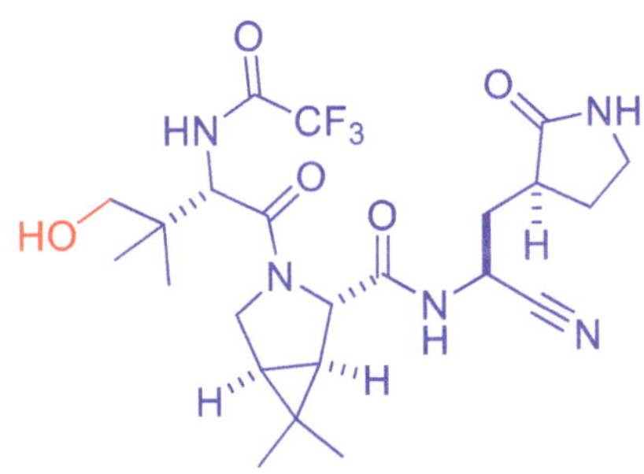

metabolite **18**

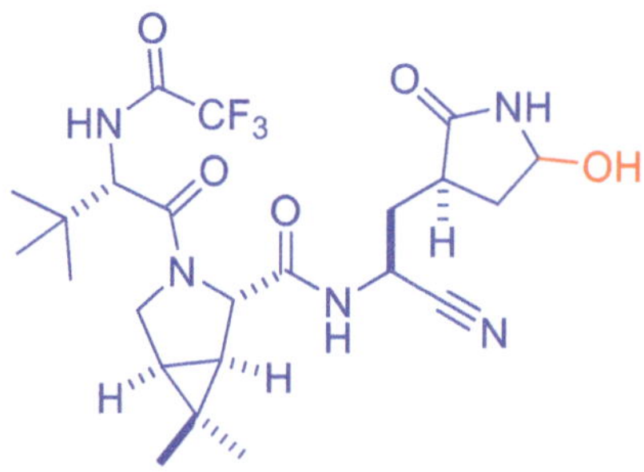

metabolite **19**

The aforementioned *in vitro* studies established a predominant role for CYP3A4 in the metabolism of nirmatrelvir (**1**). In order to overcome potentially suboptimal therapeutic exposures, nirmatrelvir (**1**) is co-dosed with the potent CYP3A4 inactivator ritonavir (**2**) as a pharmacoenhancer to boost its therapeutic concentrations.

In the past, pharmacoenhancers have been adopted for multiple purposes such as ensuring therapeutic exposure of the active product, reducing formation of toxic metabolites, changing the route of administration and increasing the cost-effectiveness of a therapy. Ritonavir (**2**) has been used as a pharmacoenhancer of several marketed protease inhibitors (e.g., darunavir and lopinavir) that are subject to metabolic clearance through CYP3A4. It is also a permeability-glycoprotein (P-gp) inhibitor, which further elevates the bioavailability of nirmatrelvir (**1**). It substantially elevates blood levels of co-administered CYP3A-dependent drugs, with increases in area-under-the-curve (AUC) blood concentrations ranging from 1.8- to 20-fold. In this case, co-administration with ritonavir (**2**) significantly improved the plasma concentration of nirmatrelvir (**1**).[15]

On the other hand, ritonavir (**2**) is an inducer of several other CYP enzyme subtypes such as CYP1A2, CYP2B6, CYP2C9, and CYP2C19, as well as UDP-glucuronyl transferases. Thankfully, since induction occurs slowly (i.e., generally 10–15 days after initiation of inducer), it is not anticipated to cause significant drug–drug interactions (DDIs) as an inducer because of its short treatment course of 5 days.[16]

5. Efficacy and Safety

Paxlovid [two 150 mg of nirmatrelvir (**1**)/one 100 mg of ritonavir (**2**)] reduced the risk of hospitalization or death by 89% in high-risk individuals who took the drug within 3 days of experiencing symptoms.

In a randomized clinical trial (EPIC-HR) with 1 : 1 Paxlovid: placebo received orally every 12 h for 5 days, it was found that those receiving Paxlovid had significantly reduced hospital admissions and deaths among those affected by COVID-19. It was found that among participants who received treatment within 3 days of beginning

COVID-19 symptoms, the risk of hospitalization or death was 89% lower than that for the placebo group, without evidence of safety concerns.[17,18]

Paxlovid's safety has also been established.

In vitro, nirmatrelvir (**1**) was remarkably selective against receptors, ion channels and other enzymes. It was also selective against other coronaviruses such as SARS-CoV-1, MERS-CoV and human coronavirus 229E using cytopathic effect (CPE) assays to measure their cellular antiviral activities. The drug was not a hERG substrate thus had a low potential for cardiotoxicity. Nirmatrelvir (**1**) was not mutagenic or clastogenic in *in vitro* genetic toxicity studies and was negative in an *in vivo* rat micronucleus assay. Finally, it was well tolerated in long-term toxicity tests in monkey with no observed adverse effects at high doses.[10]

Also in the EPIC-HR clinical trial, the incidence of adverse effects that emerged during or after the treatment period (treatment emergent adverse effects, TEAEs) was similar among recipients of Paxlovid (22.6%) and the placebo (23.9%). Most TEAEs in both treatment groups were mild to moderate (Grades 1–2) in severity. No patients in the Paxlovid group experienced an AE resulting in death (Grade 5), while there were 13 deaths among placebo recipients.[17,19]

Now that millions of doses of Paxlovid have been taken in real world. Its efficacy and safety have been well established.

6. Synthesis

The first 7 mg of nirmatrelvir (**1**) were synthesized in late July 2020. A massive scale up effort was undertaken and by late October 100 g of nirmatrelvir (**1**) was synthesized and 2 weeks later the chemists were able to scale up the synthesis to even more than 1 kg.[20]

6.1. Scale-up Route

We first review the route that Pfizer used to make their 150-g batch.[10]

The pyrrolidone-containing fragment at P_1 was prepared using known amino-ester **20**. A simple amination of **20** converted the ester to primary amide **21**, which was treated with HCl to remove the Boc protection, furnishing the P_1 fragment as **22** as its HCl salt.

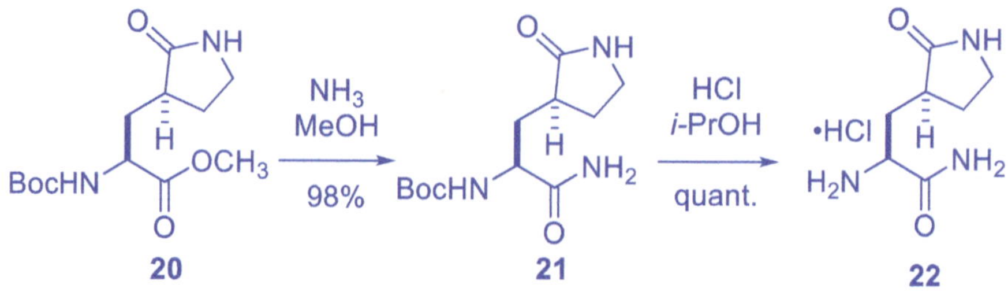

Meanwhile, dimethylcyclopropylproline ester **23** was coupled with (*S*)-*tert*-leucine **24** to assemble amides **25**. Hydrolysis of the ester functionality on **25** using lithium hydroxide then provided acid **26**, which was treated HCl to afford the "naked" amino-acid **27** as its HCl salt. Condensation of **27** with ethyl trifluoroacetate led to the formation of trifluoroacetamide **28**. Another key coupling between **28** and **22** assembled dipeptide **29**, which was readily dehydrated to deliver nirmatrelvir (**1**) as its methyl *t*-butylether (MTBE) solvate after recrystallization using a mixture of MTBE and ethyl acetate.[10]

Since each step has a high yield, the overall yield is 48% to make nirmatrelvir (**1**) from lactam **22**.

Boc protection/deprotection (twice!) and the employment of the Burgess reagent as the dehydrating reagent to make the nitrile are not amenable to commercial-scale production, which mounts to hundreds of tons of the API. Pfizer development chemists developed better synthesis for large-scale synthesis. Before disclosure of Pfizer's manufacturing route, we highlight a few examples among numerous efforts at optimizing the production of nirmatrelvir (**1**).

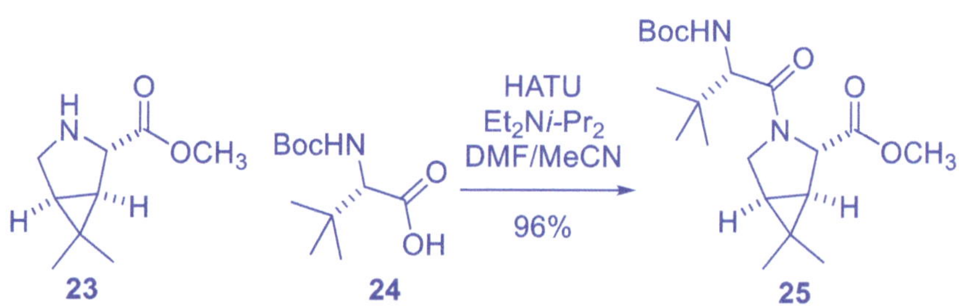

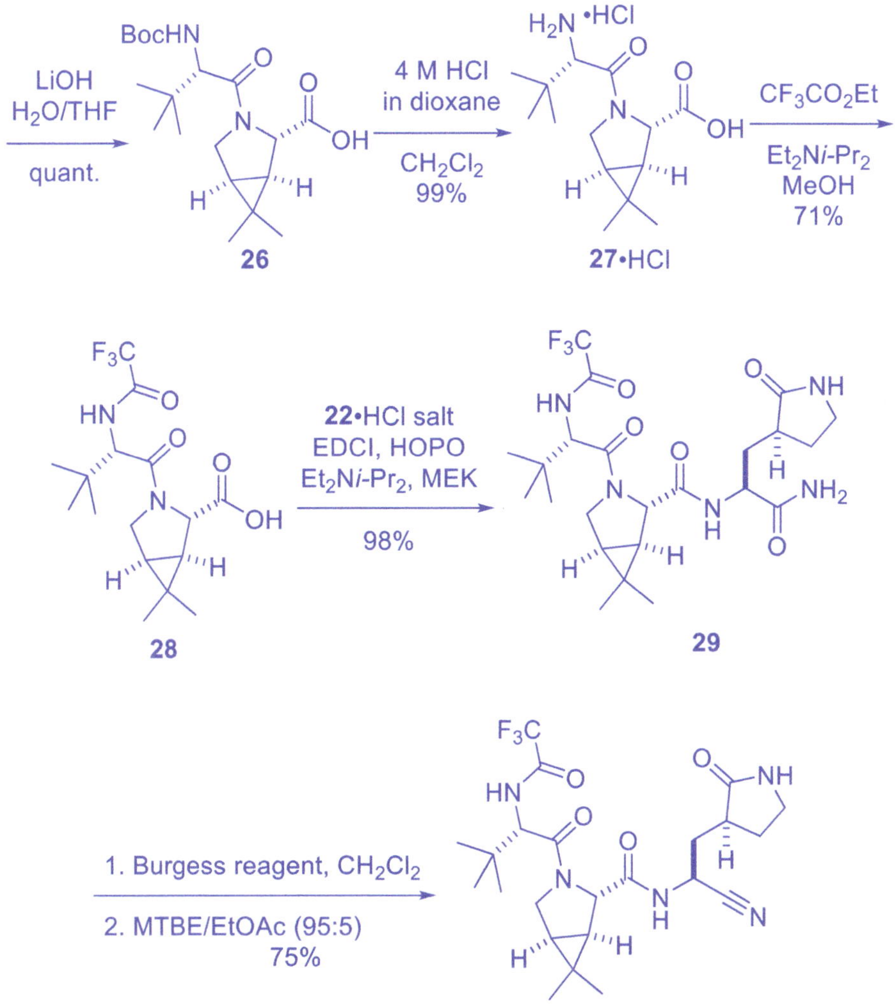

nirmatrelvir (**1**)-MTBE solvate

A group at Virginia Commonwealth University chose to use trifluoroacetic anhydride (TFAA) to replace expensive Burgess reagent, which also left behind a lot of by-product waste. As shown below, treating primary amide **29** with 2 equiv of TFAA and 2 equiv of *N*-methylmorpholine (NMM) delivered nirmatrelvir (**1**) as its methyl *t*-butylether (MTBE) solvate in 85% yield after recrystallization using a mixture of MTBE and ethyl acetate. The group also employed propanephosphonic acid anhydride (T3P) as the obvious replacement of Pfizer's coupling agents, HATU and EDCI, with good results.[21]

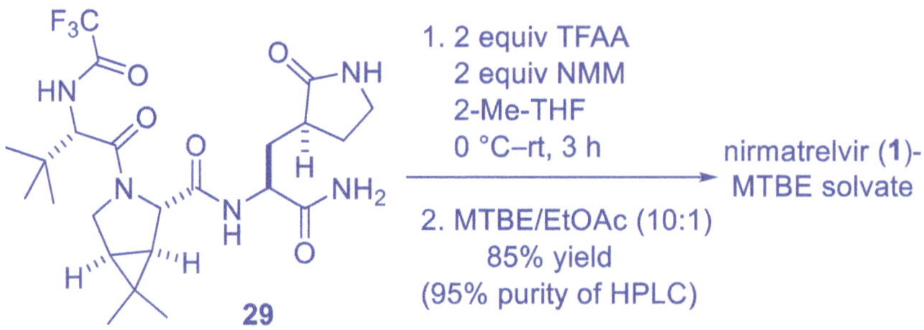

In 2023, a group in India published a preparation of nirmatrelvir (**1**) using flow chemistry. They chose to make the nitrile sooner than Pfizer did. To that end, dehydration of primary amide **21** with TFAA to give nitrile **30**, which was treated with HCl to produce the P_1 fragment as amine **31** as its HCl salt. Finally, coupling between amine **31** and acid **28** using T3P as the coupling agent delivered nirmatrelvir (**1**) directly.[21]

It could be advantageous to make the nitrile sooner than doing it at the last step.

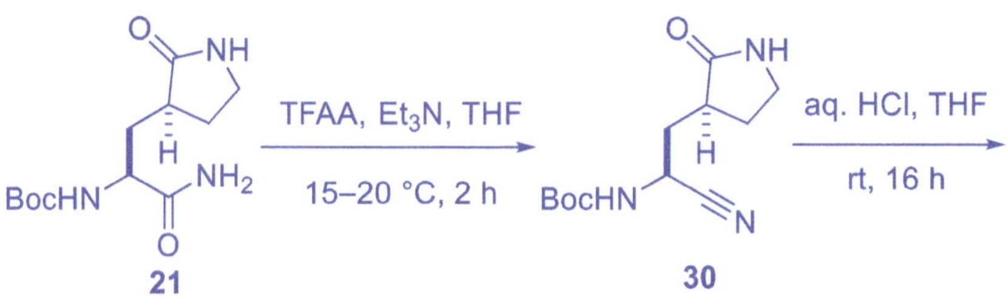

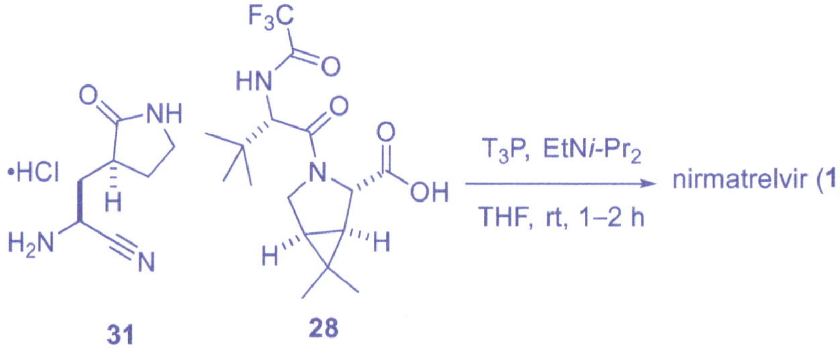

Also in 2023, a group in China published an optimized synthesis of a key intermediate of nirmatrelvir, primary amide **21**. Thus, deprotonation of dimethyl *N*-BocGlu (**32**) using LiHMDS was followed by α-cyanomethylation with bromoacetonitrile in the presence of $NdCl_3$. A one-pot Raney nickel-catalyzed hydrogenation of the nitrile group with concomitant cyclization and ammonolysis and subsequent deprotection of *N*-

Boc to deliver the target intermediate cyclic glutamine analog in three steps in high yields.[22,23]

This route reduced side reactions and undesired by-products. The operations are simple. But using LiHMDS is not my favorite feature.

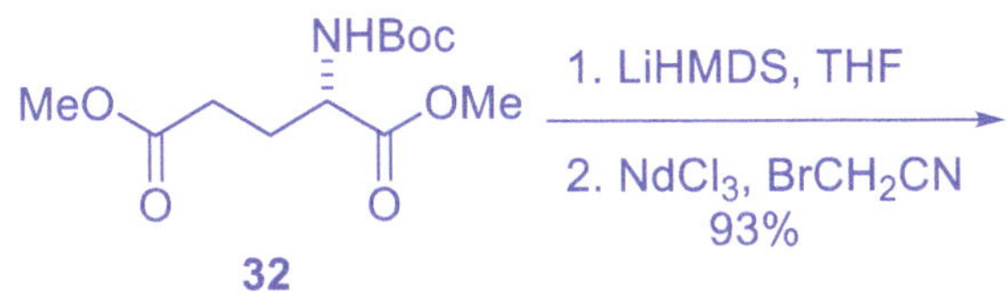

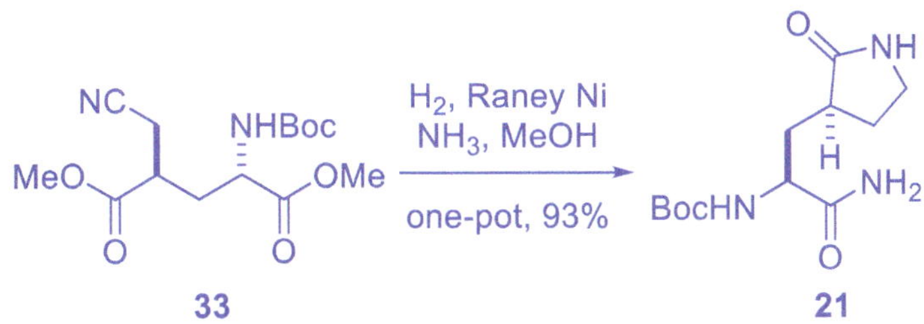

6.2. Manufacturing Route

In 2023, Pfizer Process Chemistry published their manufacturing route for nirmatrelvir (**1**), which took 17 months from first laboratory synthesis (in July 2020) to FDA emergency use authorization (in December 2021).[24]

The manufacturing route starts over 100 metric tons of starting raw materials, so the cost and even supply chain considerations were paramount, especially during the COVID pandemic.

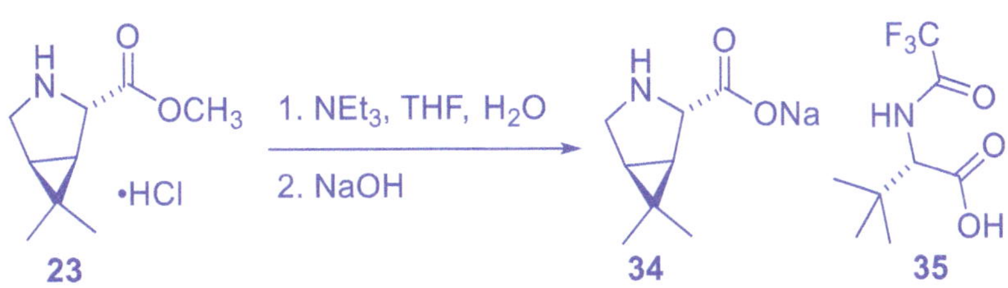

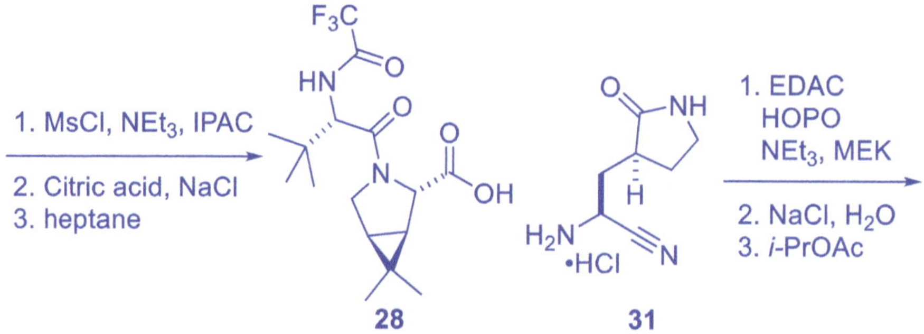

Starting material **23** was prepared by Merck/Codexis before for their HCV NS3 serine protease inhibitors program to make boceprevir (**8**). The HCl salt of **23** was at first basified with triethylamine to release the free amine and the methyl ester was converted to its corresponding sodium salt **34** by the treatment of NaOH. Treating acid **35** with mesylate gave rise to a mixed anhydride, which was coupled with **34** to assemble **28**. The coupling between acid **28** with amine **31** was accomplished by EDAC to provide adduct **29** with the aid of 2-hydroxypyridine N-oxide (HOPO), which has been known to reduce racemization. Treating primary amide with TFAA afforded nitrile **36** as the MTBE solvate, which was converted to anhydrous Form I of nirmatrelvir (**1**).[24]

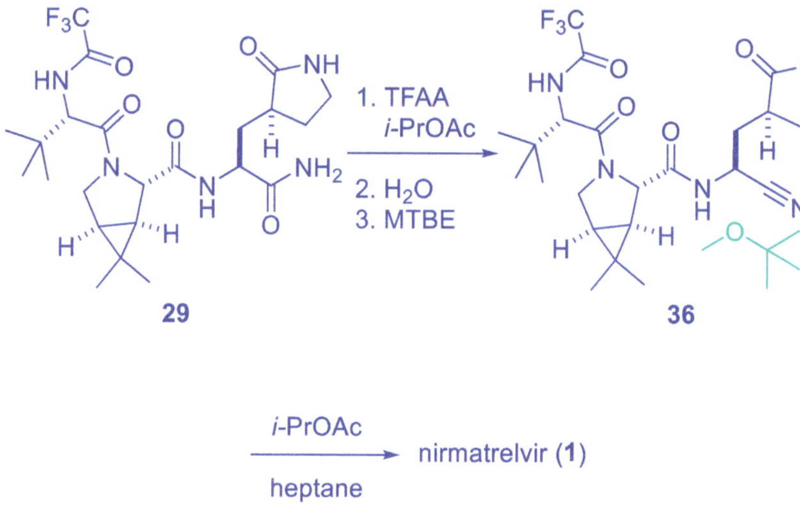

7. Summary

The discovery and development of nirmatrelvir (**1**)/Paxlovid has been an important chapter in the annals of pharmaceutical industry. The drug literally saved lives.

Pfizer's triumph has been built on decades of progress made in the field of protease inhibitors. As we see throughout this chapter, the pyrrolidone substituent at P_1, the rigid bicyclic dimethylcyclopropylproline at P_2 and the nitrile warhead at P_1' have all been precedented. Needless to say, ritonavir (**2**) as a PK enhancer was discovered in the 1990s when the first wave of HIV protease inhibitors became available.

After all, if we have seen further, it is by standing on shoulders of giants.

References

1. Focosi, D.; McConnell, S.; Shoham, S.; Casadevall, A.; Maggi, F.; Antonelli, G. Nirmatrelvir and COVID-19: development, pharmacokinetics, clinical efficacy, resistance, relapse and pharmacoeconomics. *Int. J. Antimicrob. Agents* **2023**, *61,* 106708.
2. Halford, B. The path to Paxlovid: how Pfizer scientists transformed an old drug lead into an oral COVID-19 antiviral. *ACS Cent. Sci.* **2022**, *8*, 405–407.
3. Unoh, Y.; Uehara, S.; Nakahara, K.; Nobori, H.; Yamatsu, Y.; Yamamoto, S.; Maruyama, Y.; Taoda, Y.; Kasamatsu, K.; Suto, T.; et al. Discovery of S-217622, a noncovalent oral SARS-CoV-2 3CL protease inhibitor clinical candidate for treating COVID-19. *J. Med. Chem.* **2022**, *65*, 6499–6512.
4. Gorkhali, R.; Koirala, P.; Rijal, S.; Mainali, A.; Baral, A.; Bhattarai, H. K. Structure and function of major SARS-CoV-2 and SARS-CoV proteins. *Bioinf. Biol. Insights* **2021**, *15*, 11779322211025876.
5. Cannalire, R.; Cerchia, C.; Beccari, A. R.; Di Leva, F. S.; Summa, V. Targeting SARS-CoV-2 proteases and polymerase for COVID-19 treatment: state of the art and future opportunities. *J. Med. Chem.* **2022**, *65*, 2716–2746.
6. Chen, R.; Gao, Y.; Liu, H.; Li, H.; Chen, W.; Ma, J. Advances in research on 3C-like protease (3CLpro) inhibitors against SARS-CoV-2 since 2020. *RSC Med. Chem.* **2023**, *14*, 9–21.
7. Kneller, D.W.; Phillips, G.; O'Neill, H.M.; Jedrzejczak, R.; Stols, L.; Langan, P.; Joachimiak, A.; Coates, L.; Kovalevsky, A. Structural plasticity of SARS-CoV-2 3CL Mpro active site cavity revealed by room temperature X-ray crystallography. *Nat. Commun.* **2020**, *11*, 3202.
8. Schechter, I.; Berger, A. On the size of the active site in proteases. I. Papain. *Biochem. Biophys. Res. Commun.* **1967**, *27*, 157–162.

9. Owen, D. R.; Allerton, C. M. N.; Anderson, A. S.; Aschenbrenner, L.; Avery, M.; Berritt, S.; Boras, B.; Cardin, R. D.; Carlo, A.; Coffman, K. J.; et al. An oral SARS-CoV-2 Mpro inhibitor clinical candidate for the treatment of COVID-19. *Science* **2021**, *374*, 1586–1593.

10. Hoffman, R. L.; Kania, R. S.; Brothers, M. A.; Davies, J. F.; Ferre, R. A.; Gajiwala, K. S.; He, M.; Hogan, R. J.; Kozminski, K.; Li, L. Y.; et al. Discovery of ketone-based covalent inhibitors of coronavirus 3CL proteases for the potential therapeutic treatment of COVID-19. *J. Med. Chem.* **2020**, *63*, 12725–12747.

11. Joyce, R. P.; Hu, V. W.; Wang, J. The history, mechanism and perspectives of nirmatrelvir (PF-07321332): an orally bioavailable main protease inhibitor used in combination with ritonavir to reduce COVID-19-related hospitalizations. *Med. Chem. Res.* **2022**, *31*, 1637–1646.

12. Ma, C. L.; Sacco, M. D.; Hurst, B.; Townsend, J. A.; Hu, Y. M.; Szeto, T.; et al. Boceprevir, GC-376 and calpain inhibitors II, XII inhibit SARS-CoV-2 viral replication by targeting the viral main protease. *Cell Res.* **2020**, *30*, 678–92.

13. Fu, L.; Ye F., Feng, Y.; Yu, F.; Wang, Q.; Wu, Y.; Zhao, C.; Sun, H.; Huang, B.; Niu, P.; et al. Both Boceprevir and GC376 efficaciously inhibit SARS-CoV-2 by targeting its main protease. *Nat. Commun.* **2020**, *11*, 4417.

14. Ahren, B.; Schweizer, A.; Dejager, S.; Villhauer, E. B.; Dunning, B. E.; Foley, J. E. Mechanisms of action of the dipeptidyl peptidase-4 inhibitor vildagliptin in humans. *Diabetes Obes. Metab.* **2011**, *13*, 775–783.

15. Girardin, F.; Manuel, O.; Marzolini, C.; Buclin, T. Evaluating the risk of drug-drug interactions with pharmacokinetic boosters: the case of ritonavir-enhanced nirmatrelvir to prevent severe COVID-19. *Clin. Microbiol. Infect.* **2022**, *28*, 1044–1046.

16. Martins, V.; Fazal, L.; Oganesian, A.; Shah, A.; Stow, J.; Walton, H.; Wilsher, N. A commentary on the use of pharmacoenhancers in the pharmaceutical industry and the implication for DMPK drug discovery strategies. *Xenobiotica* **2022**, *52*, 786–796.

17. Hammond, J.; Leister-Tebbe, H.; Gardner, A.; Abreu, P.; Bao, W.; Wisemandle, W.; Baniecki, M.; Hendrick, V. M.; Damle, B.; Simon-Campos, A.; et al. Oral nirmatrelvir for high-risk, nonhospitalized adults with Covid-19. *N. Engl. J. Med.* **2022**, *386*, 1397–1408.

18. Ghosh, A. K.; Mishevich, J. L.; Mesecar, A.; Mitsuya, H. Recent drug development and medicinal chemistry approaches for the treatment of

SARS-CoV-2 infection and COVID-19. *ChemMedChem* **2022**, *17*, e202200440.

19. Akinosoglou, K.; Schinas, G.; Gogos, C. Oral antiviral treatment for COVID-19: a comprehensive review on nirmatrelvir/ritonavir. *Viruses* **2022**, *14*, 2540.

20. Halford, B. Pfizer unveils its oral SARS-CoV-2 inhibitor: The antiviral candidate is the first orally administered compound to enter clinical trials that targets the virus's main protease. *Chem. Eng. News* **2021**, *99*(13), 2021.

21. Kadam, A. L.; Chiranjeevi, B.; Nunes, A. A.; Jayaraman, A.; Ahmad, S.; Aleshire, S. L.; Donsbach, K. O.; Gupton, B. F.; Nuckols, M. C.; Shanahan, C. S. Efforts to develop a cost-effective and scalable synthetic process for nirmatrelvir. *ChemRxiv* **2022**, 1–14.

22. Veeramani, K.; Shinde, M.; Eda, V. V. R.; Darapaneni, B. C.; Hindupur, R. M.; Madarapu, S. R.; Sen, S.; Oruganti, S. Alternate end-game strategies towards nirmatrelvir synthesis: defining a continuous flow process for the preparation of an anti-COVID drug. *Tetrahedron Lett.* **2023**, *116*, 154344.

23. Qin, H.; Jin, C.; Odilov, A.; Mintah Bonku, E.; Zhu, F.; Liu, Q.; Shen, J.; Aisa, H. A. Optimized synthesis of a key intermediate of nirmatrelvir. *Org. Process Res. Dev.* **2023**, *27*, 78–83.

24. Allais, C.; Ragan, J. A. Development of the commercial manufacturing process for nirmatrelvir in 17 months. *ACS Cent. Sci.* **2023**, 9, 849–857.

2

Doravirine (Pifeltro): A Third-Generation Non-Nucleoside Reverse Transcriptase Inhibitor as a Treatment of HIV-1 Infection

Jie Jack Li

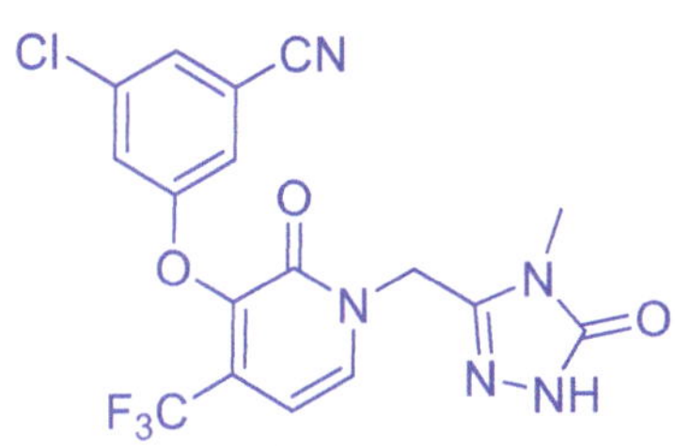

doravirine (Pifeltro, **1**)
Schering Plough/Merck, 2018
NNRTI

The coronavirus pandemic has wreaked havoc around the globe during the last few years. Yet, we must not forget that an epidemic has been going on for decades, i.e., the HIV/AIDS epidemic. From the beginning of the 1980s, AIDS is estimated to have killed more than 25 million worldwide. According to the United Nations' statistics, nearly 40 million people were living with HIV in 2021.[1] Notably, AIDS has replaced malaria and tuberculosis as the world's deadliest infectious disease. Even today, the United States sees about 40,000 new infections annually.

The FDA's approval of doravirine (Pifeltro, **1**), a third-generation non-nucleoside reverse transcriptase inhibitor (NNRTI), in 2018 was timely to contribute to the WHO's lofty goal of stopping the HIV/AIDS pandemic by 2030.

Chemistry and Pharmacology of Drug Discovery, First Edition. Edited by Jie Jack Li.

1. Background

Françoise Barré-Sinoussi and Luc Montagnier in France discovered the human immunodeficiency virus (HIV, Figure 1) in 1983 and were bestowed the Nobel Prize in 2008. The HIV encodes 15 proteins although only three of them have enzymatic activities: reverse transcriptase, protease and integrase. Nevertheless, even many nonenzymatic proteins have been successfully targeted as treatments of HIV/AIDS.

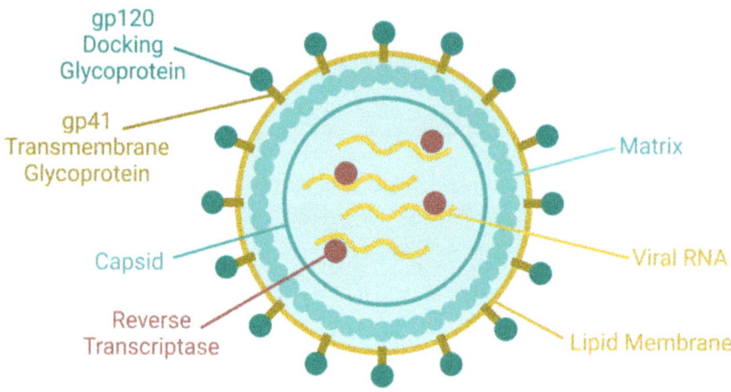

Figure 1. The structure of the HIV

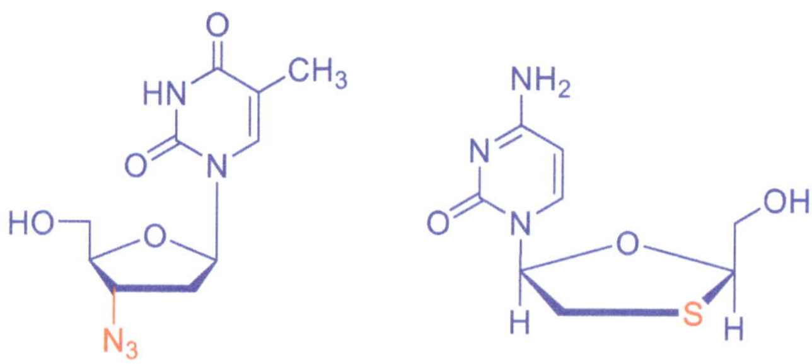

AZT (azidothymidine, Retrovir, **2**) Burroughs Wellcome, 1987 NRTI

Lamivudine (3TC, Epivir, **3**) GSK, 1995 NRTI

Since the discovery of AZT (azidothymidine, Retrovir, **2**) as the first effective treatment of AIDS, seven additional HIV-1 nucleoside reverse transcriptase inhibitors (NRTIs) have been approved by the FDA. They include GSK's Lamivudine (3TC, Epivir, **3**) and Gilead's tenofovir disoproxil (Viread, **4**) and emtricitabine (FTC, Emtriva, **5**), respectively. TNRTIs are orthosteric inhibitors binding to the active site (DNA polymerase) of the reverse transcriptase (*vide infra*), a key viral enzyme that produces double-stranded viral DNA genomes from a single-stranded viral RNA genome. In short, NRTIs function as viral DNA chain terminators.[2]

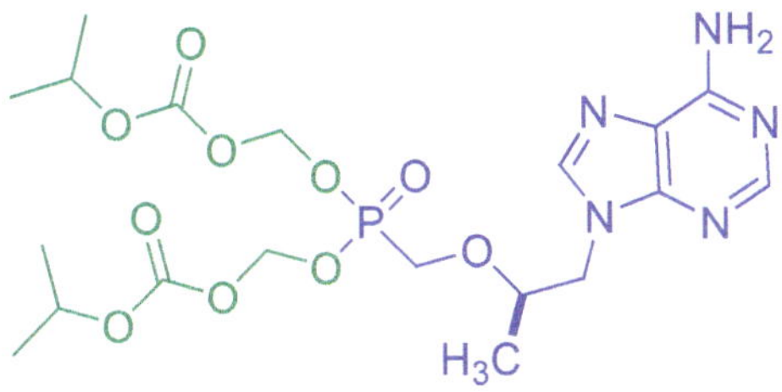

tenofovir disoproxil (Viread, **4**)
Gilead, 2001
prodrug of NRTI

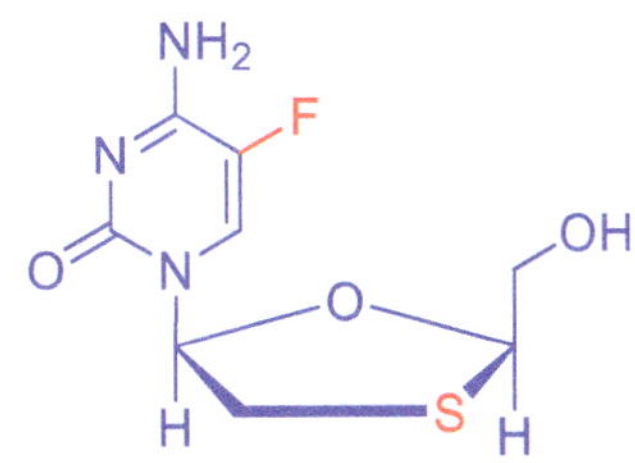

emtricitabine (FTC, Emtriva, **5**)
Gilead, 2003
NRTI

NRTIs have become the workhorse of highly active antiretroviral therapy (HAART), also known as antiretroviral therapy (ART): cocktail HIV-1 drugs that have significantly contributed to transforming AIDS from a death sentence to a chronic infection that can be managed with medicine.

HIV protease inhibitors were among the earliest drugs specifically developed for treating AIDS (AZT was initially developed as a cancer drug in the 1960s). Ten HIV protease inhibitors are on the market including ritonavir (Norvir, 1996), fosamprenavir (2005) and darunavir (Prezista, 2006). They are peptidomimetics that work as "transition state mimics." Their key hydroxyl group mimics the tetrahedral transition state of an amide bond of the polyprotein substrate being hydrolyzed by HIV protease.[3]

There are also four HIV integrase inhibitors approved by the FDA. They include raltegravir (Isentress, 2007), dolutegravir (Trivicay, 2013), elvitegravir (Vitekta, 2014), and bictegravir (Biktarvy with emtricitabine–tenofovir alafenamide, 2018). They work as integrase strand transfer inhibitors (InSTIs) to block the viral DNA from transferring its strand onto host DNA. All four HIV integrase inhibitors possess a very polar warhead as a bioisostere of diketoacid to chelate to the two magnesium ions that are important to integrase's catalytic activities.[4]

Furthermore, HIV entry inhibitors encompasses several targets. One is HIV fusion inhibitor enfuvirtide (Fuzeon) by Roche, approved in 2003. The other is CCR5 antagonist maraviroc (Selzentry) by Pfizer, approved in 2007. BMS's attachment inhibitor, fostemsavir (Rukobia) became available in 2020.[5]

More excitingly, long-acting HIV drugs are now available, which greatly enhance patience compliance. Among them, one is ViiV's Cabenuva, given one injection per month, combining an HIV integrase inhibitor cabotegravir (**6**) and a non-NRTI rilpivirine (Edurant, **7**).[6] Gilead's HIV-1 capsid protein inhibitor, lenacapavir (Sunlenca, **8**), only needs to be given as two injections per year![7]

Today, more than 30 approved drugs can efficiently suppress HIV viral load below the detectable level. Sadly, there is no effective vaccine, and no cure is available today to eradicate HIV infections globally.

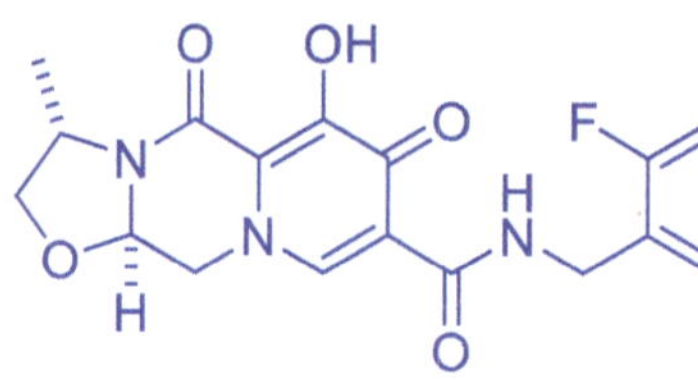

cabotegravir (**6**)
ViiV/GSK, 2020
INSTI, $t_{1/2}$ = 21–50 days

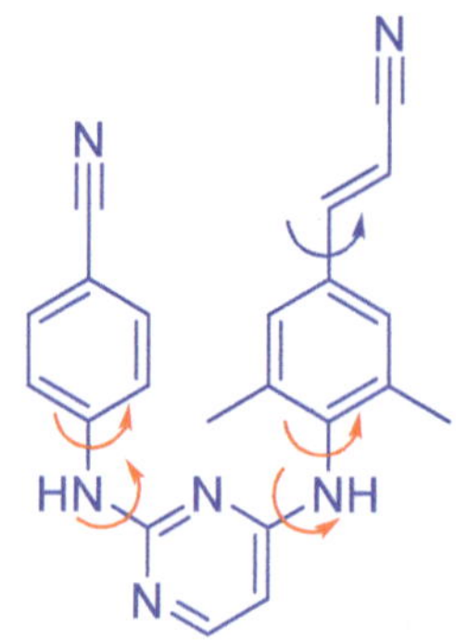

rilpivirine (Edurant, **7**)
Tibotec/Janssen, 2011
NNRTI

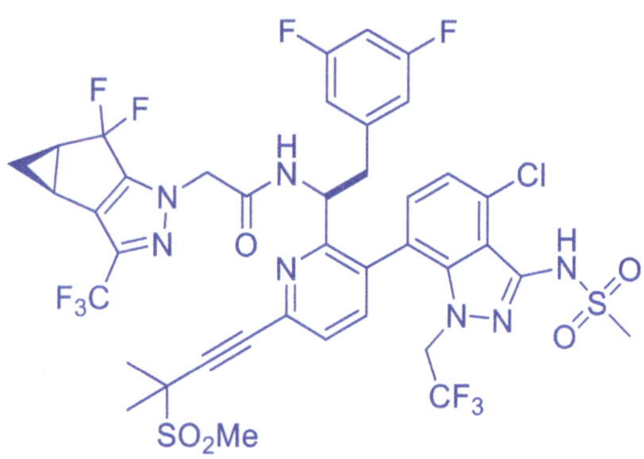

lenacapavir (Sunlenca, **8**)
Gilead, 2022
exceptionally long-lasting
($t_{1/2}$ = 7–11 weeks!)
HIV-1 capsid inhibitor

2. Pharmacology

HIV-1 reverse transcriptase is the primary enzyme responsible for the conversion of the viral single strand RNA to the double strand DNA.

In the late 1950s, Francis Crick, the codiscoverer of the double helix structure, advanced the *central dogma* in molecular biology. Central dogma states that the genetic information flows in a two-step process of transcription and translation in a unidirectional vector: DNA → mRNA → Protein (Figure 2). mRNA stands for messenger ribonucleic acid. Here comes a question. How does a retrovirus, such as HIV duplicate itself? Howard Temin and David Baltimore discovered reverse transcriptase in 1970 and won the Nobel Prize in 1975.

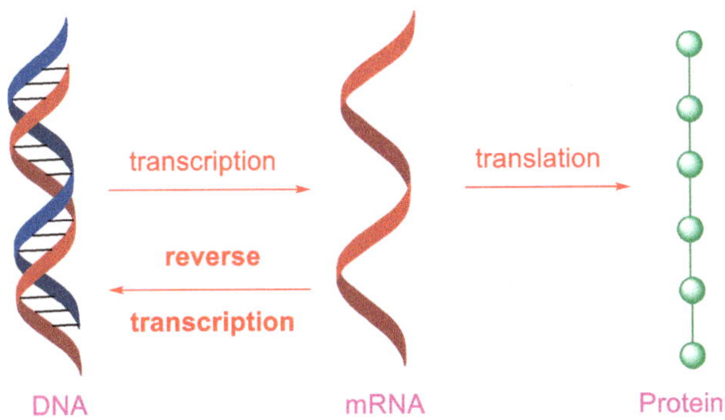

Figure 2. The central dogma of molecular biology

HIV reverse transcriptase enzyme is a component of the virion and is encoded by the *pol* gene and is manufactured in the HIV-infected cells as a *gag-pol* fusion polyprotein. HIV reverse transcriptase is an RNA-dependent DNA polymerase (RdDp). But it is not the only enzyme necessary for the translation of RNA to DNA. The other enzymes for this conversion also include DNA-dependent DNA polymerase (DdDp) and ribonuclease (RNase) H (see Figure 4, *vide infra*). In the heterodimeric protein of reverse transcriptase, the p51 subunit does not have a catalytic function; it only participates in conformational regulation for the p66 subunit. The p66 subunit, which is responsible for performing all catalytic activities, consists of the DNA polymerase domain and the RNase H domain.[8]

NRTIs as represented by **3**–**5** are *competitive* (with ATP) inhibitors. The reverse transcriptase's DNA polymerase activity is the target of *all* currently approved reverse transcriptase inhibitors. They are prodrugs that are converted to the corresponding nucleotides via phosphorylated by the host cellular kinases. From nucleosides **2**–**5**, the corresponding nucleotides generated by cellular kinases bind to the active site of the DNA polymerase and compete with their natural counterparts thus prevent insertion of the next endogenous nucleotide. Because the artificial nucleotides do not have a 3′-OH in the ribose ring, they act as chain terminators and block DNA polymerization after being incorporated into a growing strand of the pro-viral DNA during reverse transcription. The active site locates at the palm region, where DNA is elongated, as the polymerase shapes like a human right hand, so the four constituent subdomains are referred to as *fingers, palm, thumb, and connection* (Figure 3).

HIV-1 reverse transcriptase enzyme is extremely error-prone with a high error rate (1 in 1700 bases), which produces higher incidents of mutation. As shown in Figure 3, one of the most frequent single-point mutations of reverse transcriptase is K103N, i.e., amino acid lysine (K) at position-103 of the wild-type virus is mutated to asparagine (N) for the mutated virus. The use of one or two NRTIs usually led to rapid treatment failure due to mutation. Novel antiretroviral drugs with a different mechanism of action were needed to achieve sufficient virologic suppression.[9]

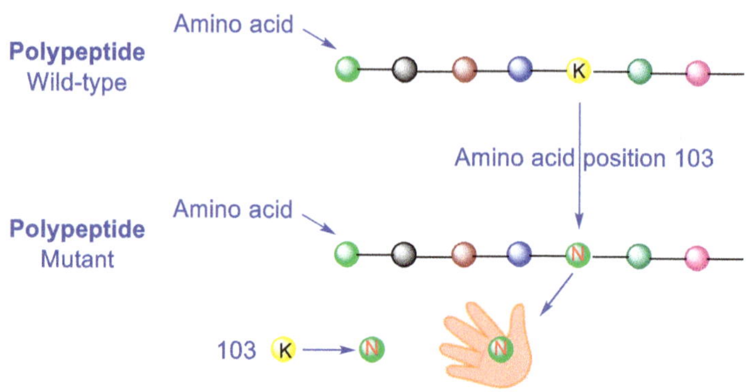

Figure 3. Single-point mutation of HIV-1 reverse transcriptase

The first-in-class NNRTI was Boehringer Ingelheim's nevirapine (Viramune, **9**). It was followed by Upjohn's delavirdine (Rescriptor, **10**) and DuPont–Merck's efavirenz (EFV, Sustiva, **11**). These three antiretroviral drugs **9–11** are classified as the first-generation NNRTIs. Unfortunately, Pfizer decided to discontinue making delavirdine (**10**) in 2018 based on a business decision and not for safety reasons.

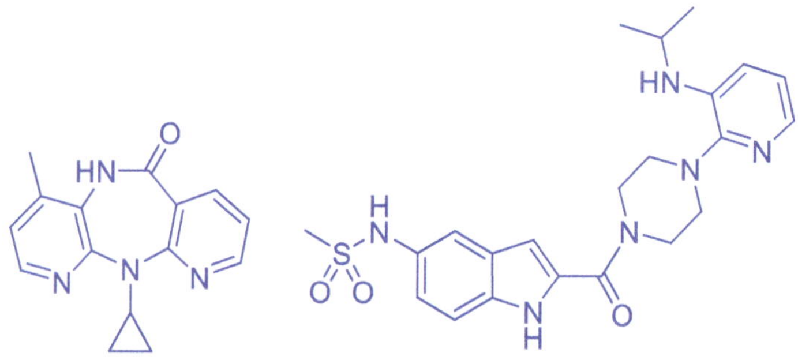

nevirapine (Viramune, **9**)
Boehringer Ingelheim, 1996

delavirdine (Rescriptor, **10**)
Upjohn/Pfizer, 1997

All NNRTIs assume a "butterfly" type of conformation in *noncompetitive* binding to the *allosteric site* of HIV-1 reverse transcriptase. In particular, they bind to a *hydrophobic domain* known as the non-nucleoside inhibitor-binding pocket (NNIBP) that is approximately 10 Å distal from the active subdomain of the DNA polymerase and 60 Å from the active site of RNase H in HIV-1 reverse transcriptase. This binding site is non-existing in the unbound RT but it is created upon binding of an NNRTI. The binding of an NNRTI to NNIBP influences the geometry of the reverse transcriptase active catalytic site and interferes with viral DNA synthesis.[10]

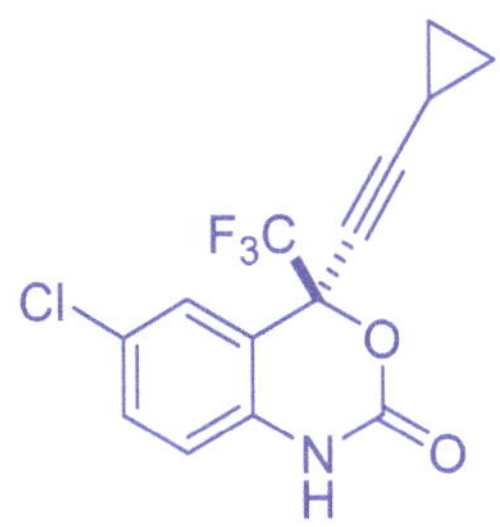

efavirenz (Sustiva, **11**)
BMS/Merck, 1998

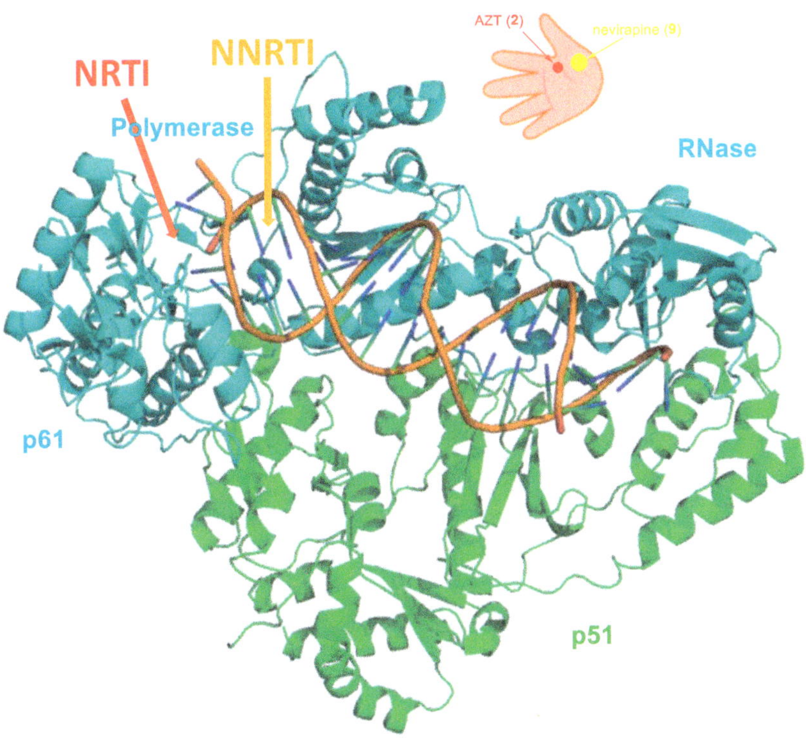

Figure 4. The three-dimensional structure of HIV-1 reverse transcriptase, drawn using PyMOL and coordinates from Protein Data Bank (PDB) file of 6UIS by Faridoon

As shown in Figure 4, the ribbon representation of the reverse transcriptase active domain illustrates its hand-like structure with AZT (**2**, red dot on the hand at the top right) in the active domain (D110, D185, and D186 as the catalytic triad) and nevirapine (**9**, yellow) in the non-nucleoside binding pocket.[11]

The allosteric binding site is not crucial to reverse transcriptase function and is not directly involved in substrate binding or viral DNA synthesis. As a consequence, point mutations can occur within the NNIBP that hinder NNRTI binding but do not interfere with reverse transcriptase's role in viral DNA synthesis. Some of these mutations make the virus resistant to NNRTI treatment. The general trends have shown that mutations create a larger allosteric site.

The first-generation NNRTIs showed a significant loss of activity with single-point mutations, especially the two most prevalent single mutants, K103N and Y181C, selected by nevirapine (**9**) and efavirenz (**11**). Second-generation NNRTIs, etravirine (Intelence, **12**) and rilpivirine (Edurant, **7**) as diarylpyrimidines, largely overcome the two mutants K103N and Y181C. Because of their conformational flexibility, these two diarylpyrimidines can "wiggle" and "jiggle" to adapt to bind the mutated allosteric site of NNIBP of the mutant reverse transcriptase. As a result, they exhibit prominent antiviral activity toward of the clinically common mutations of K103N and Y181C.[12]

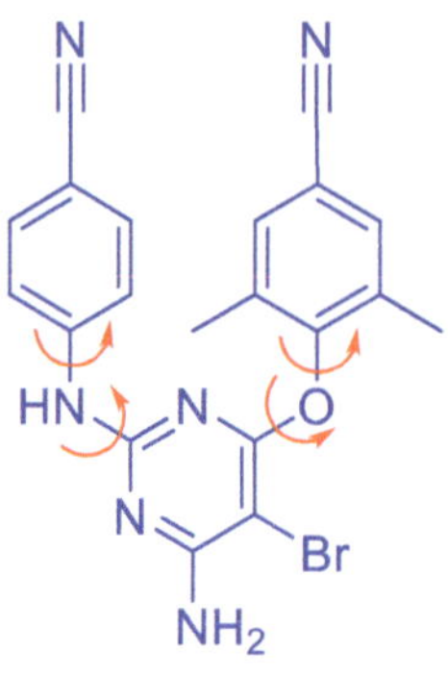

etravirine (Intelence, **12**)
Tibotec/J&J, 2008
NNRTI

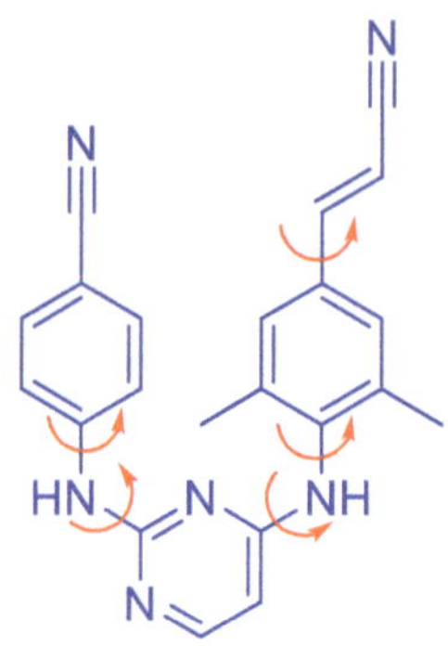

rilpivirine (Edurant, **7**)
Tibotec/Janssen, 2011
NNRTI

However, second-generation NNRTIs suffer from poor aqueous solubility and unfavorable pharmacokinetic properties although rilpivirine (**7**) has an improved bioavailability than its prototype etravirine (**12**). Furthermore, other point mutations within the NNIBP take place eventually, prominently K101P, E138R, and Y188L as shown in Table 1.[13] Some patients on either etravirine (**12**) or rilpivirine (**7**) also suffer from severe rash as a severe adverse effect (SAE). Better NNRTIs are still needed.

Third-generation NNRTIs are represented by Merck's doravirine (**1**), available since 2018 in the United States. Elsulfavirine (Elpida, **13**), a prodrug, is only approved in Russia[14] and ainuovirine (AiBangDe, **14**) has been available in China since 2021.[15]

Table 1. Key HIV-1 mutations

Generic	Approval year	Dose (mg)	Key mutation	Notes
Nevirapine (**9**)	1996	200 bid	K103N, Y181C	Liver toxicity
Delavirine (**10**)	1997	400 qd	K103N, Y181C	Discontinued
Efavirenz (**11**)	1998	600 qd	K103N, Y181C	CNS AEs
Etravirine (**12**)	2008	200 bid	K101P, Y181C	Severe rash
Rilpivirine (**7**)	2011	25 qd	E138R, Y188L	Severe rash

Source: Adapted from Wang et al. [13]

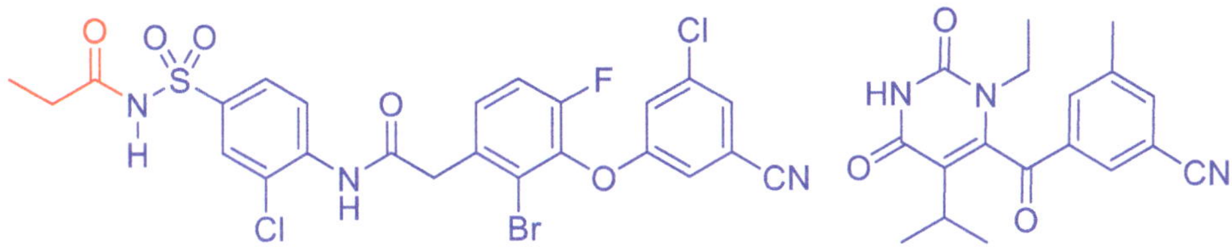

elsufavirine (Elpida, **13**)
Viriom, 2017 (Russia)
NNRTI

ainuovirine (AiBangDe, **14**)
Jiangsu Aidea, 2021 (China)
NNRTI

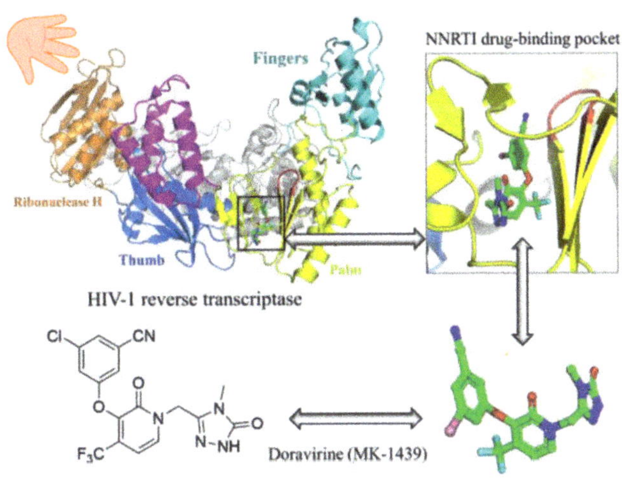

Figure 5. Doravirine (**1**) specifically targets the allosteric binding pocket of HIV-1 reverse transcriptase. Source: Reproduced with permission from Elsevier[16]

Just like all other NNRTIs, doravirine (**1**) binds to the allosteric NNIBP, which is approximately 10 Å away from the DNA polymerase active site (Figure 5). Unlike first-generation NNRTIs **2–5**, doravirine (**1**), the binding to HIV-1 reverse transcriptase is less dependent on K103 and Y181. Instead, the most relevant doravirine (**1**)-reverse transcriptase interaction is the hydrophobic interaction between its phenyl ring and the side chain of V106.[16]

Doravirine (**1**) is featured with excellent antiviral activity, favorable safety and tolerability profiles, low potential for drug resistance, and drug–drug interactions (DDIs). As a new pyridone NNRTI, doravirine (**1**) exhibits an excellent efficacy in suppressing HIV-1 viral replication (EC_{50} = 20 nM). Regarding its resistance profile, doravirine (**1**) can effectively inhibit many HIV-1 strains with NNRTI-resistant mutations including K103N, Y181C, and G190A as depicted in Table 2. Unlike its predecessors efavirenz (**11**), etravirine (**12**) and rilpivirine (**7**), doravirine (**1**) is potent against the wild-type enzyme as well as the clinically relevant single-point and double-point mutants.[17]

Table 2. Mutant Profile NNRTIs

Drug	WT (IC_{95} nM)	K103N (IC_{95} nM)	Y181C (IC_{95} nM)	K103N/Y181C (IC_{95} nM)
Efavirenz (**11**)	39	1400	90	3200
Etravirine (**12**)	33	44	240	590
Rilpivirine (**7**)	36	44	120	370
Doravirine (**1**)	19	42	25	54

Source: Côte et al. [17]/with permission of Elsevier

Doravirine (**1**)-resistance is usually associated with the combination of V106A/G190A/F227L, suggesting the limitation of doravirine (**1**) against certain HIV-1 drug-resistant strains.[18]

3. Structure–Activity Relationship (SAR)

Using a cell-based assay, Merck's initial hit from high-through screen (HTS) provided tetrazole thioacetanilide **15** as one of the two interesting hits. It was a potent inhibitor of HIV-1 reverse transcriptase polymerase (pol), with sub-micromolar activity in a cell assay and significant *in vitro* activity on the K103N mutant strain. Extensive optimization resulted in tetrazole thioacetanilide **16**, which had a reasonable oral bioavailability but suffered from a short half-life of merely 0.23 h in rat.[19]

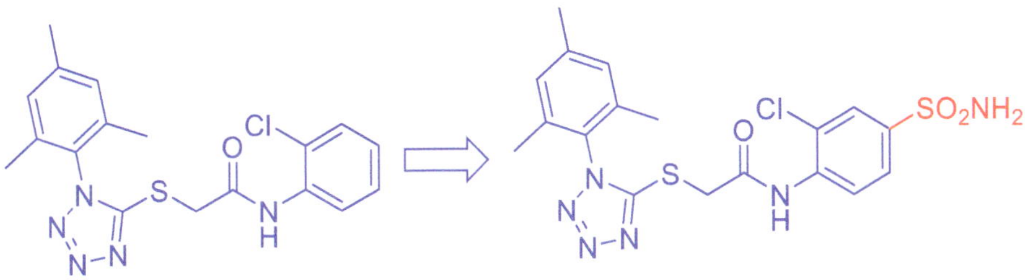

HTS hit **15**
WT HIV-1 III_B
RT pol IC_{50} = 5 nM
K103N IC_{50} = 15 nM

optimized **16**
RT pol IC_{50} = 4 nM
K103N IC_{50} = 12 nM
$t_{1/2}$ = 0.23 h (rat)
F = 33% (rat)

Inspired by Glaxo Wellcome's acetanilide *para*-sulfonamide, GW-8248 (**17**), Merck decided to use it as their starting point. Replacing the benzophenone moiety in **17** conveniently provided novel chemotype diarylether **18** although it was only moderately potent for the wild-type HIV reverse transcriptase as well as clinically relevant mutants K103N and Y181C. Further structure-activity relationship (SAR) investigation revealed that the chloro- and nitrile-substituents on **17** were crucial to its potency, probably by making all three phenyl rings A, B, and C more electron-deficient. Decorating diarylether **18** with additional chloro- and nitrile-substituents on its A- and B-ring, respectively, resulted in highly potent halogenated analog MK-4965 (**19**) where the phenylsulfonamide was replaced by an aminoindazole. It possessed high levels of potency against wild-type and key mutant viruses, excellent oral bioavailability and overall pharmacokinetics and a clean ancillary profile. It was nominated as a development candidate.[20, 21]

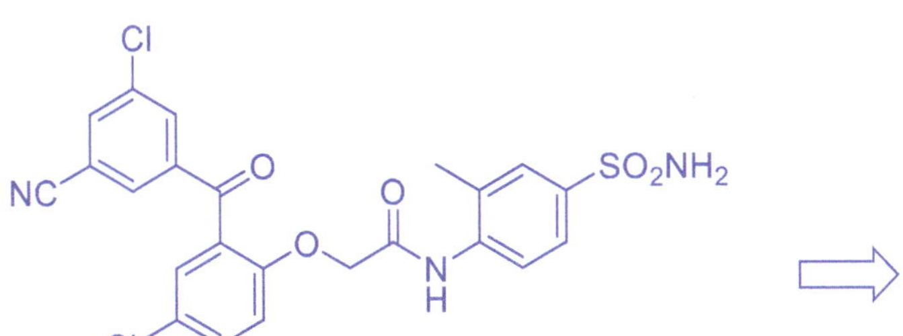

GW-8248 (**17**)

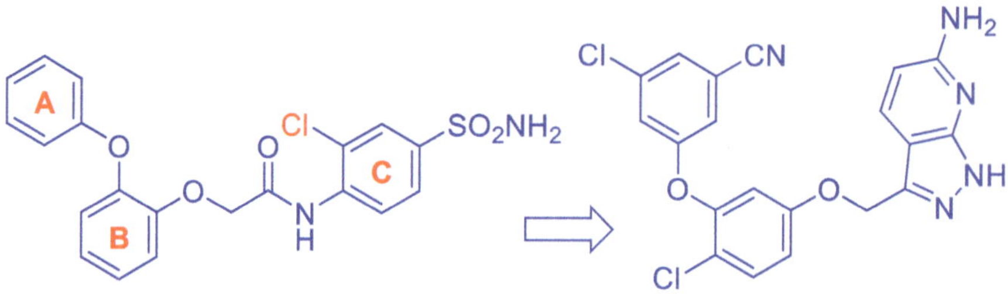

diarylether **18**
wt HIV RT IC_{50} = 52 nM
K103N IC_{50} = 190 nM
Y181C IC_{50} = 722 nM

MK-4965 (**19**)
Development candidate
wt HIV RT IC_{50} = 0.2 nM
K103N IC_{50} = 0.5 nM
Y181C IC_{50} = 0.4 nM
$t_{1/2}$ = 3.5 (rat)
F = 52% (rat)

Meanwhile, replacing the phenyl B ring with a novel pyridone core structure led to the discovery of pyridone **20** with increased polarity and binding affinity. It had better antiviral activity against HIV-1 strains with the K103N/Y181C mutation.[22]

Systemic SAR optimization of the methyl substituent revealed that the trifluoromethyl-pyridone derivative **21** was potent for both wild-type reverse transcriptase and double mutant (K103N/Y183C) enzyme inhibition. Even though **21** showed improved plasma stability, it suffered poor solubility and low oral absorption due to an esoteric, highly ordered intermolecular hydrogen bonding of the pyrazolopyridine motif. Eventually, optimization led to the replacement of pyrazolopyridine with methyl-triazolone to provide doravirine (**1**). Even though doravirine (**1**) was not as potent as MK-4965 (**19**), it had a superior DMPK profile. It became the clinical drug candidate and eventually achieved the FDA approval in 2018.[10, 16, 17]

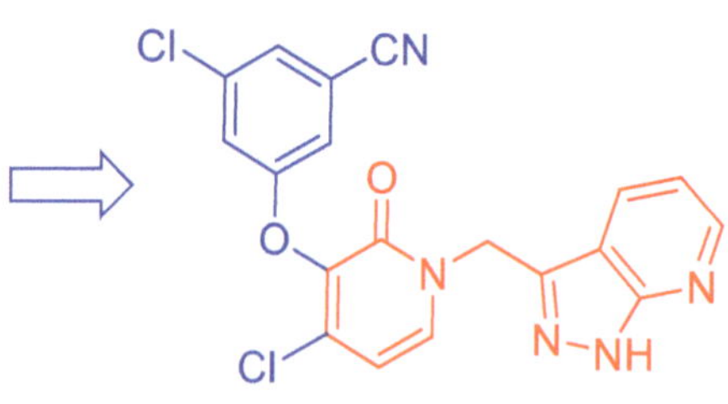

pyridone **20**
wt RT IC_{50} = 0.77 nM
K103N IC_{50} = 1.5 nM
Y181C IC_{50} = 1.0 nM
$t_{1/2}$ = 4 h (rat), F = 57% (rat)
V_d = 2.0 L/kg (rats)
Cl = 12 mL/min/kg (rat)

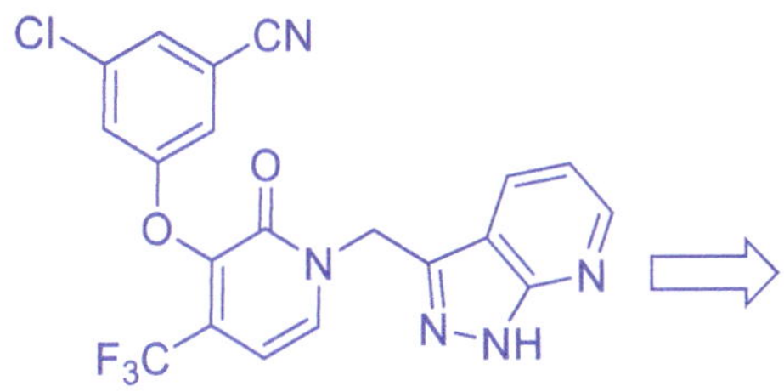

CF_3-pyridone **21**
wt RT IC_{50} = 19 nM
$t_{1/2}$ = 7 h (rat)
F = 13% (rat)

doravirine (**1**)
wt HIV RT IC_{50} = 11 nM
K103N IC_{50} = 42 nM
Y181C IC_{50} = 25 nM
K103N/Y181C IC_{50} = 54 nM
V_d = 1.4 L/kg (rhesus)
Cl = 3.5 mL/min/kg (rhesus)
$t_{1/2}$ = 4.5 h (rhesus)
F = 60% (rhesus)
AUC_{po} = 6.7 μM*h (rhesus)

Instead of a thorough tabulation of the voluminous SAR in literature, only two important optimization processes are summarized here so we can learn important lessons of drug design and medicinal chemistry.[23]

Table 3. SAR for the Optimization of the 4-Position of the Pyridone Core

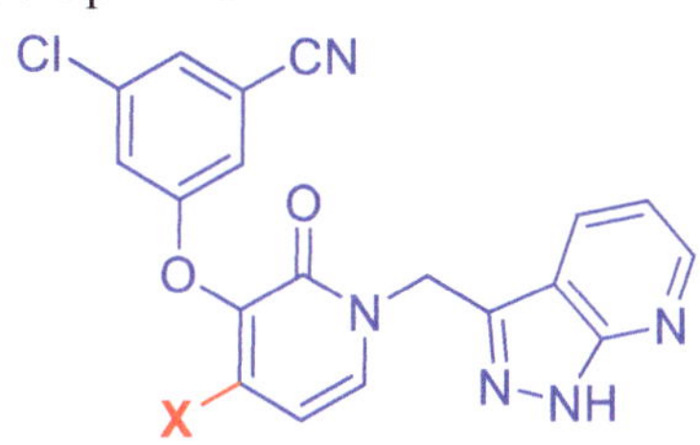

—	—	Inhibition of RT	Spread IC95	(50% NHS; nM)	–
Compound	X	(IC_{50} nM), WT	WT	K103N/Y181C	$t_{1/2}$ (h, rat)
Efavirenz (**11**)	—	2	39	3200	–
—	Me	7	12	96	0.8
20	Cl	3	8	69	2.2
21	CF_3	3	17	69	7.0
—	Br	2	4	21	1.7
—	SMe	9	15	33	1.5
—	*c*-Pr	7	12	110	1.5

Source: Adapted from Burch et al. [23]

First, let us look at the impact of the 4-position of the pyridone core on potency. As shown in Table 3, the 4-methyl pyridone derivative was tested to be a potent inhibitor of both wild type and mutant forms of HIV reverse transcriptase using the SPREAD assay in 50% normal human serum (NHS). Similarly, 4-chloropyridone **20** and 4-trifluoromethylpyridone **21** were just as potent, as were the bromo-, methylsulfanyl-, and cyclopropyl-pyridone analogs. 4-Trifluoromethylpyridone **21** stood out because it had the longest half-life in rat, suggesting potentially a higher bioavailability. As a result, 4-trifluoromethyl substituent was adopted as optimal.[17, 23]

Next, let us examine the impact of the sidechain substituents on the potency and solubility of 4-trifluoromethyl pyridones.

Table 4. Structure-activity and solubility relationship for side chain replacements

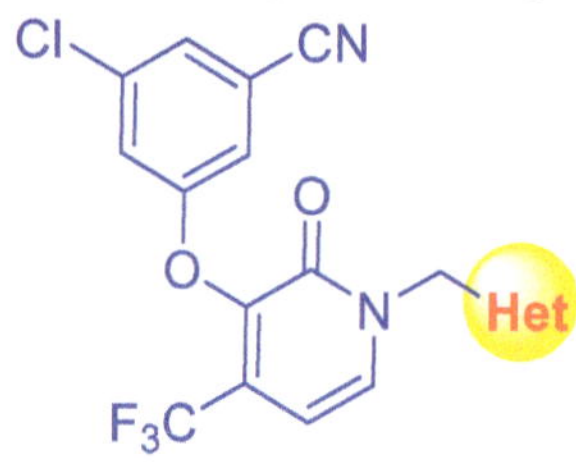

Het	Inhibition of RT (IC_{50} nM), WT	Spread IC_{95}, WT (50% NHS; nM)	Solubility (μM)
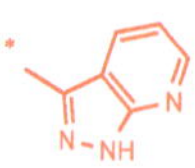 **(21)**	3	17	1.1
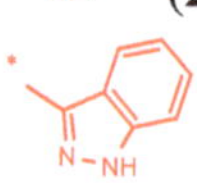	6	53	5.2
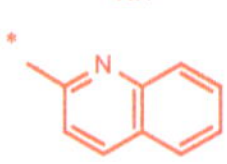	6	210	3.1
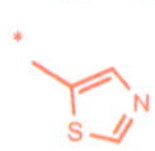	390	–	23
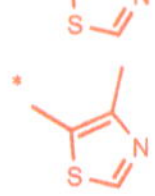	16	100	44
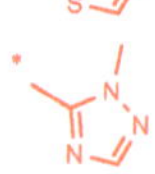	43	150	28
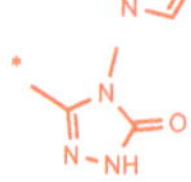 (1)	11	19	45

Because **21**'s unfortunate propensity to form highly ordered intermolecular hydrogen-bonding that resulted in poor solubility, the pyrazolopyridine sidechain was systematically optimized. As shown in Table 4, replacement sidechains were chosen to remove at least one hydrogen bond donor or acceptor to minimize the key donor–acceptor hydrogen bond with the backbone of K103. Replacing the pyrazolopyridine on **21** with indazole led to improvement of kinetic solubility in aqueous media despite being more lipophilic, presumably due to removal of the pyridine nitrogen of the donor–acceptor hydrogen bond pair. The most significant improvement of solubility came when a monocyclic thiazole was employed to replace the bicyclic heterocyclic sidechains. Although the thiazole analog lost much of its potency, installation of an additional methyl group on thiazole restored the potency. Gratifyingly, isosteric methyltrizolinone (**1**) afforded drastically improved solubility without loss of cellular potency relative to pyrazolopyridine on **21**.[17, 23]

4. Pharmacokinetics and Drug Metabolism

After oral administration, doravirine (**1**) is quickly absorbed due to its high-absolute bioavailability (64%). In healthy volunteers, the maximum plasma concentration (C_{max}) is reached in 1.5 h (range: 1–5 h) after a single dose of doravirine (**1**). The plasma half-life ($t_{1/2}$) of doravirine (**1**) is approximately 12–21 h and its clearance is approximately 3.73 L/h. The hepatic biotransformation of doravirine (**1**) is processed by cytochrome P450 (CYP450)-3A4 and P-glycoprotein (P-gp), while other compounds that induce or inhibit CYP3A may alter the plasma concentration of doravirine (**1**), thereby affecting its efficacy (Table 5).[10]

Table 5. Pharmacokinetics of NNRTIs

Generic	Approval year	Dose (mg)	$T_{1/2}$ (h)	Metabolic enzyme
Efavirenz (**11**)	1998	600 qd	40–55	CYP3A4
Rilpivirine (**7**)	2011	25 qd	50	CYP3A4
Doravirine (**1**)	2018	100 qd	11–15	CYP3A
Ainuovirine (**14**)	2021	150 qd	26	CYP2C19

Source: Adapted from Namasivayam et al.[10]

Doravirine (**1**) is rather pervious to drug metabolism, 75% of the ^{14}C-labeled drug was found in plasma unchanged. The most abundant metabolite M9 (**22**), is a product of oxidative metabolism that added an additional oxygen. The published structure of M9 (**22**), if correct, was probably a result of several steps of complicated rearrangements after metabolic hydroxylation. A glucuronide of an oxidative metabolite (M7, **23**) and an *N*-acetyl-cysteine conjugate of doravirine (M15, **24**) contributed a cumulative 2.8% of the radioactivity in this matrix. Other metabolites were less significant.[24]

Doravirine (**1**) is a P-gp substrate but P-gp efflux is not expected to play a significant role in limiting doravirine (**1**) absorption or to be involved in the elimination of doravirine (**1**). The disposition of doravirine (**1**) in human is shown below. The primary route of elimination is excretion through feces (90%) with the remaining elimination through urine (10%).[18, 24]

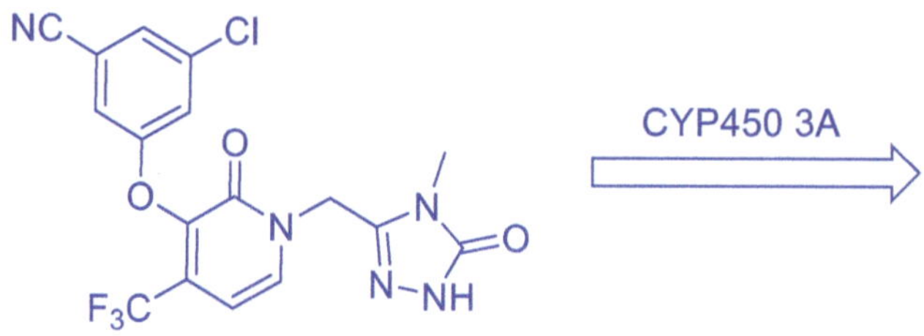

doravirine (**1**)
75% in plasma

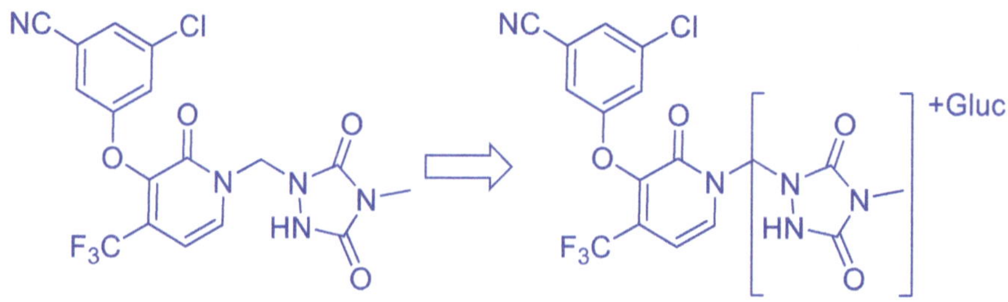

M9 (**22**)
12.9% in plasma

M7 (**23**)
~1.4% in plasma

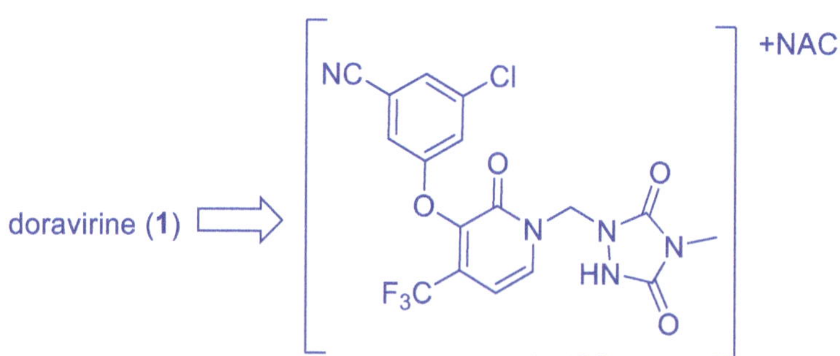

M15 (**24**)
~1.4% in plasma

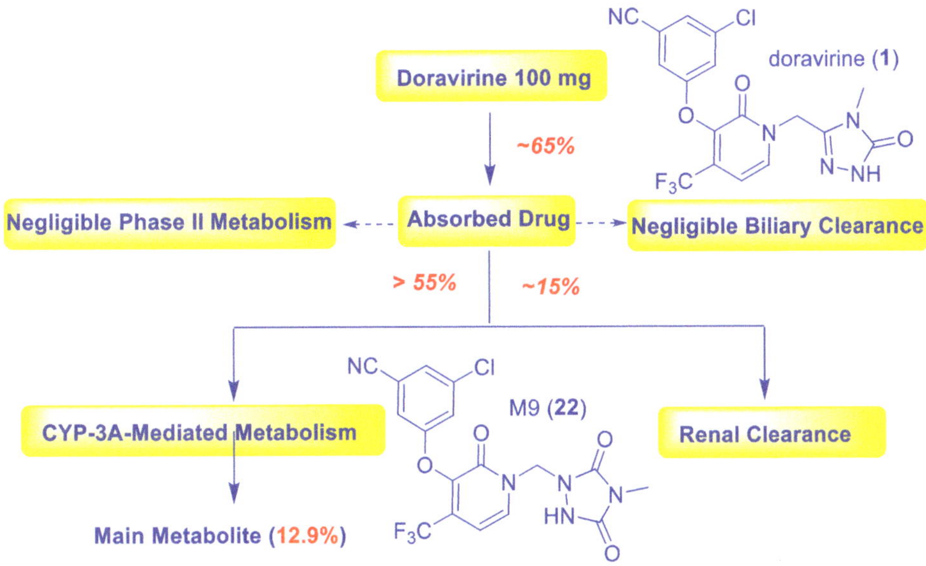

5. Efficacy and Safety

Because of the rapid development of resistance, NNRTIs cannot be applied as monotherapy in the management of HIV. The oral use of doravirine (**1**) 100 mg should be combined with other antiretroviral drugs for the treatment of HIV-1 infections in treatment-naïve adults.

In a Phase III non-inferiority trial of treatment-naive adults, 84% patients on the combination drug DOR/3TC/TDF with 100 mg doravirine (**1**) once daily co-formulated with 3TC (**3**) and TDF (**5**) achieved plasma HIV-1 RNA < 50 copies/mL at week 48. In comparison, a similar three-drug combination for EFV/3TC/TDF had 81% patients achieve the same goal. DOR/3TC/TDF was generally well tolerated with significantly fewer neuropsychiatric adverse events observed with doravirine (**1**) than with efavirenz (**11**) and a lipid profile superior to efavirenz (**11**) as assessed by change from baseline in LDL-C and non-HDL-C.[25]

Doravirine (**1**) has an IC_{50} of 88 μM against hERG, which is 30× higher than the total maximum serum concentration (C_{max}) and 126× higher than the unbound C_{max} steady-state concentrations of doravirine (**1**) at 100 mg once daily, the approved clinical dose. In clinics, a supratherapeutic dose of 1200 mg did not induce clinically meaningful differences in QTc and no issues related to delayed ventricular repolarization have been reported.[26]

6. Synthesis

Merck Process Chemistry carried out extensive optimizations to make doravirine (**1**). Herein, we only focus on the supply route and the final manufacturing route.

Generally speaking, discovery chemistry routes to make active pharmaceutical ingredients (APIs) are not amenable to process chemistry routes. This does not come as a surprise because discovery and process chemistry have different roles. Discovery chemistry enables making as many analogues as possible so the synthetic routes are often *modular*. A common intermediate can afford many derivatives to provide glimpse of SAR. In contrast, which drug to make is already known for process chemistry. Therefore, process routes are often *convergent* to make the APIs quickly, cheaply, and safely on large scales.

Merck's supply route commenced with an iridium-catalyzed *meta*-borylation of (**25**), which was superior to the palladium-catalyzed protocol that was associated with dichlorination. As important, it was discovered that inexpensive ligand 2,2′-bipyridyl (bpy) in cyclohexane (CyH) worked just as well as the expensive 4,4′-di-*tert*-butyl-2,2′-bipyridyl (dtbpy). After switching the solvent to acetone, oxidation of the resulting pinacol boronate intermediate was carried out in one-pot to give phenol **26**. Phenol **26** was then coupled with 2-chloro-3-fluoro-4-(trifluoromethyl)pyridine (**27**) to assemble diarylether **28** via a selective (for the *meta*-fluorine over the *ortho*-chlorine) S_NAr reaction. Hydrolysis of the α-chloropyridine moiety on **28** using NaOH led to pyridone **29**, which was converted to nitrile **30** using a simple S_NAr cyanation with CuCN. The crucial S_N2 coupling between pyridone **29** and triazolinone chloride **31** was straightforward to afford adduct **32**. Triazolinone chloride (**31**) was chosen over its methylated derivative due to its ready availability as it is a shared intermediate for making another Merck drug aprepitant (Emend), a substance P/neurokinin 1 receptor antagonist prescribed to treat nausea and vomiting caused by chemotherapy or surgery. Finally, selective methylation of **32** then delivered doravirine (**1**) optimally after carefully screening a variety of bases and solvents.[27]

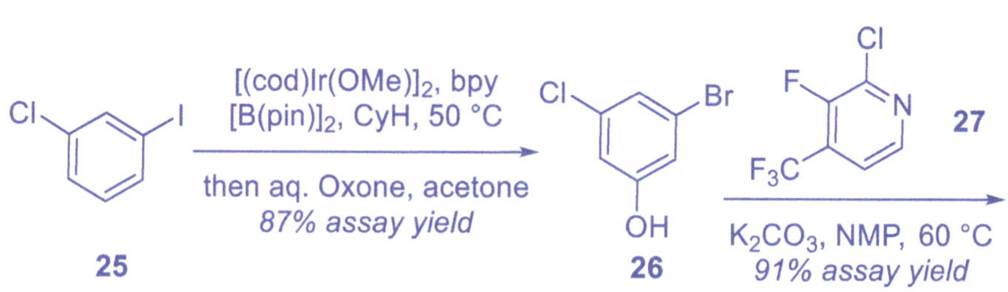

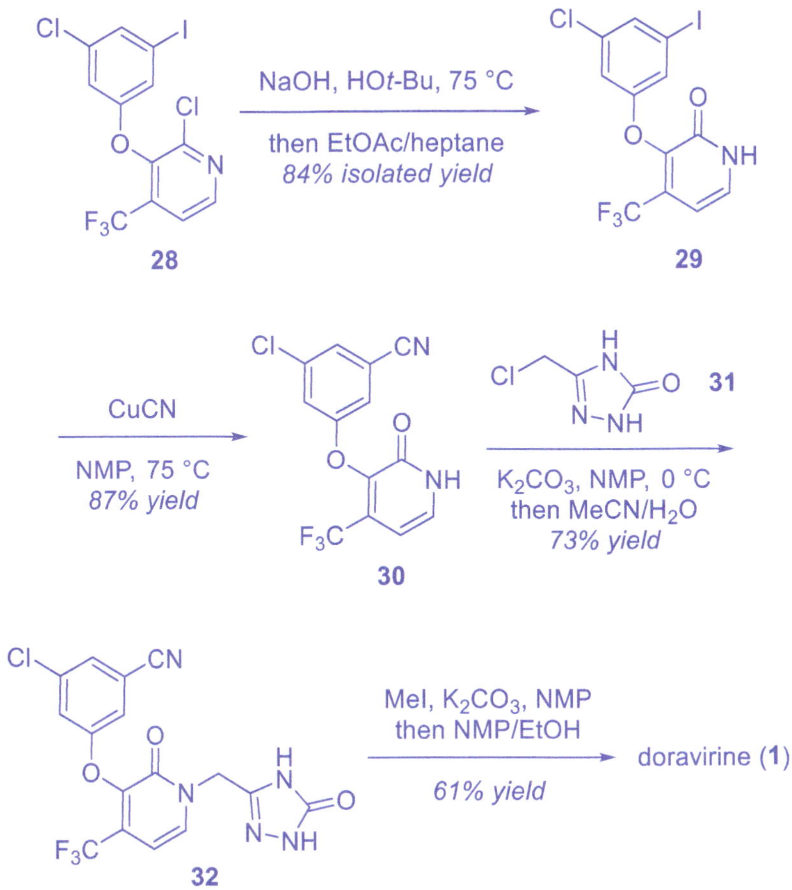

Merck's manufacturing route for making doravirine (**1**) was truly innovative. Instead of using substituted pyridines as their starting materials, they decided to build the pyridone core structure via cyclization. After extensive experimentations, flow chemistry was employed to prepare aryloxyl-pyridone **30**. As shown below, Aldol condensation between ester **33** and vinylogous ester **34** mediated by potassium *tert*-amyloxide in toluene was carried out in a flow reactor with aid of triethylamine. The unquenched aldol exit stream from the flow reactor was collected in a cooled receiver vessel to which trifluoroacetic anhydride was added synchronously to provide diene **35**. Amination, cyclization, and dehydration took place when diene **35** was treated with 28 equiv of ammonia at 60 °C to produce aryloxyl-pyridone **30** in 68% yield.[27]

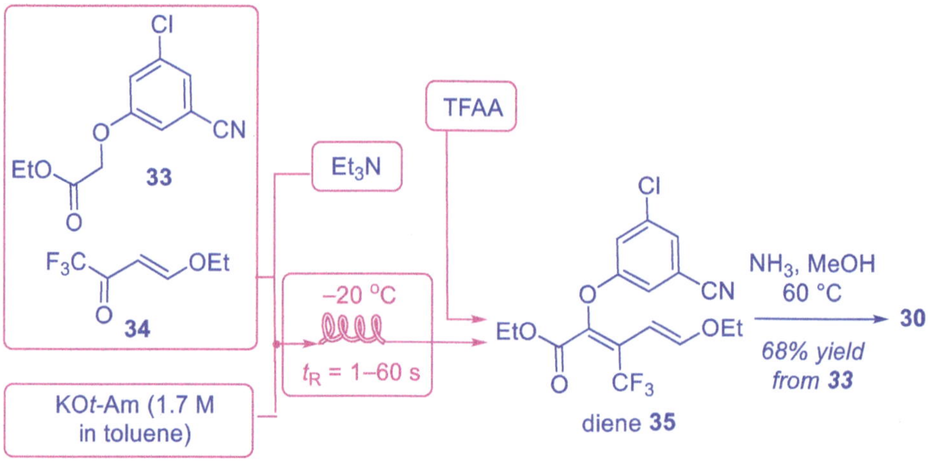

Meanwhile, a streamlined synthesis of triazolinone **40** was developed. To that end, treating phenyl chloroformate with aqueous methylamine prepared carbamate **36**. Semicarbazide **37** was generated by the addition of hydrazine in hot 2-propanol to carbamate **36** and converted without isolation to acylated adduct **38** in 81% yield over the two steps. Based-catalyzed cyclization furnished triazolinone-alcohol **39**, which was converted to triazolinone chloride **40**.[28]

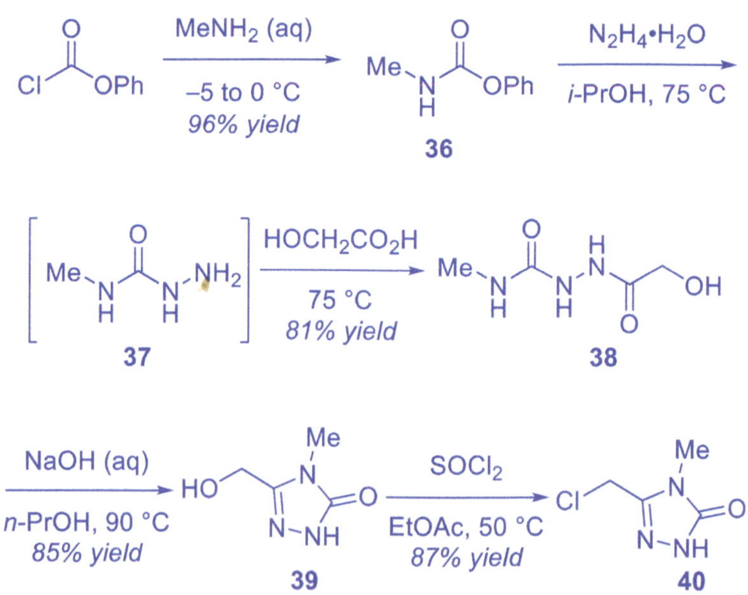

Eventually, doravirine (**1**) was assembled using an optimized S_N2 reaction between pyridone **30** and chloride **40**.[28]

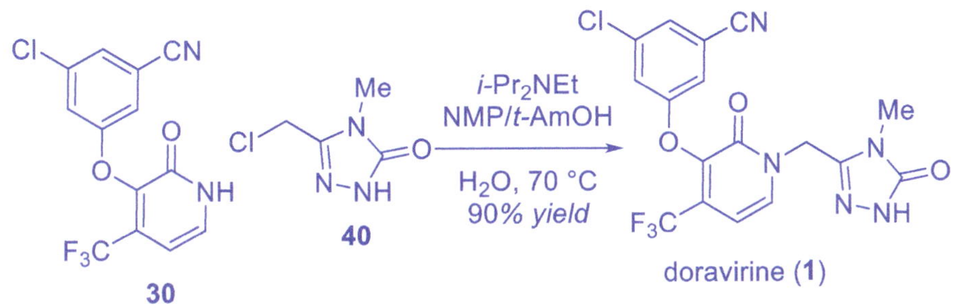

7. Summary

Although their own HTS hits did not offer viable leads, Merck did not shy away from a competitor's compound. The trick was to achieve novel intellectual properties through smart drug design of novel bioisosteres. Pyridone as the core structure was not only novel but also possessed superior physiochemical properties in comparison to the original diarylether.

Another laudable achievement of drug design was finding the monocyclic methyl-triazolone as the sidechain in place of bicyclic pyrazolopyridine. It overcame the intermolecular hydrogen bonding issue associated with pyrazolopyridine to boost the solubility. Moreover, a smaller molecule doravirine (**1**) is likely to have better physiochemical properties than the bicyclic analogues.

Finally, employing the state-of-the-art flow chemistry has contributed to the success of the robust manufacturing process.

References

1. UN/UNAIDS statistics: https://www.unaids.org/en/resources/fact-sheet, accessed on Feb. 14, 2023.
2. Amblard, F.; Patel, D.; Michailidis, E.; Coats, S. J.; Kasthuri, M.; Biteau, N.; Tber, Z.; Ehteshami, M.; Schinazi, R. F. HIV nucleoside reverse transcriptase inhibitors. *Eur. J. Med. Chem.* **2022**, *240*, 114554.
3. Ghosh, A. K.; Weber, I. T.; Mitsuya, H. Beyond darunavir: recent development of next generation HIV-1 protease inhibitors to combat drug resistance. *Chem. Commun.* **2022**, *58*, 11762–11782.

4. Wang, Y.; Gu, S.-X.; He, Q.; Fan, R. Advances in the development of HIV integrase strand transfer inhibitors. *Eur. J. Med. Chem.* **2021**, *225*, 113787.

5. Jiang, S.; Lu L., eds., *Virus Entry Inhibitors: Stopping the Enemy at the Gate. Advances in Experimental Medicine and Biology*; *Volume 1366*, Springer Nature Singapore, **2022**.

6. Taki, E.; Soleimani, F.; Asadi, A.; Ghahramanpour, H.; Namvar, A.; Heidary, M. Cabotegravir/rilpivirine: the last FDA-approved drug to treat HIV. *Expert Rev. Anti. Infect. Ther.* **2022**, *20*, 1135–1147.

7. Paik, J. Lenacapavir: first approval. *Drugs* **2022**, *82*, 1499–1504.

8. Wang, Z.; Cherukupalli, S.; Xie, M.; Wang, W.; Jiang, X.; Jia, R.; Pannecouque, C.; De Clercq, E.; Kang, D.; Zhan, P.; et al. Contemporary medicinal chemistry strategies for the discovery and development of novel HIV-1 non-nucleoside reverse transcriptase inhibitors. *J. Med. Chem.* **2022**, *65*, 3729–3757.

9. Cilento, M. E.; Kirby, K. A.; Sarafianos, S. G. Avoiding drug resistance in HIV reverse transcriptase. *Chem. Rev.* **2021**, *121*, 3271–3296.

10. Namasivayam, V.; Vanangamudi, M.; Kramer, V. G.; Kurup, S.; Zhan, P.; Liu, X.; Kongsted, J.; Byrareddy, S. N. The journey of HIV-1 non-nucleoside reverse transcriptase inhibitors (NNRTIs) from lab to clinic. *J. Med. Chem.* **2019**, *62*, 4851–4883.

11. Vite-Caritino, H.; Mendez-Lucio, O.; Reyes, H.; Cabrera, A.; Chavez, D.; Medina-Franco, J. L. Advances in the development of pyridinone derivatives as non-nucleoside reverse transcriptase inhibitors. *RSC Adv.* **2016**, *6*, 2119–2130.

12. Das, K.; Bauman, J. D.; Clark, A. D., Jr.; Frenkel, Y. V.; Lewi, P. J.; Shatkin, A. J.; Hughes, S. H.; Arnold, E. High-resolution structures of HIV-1 reverse transcriptase/TMC278 complexes: strategic flexibility explains potency against resistance mutations. *PNAS* **2008**, *105*, 1466−1471.

13. Wang, Y.; De Clercq, E.; Li, G. Current and emerging non-nucleoside reverse transcriptase inhibitors (NNRTIs) for HIV-1 treatment. *Expert Opin. Drug Metab. Toxicol.* **2019**, *15*, 1–17.

14. Al-Salama, Z. T. Elsulfavirine: first global approval. *Drugs* **2017**, *77*, 1811–1816.

15. Mitchell, M. L.; Son, J. C.; Guo, H.; Im, Y.-A..; Cho, E. J.; Wang, J.; Hayes, J.; Wang, M.; Paul, A.; Lansdon, E. B.; Kim, C. U.; et al. N1-Alkyl pyrimidinediones as non-nucleoside inhi.bitors of HIV-1 reverse transcriptase. *Bioorg. Med. Chem. Lett.* **2010**, *20*, 1589–1592.

16. Li, G.; Wang, Y.; De Clercq, E. Approved HIV reverse transcriptase inhibitors in the past decade. *Acta Pharm. Sin. B* **2022**, *12*, 1567–1590.

17. Côte, B.; Burch, J. D.; Asante-Appiah, E.; Bayly, C.; Bedard, L.; Blouin, M.; Campeau, L.-C.; Cauchon, E.; Chan, M.; Chefson, A.; et al. Discovery of MK-1439, an orally bioavailable non-nucleoside reverse transcriptase inhibitor potent against a wide range of resistant mutant HIV viruses. *Bioorg. Med. Chem. Lett.* **2014**, *24*, 917–922.

18. Hwang, C.; Lai, M.-T.; Hazuda, D. Rational design of doravirine: from bench to patients. *ACS Infect. Dis.* **2020**, *6*, 64–73.

19. Muraglia, E.; Kinzel, O. D.; Laufer, R.; Miller, M. D.; Moyer, G.; Munshi, V.; Orvieto, F.; Palumbi, M. C.; Pescatore, G.; Rowley, M.; Williams, P. D.; Summa, V. Tetrazole thioacetanilides: potent non-nucleoside inhibitors of WT HIV reverse transcriptase and its K103N mutant. *Bioorg. Med. Chem. Lett.* **2006**, *16*, 2748–2752.

20. Tucker, T. J.; Saggar, S.; Sisko, J. T.; Tynebor, R. M.; Williams, T. M.; Felock, P. J.; Flynn, J. A.; Lai, M. T.; Liang, Y.; McGaughey, G.; et al. The design and synthesis of diaryl ether second generation HIV-1 non-nucleoside reverse transcriptase inhibitors (NNRTIs) with enhanced potency versus key clinical mutations. *Bioorg. Med. Chem. Lett.* **2008**, *18*, 2959–2966.

21. Tucker, T. J.; Sisko, J. T.; Tynebor, R. M.; Williams, T. M.; Felock, P. J.; Flynn, J. A.; Lai, M. T.; Liang, Y.; McGaughey, G.; Liu, M.; et al. Discovery of 3-{5-[(6-amino-1*H*-pyrazolo[3,4-*b*]pyridine-3-yl)methoxy]-2-chlorophenoxy}-5-chloro benzonitrile (MK-4965): a potent, orally bioavailable HIV-1 non-nucleoside reverse transcriptase inhibitor with improved potency against key mutant viruses. *J. Med. Chem.* **2008**, *51*, 6503–6511.

22. Gomez, R.; Jolly, S.; Williams, T.; Tucker, T.; Tynebor, R.; Vacca, J.; McGaughey, G.; Lai, M. T.; Felock, P.; Munshi, V.; DeStefano, D.; Touch, S.; Miller, M.; Yan, Y.; Sanchez, R.; Liang, Y.; Paton, B.; Wan, B. L.; Anthony, N. Design and synthesis of pyridone inhibitors of non-nucleoside reverse transcriptase. *Bioorg. Med. Chem. Lett.* **2011**, *21*, 7344–7350.

23. Burch, J. D.; Sherry, B. D.; Gauthier, D. R. Jr.; Campeau, L.-C. Chapter 7. Discovery and development of doravirine: an investigational next generation non-nucleside reverse transcriptase inhibitor (NNRTI) for the treatment of HIV. *ACS Symp. Ser.* **2016**, *1239*, 175–205.

24. Sanchez, R. I.; Fillgrove, K. L.; Yee, K. L.; Liang, Y.; Lu, B.; Tatavarti, A.; Liu, R.; Anderson, M.S.; Behm, M. O.; Fan, Li; et al. Characterisation of the absorption, distribution, metabolism, excretion

and mass balance of doravirine, a non-nucleoside reverse transcriptase inhibitor in humans. *Xenobiot* **2019**, *49*, 422−432.

25. Orkin, C.; Squires, K. E.; Molina, J.-M.; Sax, P. E.; Wong, W.-W.; Sussmann, O.; Kaplan, R.; Lupinacci, L.; Rodgers, A.; Xu, X.; Lin, G.; Kumar, S.; Sklar, P.; Nguyen, B.-Y.; Hanna, G. J.; Hwang, C.; Martin, E. A. Doravirine/lamivudine/tenofovir disoproxil fumarate is non-inferior to efavirenz/emtricitabine/tenofovir disoproxil fumarate in treatment-naive adults with human immunodeficiency virus-1 infection: week 48 results of the DRIVE-AHEAD trial. *Clin. Infect. Dis.* **2019**, *68*, 535−544.

26. Khalilieh, S. G.; Yee, K. L.; Fan, L.; Liu, R.; Heber, W.; Dunzo, E.; Triantafyllou, I.; Hussaini, A.; Iwamoto, M. A randomized trial to assess the effect of doravirine on the QTc interval using a single supratherapeutic dose in healthy adult volunteers. *Clin. Drug Invest.* **2017**, *37*, 975−984.

27. Campeau, L.-C.; Chen, Q.; Gauvreau, D.; Girardin, M.; Belyk, K.; Maligres, P.; Zhou, G.; Gu, C.; Zhang, W.; Tan, L.; et al. A robust kilo-scale synthesis of doravirine. *Org. Process Res. Dev.* **2016**, *20*, 1476−1481.

28. Gauthier, D. R. Jr.; Sherry, B. D.; Cao, Y.; Journet, M.; Humphrey, G.; Itoh, T.; Mangion, I.; Tschaen, D. M. Highly efficient synthesis of HIV NNRTI doravirine. *Org. Lett.* **2015**, *17*, 1353−1356.

3

Cabotegravir (Vocabria): An HIV Integrase Strand Transfer Inhibitor for Treating HIV Infection

Jie Jack Li

cabotegravir (Vocabria, **1**)
ViiV/GSK, 2020
HIV Integrase Strand Transfer Inhibitor
(NNRTI), $t_{1/2}$ = 21–50 days

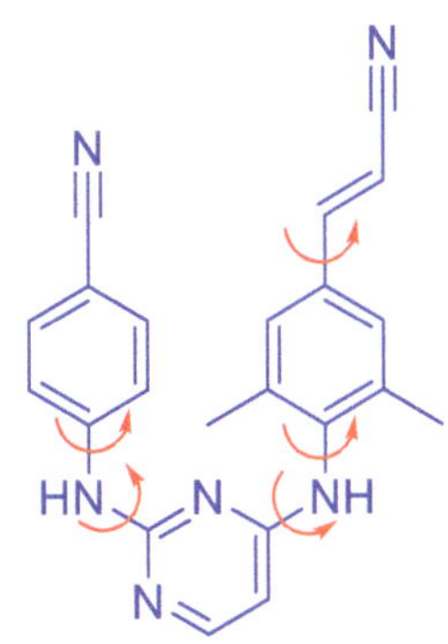

rilpivirine (**2**, Edurant)
Tibotec/J&J, 2011
NNRTI

Long-acting HIV/AIDS treatments are a great boon for patient compliance. At the end of 2022, the FDA approved Gilead's long-acting lenacapavir (Sunlenca), an HIV-1 capsid protein inhibitor. Only injections are given twice a year and lenacapavir can be used as pre-exposure prophylaxis (PrEP).

Before the approval of lenacapavir, Cabenuva was the only long-acting HIV drug on the market. Approved in January 2021 by the FDA for treating HIV infection, Cabenuva is an extended release injectable suspension of cabotegravir (**1**) and rilpivirine (**2**), given once every other month. Cabotegravir (**1**), the focus of this chapter, is an HIV integrase strand transfer inhibitor (INSTi) with a remarkable long half-life of approximately 50 days, whereas rilpivirine (**2**) is a second-generation nonnucleoside

Chemistry and Pharmacology of Drug Discovery, First Edition. Edited by Jie Jack Li.

reverse transcriptase inhibitor (NNRTI).[1,2] Cabotegravir (**1**, Vocabria) has now also been approved as the first long-acting injectable (LAI) for HIV PrEP.[3] Cabotegravir (**1**) is originated by Shionogi/GSK and developed by Viiv Healthcare, a specialized HIV company established in late 2009 by GSK and Pfizer.

The availability of highly potent antiretrovirals such as cabotegravir (**1**) with high barriers to resistance has made dual therapy, rather than the traditional triple therapy, a reality for many patients.

1. Background

We have come a long way in combating the invisible enemy HIV-1 since the early 1980s. With more than 25 FDA-approved drugs, the infection has been transformed from a death sentence to a chronic disease that can be managed using effective combination antiretroviral therapies (cARTs). One area that we are still struggling is vaccine. No safe and efficacious HIV/AIDS vaccine has emerged even though many intelligent and diligent scientists have been working on it for decades.

In Chapter 2 on Merck's doravirine, an NNRTI, I already summarized the landscape of HIV antiretroviral drugs. Here, we directly jump into the pharmacology of HIV integrase inhibitors.

2. Pharmacology

Like other INSTIs, cabotegravir (**1**) exerts its mechanism of action (MoA) by blocking integrase, an enzyme involved in inserting HIV's viral DNA into the host cellular DNA. Let us look at the drug target, HIV integrase.

2.1. HIV Integrase

HIV is a retrovirus that encodes 15 proteins, of which only three have enzymatic activities: integrase, protease, and reverse transcriptase.

2.1.1. Functions of HIV Integrase

HIV integrase catalyzes the insertion of viral DNA produced by the retro-transcription process into the genome (host DNA) of the infected human cell. This process is an essential step in the HIV viral life cycle and the key step in establishing a permanent infection. HIV integrase is an attractive target because there is no cellular homologue in human thus integrase inhibitors offer selectivity and less chances of drug resistance. The *in vivo* integration process of viral DNA into host DNA is depicted in Figure 1.

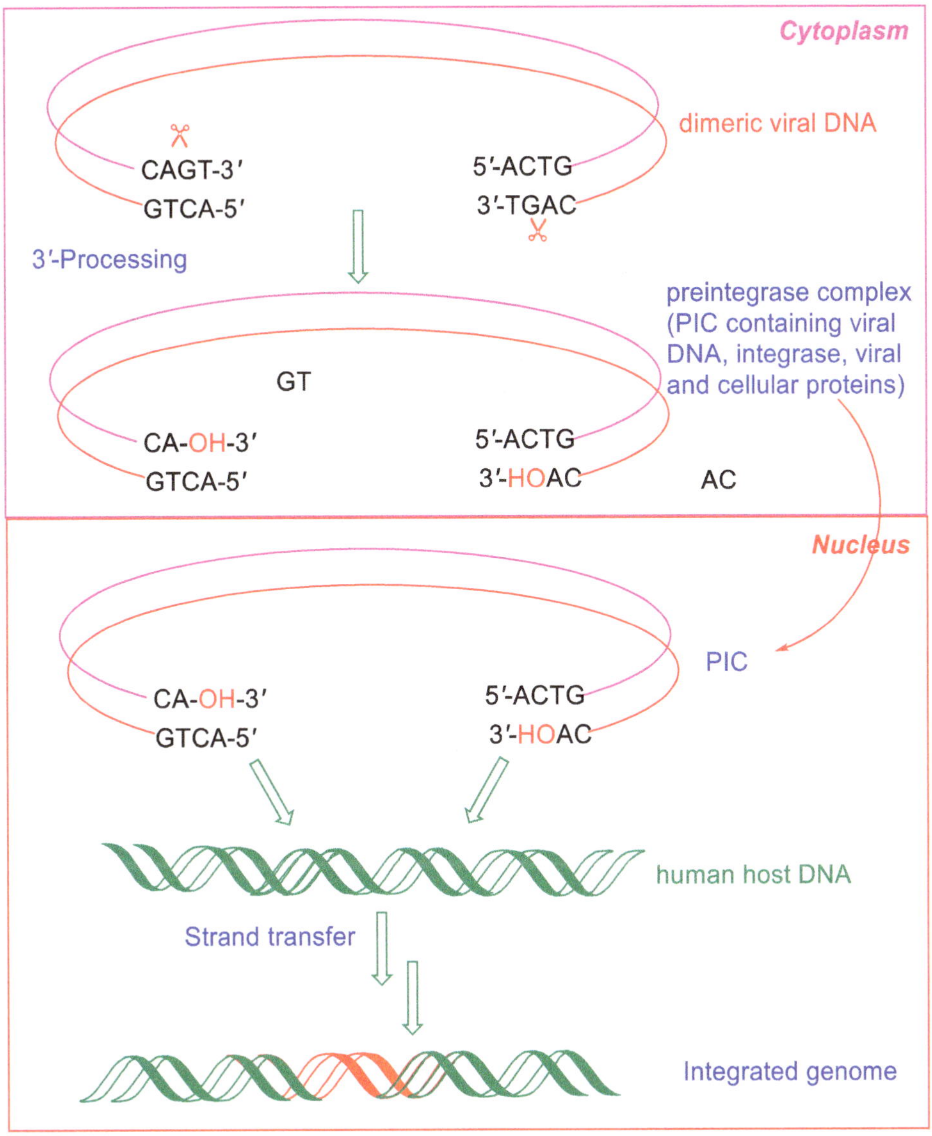

Figure 1. The *in vivo* integration process

Following reverse transcription, viral DNA is primed for integration in the cytoplasm by the integrase-mediated cleavage (a hydrolysis step) of two nucleosides from its 3′-ends, which is referred to as 3′-processing. At this point, the integrase remains bound to the viral DNA as a multimeric complex. The complex is referred to as a pre-integration complex (PIC) that bridges both ends of the viral DNA. Subsequently, the PIC migrates from cytoplasm to nucleus, where the viral DNA undergoes *strand transfer* followed by 5′-processing, a transesterification step. The result is fusion of the viral DNA into the host DNA, giving rise to integrated genome.[4]

As mentioned before, the integration process involves two integrase catalytic reactions: 3′-processing and strand transfer. Both reactions are catalyzed by two highly cooperative divalent cations centered on a phosphodiester bond. The integration process can be divided into several stages as shown in Figure 2. Stage A is when the integrase recognizes the adenine base conserved in the third position from 3′-end of viral DNA. then activates the next phosphoric ester with the two metals (Stage B to C). Stage A–C indicates the 3′-processing reaction performed in the cytoplasm. After 3′-processing, the HIV-1 PIC enters the nucleus, where integrase catalyzes the insertion of the viral DNA ends into the host chromosome.[5]

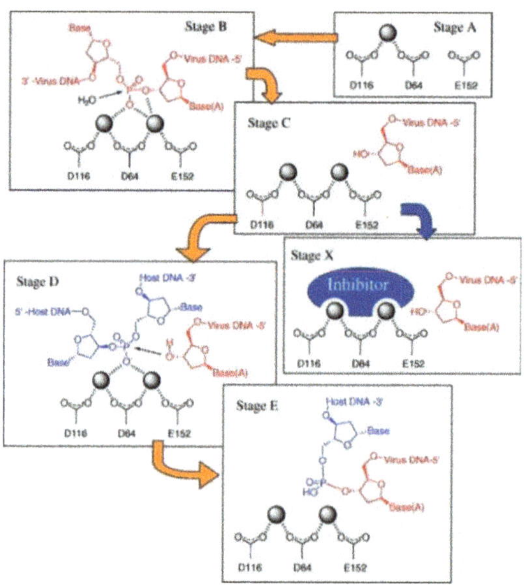

Figure 2. The two-metal-ion catalysis and inhibition mechanism. Source: Reproduced with permission Kiyama et al. [5], Elsevier

Once in the nucleus, the activated phosphoryl ester is hydrolyzed to excise the terminal dinucleotide and the recognized adenosine is exposed as the new 3′-end, giving a PIC (Stage C to D). The PIC nonspecifically binds to host DNA to activate a phosphoryl ester by the two metals (Stage D to E). The activated phosphoric ester is attacked by the recessed 3′-end in the manner of S_N2-like nucleophilic reaction, then the viral DNA and the host DNA are joined with each other. An inhibitor chelates to the two metal ions of Stage C to block the host DNA binding (Stage X). Stage C–E represents the strand transfer reaction performed in nucleus.[5]

Of course, the reality is much more complicated and nuanced. The mechanisms do not really take place sequntially and stage-wise. In fact, some steps can happen simultaneously.

2.1.2. The Structure of HIV Integrase

HIV integrase is associated with poly-nucleotidyl transferases superfamily of enzymes. It consists of 288 amino acids with a molecular weight of 30 KDa. In order to carry out the 3′-processing reaction, an integrase dimeric configuration is required. On the other hand, an integrase tetrameric configuration is needed for catalyzing the subsequent strand transfer.

Integrase's N-terminal domain (NTD) contains 50 amino acids and has the HHCC zinc-finger that binds to zinc ion (Figure 3). The HHCC zinc-finger is so-named because it contains 4 conserved residues in the form of 2 histidine (H12 and H16) and 2 cysteine (C40 and C43). The catalytic core domain (CCD) containing amino acids 51–211 is where the action is. Binding to Mn^{++} or Mg^{++}, the CCD is the enzymatic core of protein. It contains DDE motif that makes up the active site triad, also known as the *catalytic triad* with two aspartates (D64 and D116) and one glutamine (E152). The C-terminal domain (CTD) contains amino acids 212–288. It binds nonspecifically with the DNA and its linkage with CCD is essential for 3′-processing and strand transfer activities.[6]

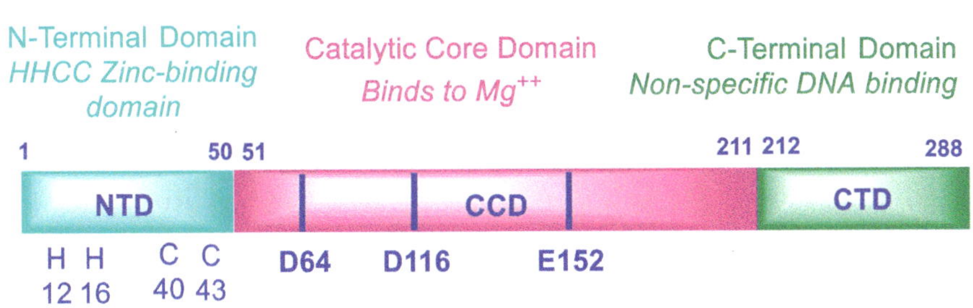

Figure 3. HIV integrase structural domains. Source: Adapted from Gill et al. [6]

Because integrase enzyme catalyzes HIV integrase inhibitors are also known as HIV INSTIs.

2.2. First-Generation HIV Integrase Inhibitors

Even though inhibitors for HIV reverse transcriptase and protease were discovered soon after the discovery of the virus, it took more than 20 years for the first HIV integrase inhibitor to appear on the market. One of the challenges was that the HIV integrase protein has shallow and solvent-exposed binding surface. Early lead structures were frequently based on catechols, hydrazides, or coumarins, all of which failed to show antiviral activity in cell culture by a mechanism that could be reliably attributed to inhibition of virus genome integration. Merck's raltegravir (**3**) was approved for marketing in 2007 as the first HIV integrase inhibitor, a culmination of considerable effort that was based on clearly defining the biochemical staging of enzyme function. The second integrase inhibitor Japan Tabacco/Gilead's elvitegravir (**4**, Vitekta) was approved in 2012.

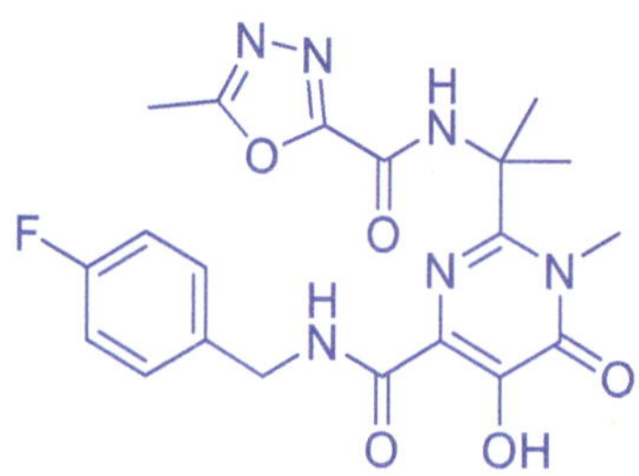

raltegravir (**3**, Isentress)
Merck, 2007
INSTI

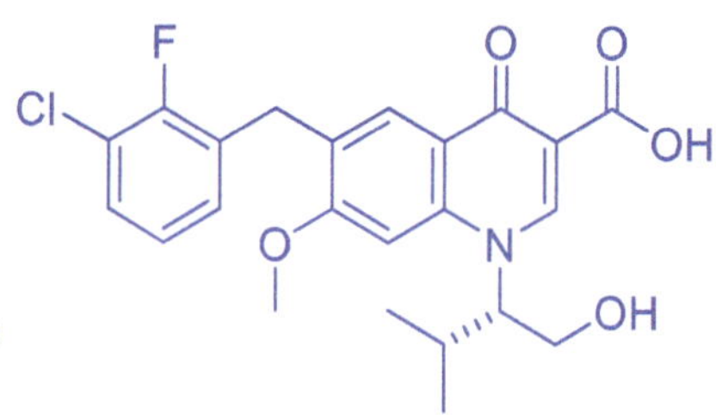

elvitegravir (**4**, Vitekta)
Japan Tabacco/Gilead, 2012
INSTI

The strand transfer step is the key enzymatic process susceptible to inhibition rather than assembly of the enzyme on viral substrate or the 3′-cleavage reaction. This mechanistic insight afforded a more effective screening assay. In 1999, using such an assay, Merck and a Japanese company Shionogi independently discovered diketoacid (DKA) derivatives (e.g., **5**) as the first specific inhibitors of HIV integrase that demonstrated antiviral activity *in cell culture*. These compounds bound to a complex of HIV and the viral DNA substrate with the DKA moiety. As a phosphate isostere, DKA binds to the two magnesium divalent ions involved in catalysis, forming a ternary complex that interferes with the binding to host cell double-stranded DNA. By replacing the carboxylic acid with a tetrazole bioisostere, Shionogi was able to obtain the first inhibitor co-crystalized integrase. That was a great contribution to the field even though the tetrazole analog never became a drug due to stability issues.[7–9]

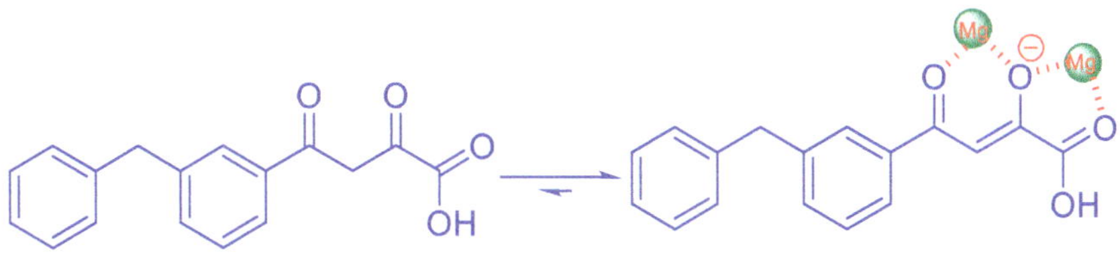

diketoacid (DKA) **5** the keto-enolate tautomer form

Merck Research Laboratories in Rome, Italy, succeeded in finding their integrase inhibitors by "inter-breeding" two drug discovery programs. In parallel to their HIV integrase program, Merck Rome also had a hepatitis C virus (HCV) inhibitors program on-going at the same time. A class of inhibitors of HCV NS5B RNA-dependent RNA polymerase (RdRp) had the dihydroxypyrimidine pharmacophore with strong metal-binding capacity even though the compounds *per se* did not inhibit HIV integration. Realizing that both HCV NS5B polymerase and HIV integrase rely on binding to the magnesium ion for their catalytic activity, Merck medicinal chemists used dihydroxypyrimidine **6** as the bioisostere of the more stable replacement of the DKA pharmacophore. Simple installation of a hydrophobic benzylamine gave rise to **7** as a very potent drug in an integrase strand transfer assay. Another major structural core change was methylation of one of the two nitrogen atom on the pyrimidine ring to convert the pharmacophore to hydroxypyrimidinone **8**. Incremental modifications to improve physiochemical properties while maintaining cell penetration and limiting protein binding delivered a drug with an exceptional potency. That became raltegravir (**3**), which has been marketed as Isentress since 2008.[8]

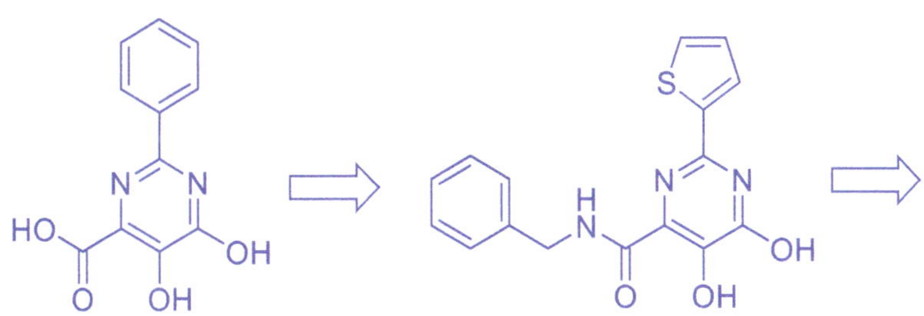

dihydroxypyrimidine **6** (HCV NS5B inhibitor)

benzylamine **7** INSTI IC_{50} = 85 nM

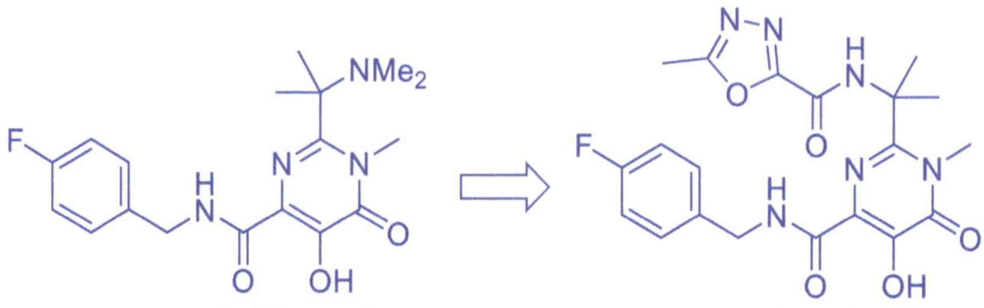

hydroxypyrimidinone **8**
INSTI IC_{50} = 230 nM

raltegravir (**3**)
INSTI IC_{50} = 10 nM

Japan Tobacco discovered their HIV integrase inhibitor elvitegravir (**4**), also from "inter-breeding" of two drug discovery programs. Their monoketo acid pharmacophore (**10**, for instance) was derived from a quinolone antibiotic scaffold originally designed for bacterial DNA gyrase activity. Even at the very beginning, they already found that simple 4-quinolone-3-carboxylic acid **10**, but not the more complicated 4-quinolone-3-glyoxlic acid **9**, had decent HIV integrase inhibitory activity. The mono-keto-acid moiety on **10** is a weaker chelator to the magnesium ions, therefore, **10** is probably more selective than diketo-acid **9** because HIV integrase belongs to a large family of DNA processing enzymes, which contain the same arrangement of three catalytically essential carboxylates. Unlike raltegravir (**3**), *para*-substitutions on the left-hand phenyl ring actually killed the activity to inhibit integrase for this series of compounds. Luckily, *ortho*- and *meta*-substitutions were fruitful, giving rise to decorated derivative **11**, which possessed both integrase inhibitory activity and antiviral activity. It turned out that the NH on the quinolone core was not essential and alkylation of **11** provided 1-hydroxyethyl analog **12**, which gained more than 10-fold of efficacy. Installation of an isopropyl group offered another 10-fold boost of efficacy. Finally, an additional 7-methoxyl substitution on the quinolone core led to elvitegravir (**4**), which showed significant improvement of inhibition of strand transfer and antiviral activity. After the FDA approval in 2012, Japan Tobacco and Gilead co-marketed it with a trade name Vitekta.[9]

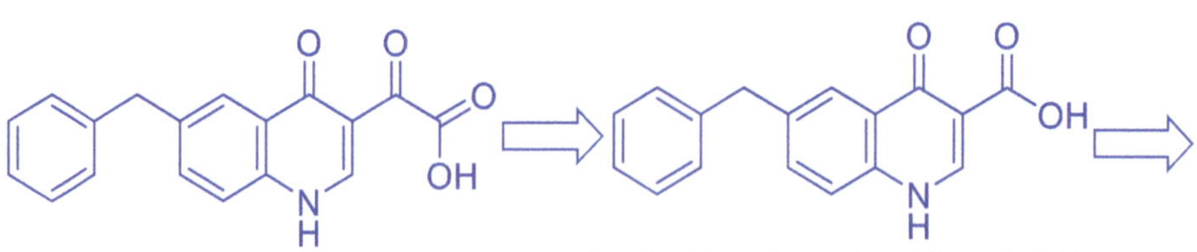

4-quinolone-3-glyoxlic acid **9**

4-quinolone-3-carboxylic acid **10**
IC_{50} = 1.6 μM, EC_{50} > 30 μM

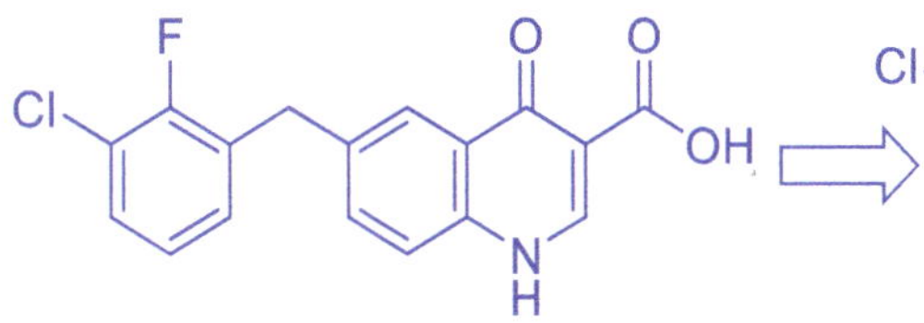

11, IC_{50} = 44 nM, EC_{50} = 805 nM

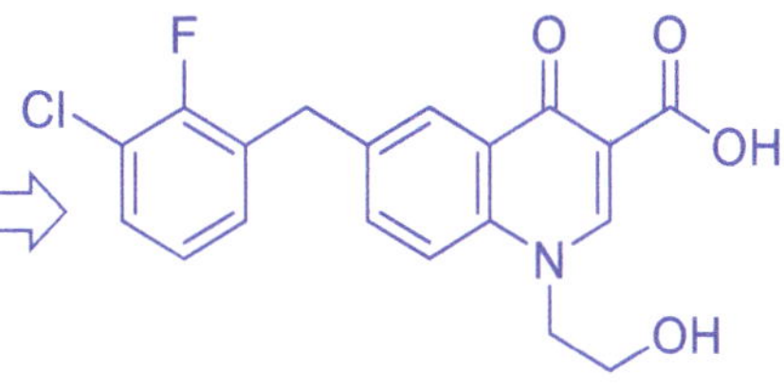

12, IC_{50} = 24 nM, EC_{50} = 76 nM

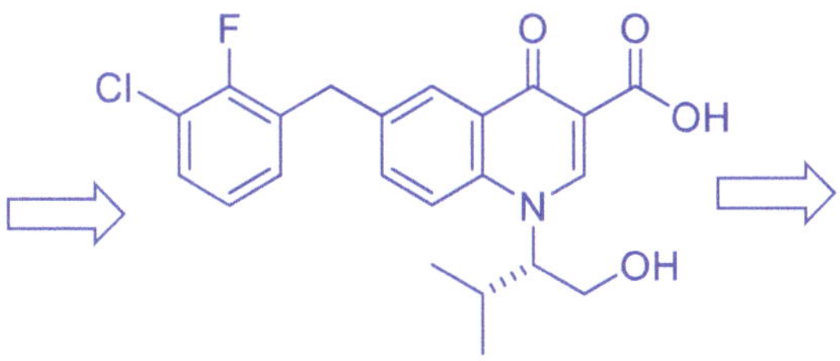

13, IC_{50} = 8.2 nM, EC_{50} = 7.5 nM

elvitegravir (**4**)
IC_{50} = 7.2 nM
EC_{50} = 0.95 nM

2.3. Second-Generation HIV Integrase Inhibitors

The first-generation INSTIs suffer from cross-resistance between each other, i.e., raltegravir (**3**) and elvitegravir (**4**). Both of them are also associated with some adverse side effects.

Second-generation integrase inhibitors doletegravir (**14**) and bictegravir (**15**) are superior to the first-generation integrase inhibitors. They include good tolerability, once-daily dosing with no need for a pharmaco-enhancer and relatively little cross-resistance that plagued the first-generation integrase inhibitors.

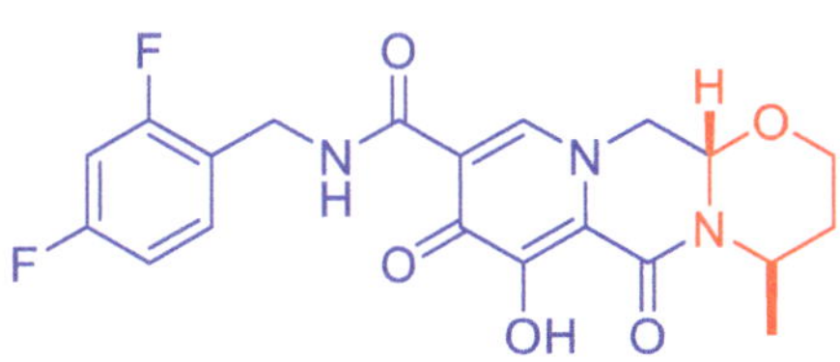

dolutegravir (**14**, Trivicay)
GSK/Shionogi, 2013
INSTI

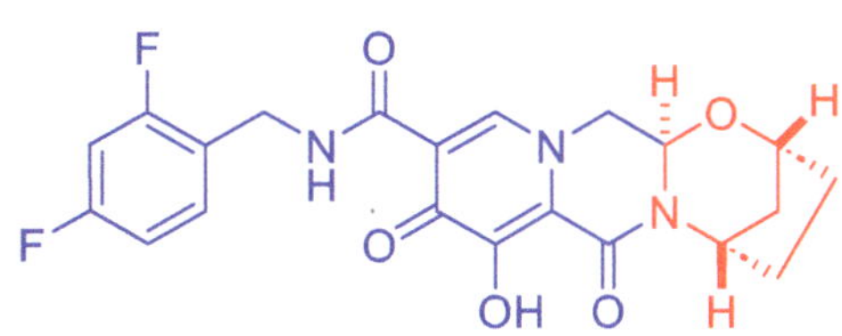

bictegravir (**15**, Biktarvy
with emtricitabine/
tenofovir alafenamide)
Gilead, 2018, INSTI

Shionogi and GSK scientists embarked on their quest of the second-generation integrase inhibitors starting with monocyclic carbamoyl pyridone **16**. Their intentional use of a triad of all oxygen-derived lone pairs to serve as the chelating donor atoms toward the two divalent metals.

Although monocyclic **16**'s enzymatic and antiviral activities are superior to raltegravir (**3**) and elvitegravir (**4**), its antiviral efficacy against resistant mutants (particularly Q148K) decreased. Nevertheless, bicyclic piperazinone **17** fixed the problem.

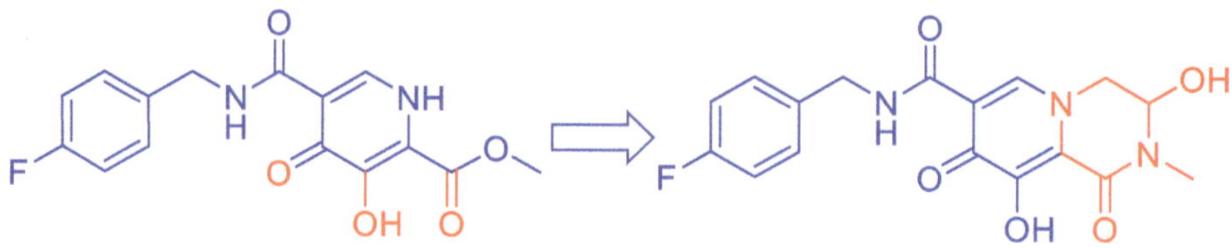

carbamoyl pyridone **16** bicyclic piperazinone **17**

If two rings are good, three rings must be better. Indeed, although racemic tricyclic hemiaminal **18** did not elevate the antiviral potency but positively modified the rat PK profile. Between the enantiomers, **18*S*** had a 43-fold loss of potency with added human serum albumin while the *R* isomer had a very modest 4-fold loss. An additional *S*-methyl substituent gave rise to dolutegravir (**14**) that was approved in 2013 with trade name Trivicay.[10–13]

Five years later in 2018, Gilead's me-too drug bictegravir (**15**), which was a very close cousin of dolutegravir (**14**), was approved. Gilead sold it as a combination drug with emtricitabine and tenofovir alafenamide with trade name Biktarvy, now a constant feature of commercials on TV every day and everywhere.

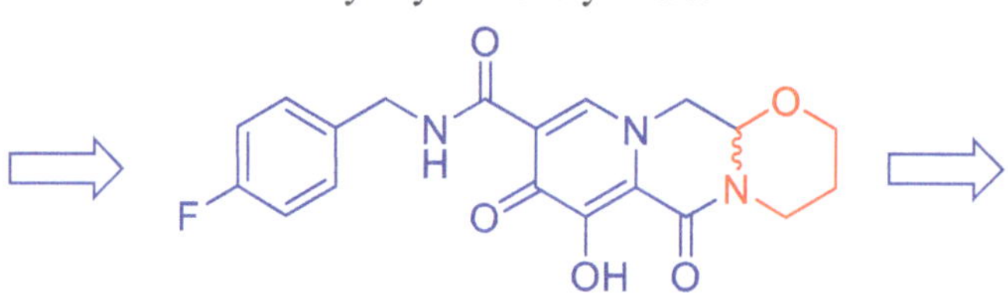

racemic tricyclic hemi-aminal **18**

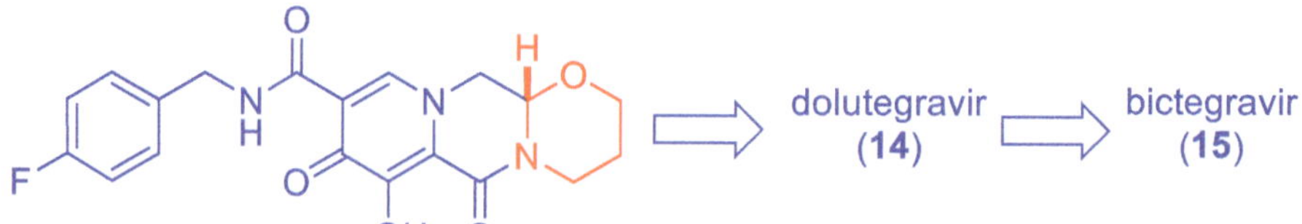

18*S*

Cabotegravir (**1**) is a new long-acting parenteral and a highly effective integrase inhibitor with a half-life of 54 days, allowing parenteral administration every other month. It features low water solubility, high activity, long half-life, and slow metabolic clearance.

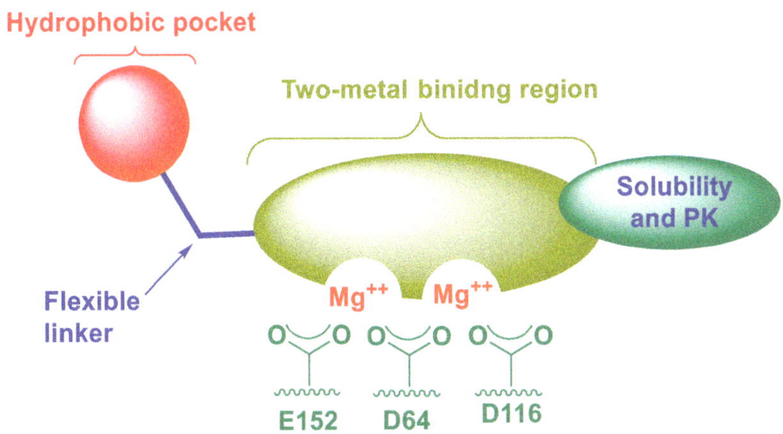

Figure 4. A typical integrase strand transfer inhibitor binding to the catalytic triad of HIV integrase

A pattern has emerged in the field of integrase inhibitors. As shown in Figure 4, a successful *competitive* (orthosteric) HIV INSTi almost always contains a chelation warhead to bind to the two metal ions. The warhead is then connected to a hydrophobic region connected by a flexible linker. The second-generation INSTIs have an additional region for optimization of pharmacokinetics and drug-like properties required for oral bioavailability.[13]

2.4. Integrase–LEDGF/p75 Allosteric Inhibitors

The five INSTIs on the market are *orthosteric* inhibitors. They are catalytic site inhibitors that specifically target the strand transfer step necessary for viral DNA insertion into the host chromatin. Due to ubiquitous drug resistance, there is always a concern using the drugs with the same MoA. Therefore, *allosteric* HIV-1 integrase inhibitors (ALLINIs) have garnered special interest because of their novel MoA.

In the life cycle of an HIV, after reverse transcription, the PIC binds to the host (cellular cofactor) lens epithelium-derived growth factor (LEDGF)/p75 protein in the nucleus. As shown in Figure 5, LEDGF is characterized by a conserved N-terminal PWWP domain. Its signature Pro–Trp–Trp–Pro motif is a chromatin reader and is able to bind simultaneously and synergistically to DNA and methyl lysines present on histone tails.

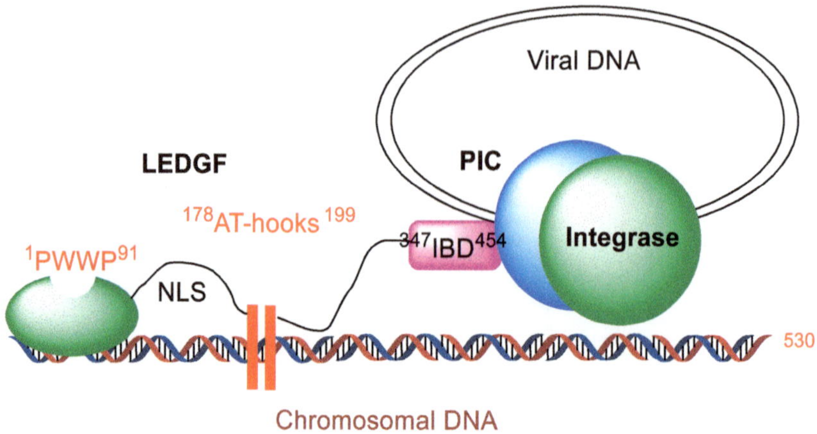

Figure 5. Allosteric HIV-1 integrase inhibitors targeting LEDGF/p75

The region following the PWWP is presumed to be a nuclear localization signal (NLS) motif, two AT-hooks (a minor-groove DNA-binding motif consisting of a Pro–Arg–Gly–Arg–Pro core that preferentially binds AT-rich sequences). LEDGF is connected by a supercoiled DNA recognition domain, which is followed by an integrase-binding domain (IBD), a compact right-handed bundle of five α-helices. LEDGF/p75 is the most studied cellular cofactor that is essential for tethering the integrase PIC to host chromatin and also for the recruitment of other cellular factors to the PIC, thereby facilitating effective integration.[14]

Most of the effective integrase-LEDGF/p75 inhibitors contain a hydrophobic moiety and a carboxylic acid functionality in common, which mimic the hot spot residues Ile365 and Asp366 within the interface of LEDGF/p75 IBD. The H-bonding interactions between the carboxylic acid and the backbone amide protons of residues E170 and H171 of integrase was critical to antiviral potency and that there was no tolerated isosteric replacement for the acid. ALLINIs elicit antiviral activity by binding to the highly conserved allosteric pocket on the integrase catalytic core that also serves as the binding pocket for LEDGF. By targeting the protein–protein interactions (PPIs) between HIV integrase and LEDGF/p75, ALLINIs induce aberrant integrase multimerization, leading to the production of replication-deficient viral particles.

Around 2014, Boehringer Ingelheim was the first to bring an ALLINI, BI 224436 (**19**), to clinical trial, which was terminated during phase I.[15] GSK described GSK1264 (**20**) that disrupts the interaction between HIV-1 integrase and the cellular factor LEDGF)/p75. Interestingly, GSK1264 (**20**) was only modestly potent to inhibit the early infection steps and had little effect on integration targeting, which is guided by the integrase–LEDGF/p75 interaction.[16]

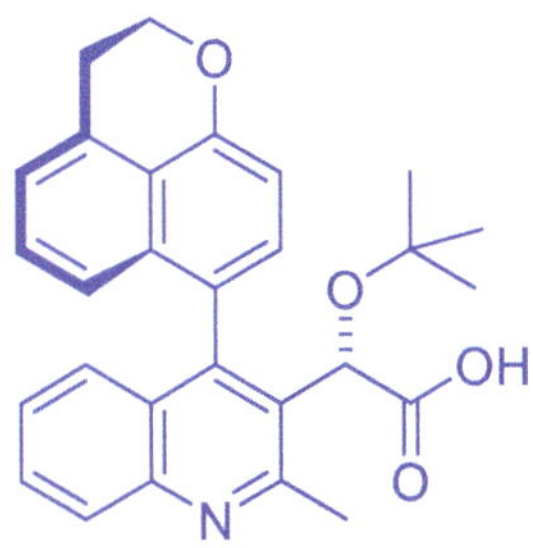

BI 224436 (**19**)

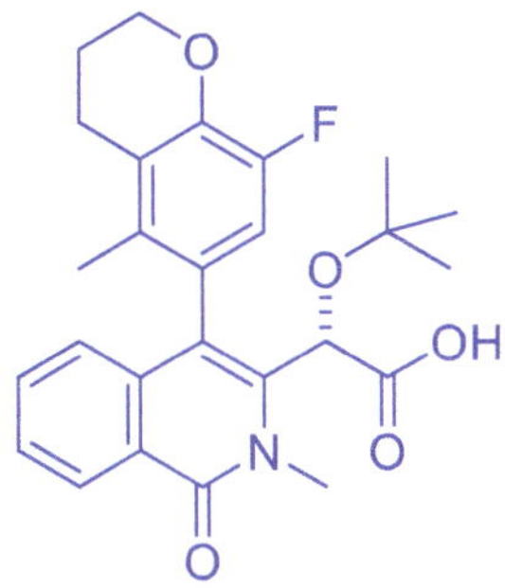

GSK1264 (**20**)

BMS carried out a drug discovery program on integrase–LEDGF/p75 allosteric inhibitors. The carried out their optimization with a specific emphasis on the inhibition of the 124/125 polymorphs such that the designed compounds showed excellent potency *in vitro* against majority of the 124/125 variants. BMS-986180 (**21**, GSK3739936) emerged as a promising preclinical lead with a good PK profile. Regrettably, findings in rat toxicology studies precluded further development of **21** in humans.[17]

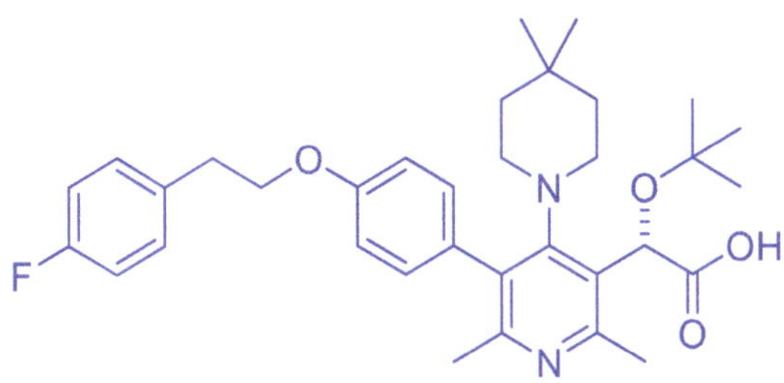

BMS-986180
(**21**, GSK3739936)
EC_{50} = 1.7 nM

Though the PPI inhibitors are less potent than the active site-directed enzyme inhibitors, allosteric inhibitors can provide new MoA to address the drug-resistance issue.

3. Structure–Activity Relationship (SAR)

In the early 2000s, Shionogi and GSK carried out extensive structure–activity relationship (SAR) investigations of carbamoyl pyridones as HIV-1 integrase inhibitors. Dolutegravir (**14**) was the fruit of their labor. Even though dolutegravir (**14**) was a very good drug, efforts were made to explore to replace its six-membered 1,3-oxazinane ring with a five-membered oxazolidine ring as represented by cabotegravir (**1**).[12]

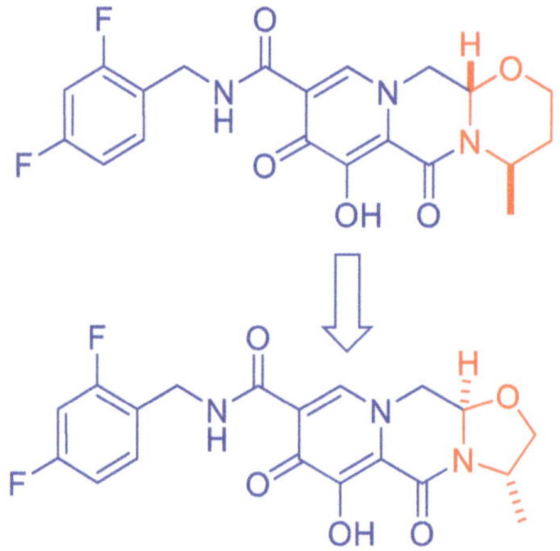

The SAR on the effects of tricyclic carbamoyl pyridones around structure **22** is summarized in Table 1. Since 3,5-difluorobenzyl motif was invariably superior to the 3-monofluorobenzyl counterpart, only the difluorobenzyl derivatives are compiled here for clarity.

Table 1. Effects of tricyclic carbamoyl pyridones

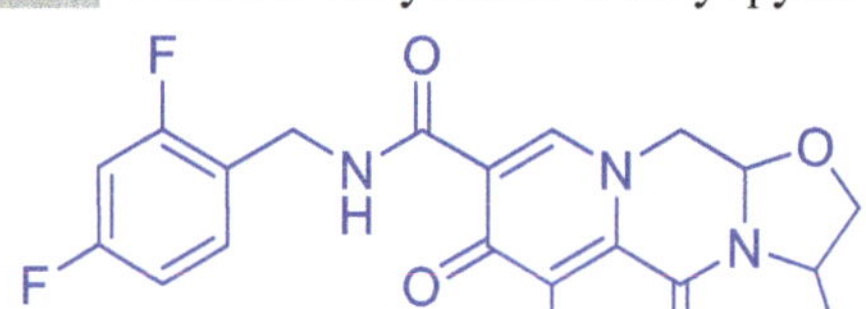

22

Compound	Structure	$^{pHIV}IC_{50}$ (nM)	$^{pHIV}PAIC_{50}$ (nM)	Q148K (FC)
1	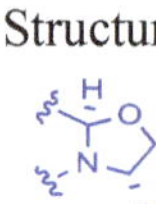	0.3	30	3.9
23	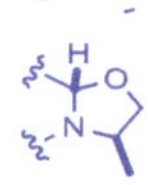	5.4	39	35
24	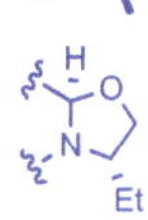	0.9	63	ND
25	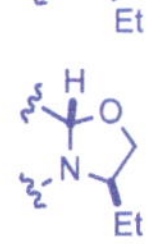	8.5	35	63
26	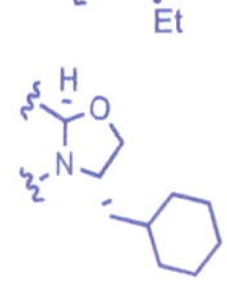	330	3100	>15

27	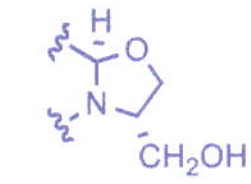	10	37	12
28	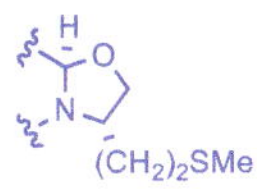	3.1	18	ND
29	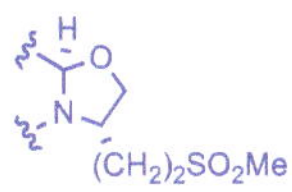	430	370	>10
30	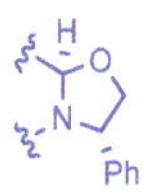	72	620	44
31	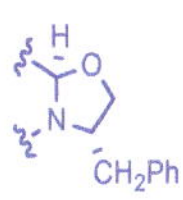	13	380	13
32	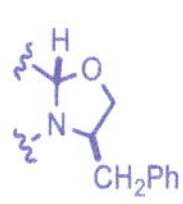	0.6	4.5	19

The potency (IC_{50}) is measured using a pseudo-typed virus assay (pHIV). Protein-adjusted potency ($PAIC_{50}$) is also listed to reflect serum protein-binding shift. Lastly, fold change (FC) is measured for the most important mutation, the Q148K mutant.

As shown in Table 1, cabotegravir (**1**) with the *S*-configuration is significantly more potent than its enantiomer **23** with the *R*-configuration. Similarly, the *S*-ethyl analog **24** is more potent than the corresponding the *R*-ethyl analog **25** as well. Once the R-substitution becomes larger, the potency falls as demonstrated by cyclohexylmethyl derivative **26**. Alcohol **27** also lost substantial potency (20×) compared to the methyl analogue **1**. However, thioether **28** retained low nM potency with less loss of potency compared to **1**, it made up for this loss with an improved protein adjusted value of $PAIC_{50}$. The sulfone **29** showed very poor activity with or without added proteins in the pHIV assay system. The rigid 3-phenyl derivative **30** was 143× less potent than the corresponding methyl analogue **1**. With an extra methylene group, the benzyl derivative **31** was somewhat more potent than **30**. Very surprisingly, even though *S*-derivatives were more potent than the *R*-derivatives when the substituents were simple methyl, ethyl, and propyl, the *R*-benzyl derivative **31** was significantly more potent than the corresponding *S*-benzyl derivative **33**.

The Q148K mutant data for the five-membered series consistently showed a sharp decline in potency against the mutant as the substituent present in the five-membered saturated ring increased in size beyond a methyl group. This was different from the six-membered series, which might explain why six-membered dolutegravir (**14**) was developed first. The methyl derivative **1** performed best within the five-membered ring series when potency, protein shift, and mutant profile were all considered. Therefore, it was not surprising that it was chosen as a drug candidate that ultimately became cabotegravir (**1**).[12]

4. Pharmacokinetics and Drug Metabolism

Preclinical pharmacokinetic parameters in Sprague–Dawley rats, beagle dogs, and cynomolgus monkeys for **14** and **1** are shown in Table 2.[12]

Both dolutegravir (**14**) and cabotegravir (**1**) are highly protein-bound (>99%). Cabotegravir (**1**) had a large protein shift (60-fold) in the presence of human serum albumin (HAS), however, again this was attenuated by its intrinsic potency and the high protein binding may have a favorable impact on pharmacokinetic properties (for example, increasing half-life).[12]

When administered as a LAI formulation, cabotegravir (**1**) exhibits an extremely long half-life of approximately 21–50 days. This prolonged half-life is a result of the poor solubility of the nanoparticles in tissue, which allows for a slow absorption rate as opposed to decreased plasma elimination. Cabotegravir (**1**) possesses a number of properties that favorably lend it to formulation as a LAI, including slow metabolism, a high melting point and low aqueous solubility.[18]

Table 2. Preclinical pharmacokinetic parameters for **14** and **1**

Compound	Species	Cl (mL/min/kg)	$T_{1/2}$ (h)	V_{dss} (L/kg)	%F
14	Rat	0.23	6.2	0.1	34
1	Rat	NR	> 18	NR	NR
14	Dog	2.2	5.2	0.3	35
1	Dog	0.34	5.7	0.14	8
14	Cyno	2.1	6.0	0.3	25
1	Cyno	0.32	4.0	0.09	6.2

Another advantage of cabotegravir (**1**) relies on its main metabolization by uridine diphosphate glucuronosyl-transferase 1A1, leading to less likely interactions with other antiretroviral drugs. To a lesser extent, cabotegravir (**1**) is metabolized hepatically by UGT1A9. It is a mild inhibitor of organic anion transporter (OAT) 1 and OAT3 and does not have any significant impact on cytochrome P450 or other UGT enzymes based on cabotegravir (**1**) plasma levels obtained from standard doses.[3]

5. Efficacy and Safety

Cabotegravir (**1**) is significantly higher than that of dolutegravir (**14**) were discovered by evaluation of antiviral activity against wild-type virus (±HSA) along with key integrase inhibitor-resistant mutants, Q148K as the most important one. Animal pharmacokinetic profiles included a key measure of the *trough drug concentration* over protein-adjusted antiviral potency ($PAIC_{50}$) along with in vitro DMPK properties. Cabotegravir (**1**) demonstrated good coverage over $PAIC_{50}$ predicting low mg unboosted once daily dosing, now validated in clinical studies. These preclinical data along with a long human $T_{1/2}$ of ~30 h in oral tablet study supports cabotegravir (**1**) as a long acting parenteral agent for once-monthly or less frequent dosing.

Table 3. Virological profile of cabotegravir (**1**) and dolutegravir (**14**)

Compound	INST IC_{50} (nM)	$^{MT4}IC_{50}$ (nM)	$^{PBMC}IC_{50}$ (nM)	100% HuS fold shift	$PAIC_{90}$ (ng/mL)
1	3.0	1.3	0.2	408	166
14	2.7	2.0	0.5	75	64

The virological profile of cabotegravir (**1**) and dolutegravir (**14**) are shown in Table 3. The biochemical assay was used to measure the potency against INST. In addition, their cellular activities were tested in both MT4 cell and peripheral blood mononuclear cell (PBMC) HIV multi-round replication assays. The fold shift (408×) for cabotegravir (**1**) is significantly higher than that of dolutegravir (**14**, 75×) when they were tested in the presence of human serum albumin (HSA). As a consequence, the $PAIC_{90}$ value for cabotegravir (**1**) is higher than that of dolutegravir (**14**).[12]

A phase 2b clinical trial, a combination of cabotegravir (**1**) and rilpivirine (**2**) was given intramuscularly every 4 weeks or every 8 weeks to adults with HIV-1 infection. The combination drug was found to be as effective as daily three-drug oral therapy of cabotegravir (**1**) plus abacavir–lamivudine at maintaining HIV-1 viral suppression through 96 weeks and was well accepted and tolerated.[19]

Thanks to their superb efficacy, cabotegravir (**1**) and dolutegravir (**14**) allow two-drug combination in place of the old three-drug combination, simplifying the treatment regimens.

Common adverse effects include injection site reactions, pyrexia, fatigue, and headache.

6. Synthesis

6.1. Discovery Route

The discovery route to make cabotegravir (**1**) was lengthy but straightforward.[11,12]

Maltol (**33**) was protected as benzylether **34**, which was converted to pyridone **35** using ammonia with the aid of NaOH. Regioselective bromination of pyridone **35** with N-bromo-succinimide (NBS) afforded 5-bromo-pyridone **36**, which was readily transformed to the corresponding ester **37** via a palladium-catalyzed carbonylation. Acetylation of **37** led to aromatization product pyridine **38**, which was oxidized to the corresponding pyridine-oxide **39** using *m*-CPBA. The Boekelheide reaction was then carried out by treating pyridine-oxide **39** with acetic anhydride at high temperature to give acetoxymethylpyridine **40**. Selective removal of the two acetates was achieved by treating **40** with sodium methoxide in methanol to reveal pyridone **41**, sparing the methyl ester. Heating **41** with 2,4-difluorobenzylamine assembled amide **42**, which was oxidized by manganese dioxide and sulfamic acid sequentially to provide acid **43**. Esterification of acid **43** with methanol was accomplished using EDAC as the coupling agent in refluxing DMF to produce the key intermediate **44**.[11]

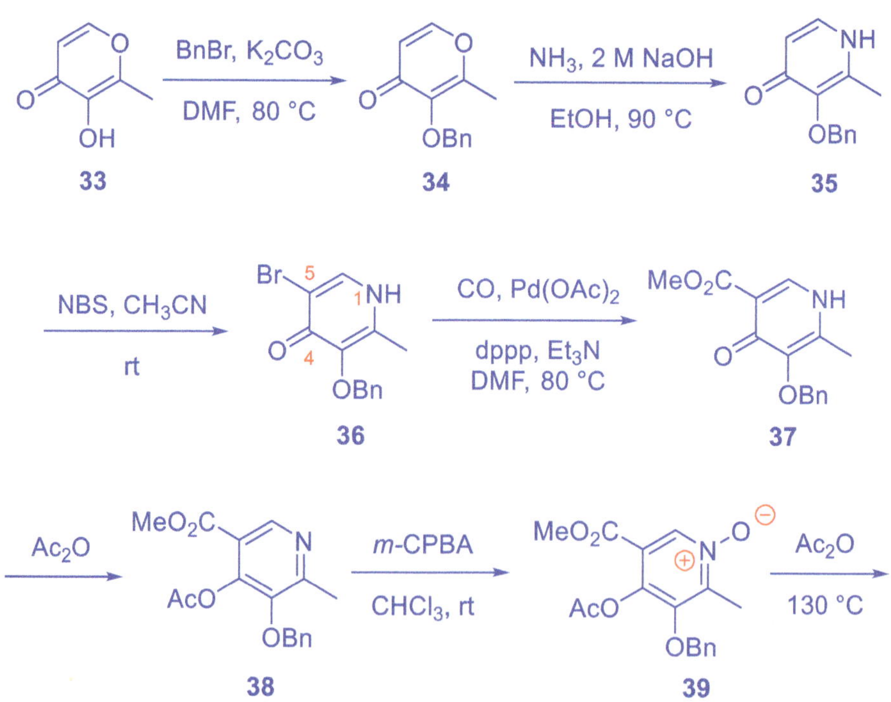

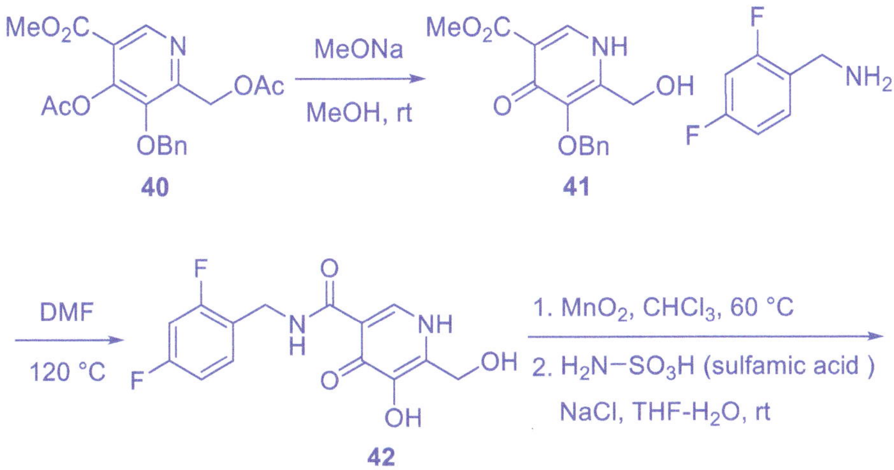

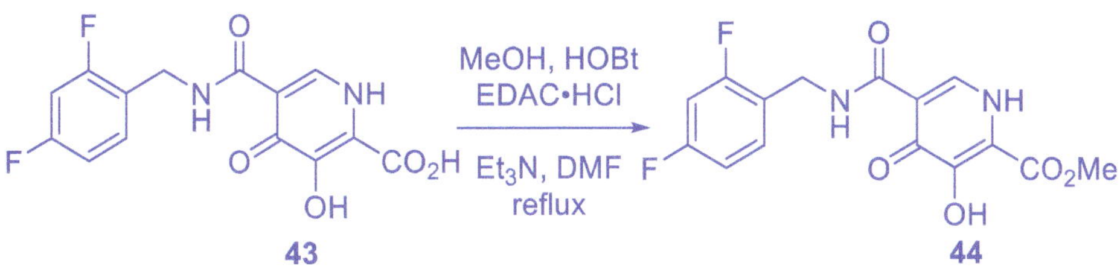

N-Allylation of pyridone **44** with allylbromide prepared terminal olefin **45**. Rather than using the dangerous ozonolysis conditions, olefin **45** was treated with sodium metaperiodate employing osmium tetraoxide (in the form of potassium osmate dihydrate) as the catalyst to make aldehyde **46**.[11] Finally, condensation of aldehyde **46** with (*S*)-2-aminopropan-1-ol was catalyzed by acetic acid to deliver cabotegravir (**1**) after palladium-catalyzed debenzylation. The condensation gave rise to cabotegravir (**1**) in a stunning 40:1 diastereomeric ratio (*dr*).[12]

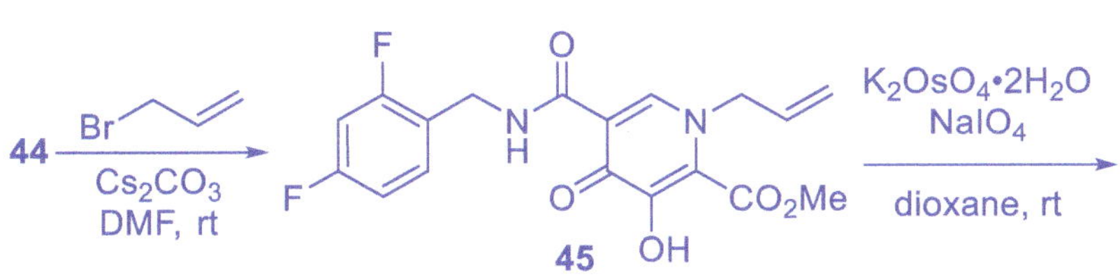

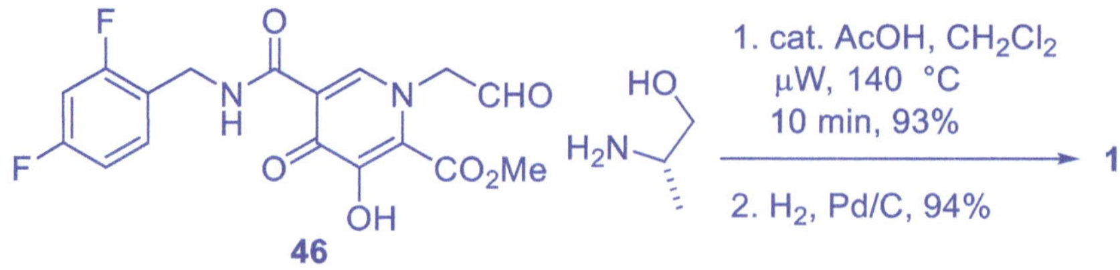

6.2. Process Route

GSK Process Chemistry group published a one-pot, four-step route to pyridone **51**.[20–22]

Condensation of β-ketoester **47** with neat DMF-DMA produced vinylogous dimethyl amide **48**, which was condensed *in situ* with aminoacetaldehyde dimethyl acetal in methanol to assemble vinylogous amide **49**. Without workup or isolation, vinylogous amide **49** was coupled with dimethyl oxalate and LiOMe in MeOH to promote the formation of pyridone **50**, which was *selectively* (10:1) hydrolyzed to mono-acidic pyridone **51**. In contrast, NaOH and KOH only gave a 3:1 selectivity.[20]

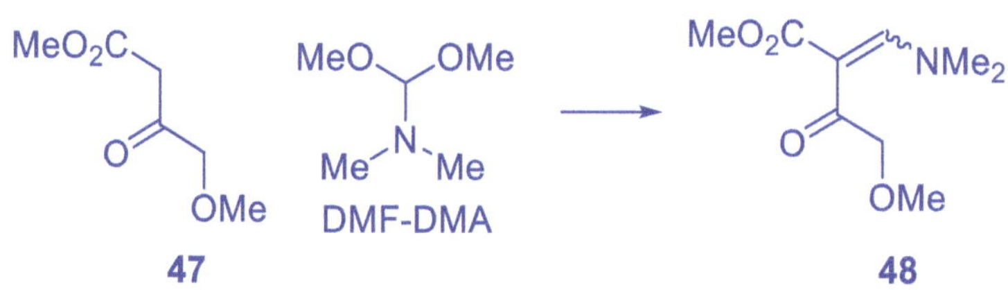

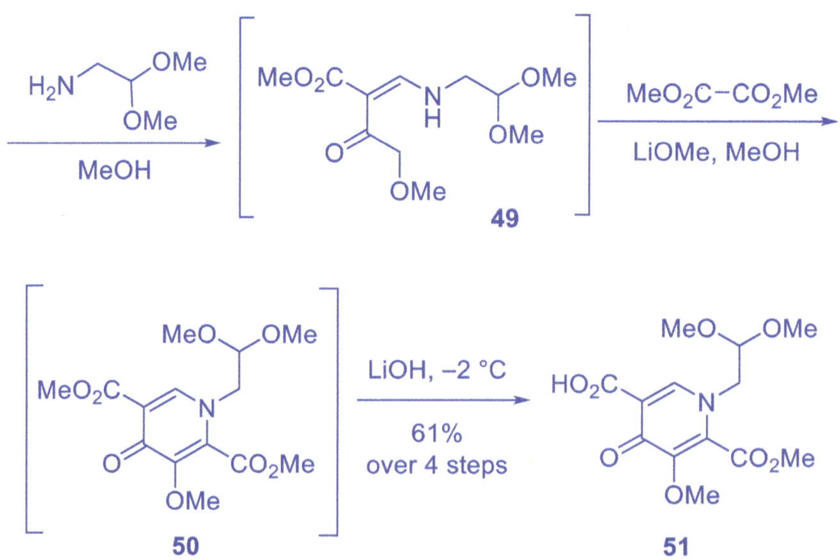

Removal of the acetal protection was facilitated by catalytic methanesulfonic acid in acetic acid and acetonitrile to reveal aldehyde **52**, which was condensed *in situ* with (*S*)-2-aminopropan-1-ol to install tricyclic pyridone acid **53** in 34:1 *dr*. It was activated with CDI and then coupled with 2,4-difluorobenzylamine to afford amide **54**. At the end, demethylation of **54** was accomplished using LiBr to deliver cabotegravir (**1**).[20]

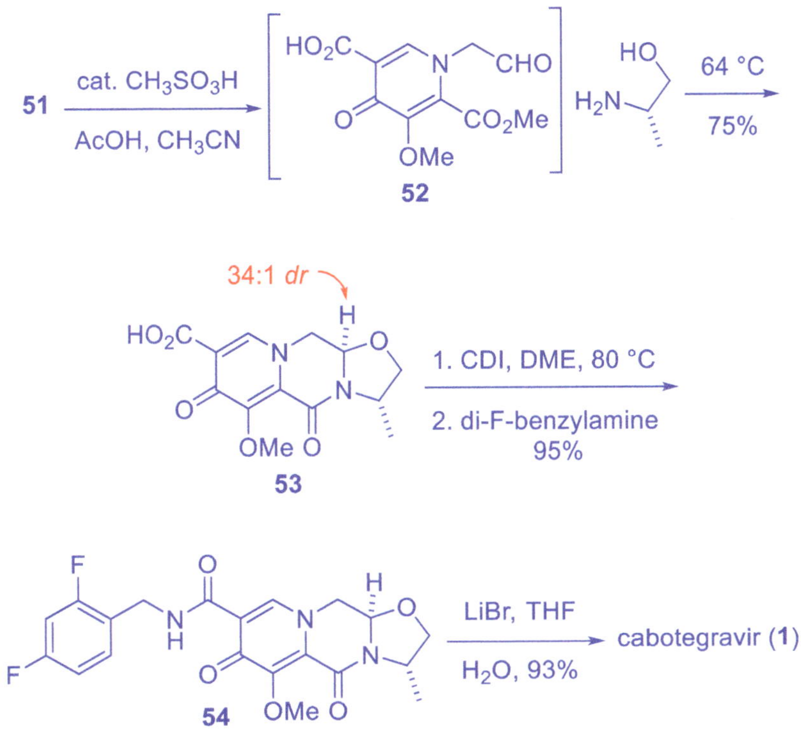

6.3. An Alternative Route

GSK Process Chemistry also disclosed an alternative route to make cabotegravir (**1**).[21,22]

Alternatively, amide formation was carried out before removal of the acetal protection. Thus, acid **51** was activated with CDI and then coupled with 2,4-difluorobenzylamine to afford amide **55**. Formic acid-promoted deprotection revealed aldehyde **56**, which was coupled with (*S*)-2-aminopropan-1-ol to deliver cabotegravir (**1**) after demethylation.[21]

It is poetic justice when a magnesium salt $Mg(OTf)_2$ was used to carried out the condensation reaction to make a drug that works by chelating also to magnesium divalent ions!

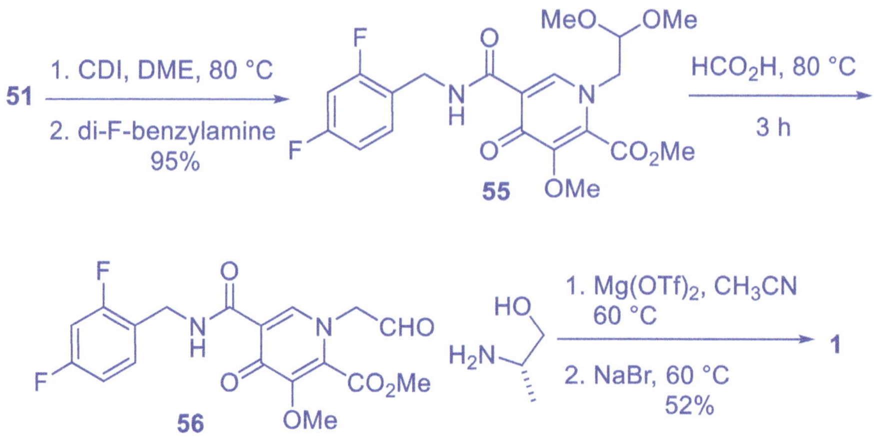

7. Summary

Cabotegravir (**1**) and doletegravir (**14**) are obtained by extensive SAR investigation from the monocyclic carbamoyl pyridones. While retaining the magnesium ion chelating functionalities, the Shionogi/GSK team discovered that the tricyclic derivatives endowed the carbamoyl pyridones with preferable DMPK profile for long-acting inhibition. Factors contributing to cabotegravir (**1**)'s long half-life include low water solubility, high activity, long half-life, and slow metabolic clearance.

Another advantage of cabotegravir (**1**) relies on its main metabolization by uridine diphosphate glucuronosyl-transferase 1A1, leading to less likely interactions with other antiretroviral drugs. The drug adds a powerful weapon in the arsenal for our war against HIV/AIDS.

References

1. Markham, A. Cabotegravir plus rilpivirine: first approval. *Drugs* **2020**, *80*, 915–822.
2. Durham, S. H.; Chahine, E. B. Cabotegravir-rilpivirine: the first complete long-acting injectable regimen for the treatment of HIV-1 infection. *Ann. Pharmacother.* **2021**, *55*, 1397–1409.
3. Durham, S. H.; Milam, A.; Waer, D.; Chahine, E. B. Cabotegravir: the first long-acting injectable for HIV preexposure prophylaxis. Ann. Phamacother. **2023**, *57*, 306–316.
4. Di Santo, R. Inhibiting the HIV integration process: past, present and the future. *J. Med. Chem.* **2014**, *57*, 539–566.

5. Kiyama, T.; Fuji, M.; Yoshinaga, T.; Sato, A.; Fujiwara, T.; Kiyama, R. A platform for designing HIV integrase inhibitors. Part 2: a two-metal binding model as a potential mechanism of HIV integrase inhibitors. *Bioorg. Med. Chem.* **2006**, *14*, 8420–8429.
6. Gill, M. S. A.; Hassan, S. S.; Ahemad, N. Evolution of HIV-1 reverse transcriptase and integrase dual inhibitors: recent advances and developments. *Eur. J. Med. Chem.* **2019**, *179*, 423–448.
7. Wang, Y.; Gu, S. X.; He, Q.; Fan, R. Advances in the development of HIV integrase strand transfer inhibitors. *Eur. J. Med. Chem.* **2021**, *225*, 113787.
8. Rowley, M. *The discovery of raltegravir, an integrase inhibitor for the treatment of HIV infection, Prog. Med. Chem.* **2008**, *46,* 1–28.
9. Shinkai, H.; Sato, M.; Matsuzaki, Y. *Elvitegravir: a novel monoketo acid HIV-1 integrase strand transfer inhibitor,* In *Antiviral Drugs,* ed., Kazmierski, W. M., Wiley: Hoboken, NJ **2011**, 197–205.
10. Kawasuji, T.; Johns, B. A.; Yoshida, H.; Taishi, T.; Taoda, Y.; Murai, H.; Kiyama, R.; Fuji, M.; Yoshinaga, T.; Seki, T.; Kobayashi, M. Carbamoyl pyridone HIV-1 integrase inhibitors. 1. Molecular design and establishment of an advanced two-metal binding pharmacophore. *J. Med. Chem.* **2012**, *55*, 8735–8744.
11. Kawasuji, T.; Johns, B. A.; Yoshida, H.; Weatherhead, J. G.; Akiyama, T.; Taishi, T.; Taoda, Y.; Mikamiyama-Iwata, M.; Murai, H.; Kiyama, R.; Fuji, M. Carbamoyl Pyridone HIV-1 integrase inhibitors. 2. Bi- and tricyclic derivatives result in superior antiviral and pharmacokinetic profiles. *J. Med. Chem.* **2013**, *56*, 1124–1135.
12. Johns, B. A.; Kawasuji, T.; Weatherhead, J. G.; Taishi, T.; Temelkoff, D. P.; Yoshida, H.; Akiyama, T.; Taoda, Y.; Murai, H.; Kiyama, R.; et al. Carbamoyl pyridone HIV-1 integrase inhibitors 3. A diastereomeric approach to chiral nonracemic tricyclic ring systems and the discovery of dolutegravir (S/GSK1349572) and (S/GSK1265744). *J. Med. Chem.* **2013**, *56*, 5901–5916.
13. Bailly, F.; Cotelle, P. The preclinical discovery and development of dolutegravir for the treatment of HIV. *Expert Opin. Drug Discovery* **2015**, *10*, 1243–1253.
14. Demeulemeester, J.; De Maeyer, M.; Debyser, Z. HIV-1 integrase drug discovery comes of age. *Top. Med. Chem.* **2015**, *15*, 1–52.
15. Fader, L. D.; Malenfant, E.; Parisien, M.; Carson, R.; Bilodeau, F.; Landry, S.; Pesant, M.; Brochu, C.; Morin, S.; Chabot, C.; et al. Discovery of BI 224436, a noncatalytic site integrase inhibitor (NCINI) of HIV-1. *ACS Med. Chem. Lett.* **2014**, *5*, 422–427.

16. Gupta, K.; Brady, T.; Dyer, B. M.; Malani, N.; Hwang, Y.; Male, F.; Nolte, R. T.; Wang, L.; Velthuisen, E.; Jeffrey, J.; et al. Allosteric inhibition of human immunodeficiency virus integrase. *J. Biol. Chem.* **2014**, *289*, 20477–20488.

17. Naidu, B. N.; Patel, M.; McAuliffe, B.; Ding, B.; Cianci, C.; Simmermacher, J.; Jenkins, S.; Parker, D. D.; Sivaprakasam, P.; Khan, J. A.; et al. Design, synthesis, and preclinical profiling of GSK3739936 (BMS-986180), an allosteric inhibitor of HIV-1 integrase with broad-spectrum activity toward 124/125 polymorphs. *J. Med. Chem.* **2022**, *65*, 4949–4971.

18. Cattaneo, D.; Gervasoni, C. Pharmacokinetics and pharmacodynamics of cabotegravir, a long-acting HIV integrase strand transfer inhibitor. *Eur. J. Drug Metab. Pharmacokinet.* **2019**, *44*, 319–327.

19. Margolis, D. A.; Gonzalez-Garcia, J.; Stellbrink, H.-J.; Eron, J. J.; Yazdanpanah, Y.; Podzamczer, D.; Lutz, T.; Angel, J. B.; Richmond, G. J.; Clotet, B.; et al. Long-acting intramuscular cabotegravir and rilpivirine in adults with HIV-1 infection (LATTE-2): 96-week results of a randomised, open-label, phase 2b, noninferiority trial. *Lancet* **2017**, *390*, 1499–1510.

20. Wang, H.; Goodman, S. N.; Mans, D.; Kowalski, M. Process for preparing carbamoylpyridone derivatives and intermediates, Int. Patent Appl. WO2011/119566 A1, WO2011119566 (2011), Sep 29, 2011.

21. Wang, H.; Kowalski, M. D.; Lakdawala, A. S.; Vogt, F. G.; Wu, Li. An efficient and highly diastereoselective synthesis of GSK1265744, a potent HIV integrase inhibitor. *Org. Lett.* **2015**, *17*, 564–567.

22. Hughes, D. L. Review of synthetic routes and final forms of integrase inhibitors dolutegravir, cabotegravir and bictegravir. *Org. Process Res. Dev.* **2019**, *23*, 716–729.

4

Lenacapavir (Sunlenca): A Long-acting HIV-1 Capsid Protein Inhibitor for Treating HIV Infection

Jie Jack Li

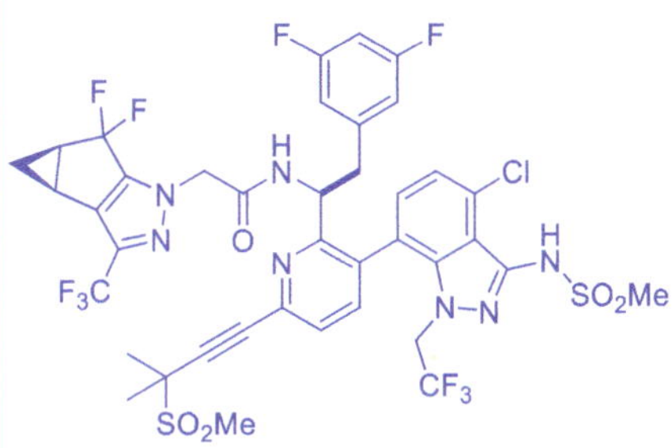

lenacapavir (**1**, Sunlenca)
Gilead, 2022
exceptionally long-lasting
($t_{1/2}$ = 7–11 weeks!)
HIV-1 capsid protein inhibitor

At the end of 2022, the FDA approved Gilead's long-acting human immunodeficiency virus-1 (HIV-1) drug lenacapavir (**1**, Sunlenca), a first-in-class HIV capsid protein inhibitor, providing a new treatment option for people suffering from multidrug resistance.[1, 2]

Several aspects about lenacapavir (**1**) are truly remarkable:

a. It is really long acting, only needs to be given two injections a year;

b. It can be used as pre-exposure prophylaxis (PrEP);

c. It is a first-in-class HIV-1 capsid protein inhibitor;

d. Its molecular weight is 968, way beyond the rule-of-5; and

e. It has 10 fluorine atoms, a record previously held by Merck's aprepitant (Emend) with "merely" seven fluorine atoms.

Chemistry and Pharmacology of Drug Discovery, First Edition. Edited by Jie Jack Li.

1. Background

Long-acting HIV-1 antiretroviral therapy, characterized by a greater than 1-month dosing interval, offers significant advantages over daily oral therapy. Before the emergence of lenacapavir (**1**), the only long-acting HIV-1 antiretroviral therapy on the market was Cabenuva, an extended release injectable suspension of a combination of cabotegravir (**2**), an HIV integrase inhibitor, and rilpivirine (**3**), a non-nucleoside reverse transcriptase inhibitor (NNRTI), given once every other month (see Chapter 3).

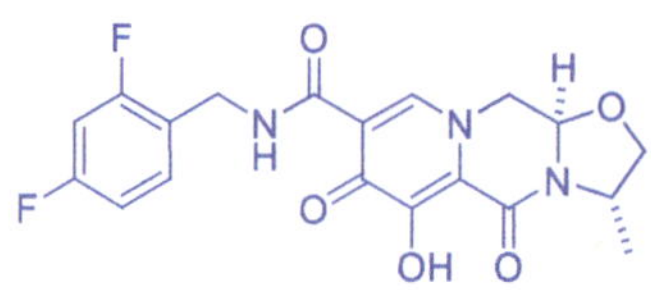

cabotegravir
(**2**, Cabenuva with rilpivirine)
ViiV/GSK, 2020
HIV Integrase Strand Transfer Inhibitor
$t_{1/2}$ = 21–50 days

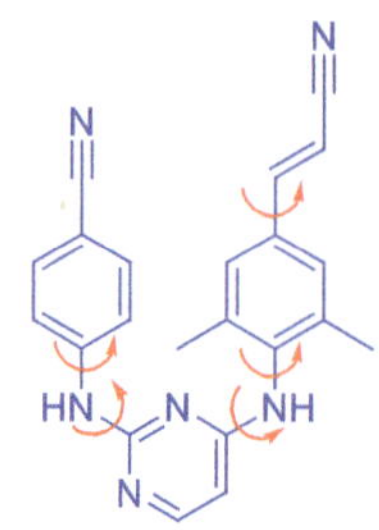

rilpivirine (**3**, Edurant)
Tibotec/J&J, 2011
NNRTI

Nowadays, the criteria for HIV drugs to enter clinical development are very high. Exceptional potency and low plasma clearance are required to meet dose size requirements; excellent chemical stability and/or crystalline form stability is required to meet formulation requirements, and new antivirals in HIV-1 therapy need to be largely free of side effects and drug–drug interactions (DDIs).

Gilead's HIV capsid protein inhibitor lenacapavir (**1**) is active *at least two points in the viral lifecycle*. Thanks to its high potency and favorable pharmacokinetics (slow sustained release and low hepatic clearance), it can be taken orally once weekly or injected subcutaneously every 6 months while maintaining supra-effective concentrations.[1]

However, lenacapavir (**1**) needs a long-acting partner if it is to deliver on the promise of being part of a complete long-acting regimen. Gilead and Merck are collaborating to develop a long-acting HIV treatment that combines lenacapavir (**1**) with Merck's islatravir (**4**), a nucleoside reverse transcriptase translocation inhibitor (NRTI). Both have long half-lives and robust virus-suppressing activity at low doses. Regrettably, with the development of islatravir (**4**) slowed recently (2023) by unexpected immunological findings, the future of this partner is far from clear.[2]

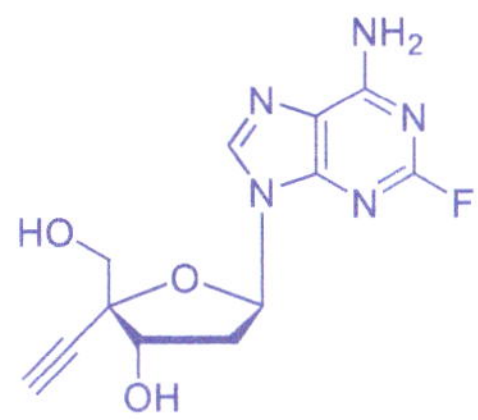

islatravir (**4**)
Merck, 2023, Ph III
exceptionally long-lasting
($t_{1/2}$ > 120 h)
and potent NRTI (safety
concerns)

2. Pharmacology

The capsid core of HIV-1 is a large macromolecular assembly that surrounds the viral genome and is an essential component of the infectious virus.

Lenacapavir (**1**) potently inhibited early steps of HIV-1 replication with half-maximal effective concentration (EC_{50} = 87 pM). It also exhibited a second, less potent (EC_{50} = 240 pM), antiviral activity during virus egress. Molecular modeling studies predicted that lenacapavir (**1**) binds to the hydrophobic pocket formed by two adjoining capsid subunits within the hexamer.

Now, let us begin learning lenacapavir (**1**)'s pharmacology by first scrutinizing the structure of HIV capsid protein.[3]

2.1. Structure of Capsid Protein

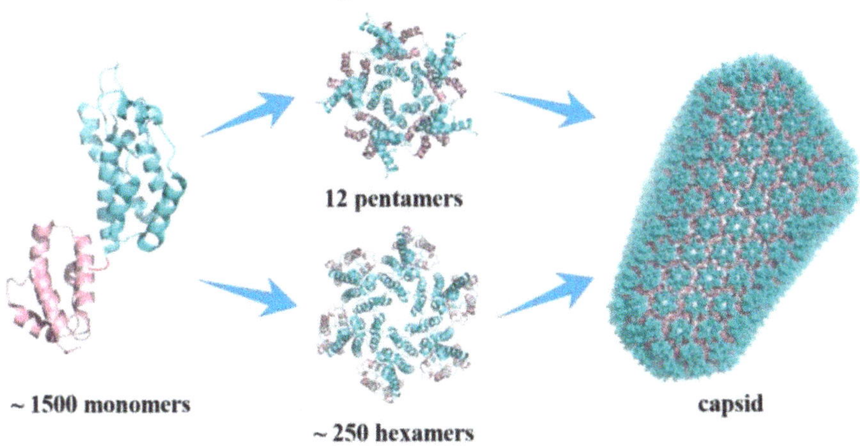

Figure 1. HIV capsid protein. Source: Zhang et al.[4]/MDPI/CC BY 4.0

A mature virion envelope is surround by a "fullerene cone"-shaped capsid shell that encapsulates two copies of the positive-strand RNA genome along with associated cellular factors and viral proteins. The HIV capsid is erected from a single protein, known as capsid protein. Consisting of 231 residues, the HIV capsid protein (CA, p24—because its molecular weight is 24 kDa) forms the shell of an electron-dense, elongated core within the virion. Capsid proteins then form pentameric and hexameric subunits, which proceed to assemble into the mature viral capsid.[5]

After the budding of the immature virion, proteolytic processing (maturation) of the *Gag polyprotein* causes the rearrangement of the capsid proteins into a conical core structure that surrounds the viral genome in the mature virus. The mature core comprises approximately 1500 capsid monomers with about 250 capsid hexamers and 12 capsid pentamers (Figure 1).[4]

Capsid consists of two domains called the N-terminal domain (NTD) and C-terminal domain (CTD), with a flexible linker connecting both. On the other hand, capsid NTD consists of seven α-helices, a β-fold, and a cyclophilin A (CypA) binding loop, while capsid CTD consists of a 310 helix, four α-helices, and an N-terminal extension chain. CypA was found to bind Gag in the capsid region back in 1993.[6]

2.2 Functions of Capsid Protein

Capsid does not have known catalytic activity although it can impact multiple viral enzymatic activities, including that of reverse transcriptase and integrase. HIV capsid protein serves critical roles in many aspects of the HIV-1 replication cycle such as reverse transcription, cytoplasmic transport, nuclear entry, and virion maturation in addition to interacting with over 20 host factors essential for infection.[6]

2.2.1. At the Early Stages of the Virus Life-Cycle

The HIV-1 capsid houses the replicative enzymes and viral genomic RNA, protecting them from antiviral factors and cellular sensors of innate immunity, allowing their traffic from entry to near integration sites before fully uncoating. Increasing evidence suggests that the capsid participates in the translocation of viral genomic material into the host nucleus for integration through partial uncoating that allows higher plasticity of the capsid.[7]

As mentioned before, capsid is generated by the cleavage of the Gag polyprotein by the protease enzyme. It is initially expressed within the Gag and Gag–Pol polyproteins and provides key interactions between proteins that are necessary for assembly of the virion. HIV-1 Gag proteins are necessary for virion assembly, virion maturation, and early stages of virus replication.

When a virus enters the cytoplasm of a cell, it undergoes a highly controlled process known as disassembly, in which capsid dissociates and releases its contents into the cytoplasm. Specifically, following fusion of the viral particle with the target cell membrane, the HIV-1 capsid core is released into the cytosol where it protects the viral genome and enzymes from host defenses while trafficking these contents along the microtubule network toward the nucleus. These replication steps require capsid to interact with human factors such as CypA, nucleoporins 153 and 358 (NUP153, NUP358), cleavage and polyadenylation-specific factor 6 (CPSF6), and other factors (Figure 2).

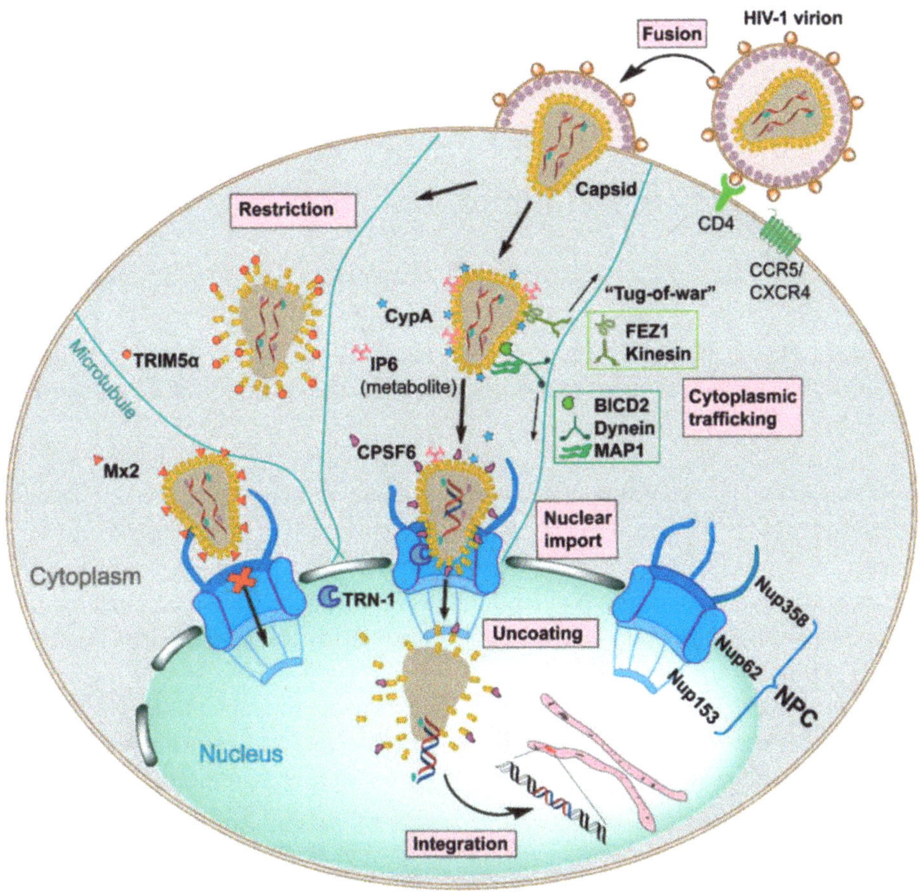

Figure 2. Schematic overview of the early stages of HIV-1 replication. Source: Zhuang and Torbett[6]/MDPI/CC BY 4.0

2.2.2. At the Late Stages of the Virus Life-Cycle

The late phase of the HIV-1 life cycle begins with transcription of the provirus, followed by export of the synthesized RNA to the cytoplasm and translation of the Gag/Gag–Pol polyproteins. In the final phase of maturation, capsid is at first expressed within the Gag/Gag–Pol polyproteins and provides key interactions between proteins that are necessary for assembly of the virion. Gag is cleaved and capsid is released, allowing the assembly of capsid into a "fullerene cone," known as the capsid core.[8]

In view of the key functions of the capsid protein, it has become a popular target for the development novel anti-HIV drugs.[9] Moreover, interfering with only a few of these capsid–capsid interactions is thus expected to reduce HIV infectiousness severely. As a comparison, each viral particle contains approximately 250 integrase proteins, of which only four need to be functionally active to achieve the irreversible integration of HIV DNA in the human genome.

2.3. Capsid Protein Inhibitors

The interactions between capsid proteins are the key determinants for the stability of the mature capsid, which is essential for the precise timing of the assembly and uncoating steps in the HIV-1 life cycle. Thus, creating small molecules that either stabilize or destabilize the capsid core is a promising strategy for the discovery of novel antiviral treatments.[6]

The NTD–CTD interprotomer pocket, a particularly therapeutically attractive capsid protein region involved in self-assembly has been defined within the past decade. This pocket within the assembled hexamer and pentamer is the binding site for host dependency factors. Disruption of the interactions of capsid with host factors could interfere with HIV-1 replication. Moreover, the NTD–CTD interface is also the binding site of extensively studied small molecule modulators such as lenacapavir (**1**) and PF-3450074 (PF74, **5**).

In 2010, Pfizer reported that PF74 (**5**) exhibited antiretroviral activity. It is a peptidomimetic compound built around a phenylalanine core and capped with indole-3-acetic acid and aniline moieties at the amino and carboxylate ends, respectively.

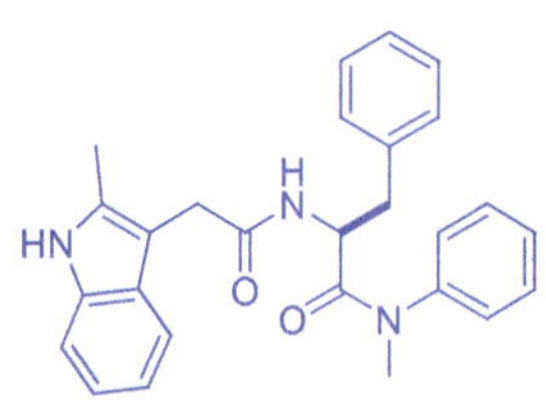

PF3450074 (PF74, **5**)
Agouron phenotypic screen
Pfizer discontinued (2020 poster)
EC_{50} = 570 nM
1.25 μM in cell, $t_{1/2}$ = 3 min
human predCL = 1.2 L/h/kg
in vitro and *in vivo* POC
It was an accelerator in Gilead's assay

Cocrystal structures revealed that (*S*)-PF74 (**5**) binds to a novel binding pocket in the NTD of the protein. This hydrophobic cavity, also known as the phenylalanine-glycine (FG) binding pocket, is the same binding site of lenacapavir (**1**). (*S*)-PF74 (**5**) also binds to the cellular HIV-1 cofactors that mediate nuclear import of pre-integration complexes (PICs). Since (*S*)-PF74 (**5**) interferes with both capsid assembly and disassembly, it inhibits both the early (virion uncoating) and late (viral core assembly) events in the viral replication cycle.[10]

The (*S*)-PF74 (**5**) inhibits viral replication via two distinct mechanisms of action: by disrupting CA–CA interactions, they could alter the overall core stability and impact viral assembly and uncoating; by competing against NUP153 and CPSF6 for capsid binding, they could block viral nuclear entry and/or productive integration. In addition to (*S*)-PF74 (**5**), many capsid protein inhibitors, especially the ones binding to the FG pocket, have been reported.[11]

Clinical trials of (*S*)-PF74 (**5**) was carried out but was terminated during phase I.

2.4. Lenacapavir

Employing (*S*)-PF74 (**5**) as the starting point and drawing on X-ray crystallographic information, Gilead designed lenacapavir (**1**) to bind tightly at a conserved interface between capsid protein monomers. Lenacapavir (**1**) reduces viral replication at both the early and late stages of the HIV life cycle by binding at the interface between two capsid proteins. This binding interferes with capsid-mediated interactions between proteins that are essential for multiple phases of the viral replication cycle (Figure 3).[7]

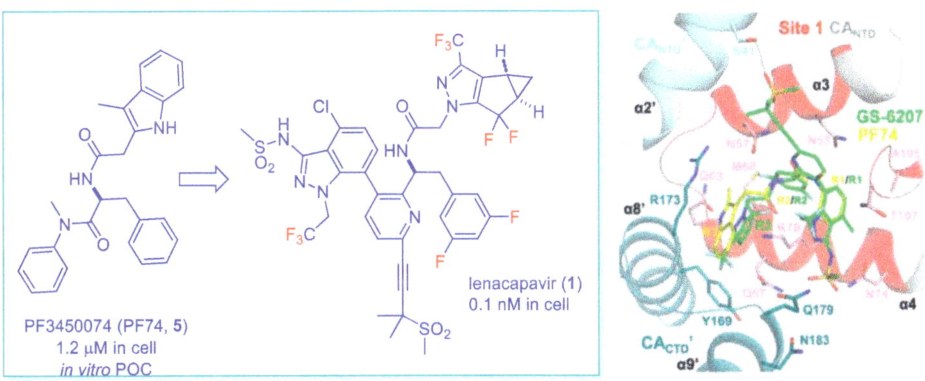

Figure 3. The X-ray crystal structure of the HIV-1 capsid hexamer bound to PF-3450074 (**5**, PDB code: 4XFZ) and lenacapavir (**1**, PDB code: 6V2F). Source: McFadden et al.[7]/Springer Nature/CC BY 4.0

Just like (*S*)-PF74 (**5**), lenacapavir (**1**) inhibits HIV-1 replication mainly by binding to the FG binding site, stabilizing and thereby preventing capsid disassembly in infected T cells.

It is a tight inhibitor (with picomolar potency) that binds two contiguous capsid subunits while promoting distal intra- and inter-hexamer interactions that stabilize the curved capsid lattice. A crystal structure of lenacapavir (**1**) bound to a cross-linked capsid hexamer shows extensive hydrophobic and electrostatic interactions, *seven hydrogen bonds,* and *two cation–π interactions* between lenacapavir (**1**) and residues of binding site 1 (Figure 4).[3]

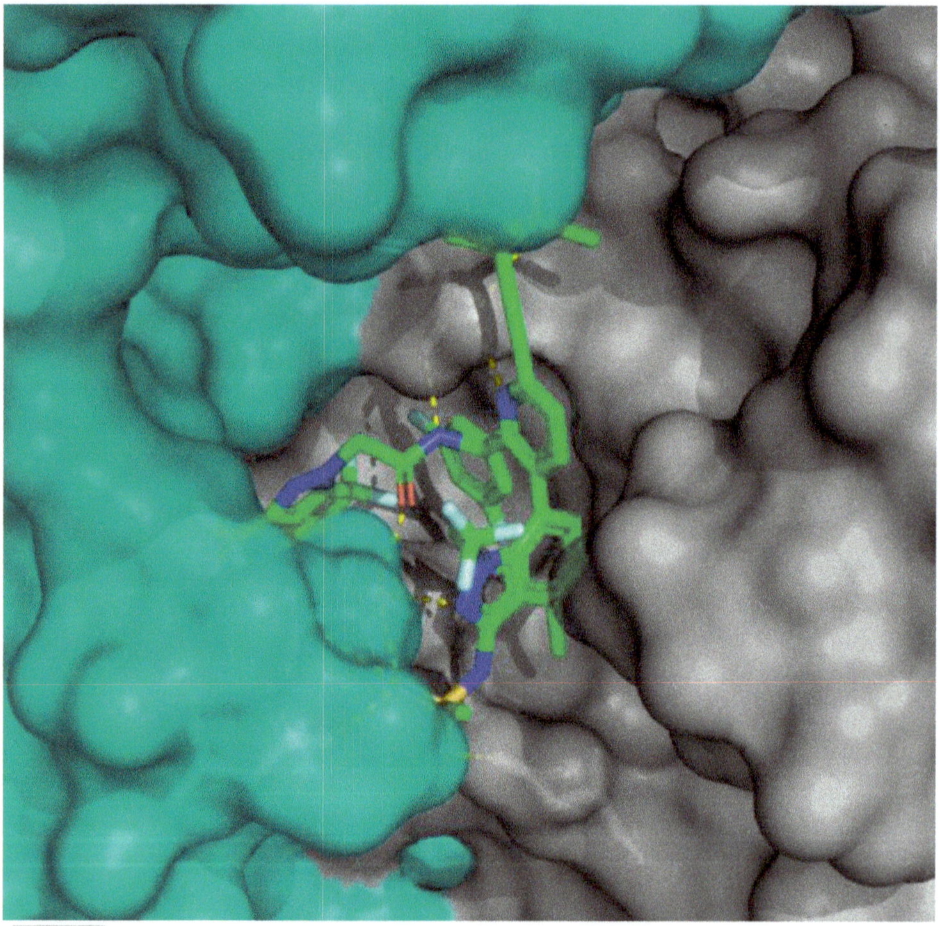

Figure 4. The X-ray crystal structure of the HIV-1 capsid hexamer bound to lenacapavir (**1**, PDB code: 6V2F). Source: Link et al. [3]/with permission of Springer Nature

Inspired by lenacapavir (**1**)'s spectacular success, "me-too" drugs soon emerged. ViiV's capsid inhibitor ViiV-1 (**6**) also binds to the now-familiar hydrophobic FG pocket.[7] In the same vein, BMS/ViiV in 2023 published their potent long-acting inhibitors targeting the HIV-1 capsid based on a versatile quinazolin-4-one scaffold. The development candidate GSK878 (**7**) has an EC_{50} value of 39 pM.[12]

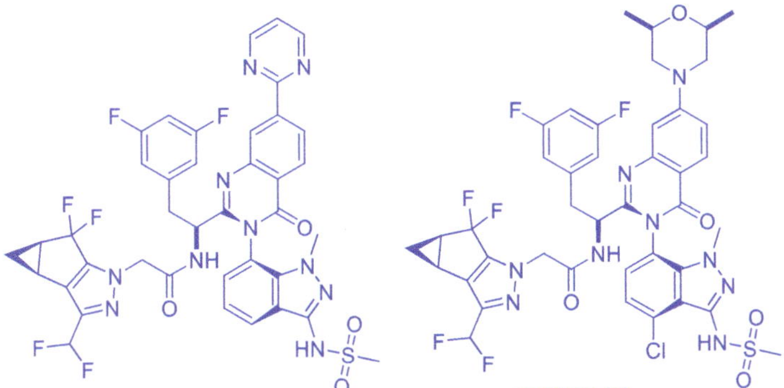

ViiV-1 (**6**)

GSK878 (**7**)
Stable axial chirality
HIV-1 EC_{50} = 39 pM

3. Structure–Activity Relationship (SAR)

Gilead has not yet published any SAR for this series of capsid inhibitors at the time of writing of this review (2023). In lieu of SAR, the progression of the project is summarized here according to John Link's seminar presentation at the 2023 Steamboat Medicinal and Bioorganic Chemistry Conference in Steamboat, Colorado.

In the hands of Gilead scientists, (*S*)-PF74 (**5**) was tested as an *accelerator*, not an inhibitor, of capsid assembly. Its EC_{50} in MT-4 cells was 1.24 μM whereas its cytotoxicity concentration, CC_{50} = 32 μM, providing a therapeutic window of 26.

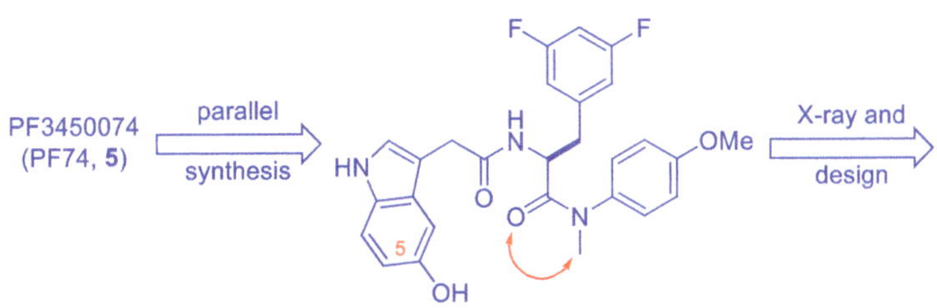

dipeptide **8**
EC_{50} = 52 nM (25-fold)
PA EC_{50} = 370 nM (35-fold)
iEC_{50} = 14 nM
Metabolite ID
with 5-OH, glucuronide + amide cleavage
w/o 5-OH, amide cleavage + multiple oxidations

In order to quickly evaluate the SAR of the indole motif, a parallel synthesis was carried out to prepare an array of derivatives at Gilead. Among them, dipeptide **8** with a 5-OH-substituted indole showed a 25× boost of potency (EC_{50} = 52 nM). However, the presence of the phenol functionality on dipeptide **8** led to extensive phase II metabolism through formation of the corresponding 5-OH glucuronide, accompanied by amide bond cleavage.

Scrutinizing the X-ray crystal structure of binding pose of dipeptide **8** to capsid protein, it was realized that the amide bond could be rigidified to afford pyridine **9**, which only had one amide bond left. Another parallel synthesis led to carboxamide **10** with its primary amide forming a hydrogen bond with histidine-75 of the capsid protein, which became known as a potency driver.

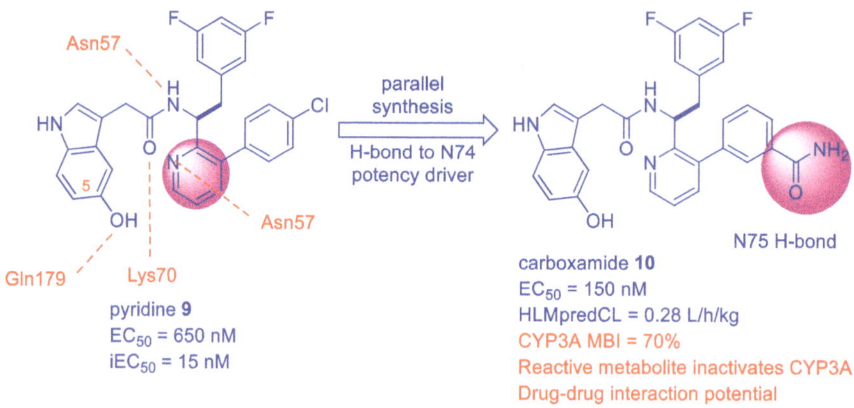

While carboxamide **10** was more potent than its precursor, it saw significant (70%) metabolism-based inhibition (MBI). Not only did this metabolism pose potential toxicity issue, but it also had DDI potential. As shown below, it is reasonable to assume that amide **10** was initially oxidized at the "benzylic" position by CYP3A to the corresponding alcohol **11**. This unstable alcohol then underwent an elimination process to provide enamide **12** as the proposed metabolite, which then potentially formed a covalent bond with a nucleophile of CYP3A to inactivate the enzyme.

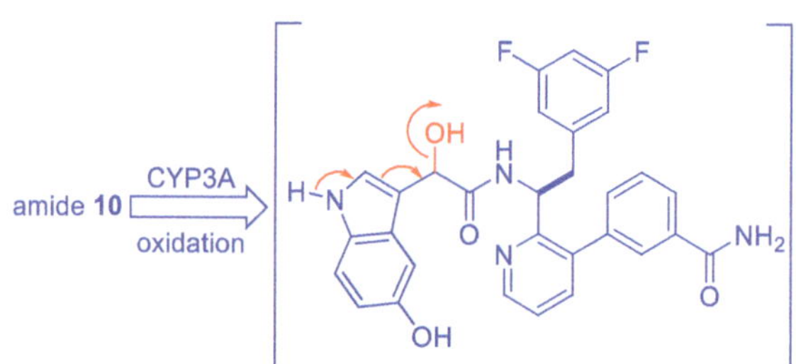

alcohol **11**

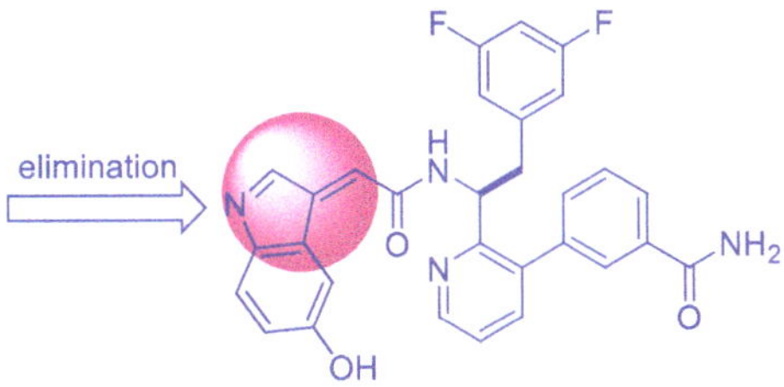

enamide **12**
proposed reactive metabolite
inactivated CYP3A

After extensive exploration, it was found that cyclohexyl ring-fused pyrazole **13** did not suffer MBI (<1%) as did amide **10**. But it was extensively metabolized/cleared. In order to explore the metabolic "soft spots," the suspected cyclohexyl ring was removed to afford the "naked" pyrazole **14**, which completed lost its potency in MT-4 cells but its metabolic stability was significantly improved—indicating that the cyclohexyl ring on pyrazole **13** contained indeed the "soft spots."

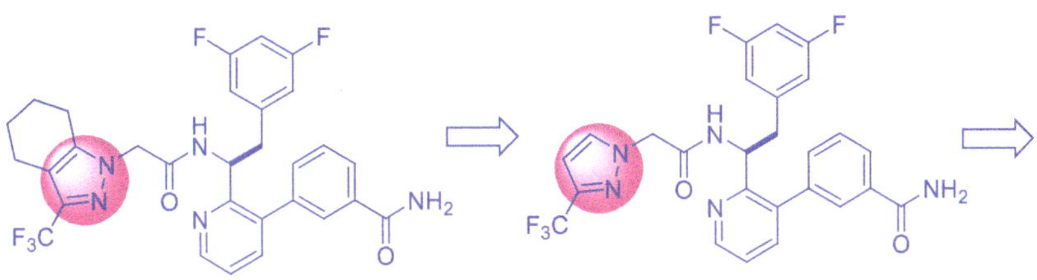

pyrazole **13**
EC_{50} = 260 nM (MT4)
MBI < 1%
HLMpredCL = 1.2 L/h/kg

pyrazole **14**
EC_{50} > 40 μM (MT4)
HLMpredCL < 0.17 L/h/kg
potency lost, but stability improves

In an attempt to block the metabolic "soft spots," tetrafluorocyclohexyl **15** was prepared, which eventually morphed into the tricyclic cyclopropyl **16**. It enjoyed both good potency and stability.

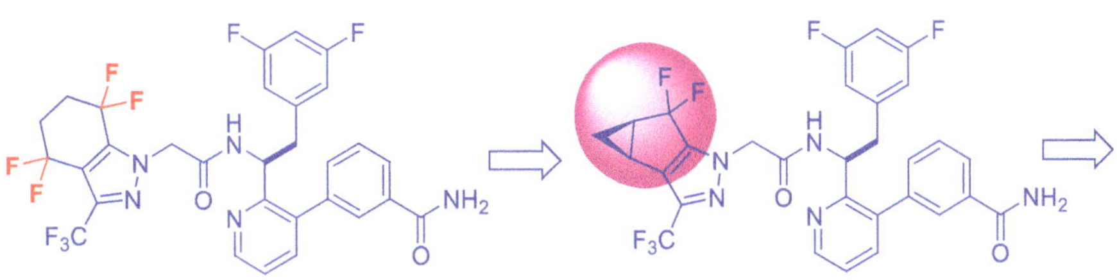

tetrafluorocyclohexyl **15**

cyclopropyl **16**
EC_{50} = 167 nM
HLMpredCL < 0.17 L/h/kg

Gilead's efforts to enhance potency led to modifications on the right-hand portion of the molecules. Aminoindazole **17** was nearly 10× more potent than its precursor **16** although it had metabolism issues as well, most likely phase II metabolism associated with the "naked" amine. Indeed, sulfonylation of **17** gave rise to sulfonamide **18**, which was tested to be more robust toward metabolism.

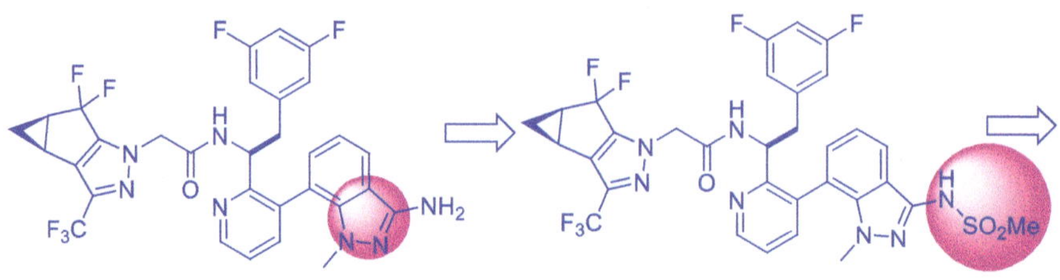

aminoindazole **17**
EC_{50} = 19 nM
HLMpredCL < 0.17 L/h/kg
HLM+UDPGA = 0.87 L/h/kg

sulfonamide **18**
EC_{50} = 12 nM
HLM+UDPGA = 0.26 L/h/kg

To further enhance the drug's affinity to the target protein, it was decided to extend the molecule by making alkynol **19**, which further morphed into sulfone **20** (EC_{50} = 200 pM). Now the potency of this series of drugs had entered the enviable picomolar territory. Fine-tuning of sulfone **20**, especially at the right-hand portion, led to GS-CA1 (**21**) as a potential drug candidate. The EC_{50} of GS-CA1 (**21**) was 240 pM whereas its cytotoxicity concentration was CC_{50} > 50 μM, providing a therapeutic window of >208,300! When administered subcutaneously to a mouse model, one dose of GS-CA1 (**21**) was able to maintain compound levels above its EC_{50} in the blood plasma for 56 days, outperforming the long-lasting NNRTI, rilpivirine (**3**), in both potency and selectivity.[13]

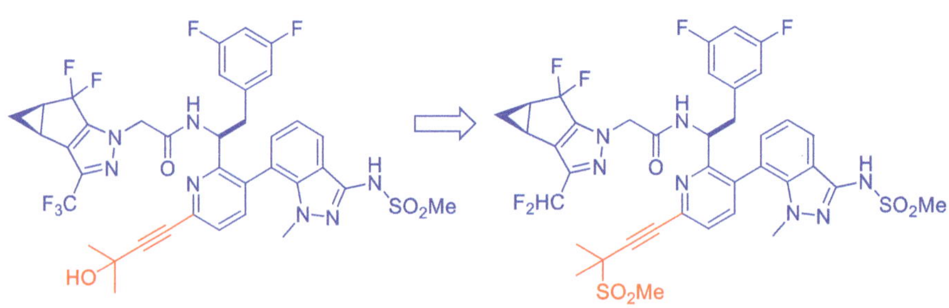

alkynol **19**
EC_{50} = 10 nM
HLM+UDPGA < 0.09 L/h/kg

sulfone **20**
EC_{50} = 0.2 nM
HLM+UDPGA < 0.09 L/h/kg

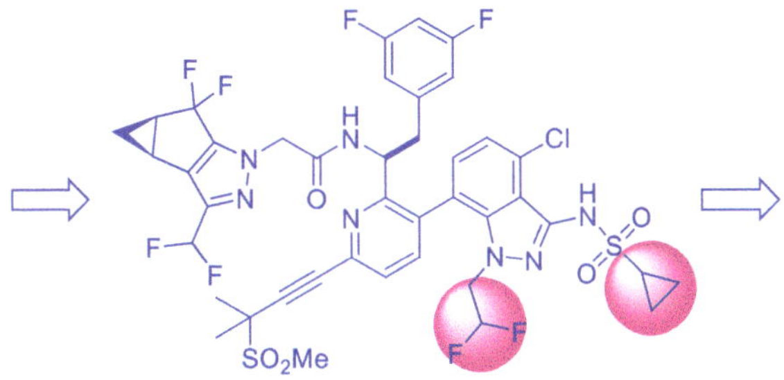

GS-CA1 (**21**)
EC_{50} = 0.24 nM

Obsession to obtain the perfect drug led Gilead to eventually "settling" at lenacapavir (**1**) as their drug development candidate. The decision was apparently a wise one as witnessed by its approval at the end of 2022.

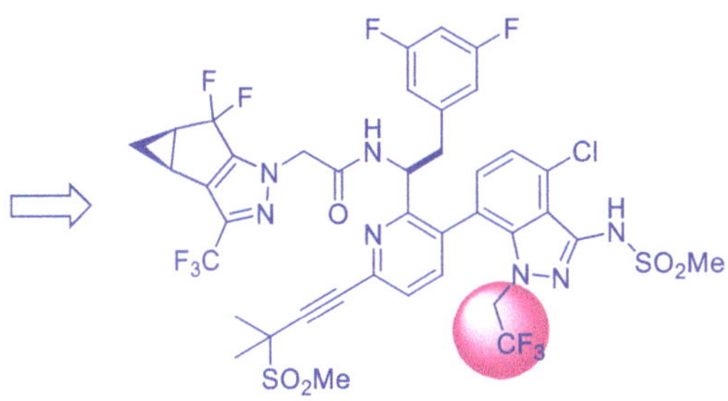

lenacapavir (**1**)
EC_{50} = 0.1 nM
predCL = 0.09 L/h/kg
IVIVC DCL = 0.035 L/h/kg
V_{ss} = 2.86
MRT = 10.5

In summary, low hepatic clearance is an essential attribute for a long-acting agent. Thus, along with enhancement of potency, the design process that culminated in lenacapavir (**1**) also focused on *blocking metabolically labile sites through incorporation of electron-withdrawing groups (halogens and sulfonyls), metabolically stable ring systems (cyclopropane and pyrazoles) and rigidifying elements*. Many invaluable lessons in drug design can be learned from this process.

4. Pharmacokinetics and Drug Metabolism

Tritiation of lenacapavir (**1**) was necessary to accurately measure the low turnover of lenacapavir (**1**) in primary human hepatocytes and showed a predicted rate of hepatic clearance of 0.01/h*kg, or 0.8% of the hepatic extraction.[3]

Not surprisingly, little metabolism was observed for this very robust molecule. Lenacapavir (**1**) has a half-life of 7–11 weeks when administered subcutaneously, and it can be administered orally. It can be used as a long-acting agent for pre-exposure prophylaxis (PrEP).

5. Efficacy and Safety

Lenacapavir (**1**) has two mechanisms of action at separate stages of the viral life cycle, thus posing a barrier to resistance that may be intrinsically higher.

The FDA approved lenacapavir (**1**) on the basis of the phase II/III CAPELLA trial in 72 heavily treatment-experienced subjects with multiclass-resistant HIV-1. The trial met its primary endpoint, with 88% of lenacapavir (**1**) recipients having a decrease of at least 0.5 log10 copies per milliliter in the viral load at day 15, compared with 17% of placebo recipients.[1]

For long-acting drugs, their safety is of utmost importance because they stay in the system for over 1 month. Gilead was scrupulous to ensure the safety profile to be nearly perfect for their drug candidate. Their labor paid off at the end. No serious adverse events related to lenacapavir (**1**) were identified in clinical trials.[1]

6. Synthesis

A synthesis of lenacapavir (**1**) was published as the supplemental information (SI) of Gilead's 2020 *Nature* article.[3] Despite its molecular complexity, a retrosynthetic excise may dissect lenacapavir (**1**) into three simpler and more manageable building blocks: fragment A, B, and C.

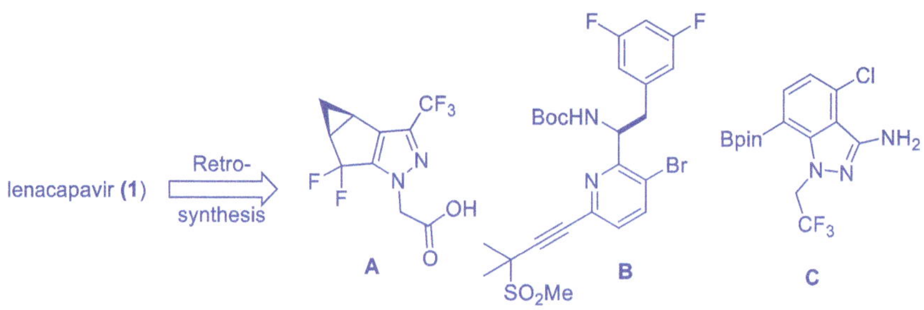

6.1. Fragment A

Preparation of fragment A commenced with a Simmons–Smith cyclopropanation. Thus, cyclopent-3-en-1-ol (**22**) was treated with diethylzinc and diiodomethane to install the cyclopropane onto **22**. Subsequent Dess–Martin periodinane (DMP) oxidation offered ketone **23**. Treatment of ketone **23** with lithium hexamethyldisilazide was followed by addition of ethyl-trifluoroacetate to generate enolate **24**, which was used *in situ* to react with ethyl aminoglycinate to forge pyrazole **25**. Oxidation of the "benzylic" methylene was accomplished using sodium chlorite and 2-hydroxyisoindoline-dione as co-oxidants to provide ketone **26**, which was then protected as dithiolane **27**.

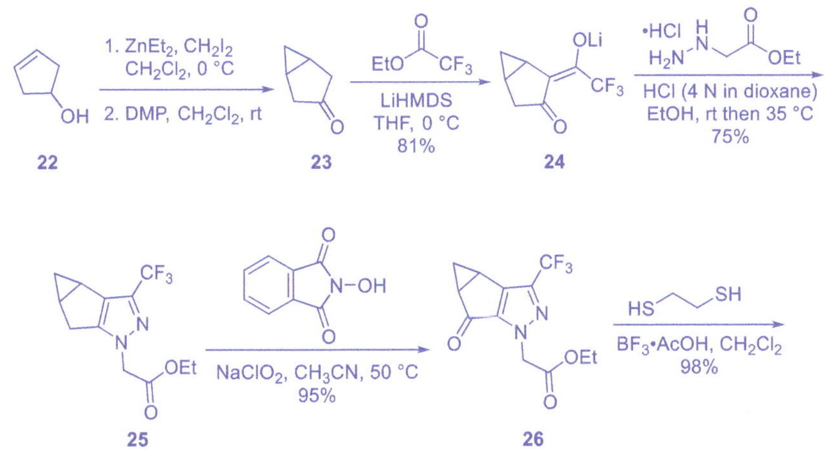

Subsequently, treatment of dithiolane **27** with 1,3-dibromo-5,5-dimethylhydantoin (DBDMH, **28**) and HF•pyridine converted dithiolane **27** to the corresponding difluoro-derivative **29**. Lithium hydroxide-mediated hydrolysis of the ethyl ester on **29** revealed carboxylic acid **30**. Chiral supercritical fluid chromatography (SFC) separation then delivered fragment A with the desired configuration.[3]

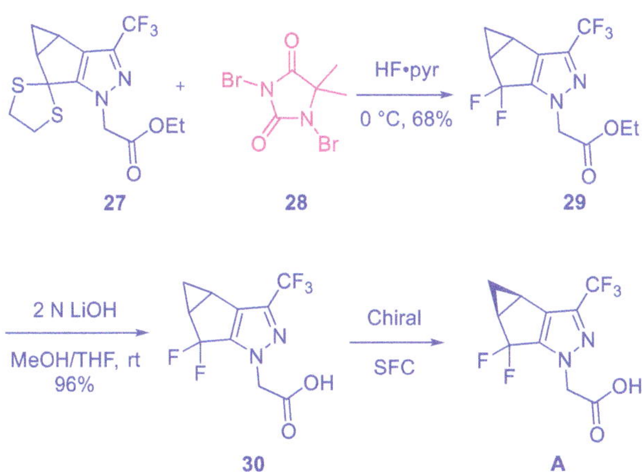

6.2. Fragment B

Preparation of fragment B began with making terminal alkyne **33** by CuCl-mediated coupling reaction between propargyl chloride **31** and sodium methanesulfinate (**32**).

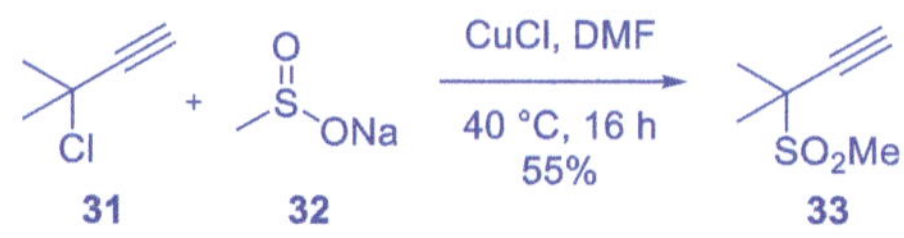

Meanwhile, 3,6-dibromo-2-methylpyridine (**34**) underwent a Wohl–Ziegler reaction to afford the corresponding "benzyl" bromide, which was subsequently converted to aldehyde **35** under the influence of silver nitrate. Condensation of aldehyde **35** with Ellman's chiral auxiliary (*S*)-sulfinamide **36** gave rise to imine **37**, which was immediately coupled with benzylzinc **38** to assemble adduct **39**. Removal of the chiral auxiliary was carried out using strong acid to expose the "naked" amine **40**, which was protected *in situ* as the corresponding Boc derivative **41**. A Sonogashira coupling reaction between terminal alkyne **33** with **41** *chemoselectively* took place with the α-bromide to install fragment B. For the two bromides on pyridine **41**, the α-bromide is exponentially more reactive than the γ-bromide because the α-bromide is activated by the neighboring nitrogen atom.

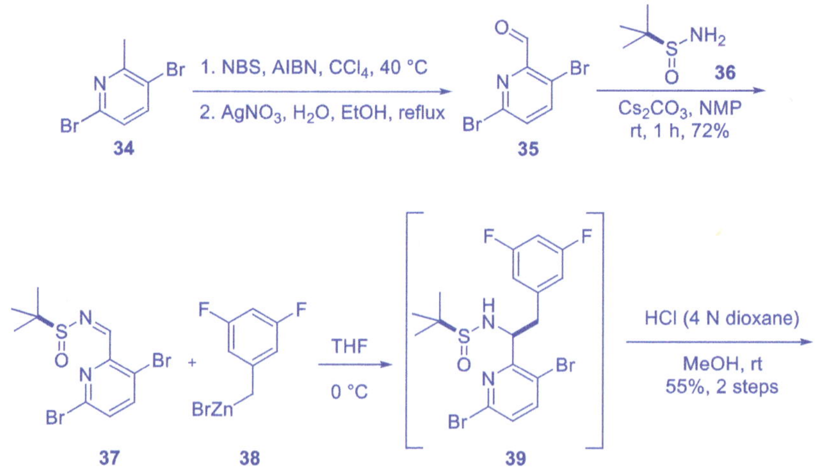

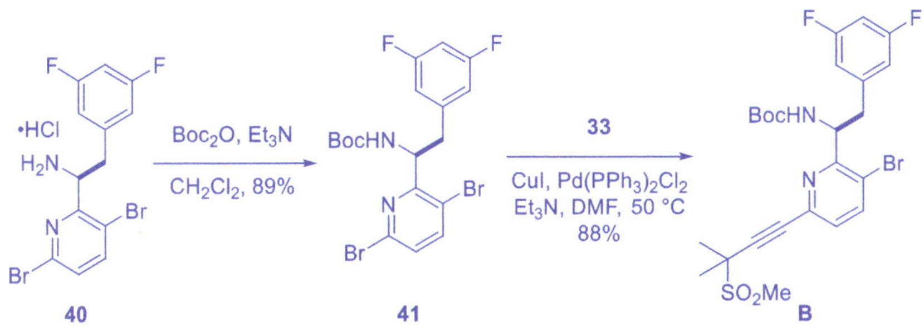

6.3. Fragment C

Production of fragment C employed 1-bromo-4-chloro-2-fluorobenzene (**42**) as the starting material. Deprotonation of **42** with lithium diisopropylamide took place selectively at the most acidic C-3 position. Quenching the anion with DMF provided the corresponding aldehyde, which was promptly coupled with sulfamic acid (hydroxylamine-O-sulfonic acid), and the adduct immediately collapsed to offer nitrile **43**. Heating nitrile **43** with hydrazine hydrate in ethanol gave the S_NAr adduct by replacing the fluorine. The adduct then simultaneously underwent an intramolecular cyclization to afford aminoindazole **44**. Alkylation of aminoindazole **44** selectively attached the trifluorethyl group at the most acidic 1H position to prepare **45**. A Miyaura coupling of **45** with bis(pinacolato)diboron then delivered fragment C as a boronate.

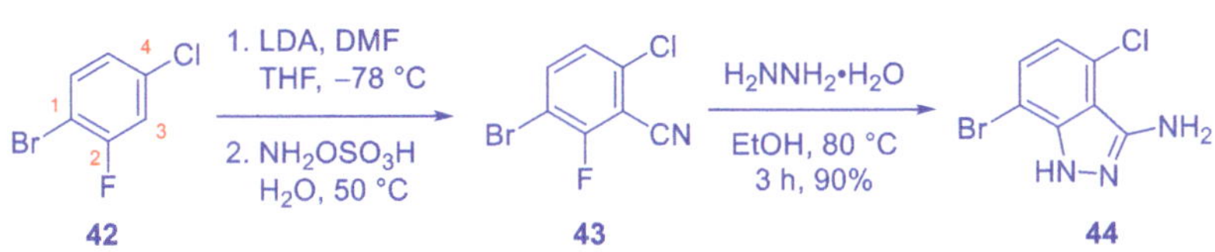

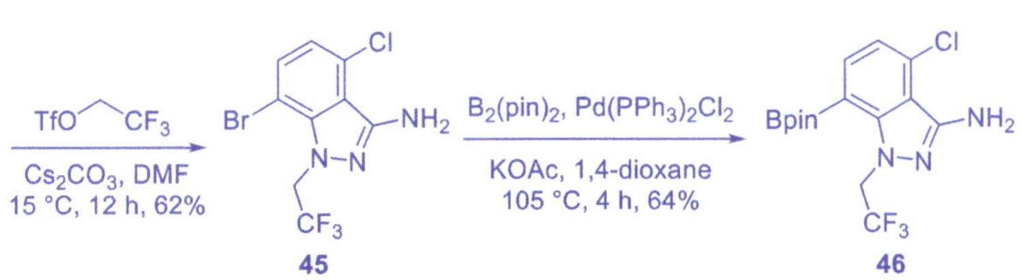

6.4. Assembly of Lenacapavir (1)

The final assembly of lenacapavir (**1**) began with a Suzuki coupling between the bromide on fragment B with the boronate on fragment C to produce adduct **47**. Sulfonylation of **47** with mesyl chloride could not stop at the mono-sulfonylation stage, giving rise to bis-sulfonamide **48**. Removal of the Boc protection on **48** with trifluoroacetic acid exposed the primary amine on **49**, which was then coupled with fragment A using HATU as the coupling agent. Treatment of the adduct with sodium hydroxide then removed one of the two sulfonamides to deliver lenacapavir (**1**).

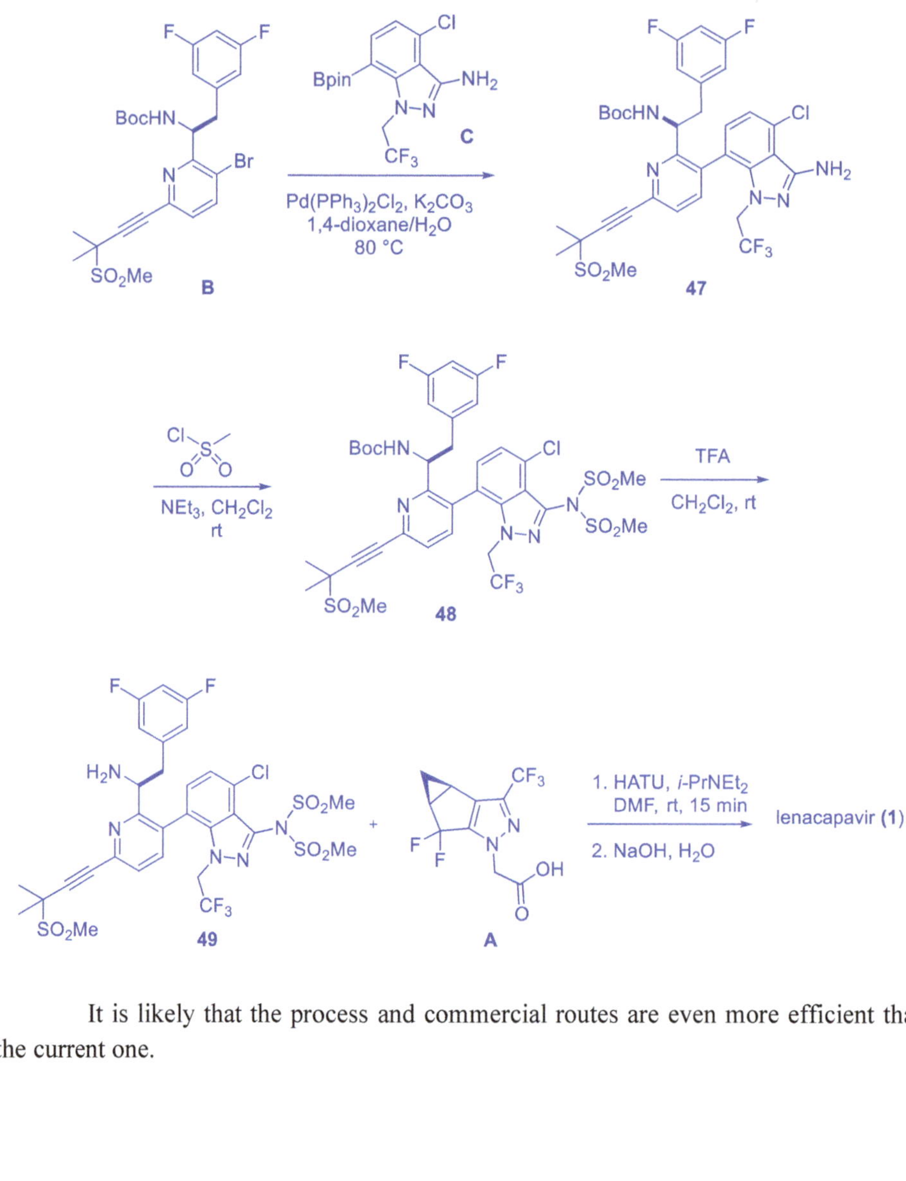

It is likely that the process and commercial routes are even more efficient than the current one.

7. Summary

We have come a long way on the road to conquest HIV, the invisible enemy. Impressively, nearly half of all antiviral drugs on the market are used for the treatments of AIDS. The other class of major antiviral drugs are anti-hepatitis C virus (HCV) drugs. The lessons that we learned from our war against HIV/AIDS and HCV have greatly aided our endeavor on discovering drugs to treat SARS-CoV-2.

Long-acting HIV drugs are at the frontier of the AIDS field. Now with the availability of lenacapavir (**1**), patient compliance will be greatly improved now that it only needs to be given twice a year. PrEP is another significant advantage of this drug.

As we advance the field of antiviral drugs against HIV, it is not a stretch of imagination that 1 day when eradication of this once deadly disease becomes a reality!

References

1. Mullard, A. FDA approves first-in-class HIV capsid inhibitor. *Nat. Rev. Drug Discovery* **2023**, *22*, 90.
2. Orkin, C. Lenacapavir in first-line therapy. *Lancet HIV* **2023**, *10*, e2–e3.
3. Link, J. O.; Rhee, M. S.; Tse, W. C.; Zheng, J.; Somoza, J. R.; Rowe, W.; Begley, R.; Chiu, A.; Mulato, A.; Hansen, D.; et al. Clinical targeting of HIV capsid protein with a long-acting small molecule. *Nature* **2020**, *584*, 614–618.
4. Zhang, X.; Sun, L.; Xu, S.; Shao, X.; Li, Z.; Ding, D.; Jiang, X.; Zhao, S.; Cocklin, S.; Clercq, E. D.; et al. Design, synthesis, and mechanistic study of 2-pyridone-bearing phenylalanine derivatives as novel HIV capsid modulators. *Mol. Ther.* **2022**, *27*, 7640.
5. Momany, C.; Kovari, L.; Prongay, A.; Keller, W.; Gitti, R.; Lee, B.; Gorbalenya, A.; Tong, L.; McClure, J.; Ehrlich, L.; Summers, M.; Carter, C.; Rossmann, M. Crystal structure of dimeric HIV-1 capsid protein. *Nat. Struct. Biol.* **1996**, *3*, 763–770.
6. Zhuang, S.; Torbett, B. E. Interactions of HIV-1 capsid with host factors and their implications for developing novel therapeutics. *Viruses* **2021**, *14*, 417.
7. McFadden, W. M.; Snyder, A. A.; Kirby, K. A.; Tedbury, P. R.; Raj, M.; Wang, Z.; Sarafianos, S. G. Rotten to the core: antivirals targeting the HIV-1 capsid core. *Retrovirology* **2021**, *18*, 41.
8. Freed, E. O. HIV-1 assembly, release and maturation. *Nat. Rev. Microbiol.* **2015**, *13*, 484–496.

9. Zhang, X.; Xu, S.; Sun, L.; Ding, D.; Tao, Y.; Kang, D.; Liu, X.; Zhan, P. HIV-1 capsid inhibitors: a sword to destroy the virus. *Future Med. Chem.* **2022**, *14*, 605–607.

10. Blair, W. S.; Pickford, C.; Irving, S. L.; Brown, D. G.; Anderson, M.; Bazin, R.; Cao, J.; Ciaramella, G.; Isaacson, J.; Jackson, L.; et al. HIV capsid is a tractable target for small molecule therapeutic intervention. *PLoS Pathog.* **2010**, *6*, e1001220.

11. Sun, L.; Zhang, X.; Xu, S.; Huang, T.; Song, S.; Cherukupalli, S.; Zhan, P.; Liu, X. An insight on medicinal aspects of novel HIV-1 capsid protein inhibitors. *Eur. J. Med. Chem.* **2021**, *217*, 113380.

12. Gillis, E. P.; Parcella, K.; Bowsher, M.; Cook, J. H.; Iwuagwu, C.; Naidu, B. N.; Patel, Manoj; Peese, K.; Huang, H.; Valera, L.; et al. Potent long-acting inhibitors targeting the HIV-1 capsid based on a versatile quinazolin-4-one scaffold. *J. Med. Chem.* **2023**, *66*, 1941–1954.

13. Yant, S. R.; Mulato, A.; Hansen, D.; Tse, W. C.; Niedziela-Majka, A.; Zhang, J. R.; Stepan, G. J.; Jin, D.; Wong, M. H.; Perreira, J. M.; Singer, E. A highly potent long-acting small-molecule HIV-1 capsid inhibitor with efficacy in a humanized mouse model. *Nat. Med.* **2019**, *25*, 1377–1384.

5

Fostemsavir (Rukobia): An HIV-1 gp120-Directed Attachment Inhibitor for Treating AIDS

Tao Wang and
Xiang Li

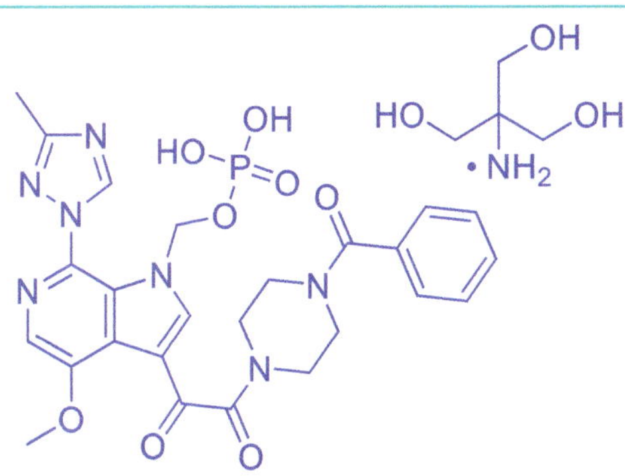

fostemsavir (**6**, Rukobia)
BMS-ViiV, 2020
HIV-1 gp120-directed
attachment inhibitor

1. Background

Since the start of the AIDS outbreak four decades ago, HIV-1 has infected roughly 84 million people worldwide and caused the deaths of approximately 40 million individuals.[1] The virus is spread through bodily fluids, including blood and genital secretions. If left untreated, HIV-1 can progress to AIDS, which can prove fatal for adults within 2–10 years. Infants who are infected during gestation, perinatal or lactation period via vertical transmission experience rapid disease progression, resulting in a 50% mortality rate within 24 months.[2]

Chemistry and Pharmacology of Drug Discovery, First Edition. Edited by Jie Jack Li.

The Fast-Track Strategy of UNAIDS targets the end to AIDS by 2030, with the objective of achieving 95–95–95: 95% of those living with HIV know their status, 95% are aware of their treatment options, and 95% receive viral load suppression therapy. The goal is set to decrease annual new adult HIV infections to 200,000 and to eradicate discrimination.[3]

However, the HIV epidemiological data of 2021 suggests that the goals set by UNAIDS are still far to reach. Globally, an estimated 1.5 million people are infected with the virus annually, including 160,000 children. And in each year, 650,000 individuals (ranging from 510,000–860,000) die of AIDS and related complications. As of the end of 2021, approximately 38.4 million people worldwide are living with AIDS, with two-thirds (25.6 million) residing in the African region.[1]

The AIDS epidemic posts severe economic and social issues in Africa. Reports state that certain areas in South Africa have an alarming prevalence rate of 35% among all adults aged 15–59. Age and gender stratification reveals that females aged 35–39 have a prevalence rate as high as 59%.[4]

Although worldwide AIDS-related deaths and infections have decreased stately, epidemiological trend in Eastern Europe and Asia, continue to worsen, increasing by 46% and 49% since 2010[5]. Approximately daily infections and AIDS-related fatalities in these regions are 440 and 130, respectively.

Clearly, unmet clinical needs persist in HIV therapy, prevention, and cure. Ongoing research aims to enhance treatment efficacy by developing new drugs with novel mechanisms, optimizing current regimens, and discovering methods to eliminate latent HIV reservoirs. Such advances will offer treatment alternatives for patients who have developed drug resistance or are intolerant to existing therapies and will reduce transmission. Potentially a functional cure for HIV can be achieved.

Before 2020, HIV-1 treatment involved six main antiretroviral categories that target different aspects of the HIV life cycle. Entry inhibitors prevent the binding of HIV envelope glycoprotein gp120 to co-receptor CCR5 (1) or cell fusion mediated by HIV gp41 (2). Nucleoside reverse transcriptase inhibitors (NRTIs) (3) and non-nucleoside reverse transcriptase inhibitors (NNRTIs) (4) inhibit the conversion of viral RNA to cDNA. Integrase Strand Transfer Inhibitors (INSTIs) (5) and protease inhibitors (PIs) (6) are antiretroviral drugs that inhibit the integration of proviral DNA into the host genome and the processing of gag and gag-pol precursors by protease, respectively.[6]

Although antiretroviral therapy (ART), a combination of antiviral medications that target various molecular targets, can partially relieve HIV-1 progression, it also results in several deleterious side effects such as psychiatric symptoms, renal function and bone density changes, elevated cardiovascular risk, and altered body weight.[7]

More important, the emergence of drug-resistant and multidrug-resistant strains cause ART failure, and lead to a heightened risk of HIV disease progression and mortality. According to the WHO report, 10% of adults with HIV who received treatment are resistant to NNRTIs.[8] Additionally, patients who have undergone antiretroviral therapy in the past are three times more likely to develop resistance to the NNRTI drug class compared to those who are currently receiving treatment for the first time. The prevalence of simultaneous resistance to six classes of antiviral medications can be as high as 5–10% among different groups of antiretroviral-treated patients in North America and Europe. Consequently, a significant number of these patients will be left without any viable treatment options.[9]

Efficient strategies for confronting HIV drug resistance require a global effort. These tactics comprise routine viral load monitoring, genotyping for resistance detection, changing regimens as necessary, and selecting the most potent antiretroviral drug combination and improved adherence to achieve long-term viral suppressing success.

Correspondently, research priorities include:

(a) Improving the safety and efficacy of antiretroviral therapy;
(b) Enhancing medication adherence by reducing treatment frequency, simplifying regimens, or utilizing long-acting medications to overcome adherence challenges;
(c) Developing new medicines with innovative mechanisms of action; and
(d) Preventing HIV transmission, via microbicide or vaccine.

Given the current landscape of HIV-1 therapies, we focus on the salvage therapy of the group comprising 6% of the patient population who have undergone five unsuccessful rounds of antiretroviral therapy need immediate access to new treatments as their options are exhausted.

To address this need, we discovered HIV-1 Adhesion Inhibitors (HIV AIs), a novel class of mechanistic antiviral drugs for HIV-1, inhibit viral infection by interfering with HIV's cellular adhesion process.[10] Among them, fostemsavir (**6**), the prodrug of temsavir (**5**) became the first of its kind to receive FDA approval in July 2020, is currently available under the trade name of Rukobia™ by ViiV/GSK.

Rukobia™ can be used in combination with drugs of distinct mechanisms. No evidence of cross-resistance with these treatments has been found to date.[11]

2. Pharmacology

The HIV-1 attachment inhibitor (HIV AI) belongs to a distinct mechanistic class of entry inhibitors for HIV-1. It specifically targets gp120, the membrane surface glycoprotein of

HIV-1, and prevents gp120's interaction with $CD4^+$ T cells, thereby halting the HIV-1 life cycle during the initial stage of cis-infection mode (Figure 1a).[11]

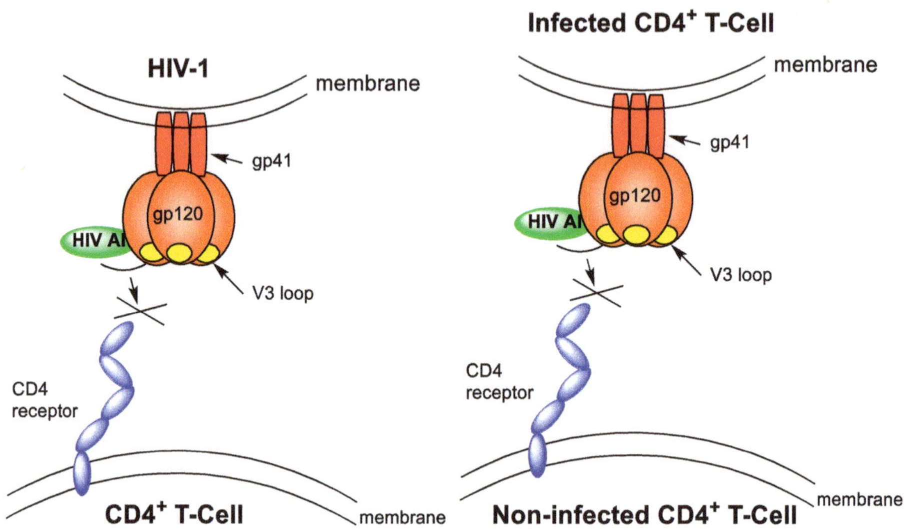

Figure 1. HIV AI blocks gp120 of HIV-1 binding to CD4 receptor (a) and HIV AI blocks gp120 of the infected CD4 T cell binding to CD4 receptor of non-infected T cell (b).

The second mode of infection is cell-to-cell fusion. During this process, an infected $CD4^+$ T-cell transmits the virus to a non-infected $CD4^+$ T-cell. The gp120 protein on the infected $CD4^+$ T-cell membrane and the CD4 receptor on the noninfected $CD4^+$ T-cell form an immunological synapse, which allows their membranes to fuse together and transfer the viral contents (Figure 1b). HIV AI also blocks cell-to-cell HIV-1 infection by inhibiting gp120.

3. Structure–Activity Relationship (SAR)

In 1998, Bristol-Myers Squibb revolutionized its drug discovery process by randomly testing its chemical inventory using high-throughput screening. One of the first projects adopted a pseudo-type of cell-based assay to identify compounds that could block HIV-1 infection. After reviewing approximately 100,000 compounds within 6 months, several hits emerged. Upon careful analysis and examination, a single hit, BMS-216 (Figure 2, **1**), appeared to have a unique mechanism of action. The compound displayed an EC_{50}

value of 153 nM, regardless of the HIV-envelope used, either the CCR5-dependent JRFL strain or the CXCR4-dependent LAI strain. It was generally non-cytotoxic, with a CC_{50} value of 339 μM in uninfected cells.[12]

Efforts were made to promptly establish the structure–activity relationship (SAR) of the indole, piperazine, and benzoyl fractions. Initial results showed that a small substitution on the 4-position of the indole significantly enhances the antiviral activity. The lead compound, 4-F indole derivative BMS-705 (**2**), reduced the EC_{50} value to 2.6 nM.[10] However, due to its low water solubility and metabolic instability, **2** was deemed "impossible" to be formulated for oral dosing.

Theoretically, increasing polarity and reducing electronic potential could improve aqueous solubility and metabolic stability. Replacement of a carbon atom with a nitrogen atom of aromatic ring should suffice to achieve the goal.[13] Accordingly, 4-/5-/6-/7-azaindole analogs of BMS-216-Me were synthesized, resulting in >26-fold improvement in aqueous solubility and >2.3-fold enhancement of metabolic stability.[13]

The further installation of a MeO- group at the 4-position of the 7-azadole led to BMS-806 (**3**).[11,14] It progressed to Phase I clinical studies but failed to deliver the desired exposure. This was likely due to **3**'s moderate metabolic stability ($t_{1/2}$ 47 min in HLM) and permeability (Caco-2 Pc 51 nm/sec) measured *in vitro*.

Given lipophilicity's positive impact on permeability, the polarity that is against lipophilicity was reduced by moving the N atom from the 7- to the 6-position and simultaneously adding a methoxy group at position 7. This operation hypothetically should also level the electron potential to retain metabolic stability, compared with **3**. After removing the 2-methyl group from piperazine, BMS-043 (**4**) was created, which provided $t_{1/2}$ of over 100 min in HLM and Caco-2 Pc 178 nm/sec.[13] In 2004, **4** completed Phase II clinical trials and demonstrated the proof-of-concept (POC).[15,16] However, only approximately 68% of patients experienced a desired viral load drop, and 1.8 g of **4** was required co-dosing with a high-fat meal. Further optimization of activity was necessary.

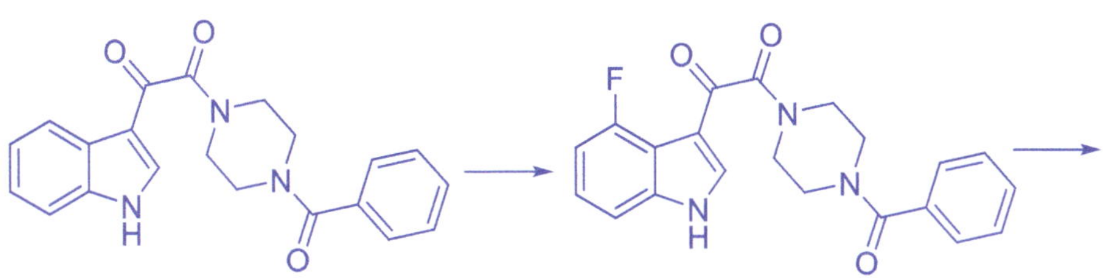

BMS-216, **1** BMS-705, **2**

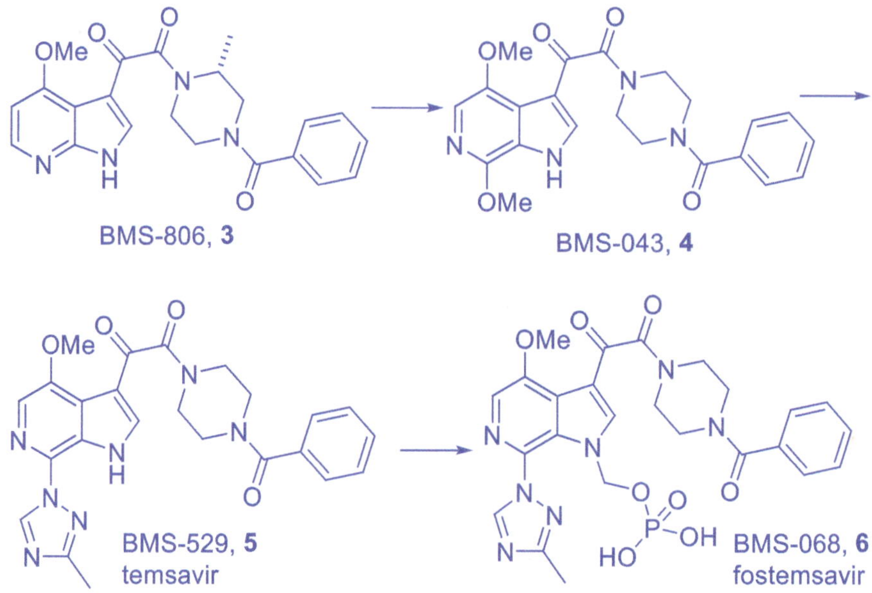

Figure 2. Optimization from BMS-216 to fostemsavir

The focus shifted to modifying position 7 of the 6-azaindole. It was discovered that a sp2 hybridized atom extension at C-7 was critical for potency. Extensive research was conducted on amides, aromatic rings, C-linked heteroaryl, and N-linked heteroaryl substituents. Additionally, in a subseries with the same heteroaryl ring, the relative strength of antiviral activity depended on the arrangement of atoms, which preferred coplanar between the heteroarene and (aza)indole core. In detail, the optimal compound in the subseries was the structure that held assembly on both flanking sides of the heteroaryl, labeled as α and α′ in Figure 3a, which facilitated coplanarity. This structural arrangement ensured that at the α-position of the C-7 heteroaryl, the heteroatom (O, S, or N) and the NH of the pyrrole ring of the 6-azaindole could form an internal hydrogen bond. Meanwhile, the heteroaryl's α′-position contained a hydrogen that could form an H or pseudo-H bond with the C-6 azaindole nitrogen, as shown in Figure 3a. The activities of all members of the furan (EC_{50}: **7** 0.02 nM, **8** 0.11 nM) and pyrimidine (EC_{50}: **9** 0.11 nM, **10** 0.51 nM, **11** 1.28 nM) subseries were compared head-to-head, as illustrated in Figure 3a and b, which clearly supported the assumption. Coplanarity was affected by several factors. The strength of an internal H-bond was determined by the substitution on the C-7 heteroaryl ring. The first, in general, electron-donating ring substitutions (such as an amino group) increased the electron density of the α-heteroatom, resulting in a stronger H-bond. Conversely, an electron-withdrawing functionality (such as an alkoxy group at the alpha-position of the ring nitrogen atom)[17] decreased the electron density of the alpha-heteroatom, leading to a weaker H-bond. The greatest impact was expected from substitutions at the adjacent position to the α-heteroatom. Secondly, the bulkiness of the substitution at the α- or α′-position of the C-7 heterocycle was also a significant determinator, as steric repulsion (e.g., **8** and **10**) was an obvious negative factor. The third factor was the location of the in-ring heteroatom. A heteroatom locked in the α-position experienced the long-pair repulsion from the nitrogen atom of 6-azaindole (e.g., **11**), causing the aromatic ring at C-7 to move out of the plane of the azaindole.

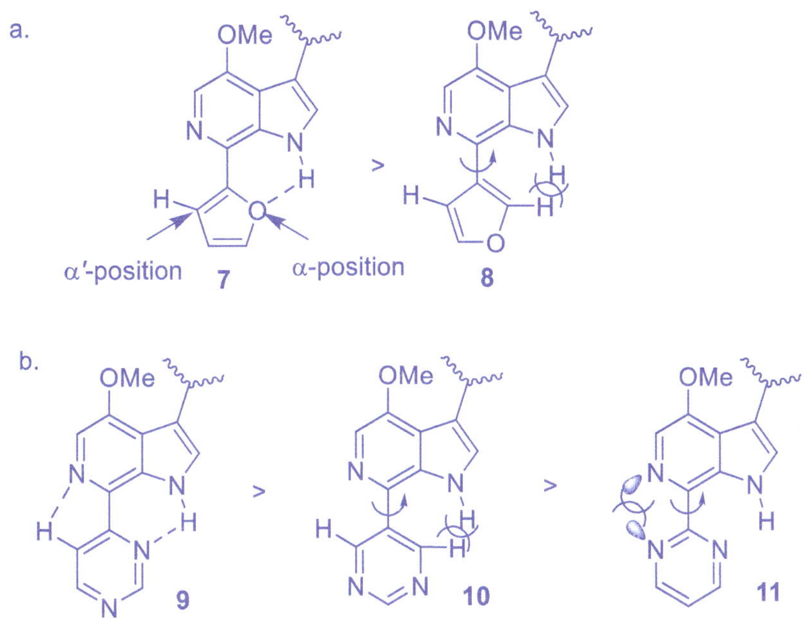

Figure 3. Coplanarity model with C-linked heteroarenes

It was observed that the permeability was correlated with the coplanarity between the heteroarene and azaindole core. For example, the Caco-2 Pc value favored furan **7** (86 nm/sec) over **8** (46 nm/sec), and 6-pyrimidinyl **9** (378 nm/sec) over 2-pyrimidinyl **11** (106 nm/sec) and 5-pyrimidinyl **10** (<15 and 46 nm/sec).

The liver microsome stability in six-membered ring extensions was discovered to be orthogonal to the coplanarity. For instance, the HLM $t_{1/2}$ values of pyrimidines **9**, **11**, and **10** were 15, 32, and over 100 min, respectively. A plausible explanation would be that the out-of-plane aromatic substitution had stronger electron-withdrawing effect, which led to lower electronic potential, in turn, better metabolic stability. The relationship between metabolic stability and coplanarity in the five-membered heteroaryl subscrics was unclcar (rcsults not shown).

However, despite optimization efforts on the C-linked heteroaryl series, a desired balance of antiviral activity, metabolic stability, and permeability could not be achieved, even after surveying hundreds of newly synthesized compounds. Then, the relationships between (aza)indole and its C-7 aryl/heteroaryl substituent (activity, metabolic stability, and permeability) were tested in the N-linked heteroaryl series against coplanarity. This turned out to be a critical design for the success.

Figure 4 illustrates the coplanarity setup of the N-linked heteroaryl 4-methoxy-6-azaindole series, which requires a five-membered heteroaryl ring with a nitrogen at the α-position and a CH at the α′-position to maximize coplanarity. At least two nitrogen atoms were required in the ring. One nitrogen atom connected the heteroaryl group to the 6-azaindole at the C-7 position. The second nitrogen atom was positioned at the α-position to form an internal hydrogen bond with the azaindole's pyrrole NH. At the α′-position, steric or electronic repulsion would be avoided with a non-substituted carbon atom.

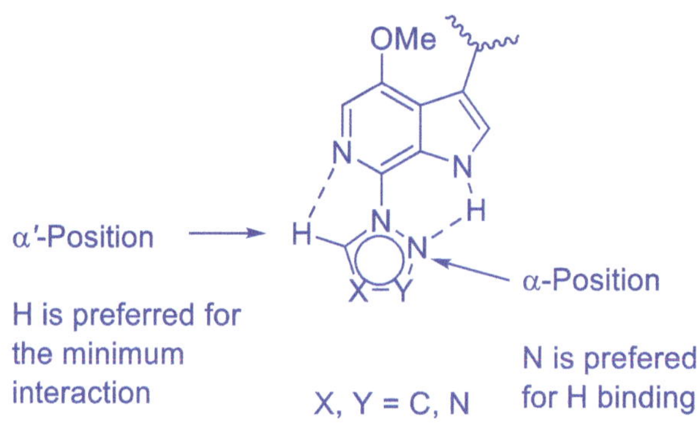

Figure 4. Coplanarity model with N-linked heteroarenes

Several hypotheses on physicochemical properties have been proposed. The first hypothesis suggests that the C–N bond of a N-linked heteroaryl would be shorter than the C–C bond of a C-linked heteroaryl, which would bring the heteroaryl closer to the azaindole ring and reinforce coplanarity. Secondly, the polarity and electronic potential of 6-azaindole in a C-linked heteroaryl series are further reduced in the corresponding N-linked heteroaryl series. Additionally, the lipophilicity of the N-linked heteroaryl series increases contender due to electronic withdrawing effect, in comparison with the C-linked heteroaryl contender. Overall, the N-linked heteroaryl series offers higher permeability and metabolic stability compared to its C-linked heteroaryl partner. Design in Figure 4 resulted in the synthesis of C-7 N-linked pyrazoles, as well as 1,2,3- and 1,2,4-triazoles.

After extensive efforts, 1,2,4-triazole derivative BMS-529 (temsavir, **5**) was discovered with an EC_{50} of 0.1 nM (0.10 ± 0.04 nM, n = 177), exhibiting a much broader spectrum of anti-HIV activity than the POC compound **4** did. The preclinical profile was comparable to that of **4**, including all physicochemical properties and safety parameters. The water solubility of the compound remained low at 0.022 mg/mL at pH = 7.4. This caused lower-than-dose-proportional increases in exposure during dose-escalation PK studies. It was concluded that the solubility was due to kinetic factors, which limited absorption.

The dissolution rate was significantly increased using the phosphonooxymethyl prodrug strategy. After an individual takes a phosphonooxymethyl prodrug orally, the gut membrane-bound alkaline phosphatase cleaves the phosphate bond, releasing formaldehyde and the corresponding parent molecule. When the parent molecule is highly permeable, it will promptly enter the bloodstream. This was observed in the case of BMS-068 (fostemsavir, **6**).[17,18] **6** converted to **5** in the intestinal tract, and **5** was rapidly absorbed. Only a negligible quantity of **6** was detected in the early stages after oral administration (Figure 5).

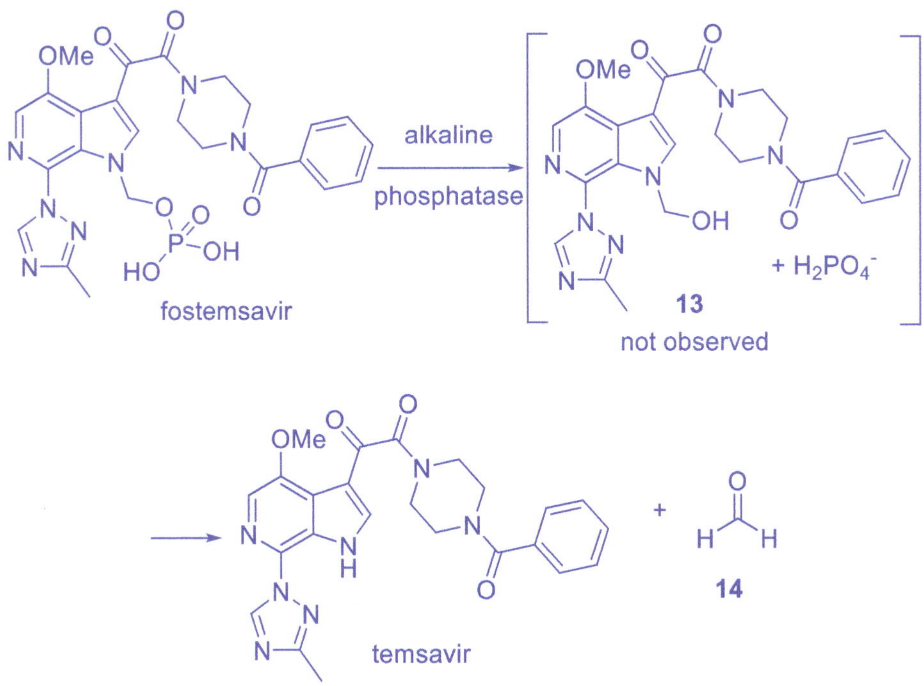

Figure 5. Fostemsavir converted to temsavir in the presence of alkaline phosphatase

4. Pharmacokinetics and Drug Metabolism

4.1. Pharmacokinetics of Temsavir

In rat PK studies, temsavir (**5**) was compared to the benchmark BMS-043 (**4**). Table 1 lists some of the results. **5** had a 7-fold better AUC after p.o. dosing at 5 mg/kg. Its intrinsic clearance was 1/10 of **4**'s, while its volume of distribution was 1/3 of **4**'s, which together displayed a 1.5-fold longer half-life. Oral bioavailabilities were similar, with **5** at 82% and **4** at 90%.

Table 1. PK profile of BMS-043 and temsavir in rat

IV 1 mg/kg PO 5 mg/kg	AUC p.o. 24 h (μM*h)	CL i.v. (mL/min/kg)	Vss i.v. (L/kg)	$t_{1/2}$ i.v. (h)	*F* (%)
BMS-043 (**4**)	15 ± 6.3	13 ± 4.0	1.1 ± 0.22	2.4 ± 0.33	90
BMS-529 (**5**)	111 ± 25	1.3 ± 0.19	0.36 ± 0.098	4.3 ± 1.1	82

The next evaluations were conducted on higher species, including dogs, monkeys, and chimpanzees (Table 2). **5** showed an advantage over **4** in term of AUC in dogs (29 vs 18 μg*h/mL), but not in monkeys and chimpanzees. The clearance rates of both compounds were similar in dogs, with higher rates for **5** in monkeys and chimpanzees. However, **5**'s exceptional *in vitro* metabolic stability in microsomes produced better oral bioavailability in dogs and monkeys. **5** was advanced to clinical trials, given its well-balanced anti-HIV-1 activity spectrum, metabolic stability, permeability, and *in vivo* exposure.

Table 2. PK profiles of temsavir and BMS-043 in higher species

	Temsavir (**5**)	BMS-043 (**4**)
Oral *F* (%)		
Dog	89	57
Monkey	64	60
Chimpanzee	16 (susp)	25 (susp)
Oral AUC at 5 mg/kg (μg*h/mL)		
Dog	29	18
Monkey	6.8	12
Chimpanzee	2.1	7.1
Total CL (mL/min/kg)		
Dog	2.6 (low)	2.4 (low)
Monkey	7.5 (low)	4.3 (low)
Chimpanzee	6.6 (inter)	3.2 (low)

In humans, the clearance of **5** was identified 51% via renal path.[19]

4.2. Pharmacokinetics of Fostemsavir (**6**)

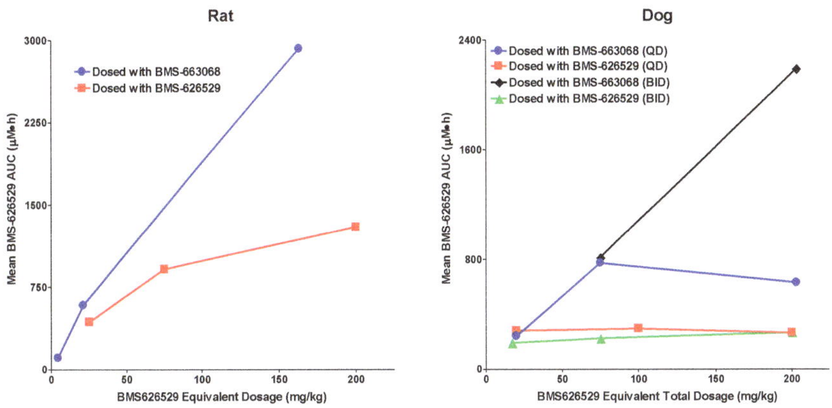

Figure 6. Plasma exposure profiles of temsavir (BMS-626529) as temsavir and fostemsavir (BMS-663068) at escalating doses in the rat (a) and dog (b)

At pH 7.4, temsavir (**5**, BMS-626529)'s aqueous solubility in a crystalline form was 0.022 mg/mL, which classified it in the BSC II category due to its low aqueous solubility and high Caco-2 permeability. When an un-optimized suspension was compared with aqueous solution, its relative oral bioavailability was 52%, indicating its absorption was limited by dissolution or solubility. In the additional dose escalation studies conducted on rats and dogs (red and green lines in Figure 6), the exposure increased at a lower rate than the escalated dose, which confirmed the concern of absorption being limited by dissolution or solubility.

Fostemsavir (**6**, BMS-663068) was a phosphonooxymethyl prodrug designed to enhance the dissolution rate of **5**. **6** increased **5**'s aqueous solubility from 22 µg/mL at pH 7.4 to over 11 mg/mL at pH 1.5–8.2 at room temperature. **6** was stable in both its solid state and acidic or neutral solutions for more than 24 h at 37 °C. As previously described, alkaline phosphatases located at the brush border membranes of the intestinal lumen hydrolyzed **6** into **5** which was rapidly absorbed into the bloodstream. *In vitro* studies showed that **6** is converted to **5** by human placental alkaline phosphatase and hepatocytes, which suggested a similar process occurring *in vivo*. Good to excellent oral bioavailability (80–122%) of **5** was observed after administering low doses of **6** solutions to rats, dogs, and monkeys. Minimal **6** exposure was detected in the blood.

Figure 6a's blue line compares the AUC exposure of **5** dosed with **5** or **6** in rats. At lower dosages (e.g., ≤25 mg/kg), both compounds had similar AUCs because they could be formulated as a solution (PEG-400/ethanol/0.1 N NaOH for **5** and water for **6**. However, at a higher dosage (e.g., 200 mg/kg), **5** could only be formulated as a suspension due to its poor water solubility, while **6**'s salt's high solubility allowed for an aqueous solution formulation, to offer superior exposure of **5**.

Figure 6b describes single dose toxicokinetic and tolerability studies in dogs. The dogs were dosed with **6** in API capsules or water solution at doses of 25, 92, and 250 mg/kg daily (molar equivalent to 20, 75, and 203 mg/kg of **5**, or BID at doses of 46 and 125 mg/kg (equivalent to QD of **5** as the two highest single QD doses). **5** was given at doses of 15, 75, or 200 mg/kg QD or BID at doses of 37.5 and 100 mg in PEG400/Ethanol/0.1 N NaOH.

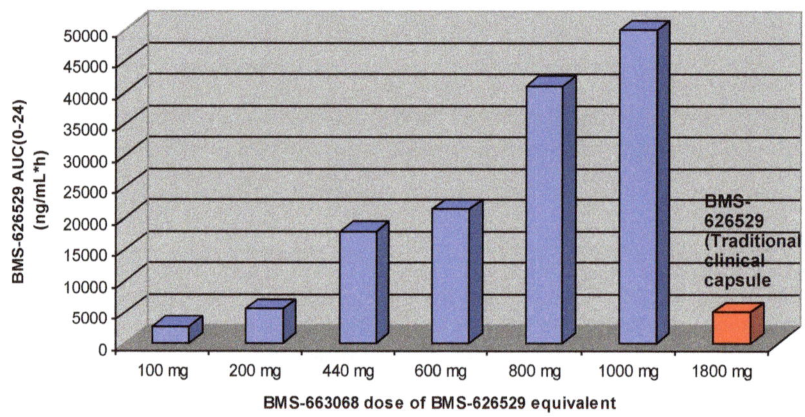

Figure 7. Plasma exposure of temsavir following administration of fostemsavir to NHVs in a phase 1 clinical trial compared with the exposure of temsavir following administration of temsavir in a capsule formulation

In a Phase I clinical study, **6** was found to provide significantly greater systemic exposure of **5** compared to **5** alone. The study involved a single ascending dose (SAD) of **6** administered in a standard capsule formulation at 5 equiv dosages ranging from 100 to 1000 mg. The systemic exposure of **5** increased more than proportionally with dose, as shown in Figure 7. When compared to **5** given in a traditional capsule at a dose of 1.8 g (red bar in Figure 7), **6** at any dosage greater than 200 mg resulted in better plasma concentration. This indicated that the prodrug approach effectively overcame the issue of solubility-limited absorption. Subsequently, an extended-release formulation was developed for twice-daily dosing, which was approved by the FDA in 2020.

4.3. Metabolism of Temsavir

During human clinical trials, temsavir (**5**) was extensively metabolized with less than 2% excreted in the urine. The primary metabolic pathways were through esterases (36.1%) and CYP3A4 enzymes (21.2%).[19]

4.4. Potential Drug–Drug Interaction of Temsavir

Temsavir (**5**) inhibited the transporters OATP1B1, OATP1B3, and BCRP. **5** was found to be a substrate of P-gp and BCRP. **5** did not induce the expression of enzymes and transporters, including CYPs, UGTs, P-gp, MRP2, BSEP, NTCP, OAT1, and OAT2. Based on preclinical and clinical data, there were no significant drug–drug interactions with common metabolic enzymes and transporters. Therefore, there is no need for dosage adjustment for patients with either renal or hepatic impairment.[19]

5. Efficacy and Safety

The Phase IIb clinical trial AI43811 (NCT01384734) evaluated the safety and efficacy of fostemsavir (**6**) in antiretroviral-experienced patients. The trial was randomized and active-controlled, with four arms testing different dosages and dosing schedules of **6** in combination with raltegravir 400 mg BID and tenofovir disoproxil fumarate 300 mg QD. The study involved four arms: The subjects were randomized equally among the four arms, with a total of 251 participants. Arm 1 received **6** at a dose of 400 mg twice daily; Arm 2 received **6** at a dose of 800 mg twice daily; Arm 3 received **6** at a dose of 600 mg once daily; and Arm 4 received **6** at a dose of 1200 mg once daily. The reference arm received ritonavir (300 mg)-boosted atazanavir (100 mg) once daily, along with raltegravir 400 mg twice daily and tenofovir disoproxil fumarate 300 mg once daily. The observed analysis showed that HIV-1 RNA levels were below 50 copies/ml in 77–95% of the **6** arms and 88% of the ATV/r arm, respectively. At the end of week 48, the virological response rates for the **6** arms ranged from 74% to 100%, while the reference arm showed a rate of 96% in patients with a baseline viral load of less than 100,000 copies/mL, and 60% to 91% compared to 71% in patients with a baseline viral load of 100,000 copies/mL or more. The median $CD4^+$ T-cell count in the **6** arms increased by 145–186 cells/μL compared to 142 cells/μL in the ATV/r arm.

There were no significant safety concerns associated with **6**, and it was generally well-tolerated with no discontinuations in the **6** arms.[20]

6. Synthesis

6.1. Discovery/Scale-Up Route

Scheme 1 shows the original synthesis route developed by the discovery medicinal chemistry team. 5-bromo-2-chloro-3-nitropyridine **15** was treated with an excess amount of vinyl Grignard agent (3–4 equiv) at low temperature (–78 to –20 °C) via a Bartoli-type reaction to generate 4-bromo-7-chloro-6-azaindole **16**. Subsequently, CuBr- or CuI-mediated radical coupling of methoxide with **16** produced 4-methoxy-7-chloro-6-azaindole **17**. Acylation of the 6-azaindole at the C-3 position with methyl or ethyl 2-chloro-2-oxoacetate and aluminum chloride gave oxoacetate **18**. The ester group of **18** was then removed by potassium carbonate in methanol and water, resulting in the formation of potassium salt **19**. Subsequently, diamide **21** was obtained by coupling compound **19** with piperazine **20** using a coupling agent such as DEBPT, TBTU, or HATU to furnish diamide **21**, which served as the key intermediate for further derivatization at the C-7 position of 6-azaindole. Finally, the synthesis of temsavir (**5**) was completed by coupling compound **10** with azole **11** under copper(0) catalyst.

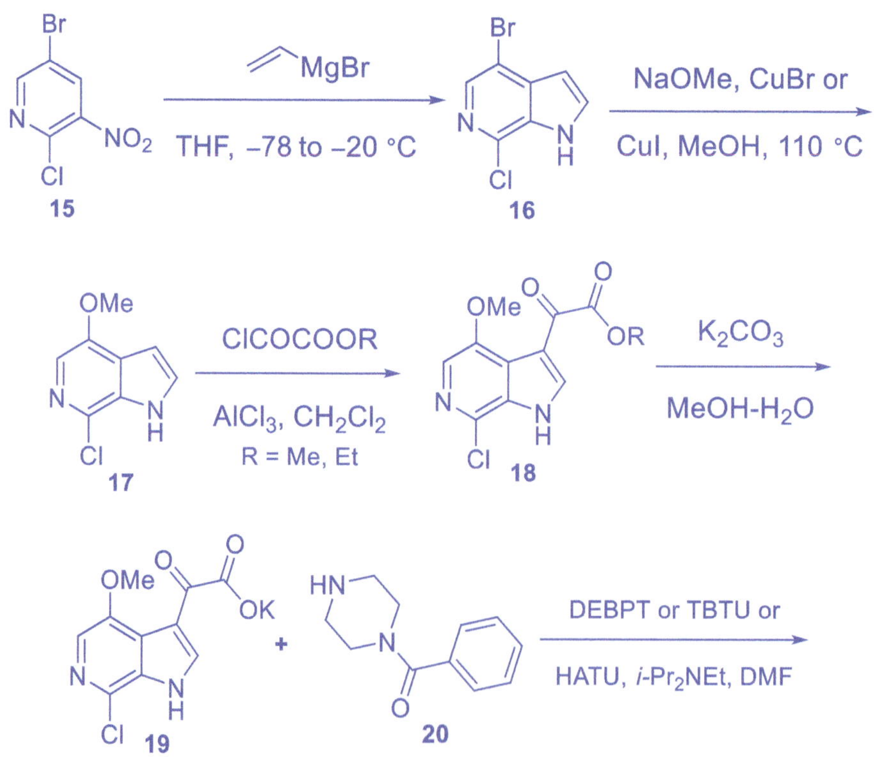

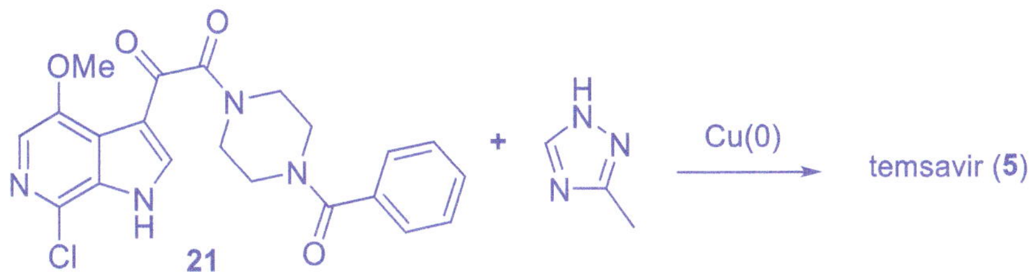

Scheme 1. Discovery synthetic route towards temsavir (5)

A second synthetic pathway involved installing 3-methyl-1,2,4-triazole at the C-7 position of 4-methoxy-6-azaindole before acylation at the C-3 of compound **22** to give compound **23** as shown in Scheme 2. After hydrolysis of ester **23**, the resulting salt **24** was coupled with benzoyl piperazine **20** to produce **5**.

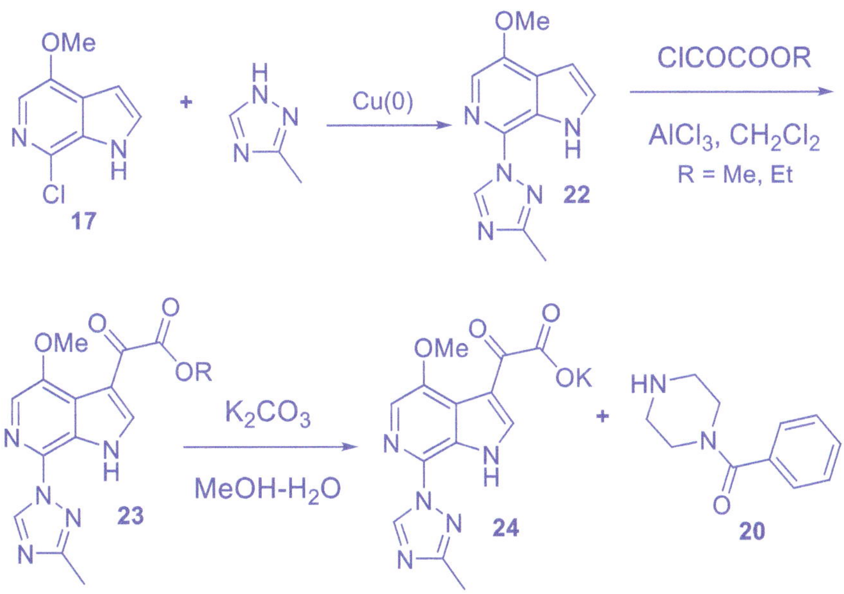

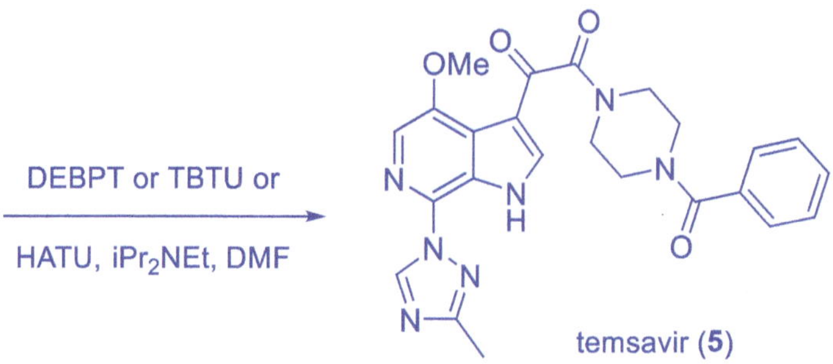

Scheme 2. Alternative discovery synthetic route toward temsavir

Compound **5** had a solubility-limited absorption issue which prevented it from achieving the desired exposure at higher doses, reaching sufficient safety multiples in preclinical safety studies, and owning a broad spectrum against HIV-1 viruses in clinical studies. Among all the chemical modifications or physical formulations, the phosphonoxymethyl prodrug approach proved to be the most effective. Scheme 3 illustrates its preparation. **5** was treated with an excess of NaH (5 equiv), followed by iodine (1 equiv), and freshly prepared di-*tert*-butyl chloromethyl phosphate (10 equiv) to yield phosphenoxy di-*t*-butyl ester **25**. The *t*-butyl ester was then removed by thermal treatment at 40 °C in a mixed solution of acetone and water to form phosphoric acid fostemsavir (**6**). Monotromethamine (tris) was added to form fostemsavir-tris (**26**), which crystallized and precipitated from the solution. This salt was later marketed as Rukobia™.

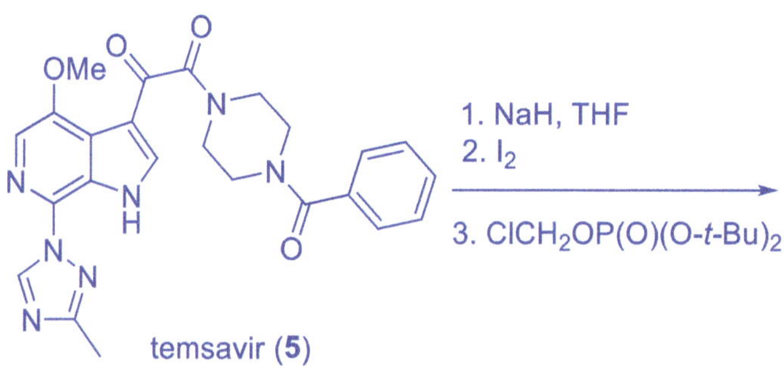

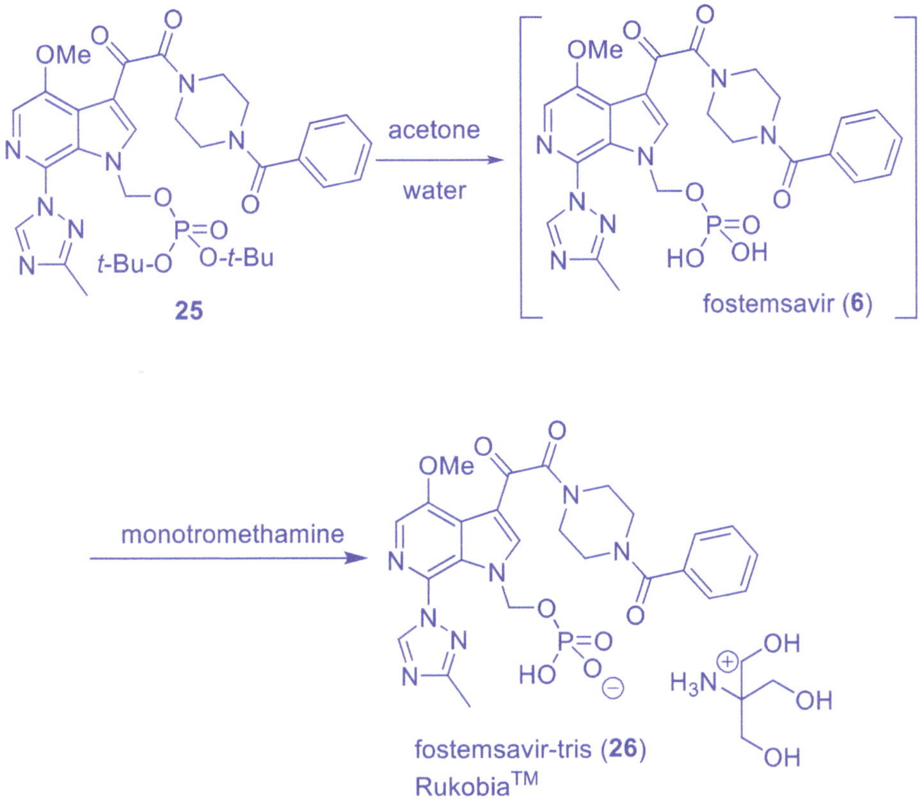

Scheme 3. Preparation of fostemsavir-tris from temsavir

6.2. Non-Discovery Chemistry Route

One of the later processes (Scheme 4) began with pyrrole **27**. An acetyl group was added to its meta-position using acetyl chloride and aluminum chloride. The alpha position of the acetyl was then chlorinated using NCS to produce compound **28**, which reacted with amidate **29** to yield amide **30**. The amide bond was cleaved under acidic conditions, and the ketone group was protected by ethylene glycol, in the form of compound **31**. Compound **32** was constructed by bringing the N atom and pyrrole's C-2 carbon together through a methylene group originated from paraformaldehyde. Tosyl group removal via E2 elimination and mesyl group-assisted aromatization led to the formation of 4-methoxy-6-azaindole **33**. Bromination in four steps installed bromine at position 7 of 6-azaindole, providing a crucial intermediate **34**. The following processes set up the glyoxylic functional group at the C-3 location, and hydrolysis of the resulting ester to acid **35** which further reacted with benzoyl piperazine **20** to furnish diamide **36**. Coupling

3-methyl-1,2,4-triazole with the C-7 bromine of **36** via a radical mechanism generated temsavir (**5**). **5** was then alkylated with phosphonooxymethyl chloride to produce diphosphate **25**. Hydrolysis of both *tert*-butyl groups led to fostemsavir (**6**), which *in situ* complexed with tris to form fostemsavir-tris (**26**).

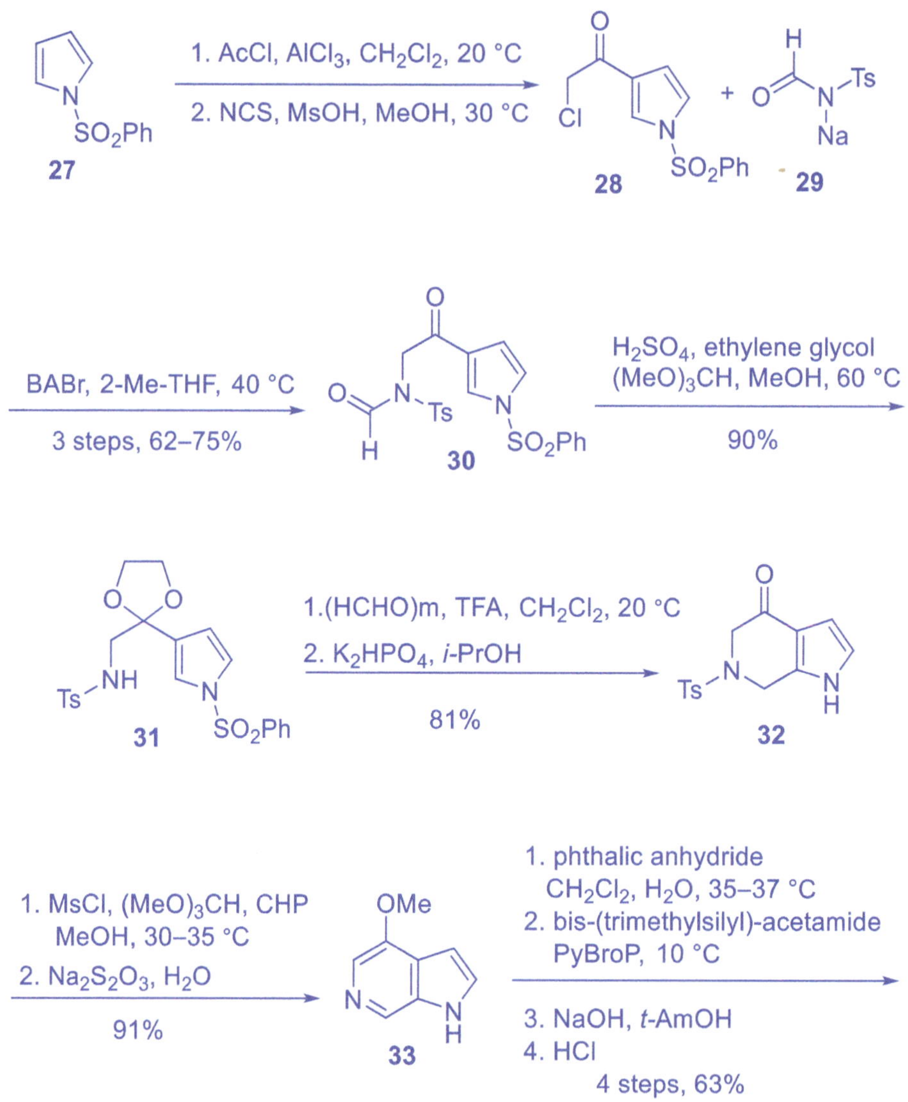

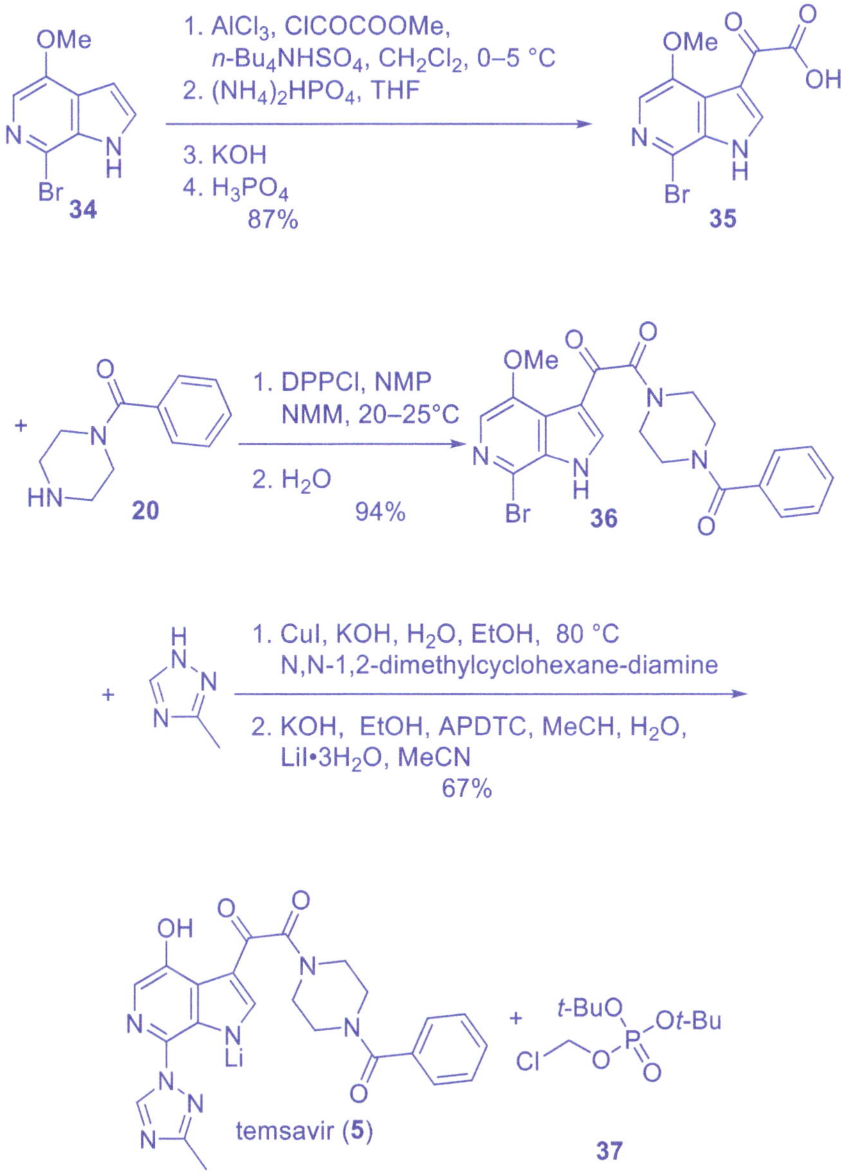
OMe
34
1. AlCl3, ClCOCOOMe, n-Bu4NHSO4, CH2Cl2, 0–5 °C
2. (NH4)2HPO4, THF
3. KOH
4. H3PO4
87%
35
20
1. DPPCl, NMP NMM, 20–25°C
2. H2O
94%
36
1. CuI, KOH, H2O, EtOH, 80 °C N,N-1,2-dimethylcyclohexane-diamine
2. KOH, EtOH, APDTC, MeCH, H2O, LiI•3H2O, MeCN
67%
temsavir (5)
t-BuO
Ot-Bu
37

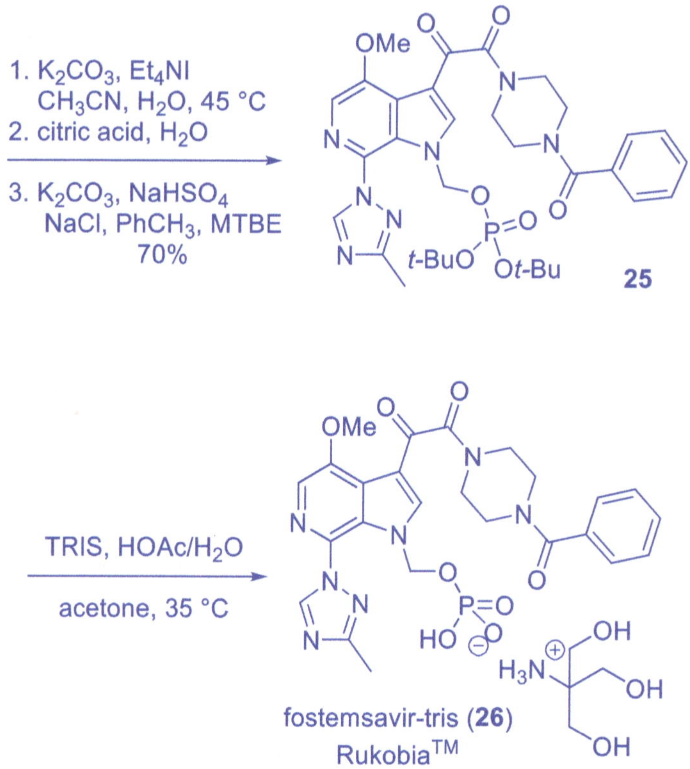

Scheme 4. Development route toward fostemsavir-tris

7. Summary

Rukobia^{TM} could benefit from significant improvements. Patient adherence is impaired by the need for twice-daily administration. Rukobia^{TM} has limited efficacy against different HIV-1 genotypes, with genotype B exhibiting the highest effectiveness while genotype A/E is the least effective. It is important to note that A/E is now one of the major HIV-1 subtypes and is reported to be the second most prevalent subtype in China. Furthermore, Rukobia^{TM} is unable to cross the blood-brain barrier to neutralize HIV particles that enter the brain.

In addition to therapy, HIV-1 attachment inhibitors can be linked with various functional effector moieties, such as enzymes, antibodies, cytokines, and toxins. In such cases, HIV AIs behave as a delivery tool. This tool is capable of not only targeting the HIV virus but also attacking infected cells, taking advantage of immune effects or toxins, offering the HIV cure.

HIV AIs can also be used as sensitizers for the broad-spectrum neutralizing antibodies. Their combination would lead to the cure for HIV-1.

HIV AIs can be further modified to enter the lymph nodes and brain, which provides opportunities to eradicate HIV-1 in these reservoirs.

Finally, topical application of HIV AIs can serve as a prophylactic to protect healthy individuals against HIV infection by blocking HIV entry through mucous membranes and skin. The intervene of disease transmission is always the best choice to limit infection from spreading.

In conclusion, despite the scientific and technical challenges remain high, the optimization of Rukobia™ presents significant opportunities for the treatment, cure, and prevention of HIV at all the levels.

References

1. *World Health Organization HIV-AIDS Fact Sheets*. https://www.who.int/news-room/fact-sheets/detail/hiv-aids (accessed 141 July 2023).
2. Organization, W. H. HIV in pregnancy: a review. **1999**.
3. *90-90-90: AN AMBITIOUS TREATMENT TARGET TO HELP END THE AIDS EPIDEMIC*. https://www.unaids.org/en/resources/909090 (accessed 141 July 2023).
4. Haeuser, E.; Serfes, A. L.; Cork, M. A.; Yang, M.; Abbastabar, H.; Abhilash, E. S.; Adabi, M.; Adebayo, O. M.; Adekanmbi, V.; Adeyinka, D. A.; et al. Mapping age- and sex-specific HIV prevalence in adults in sub-Saharan Africa, 2000–2018. *BMC Med.* **2022**, *20* (1), 488. 10.1186/s12916-022-02639-z.
5. Publications, U. N. *In Danger: UNAIDS Global AIDS Update 2022*; UN, **2022**.
6. Menendez-Arias, L.; Delgado, R. Update and latest advances in antiretroviral therapy. *Trends Pharmacol. Sci.* **2022**, *43* (1), 16–29. 10.1016/j.tips.2021.10.004 From NLM Medline.
7. Montessori, V.; Press, N.; Harris, M.; Akagi, L.; Montaner, J. S. Adverse effects of antiretroviral therapy for HIV infection. *CMAJ* **2004**, *170* (2), 229–238. From NLM Medline.
8. Organization, W. H. HIV drug resistance report 2021. **2021**.
9. Cozzi-Lepri, A.; Phillips, A. N.; Ruiz, L.; Clotet, B.; Loveday, C.; Kjaer, J.; Mens, H.; Clumeck, N.; Viksna, L.; Antunes, F.; et al.

Evolution of drug resistance in HIV-infected patients remaining on a virologically failing combination antiretroviral therapy regimen. *AIDS* **2007**, *21* (6), 721–732. 10.1097/QAD.0b013e3280141fdf From NLM Medline.

10. Wang, T.; Ueda, Y.; Zhang, Z.; Yin, Z.; Matiskella, J.; Pearce, B. C.; Yang, Z.; Zheng, M.; Parker, D. D.; Yamanaka, G. A.; et al. Discovery of the human immunodeficiency virus type 1 (HIV-1) attachment inhibitor temsavir and its phosphonooxymethyl prodrug fostemsavir. *J. Med. Chem.* **2018**, *61* (14), 6308–6327. 10.1021/acs.jmedchem.8b00759 From NLM Medline.

11. Wang, T.; Kadow, J. F.; Meanwell, N. A. Innovation in the discovery of the HIV-1 attachment inhibitor temsavir and its phosphonooxymethyl prodrug fostemsavir. *Med. Chem. Res.* **2021**, *30* (11), 1955–1980. 10.1007/s00044-021-02787-6 From NLM PubMed-not-MEDLINE.

12. Meanwell, N. A.; Krystal, M. R.; Nowicka-Sans, B.; Langley, D. R.; Conlon, D. A.; Eastgate, M. D.; Grasela, D. M.; Timmins, P.; Wang, T.; Kadow, J. F. Inhibitors of HIV-1 attachment: the discovery and development of temsavir and its prodrug fostemsavir. *J. Med. Chem.* **2018**, *61* (1), 62–80. 10.1021/acs.jmedchem.7b01337 From NLM Medline.

13. Kozal, M.; Aberg, J.; Pialoux, G.; Cahn, P.; Thompson, M.; Molina, J. M.; Grinsztejn, B.; Diaz, R.; Castagna, A.; Kumar, P.; et al. Fostemsavir in adults with multidrug-resistant HIV-1 infection. *N. Engl. J. Med.* **2020**, *382* (13), 1232–1243. 10.1056/NEJMoa1902493 From NLM Medline.

14. Chen, K.; Risatti, C.; Simpson, J.; Soumeillant, M.; Soltani, M.; Bultman, M.; Zheng, B.; Mudryk, B.; Tripp, J. C.; La Cruz, T. E.; et al. Preparation of the HIV attachment inhibitor BMS-663068. Part 2. Strategic selections in the transition from an enabling route to a commercial synthesis. *Org. Process Res. Dev.* **2017**, *21* (8), 1110–1121. 10.1021/acs.oprd.7b00121.

15. Fox, R. J.; Tripp, J. C.; Schultz, M. J.; Payack, J. F.; Fanfair, D. D.; Mudryk, B. M.; Murugesan, S.; Chen, C.-P. H.; La Cruz, T. E.; Ivy, S. E.; et al. Preparation of the HIV attachment inhibitor BMS-663068. Part 1. Evolution of enabling strategies. *Org. Process Res. Dev.* **2017**, *21* (8), 1095–1109. 10.1021/acs.oprd.7b00134.

16. Bultman, M. S.; Fan, J.; Fanfair, D.; Soltani, M.; Simpson, J.; Murugesan, S.; Soumeillant, M.; Chen, K.; Risatti, C.; La Cruz, T. E.; et al. Preparation of the HIV attachment inhibitor BMS-663068. Part 4.

Synthesis of the 6-azaindole core. *Org. Process Res. Dev.* **2017**, *21* (8), 1131–1136. 10.1021/acs.oprd.7b00152.

17. Fox, R. J.; Cohen, B.; La Cruz, T. E.; Simpson, J. H.; Freitag, A.; Saurer, E.; Tripp, J. C.; Chen, C.-K.; Beutner, G. L.; Rosso, V. W.; et al. Preparation of the HIV attachment inhibitor BMS-663068. Part 8. Installation of the phosphonoxymethyl prodrug moiety. *Org. Process Res. Dev.* **2017**, *21* (8), 1166–1173. 10.1021/acs.oprd.7b00135.

18. Gallagher, W. P.; Soumeillant, M.; Chen, K.; Fox, R. J.; Hsiao, Y.; Mack, B.; Iyer, V.; Fan, J.; Zhu, J.; Beutner, G.; et al. Preparation of the HIV attachment inhibitor BMS-663068. Part 7. Development of a regioselective Ullmann–Goldberg–Buchwald reaction. *Org. Process Res. Dev.* **2017**, *21* (8), 1156–1165. 10.1021/acs.oprd.7b00191.

19. Nettles, R. E.; Schürmann, D.; Zhu, L.; Stonier, M.; Huang, S.-P.; Chang, I.; Chien, C.; Krystal, M.; Wind-Rotolo, M.; Ray, N.; et al. Pharmacodynamics, safety, and pharmacokinetics of BMS-663068, an oral HIV-1 attachment inhibitor in HIV-1–infected subjects. *J. Infect. Dis.* **2012**, *206* (7), 1002–1011. 10.1093/infdis/jis432 (acccessed 18 Dec 2023).

20. Hoeger, K.; Davidson, K.; Kochman, L.; Cherry, T.; Kopin, L.; Guzick, D. S. The impact of metformin, oral contraceptives, and lifestyle modification on polycystic ovary syndrome in obese adolescent women in two randomized, placebo-controlled clinical trials. *J. Clin. Endocrinol. Metabol.* **2008**, *93* (11), 4299–4306. 10.1210/jc.2008-0461 (acccessed 12 Dec 2023).

6

Oteseconazole (Vivjoa): A CYP51 Inhibitor for Treating Recurrent Vulvovaginal Candidiasis

Charles L. Lail III and
Timothy J. Hagen

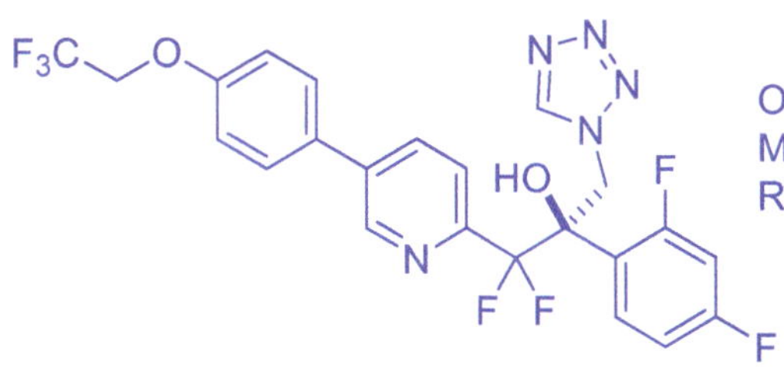

Oteseconazole (Vivjoa, **1**)
Mycovia Pharm., 2020
RVVC

1. Background

It is reported that around 9% of all women will develop recurrent vulvovaginal candidiasis (RVVC) in their lifetime, with around 75% of women having at least one case of vulvovaginal candidiasis in their lifetime.[1,2] RVVC is defined by the Center for Disease Control and Prevention (CDC) as the development of at least three cases of vaginal yeast infections caused by a fungus from the genus *Candida* in a 12-month period. RVVC has side effects of burning, itching, pain, and vaginal discharge. Oteseconazole (**1**, Vivjoa) was developed by Mycovia Pharmaceuticals Inc. as a breakthrough in the treatment of RVVC. Oteseconazole (**1**) was patented on August 18, 2020, and FDA approved on April 28, 2022.[3] Oteseconazole (**1**) is an orally administered

Chemistry and Pharmacology of Drug Discovery, First Edition. Edited by Jie Jack Li.

150-mg pill. Functionally, oteseconazole (**1**) is a metalloenzyme-inhibitor that inhibits the CYP51 enzyme that is found in most members of the genus *Candida*.

The standard drug used to treat *Candida* fungal infections before oteseconazole (**1**) is fluconazole (**14**). Structurally, fluconazole (**14**) has a 1,2,4-triazole ring that is responsible for its activity. Fluconazole (**14**) works quite well; however, there are many species of *Candida* fungi that have developed resistance to fluconazole (**14**). Thus, there is a continuing need for stronger treatment methods that remain highly potent and can counter antibiotic resistance. Mycovia Pharmaceuticals sought to solve this problem by using an extra nitrogen at position four in the azole ring, making oteseconazole (**1**) a tetrazole. The tetrazole unit shows a marked selectivity for *Candida albicans* CYP51 over several of the common human CYP enzymes, something that had been a struggle for previous generations of azole drugs (Figure 1).

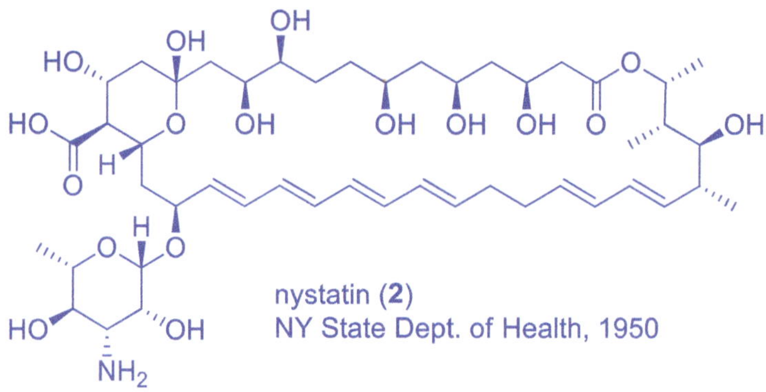

nystatin (**2**)
NY State Dept. of Health, 1950

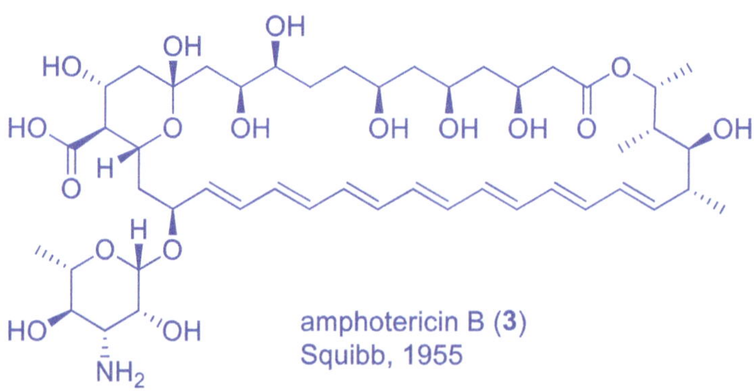

amphotericin B (**3**)
Squibb, 1955

Figure 1. Structure of nystatin (**2**) and amphotericin B (**3**)

There are four major classes of antifungals available today: polyenes, allylamines, echinocandins, and azoles. Each class acts through a different mechanism and each class has its own strengths and weaknesses. The first group, the polyenes, was started in 1951 with the discovery of nystatin (**2**). About 5 years later, a molecule known as amphotericin B (**3**) was developed, which became the first broad spectrum antifungal treatment. As the name implies, polyenes are distinguishable by structures, which contain many conjugated double bonds. Nystatin (**2**) has six double bonds (with one single bond breaking the conjugated chain) and amphotericin B (**3**) has seven conjugated double bonds. The polyenes are typically applied intravenously due to their larger molecular weights; nystatin (**2**) has a molar mass of 926.1 g/mol, while amphotericin B (**3**) has a molar mass of 924.1 g/mol. The polyenes are commonly believed to complex directly to ergosterol itself, thus interrupting the synthesis of molecules necessary for the maintenance of plasma membranes.[4–6] This would lead to direct terminal lysis of the fungal cells. The polyenes thus have a different mode of action to the other three categories in that they do not inhibit a specific enzyme, rather, they inactivate the final product of the ergosterol pathway. While the polyenes are beneficial due to their strong broad-spectrum treatment properties, their drawbacks are the lack of oral bioavailability and known nephrotoxicity associated with potassium wasting.[7] For these reasons, amphotericin B (**3**) typically fits into the therapeutic arsenal as being the drug of choice only in cases of severe, systemic fungal infections that could not be effectively treated by drugs with more mild side effects (Figure 2).

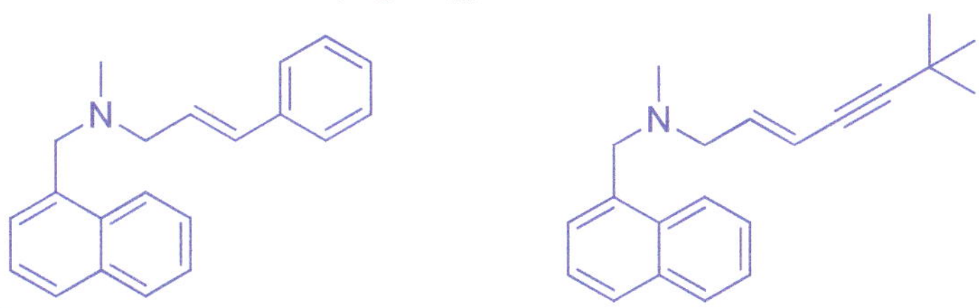

naftifine (**4**)
Sandoz, 1976

terbinafine (**5**)
Novartis, 1991

Figure 2. Structure of naftifine (**4**) and terbinafine (**5**)

The second group, the allylamines, was introduced in the late 1970s. The first of the allylamines, naftifine (**4**), was patented in 1976.[8] The allylamines all bear the same functional group—a tertiary amine adjacent to an allyl group. The amine is typically connected to a fused aromatic system, such as naphthalene, as in the structures of naftifine (**4**) and terbinafine (**5**). Other ring systems than naphthalene have been employed. The mode of action for allylamines is inhibition of squalene epoxidase.[9,10] Of the four major classes of antifungals, allylamines act the earliest in the ergosterol synthesis pathway. The allylamines are generally quite orally bioavailable and can be administered as oral pills.[11,12] In humans, the C_{max} of terbinafine (**5**) is reached in under 2 h.[12] Allylamines have the distinct advantage of being both fungistatic and fungicidal, that is, they both inhibit further growth of fungi and kill fungi.[13] The allylamines also generally avoid any interactions with cytochrome P450 and therefore have very few drug–drug interactions.[13] The primary downside of the allylamines is that they undergo heavy metabolism which limits their effectiveness; for example, terbinafine (**5**) can be metabolized into 15 different metabolites, each of which is inactive (Figure 3).[12]

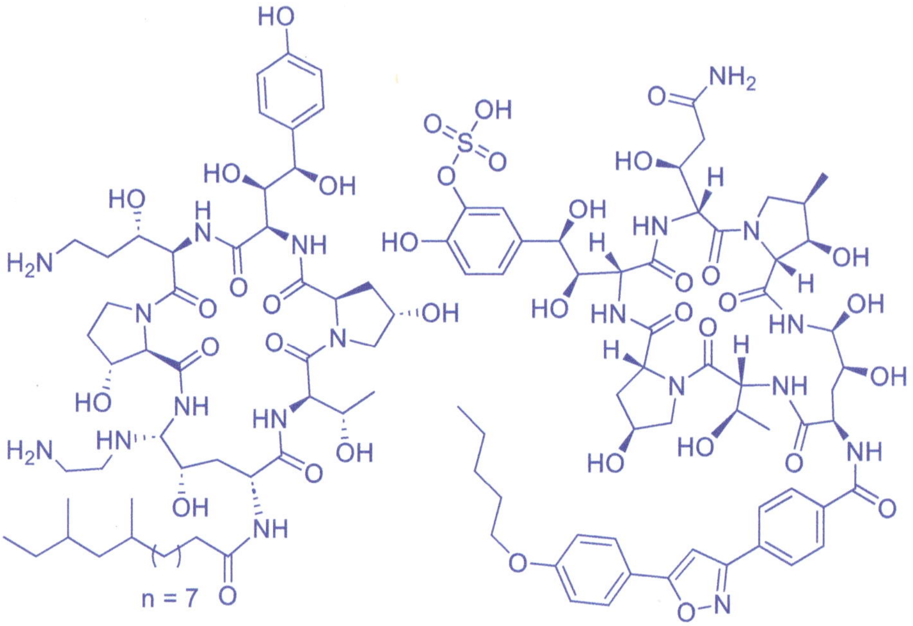

Figure 3. The structures of caspofungin (**6**) with molecular weight of 1093.31 g/mol, and micafungin (7) with molecular weight of 1270.28 g/mol

The third group, the echinocandins, is the most recently developed class. They have their beginning in 1992 with the development of caspofungin (**6**). The other popular echinocandin is micafungin (7), developed in 2004 by Astellas Pharmaceuticals. They are the only class of molecules out of the four that do not work inside of the ergosterol synthesis pathway. Echinocandins are responsible for the noncompetitive inhibition of the 1,3-β-D-glucan synthase enzyme. As the name implies, this enzyme is responsible for the synthesis of 1,3-β-D glucans. The 1,3-β-glucans are essential building blocks for the structural integrity of cell walls. Fungal infections can actually be detected by elevated levels of 1,3-β-glucans.[14] Most species of fungi have these glucans; 1,3-β-glucans are typically responsible for about 30–55% of the cell wall composition.[15] Structurally speaking, 1,3-β-glucans are polymers of glucose that are connected by a 1,3-linkage; there is typically some branching off of the 6 position, but is not a requirement.[15] These polymers are typically about 1,500 units long and are responsible for both the elasticity and strength of the cell walls.[16] Echinocandins have used 1,3-β-glucans as a target because they are simultaneously essential in fungi and completely absent in mammals.[16,17] The echinocandins are not broad-spectrum antifungals. They are typically quite good at treating fungi from *Candida* and *Asperigillus* but struggle against fungi such as *Cryptococcus neoformans*.[17,18] Echinocandins are fungicidal against *Candida* spp., but are fungistatic against *Asperigillus* spp.[19,20] The weakness of the echinocandins is that, since they are so large, they are typically required to be administered as an intravenous injection rather than the preferred model of an oral pill.[20]

The last group, the azoles, is the largest group on the list. The azoles can be roughly divided into generations depending on the number of nitrogen atoms in the ring. The first generation of azoles utilized an imidazole ring, the second generation utilized a triazole ring (usually a 1,2,4-triazole ring), and the most current azole drugs seem to be utilizing tetrazole rings. The origin of azole drugs begins with the discovery of the mild antifungal properties of benzimidazole, reported by D. W. Woolley in 1943.[21] He noted the effect of benzimidazole and several substituted benzimidazole derivatives, such as chlormidazole (**8**), against *Saccharomyces cerevisiae*. The next development occurred in 1959 when chlormidazole (**8**), a substituted benzimidazole, was developed and marketed as a 5% topical ointment for antifungal treatment. This became the first drug developed and sold as an azole antifungal.[22] The next advancement for azoles occurred in 1969 when three azole antifungals were reported: clotrimazole (**9**), miconazole (**11**), and econazole (**12**).[22]

Structurally speaking, miconazole (**11**) and econazole (**12**) are quite similar. They are only different by miconazole (**11**) having an extra chlorine atom. Miconazole (**11**) became the more popular of the two and is thus seen as the "parent compound" after which structure many other azoles were developed. Both miconazole (**11**) and econazole (**12**) are successful at treating superficial fungal infections of various types; however, miconazole (**11**) can be used to treat systemic fungal infections, while econazole (**12**) is bound too tightly by serum proteins to be useful against systemic fungal infections.[22] Drugs similar to miconazole (**11**) and econazole (**12**) include: dapaconazole, isoconazole, tioconazole, fenticonazole, sertaconazole, sulconazole, and oxiconazole. Each of these compounds can be synthesized from the intermediate 1-(2,4-dichlorophenyl)-2-(1*H*-imidazol-1-yl)-ethanone. This intermediate compound is generally reduced from a ketone to an alcohol, and the alcohol is then alkylated. Miconazole (**11**) and its derivatives are usually marketed as racemic mixtures. Miconazole (**11**) had some early success as a intravenously administered drug; however, its serum concentration was reported to decrease rapidly.[22] In addition, the developers used a 10% solution of Cremophor EL® to make the compound soluble; subsequently, there were toxicity concerns. There is still debate over whether the toxicity was from miconazole (**11**) itself, or whether it was from the Cremophor EL.[22–25] Either way, miconazole (**11**) and each of its derivatives is generally applied topically (Figure 4).

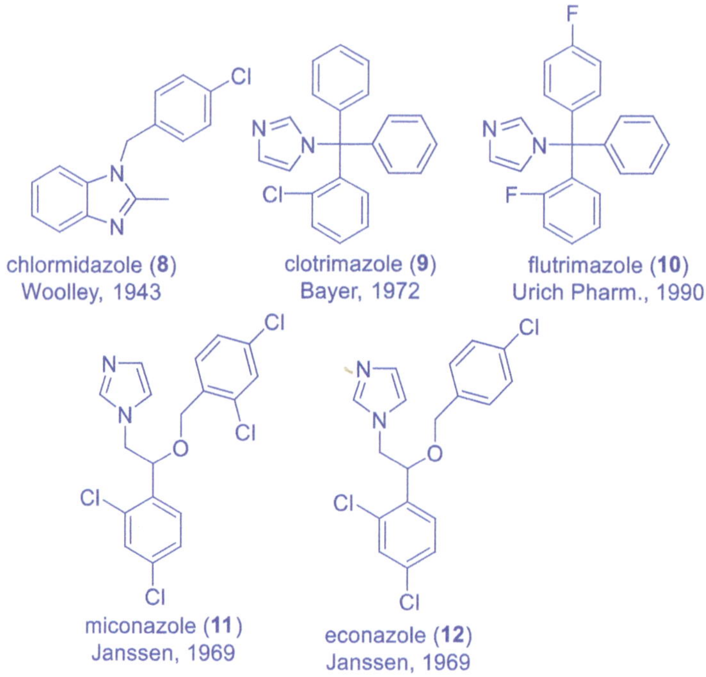

Figure 4. First generation azoles chlormidazole (**8**), clotrimazole (**9**), flutrimazole (**10**), miconazole (**11**), and econazole (**12**)

Clotrimazole (**9**) has a different structure compared to the miconazole (**11**) derivatives. It still contains the imidazole moiety but has a central quaternary carbon that is bonded to two phenyl rings and a 2-chlorophenyl ring. There is only one real derivative of clotrimazole (**9**) named flutrimazole (**10**). As the prefix implies, flutrimazole (**10**) uses fluorine instead of the chlorine from clotrimazole (**9**). Clotrimazole (**9**) managed slightly better success than did the miconazole (**11**) derivates in that it has very good broad spectrum activities that even rival those of amphotericin B (**3**).[22] Early data showed clotrimazole (**9**) to be orally bioactive, however, fast metabolism in the liver and toxicity concerns have limited clotrimazole (**9**) to a topical cream.[26]

The biggest breakthrough for the first generation of azoles was the development of ketoconazole (**13**) by Janssen Pharmaceutica in 1977. Ketoconazole (**13**) keeps the imidazole and the 2,4-dichlorophenyl ring but makes major changes to the rest of the molecule. First, the ether linkage is turned into a dioxolane ring, and then an *N*-acyl 4-piperazinyl-phenyl substituent is attached to the dioxolane ring by an ether linkage. Ketoconazole (**13**) is sold as a racemic mixture. The biggest achievement of ketoconazole (**13**) is that it is orally bioactive against systemic fungal infections. This achievement made ketoconazole (**13**) the "gold-standard" of azole drugs for nearly a decade. Ketoconazole (**13**) has very broad-spectrum activity making it similar in properties to miconazole (**11**); the only exception is that ketoconazole (**13**) is inactive against *Aspergillus*.[22,25,27] Despite its major breakthrough of being the first azole to be orally bioactive, the drug does have several negatives that have resulted in the assignment of a black-box label from the FDA in June 2013.[28] The major short-comings of ketoconazole (**13**) are: oral absorption varied widely on an individual basis and was dependent on pH, an inability to cross the blood-brain-barrier, strong cases of hepatotoxicity, dose-correlated reduction in testosterone and cortisol, and numerous drug–drug interactions from inhibition of Cytochrome P450.[22–25,29]

The next breakthrough in the development of the azoles comes with the development of fluconazole (**14**) by Pfizer, which was given FDA approval in 1990. Fluconazole (**14**) began the second generation of azoles, which uses 1,2,4-triazole rings instead of imidazole rings. Fluconazole (**14**) represents a large increase in progress over the imidazole-generation, particularly from a pharmacokinetic perspective. Fluconazole (**14**) has around 100% oral absorption with about 90% oral bioavailability. Another improvement over previous drugs is that 64% of fluconazole (**14**) clears the kidneys unmetabolized and is excreted through the urine.[30] Its water solubility and bioavailability have led fluconazole (**14**) to be an invaluable tool in the treatment of cryptococcal meningitis.[23] Fluconazole (**14**) is also one of the least expensive antifungal drugs on the market today. The downsides of fluconazole (**14**) are that it is not active against filamentous fungi such as *Aspergillus* spp. and that there is an ever increasing amount of resistance to fluconazole (**14**) treatment.[23,26] Fluconazole (**14**) is still one of the most commonly prescribed antifungal drugs on the market today.

Voriconazole (**15**), also developed by Pfizer, was FDA approved in 2002. The oral form of voriconazole (**15**) has equal bioavailability with fluconazole (**14**).[25,31] It has good broad-spectrum activity including the treatment of pathogens from *Candida, Asperilligius, C. neoformans*, and several types of mold pathogens such as *Scedosporium* spp. Or *Fusarium* spp.[31,32] Voriconazole (**15**) has also demonstrated good activity (10–100× greater) against fluconazole (**14**)-resistant strands of *C. albicans*.[31,32] Voriconazole's (**15**) primary uses today are prophylactic treatment in stem-cell transfer patients, and in the general treatment of Aspergillosis, against which it has actually outperformed amphotericin B (**3**).[25,33] There are two primary downfalls of voriconazole (**15**). First is its inhibition of three human CYP450 enzymes (2C9, 2C19, and 3A4) that has caused problems.[25,32,34] Second, in addition to having similar side-effects as the other azoles, voriconazole (**15**) uniquely bears side-effects such as skin reactions, hallucinations, and mental confusion.[25,32] Voriconazole (**15**) is also highly metabolized by the liver, with at least nine different metabolites being structurally confirmed; studies have shown that only 5% of voriconazole (**15**) is excreted unchanged.[31] Voriconazole (**15**) remains useful as it is able to treat several pathogens that are not treatable by other antifungals, but the possible side effects must be carefully monitored.

Itraconazole (**16**), a structural derivative of ketoconazole (**13**), was FDA approved in 1992. It has broad-spectrum activity and is useful in the treatment of aspergillosis, histoplasmosis and blastomycosis.[25] It is also active against *C. neoformans*, *Torulopsis* spp., and dematiaceous fungi.[11,35] Itraconazole (**16**) has similar problems as ketoconazole (**13**), in that its absorption is highly variable from person to person and additionally reliant on the presence of food and the pH of the stomach.[25] The half-life of itraconazole (**16**) also varies depending upon the dosage and the duration of the treatment.[36] Another drawback is that the compound is so highly metabolized in the liver that over 30 different metabolites have been reported, most of which are biologically inactive.[11,27] Despite the heavy metabolism, itraconazole (**16**) has a much better safety profile than most of the other azoles, with the exception of being contraindicated for pregnant women.[11,37]

Posaconazole (**17**), a structural derivative of itraconazole (**16**), was developed at the Schering–Plough Research Institute. The chemists were working with a lead compound and were surprised by abnormally high activity of one analog of their lead compound. The researchers believed that an active metabolite of the compound was responsible for better activity than the lead compound. This led to the development of Posaconazole (**17**) in 2004.[24,38] The deviations in structure include the changing of the 1,3-dioxolane ring to a furan ring, replacing the 2,4-dichlorophenyl ring with a 2,4-difluorophenyl ring, and the inclusion of an alcohol on the triazolone ring substituent. Posaconazole (**17**) is another broad-spectrum antifungal drug. It has strong activity against *C. albicans, C. neoformans*, and *Aspergillus* spp., among others.[31] Posaconazole (**17**) specialized in the treatment of *C. neoformans*, against which it significantly outperforms fluconazole (**14**) and amphotericin B (**3**).[31] Posaconazole (**17**) is both fungistatic and fungicidal against *C. neoformans*. Posaconazole (**17**) shares the same problems as ketoconazole (**13**) and itraconazole (**16**) in that absorptions varies widely upon food intake and gastric pH.[25] It has low oral bioavailability (10–40%), and it is primarily metabolized in 15–20 h by glucuronidation, although most of the drug passes through the body unmetabolized.[34,39] Unlike most of the other azole drugs, posaconazole (**17**) has linear pharmacokinetics.[39] Posaconazole (**17**) has a better safety profile than voriconazole (**15**) and itraconazole (**16**). It avoids the intense metabolism problems of itraconazole (**16**), and has much better interactions with the human CYP enzymes, with the exception of 3A4 (Figure 5).[25]

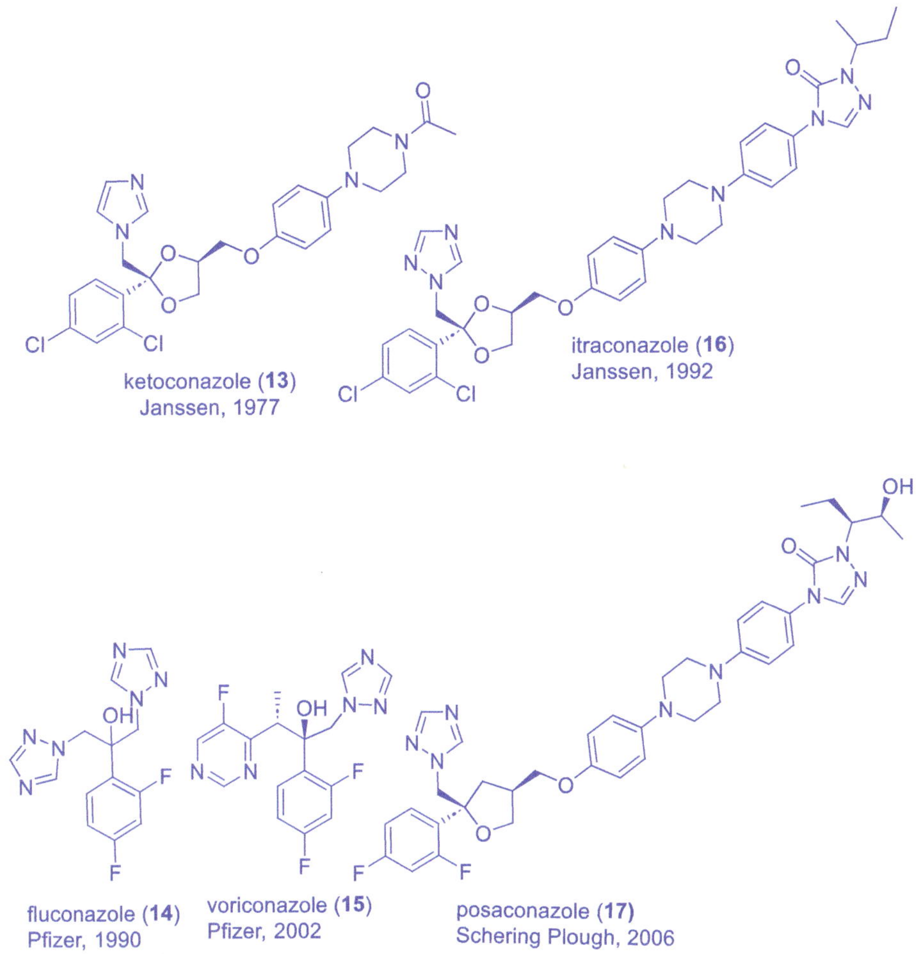

Figure 5. The structures of ketoconazole (**13**) and 1,2,4-triazole drugs (**14–17**)

The third generation of azoles is the tetrazoles. The tetrazole ring has been gaining popularity in its use in antifungal medications.[40] Some drugs, such as the subject of the review, have chosen to use the tetrazole instead of the triazole, while other antifungals have opted to add tetrazole rings in addition to the 1,2,4-triazole ring. The primary advantage of the tetrazole antifungals is that the show marked selectivity between antifungal CYP51 and human CYP enzymes, specifically CYP3A4, which is typically an area of strong inhibition for the azole antifungals.[24,41]

One of the most frequently mentioned problems with the azole class when compared to the polyenes, allylamines, and echinocandins, is that the azoles typically inhibit several of the human CYP450 enzymes. This generally leads to a laundry list of contraindications as to which other drugs they may not be co-administered with. As fungal infections tend to happen in patients who are already diseased, this represents a significant flaw with the azole class. Below is a table that shows the degree of interference with 6 CYP enzymes caused by several antifungal drugs. The azoles are typically active against 2C9, 2C19, and especially active against 3A4. This is one area that oteseconazole (**1**) will look to significantly improve upon (Table 1).

Inhibitor	**Class**	**1A2**	**2C9**	**2C19**	**2D6**	**2E1**	**3A4**
Micafungin (7)[42–44]	Echinocandin	None	None	None	None	None	13.5
Miconazole (**11**)[42–44]	Azole – Imidazole	2.9	2	0.33	6.46	>10	0.07
Ketoconazole (**13**) [45]	Azole – Imidazole	13	>100	28	17	90	0.2
Fluconazole (**14**)[42–44]	Azole – Triazole	None	30.3	12.3	None	None	13.1
Voriconazole (**15**)[42–44]	Azole – Triazole	None	8.4	8.7	None	None	10.5
Itraconazole (**16**)[42–44]	Azole – Triazole	None	>10	>10	None	None	0.03

Table 1. A summary of human CYP450 enzyme inhibition of several antifungal drugs. All values are IC_{50} (μM)

In summary, there are only four main classes of antifungal drugs. Each group has specific pros and cons. All four categories work by targeting structural molecules responsible for the formation and maintenance of cell walls or plasma membranes. Polyenes are the best broad-spectrum compounds but have toxicity concerns and are not orally bioavailable. Allylamines are orally bioavailable and have much lower drug–drug interactions, but suffer from rapid, inactivating metabolism. Echinocandins are specific in their treatment of fungi but are not orally bioavailable. Finally, azoles are generally orally bioavailable and are fairly broad-spectrum, but have been known to cause drug–drug interactions by off-site inhibition of Cytochrome P450. Oteseconazole (**1**) was designed as an azole drug that would be both orally bioavailable and have much fewer drug–drug interactions. It is less broad-spectrum than other azoles but is much more specific than the previous imidazoles and triazoles.

2. Pharmacology

Oteseconazole (**1**) is a metalloenzyme inhibitor that targets the *Candida* CYP51 enzyme. The CYP51 enzyme, which belongs to the cytochrome P450 monooxygenase superfamily, is a sterol 14α-demethylase enzyme, meaning that its primary function is to catalyze the stereo- and regio-selective removal of a methyl group at position 14 of the lanosterol chemical scaffold.[46,47] CYP51 removes a methyl group from lanosterol (**18**) to form 4,4-dimethyl-5α-cholesta-8,14,24-trien-3ß-ol (**19**), which is abbreviated FF-MAS (follicular fluid meiosis-activating sterol). FF-MAS (**19**) will undergo eight further reactions by a total of nine different enzymes to become cholesterol (**20**) in mammals and ergosterol (**21**) in fungi.[48] Mutation of the *erg11* gene that encodes for CYP51 has shown a buildup of 14α-methyl-ergosta-8,24(28)dien-3β,6α-diol later down the enzymatic pathway. This diol has been shown to be toxic to fungal cells (Figure 6).[48]

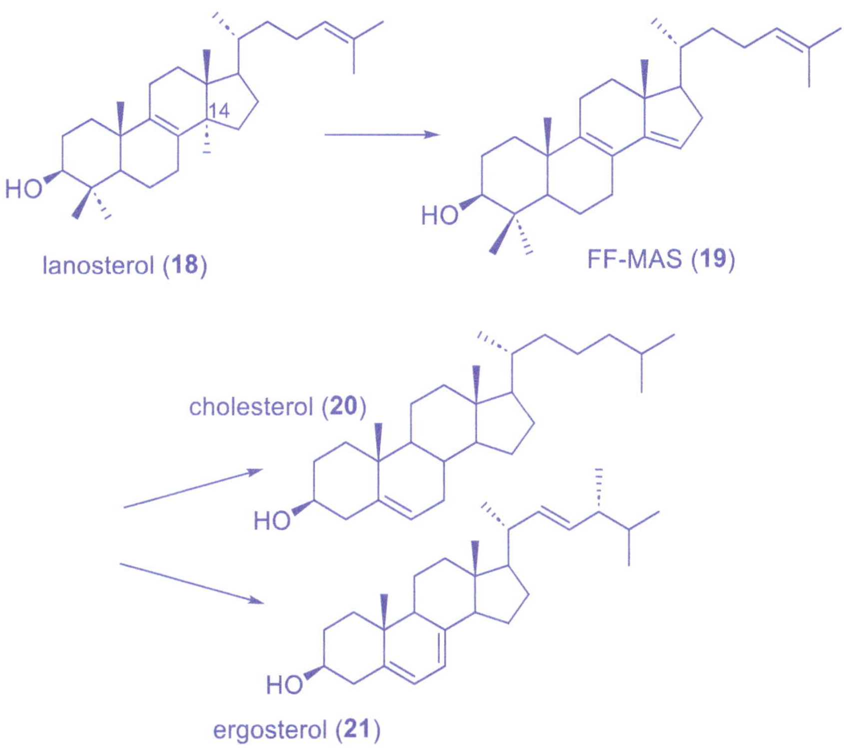

Figure 6. The function of CYP51 is to catalyze the stereo- and regio-selective removal of the methyl group at position 14. From there, cells can redox the structure into a number of steroid derivatives such as cholesterol (**20**) in mammals and ergosterol (**21**) in fungi[47]

Cholesterol and ergosterol have long been known to be key components of cellular membranes. Sterols have been shown to primarily govern the fluidity and permeability of the cellular membranes.[48] This makes the CYP51 enzyme a very attractive target in the treatment of fungal diseases.

Oteseconazole (**1**) uses its tetrazole ring to bind to the heme atom of the CYP51 enzyme. The nitrogen atom at the 4-position of the tetrazole ring is responsible for coordinating to the iron atom of the CYP51 enzyme. The heme atom generally participates in the electron transfer process by varying its oxidation state. The coordination of the basic nitrogen atom to the Lewis acid iron severely limits the iron's ability to switch oxidation state (Figure 7).

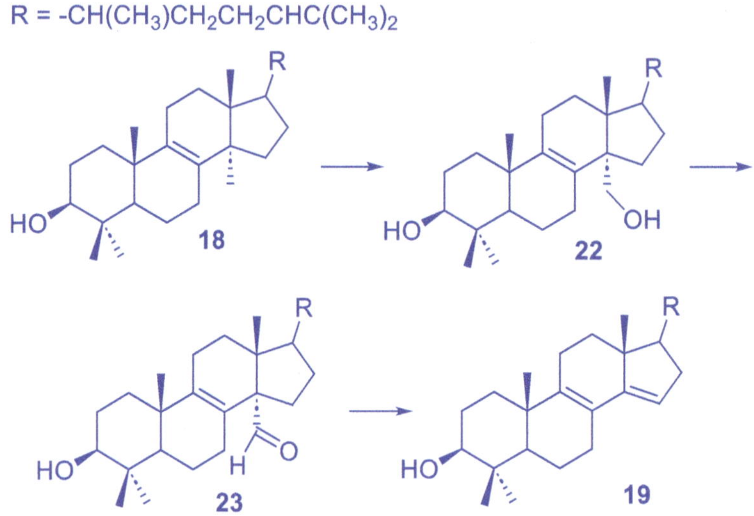

Figure 7. The proposed step-wise mechanism of CYP51[24,47]

3. Structure–Activity Relationship (SAR)

The development of oteseconazole (**1**) began by using the structure of voriconazole (**15**) and optimizing each structural region. The first area to be optimized was the metal-binding group (MBG). The MBGs were evaluated for heme affinity with an *in silico* model. Each MBG group was optimized using the PM3 method and geometry optimized in the heme-bound position. The enthalpy values for each MBG group were calculated, but not reported in the literature. Several MBGs were chosen to take into chemical synthesis based upon their predicted affinity with the heme iron atom. The developers were clear that they wanted groups that demonstrated high affinity for the heme iron but did not have too high affinity for the iron. This choice was intended to avoid as many nonspecific interactions as possible. A total of five compounds (**24**–**28**) were synthesized with different MBGs using the chemical structure of voriconazole (**15**) as a starting scaffold. Each compound in the SAR was evaluated based upon its MIC_{50} (μg/mL) against *C. albicans* and its IC_{50} (μM) against human CYP3A4. Instead of screening every molecule against each of the common CYP enzymes that are responsible for drug–drug interactions, CYP3A4 was chosen as a representative due to its history of strong interactions with common azole drugs. The best MBG was the 1-imidazole **27**. However, it was not further utilized because it was mildly basic (pK_a = 6.8) and it showed high amounts of human CYP3A4 inhibition (IC_{50} = 0.8 μM). The 1-tetrazole group **25** was selected for further study due to its low basicity (pK_a = –1.1) and its low levels of human CYP3A4 inhibition (IC_{50} = 32 μM) (Table 2).

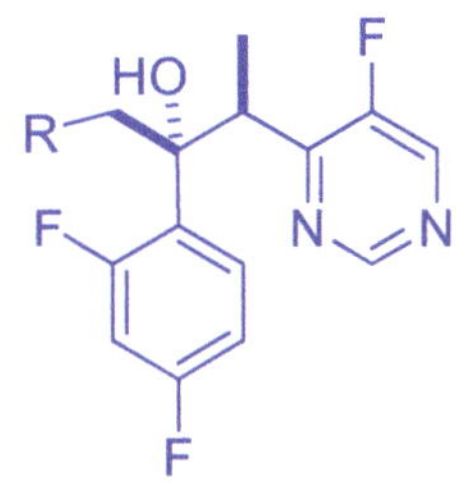

No.	R group	MIC_{50} *C. albicans*[a]	IC_{50} CYP3A4[b]	Selectivity ratio[c]
24	1-(1,2,3-Triazole)	>16	8.0	0.5
25	1-Tetrazole	1	32	32
26	4-(1,2,4-Triazole)	>16	51	3.2
27	1-Imidazole	0.5	0.8	1.6
28	2-Tetrazole	8	46	5.8

Table 2. All SAR information taken from Hoekstra et al.[49]

[a] Units are μg/mL

[b] Units are μM

[c] Ratio is equal to IC_{50} CYP3A4/MIC_{50} *C. albicans*

The next portion of the SAR uses the tetrazole ring as its MBG and optimizes the 3-fluoropyrimidine ring. In addition to changing the ring, the methyl group was replaced with a difluoromethyl linker fragment that is more metabolism resistant. The 3-fluoropyrimidine ring was changed to a 2,5-disubstituted pyridine ring for compounds **29–31**. These compounds gave a slight increase in potency while demonstrating much better selectivity than the 3-fluoropyrimidine ring, by comparison of structures **25** and **29**. As a point of clarity, compounds **29–34** were tested as racemic mixtures. A homology docking model based on the structure of *Trypanosoma cruzi* highlighted a tyrosine residue that could provide a favorable π-interaction if an aromatic group were placed in the 5-position of the pyridyl ring. This gives rise to compounds **32–34**. The additional phenyl group was good increases to both the potency and the selectivity. The best compound from the second phase was the racemic compound **34** (Table 3).

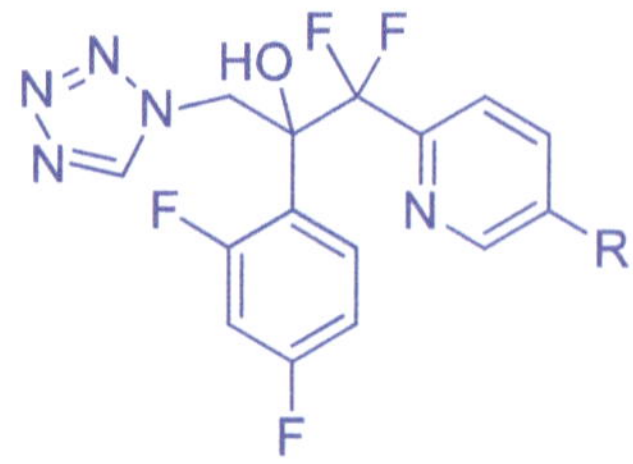

No.	R group	MIC_{50} *C. albicans*[a]	IC_{50} CYP3A4[b]	Selectivity ratio[c]
29	H	0.25	136	544
30	Cl	0.25	74	296
31	OCH_2CF_3	0.06	72	1200
32	4-F-Ph	0.016	53	3313
33	4-CN-Ph	≤0.016	16	1000
34	4-CF_3-Ph	≤0.016	>60	3750

Table 3. All SAR information taken from Hoekstra et al.[49]

[a] Units are μg/mL

[b] Units are μM

[c] Ratio is equal to IC_{50} CYP3A4/MIC_{50} *C. Albican*

The third phase of the SAR, compounds **35-37** and **1**, re-institutes the same stereochemistry as the voriconazole (**15**) starting scaffold with respect to the alcohol and the MBG. Each compound in this last group was synthesized as a racemic mixture, and then separated via chiral chromatography. The single enantiomer **35** gave decisively better results than did the racemic mixtures **34**. Note the increase in both activity and selectivity between compounds **34** and **35**. Compounds **35–37** and **1** were more potent than **32–34** and featured selectivity profiles that were much better than any previous compounds. The idea for the ether group in compounds **37** and **1** came from the homology model, which showed a serine residue toward the *meta/para* positions of the phenyl ring. An ether group was added in the para position to hydrogen bond with the serine residue. Compounds **37** and **1** were the best compounds in the Table 4 with low MIC_{50} values and excellent selectivity for *C. albicans* CYP51 over the human CYP3A4.

Compound **1** was chosen over **37** due to three reasons. First, compound **37** showed inhibition of human CYP2C9 (IC_{50} = 99 μM) and CYP2C19 (IC_{50} = 72 μM) in addition to the CYP3A4 data. Second, **1** was extremely resistant to general liver metabolism. The researchers incubated **1** with both human and animal liver microsomes for 2 h and recovered 99% of compound **1**. The oral-half life in humans was also found to be >>24 h. Third, **1** was found to be orally efficacious from dosing in mouse and guinea pig testing.

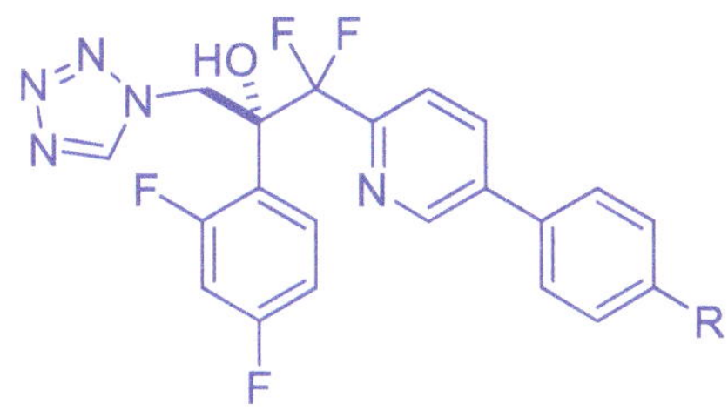

No.	R group	MIC50 *C. albicans*[a]	IC50 CYP3A4[b]	Selectivity ratio[c]
35	CF_3	≤0.001	54	54,000
36	Cl	≤0.001	36	36,000
37	OCF_3	≤0.001	79	79,000
1	OCH_2CF_3	≤0.001	65	65,000

Table 4. All SAR information taken from Hoekstra et al.[49]

[a] Units are μg/mL

[b] Units are μM

[c] Ratio is equal to IC_{50} CYP3A4/MIC_{50} *C. albicans*

4. Pharmacokinetics and Drug Metabolism

Considering that oteseconazole (**1**) is a drug that is intended to treat recurring fungal infections, it is beneficial that oteseconazole (**1**) is extremely resistant to general liver metabolism. Preclinical data showed that when oteseconazole (**1**) was incubated with both human and animal liver microsomes for 2 h over 99% of the starting material was recovered completely un-metabolized.[49] Oteseconazole (**1**) is reported to have a human half-life of approximately 138 days in humans.[50] During Phase 3 clinical trials, the average blood plasma concentrations were found: 1396.9 μg/L at the end of Week 2 (induction period), 2679.3 μg/L at the end of Week 14 (maintenance period), and 873.4 μg/L by week 50 (end of study).[51] The median t_{max} is approximately 4–5 h, depending on dose size.[52] At the end of the maintenance period (week 14), the C_{max} is 2.78 μ/mL, while the C_{min} is 2.53 μg/mL. The $AUC_{24\ h}$ is 64.2 h μg/mL.[50] If oteseconazole (**1**) is consumed with a high-fat, high-calorie meal (800–1000 Calories; 50% fat) the C_{max} is increased by 45% and the $AUC_{0\text{–}72\ h}$ is increased by 36%.[50] Several of the previous azole antifungal drugs possessed a similar ability. There was no difference observed when oteseconazole (**1**) was consumed with a low-fat, low-calorie meal. Oteseconazole (**1**) is primarily excreted through feces (56%) and secondarily through urine (26%). There was no difference in the pharmacokinetic profile regarding sex, race/ethnicity or mild to moderate renal impairment. Oteseconazole (**1**) is not a P-glycoprotein (Pgp) substrate.[50]

5. Efficacy and Safety

5.1. Efficacy

The third phase of clinical trials directly compared the efficacy of oteseconazole (**1**) to fluconazole (**14**) and placebo. The study was conducted on 219 women with a history of vulvovaginal candidiasis infection at 38 different sites within the United States.[51] The patients were randomly split into two groups in a 2:1 ratio, with the larger group taking oteseconazole (**1**). The larger group was induced with oteseconazole (**1**) according to the following dosing regimen: on Day 1 consume 600 mg and on Day 2 consume 450 mg. The smaller group was induced with fluconazole (**14**) according to the following dosing regimen: on Days 1, 4, and 7 consume 150 mg of fluconazole (**14**). Following the induction phase, the maintenance phase for the oteseconazole (**1**) group consisted of taking 150 mg of oteseconazole (**1**) once per week for 11 weeks. The maintenance phase for the fluconazole (**14**) group consisted of taking 150 mg of placebo once per week for 11 weeks. No drugs were administered after the eleventh week. The patients were then monitored until the end of week 50 (Tables 5 and 6).

	Oteseconazole (1)	**Fluconazole (14)/Placebo**
Resolution of current VVC infection after Day 14	93.2%	95.8%
Occurrence of VVC infection during maintenance period	5.1%	42.2%

Table 5. Efficacy of oteseconazole (**1**) in Phase 3 Clinical Trials[51]

Isolate (# occurrences)	**Oteseconazole (1) MIC_{50} (μg/mL)**	**Fluconazole (14) MIC_{50} (μg/mL)**
Candida albicans (266)	0.004	0.25
Candida glabrata (42)	0.030	2.00
Candida parapsilosis (19)	0.008	0.25
Candida tropicalis (11)	0.008	0.50

Table 6. MIC_{50} of oteseconazole (**1**) and fluconazole (**14**) against clinical isolates[51]

As another source of direct comparison between oteseconazole (**1**) and fluconazole (**14**), during the Phase 3 Clinical Trials, researchers isolated the specific strains of fungal infection that occurred during the maintenance period. They then determined the MIC_{50} (μg/mL) of oteseconazole (**1**) and fluconazole (**14**) directly against the isolated fungal strains. Oteseconazole (**1**) was between 31- and 67-times more effective against these isolates than fluconazole (**14**).

5.2. Safety

The use of oteseconazole (**1**) is contraindicated in three groups of people: (1) females of reproductive potential, (2) pregnant and lactating women, and (3) patients with known hyper-sensitivity to oteseconazole (**1**).[50] Further, oteseconazole (**1**) carries warnings for an additional three groups of people: (1) those with renal impairment (with or without dialysis), (2) those with hepatic impairment, and (3) those who are using breast cancer resistance protein substrates (BCRPSs).

6. Synthesis

The first reported synthesis of oteseconazole (**1**) is a five-step pathway by Hoekstra et al. in 2014.[49] It begins by reacting ethyl difluorobromoacetate with 2,5-dibromopyridine **38** in the presence of copper and DMSO to achieve an 81% yield of compound **39**. The next step is addition of 2,4-difluorobromobenzene **40** with *n*-butyl lithium at –70 °C for 2 h. This provided compound **41** in 43% yield. Compound **41** was then treated with methyldiazenide at 0 °C and allowed to warm to room temperature. This solution was then stirred for 2 h to afford compound **42** in 59% yield. Compound **42** then underwent Suzuki coupling with 4-(2,2,2-trifluoroethoxy)phenylboronic acid **43** in the presence of sodium carbonate using Pd(dppf)Cl_2 as catalyst. After 4 h under argon at 75 °C, the product **44** was obtained in 71% yield. This compound was then reacted directly with 1*H*-tetrazole **45** in the presence of potassium carbonate to obtain a racemic mixture of compound **1** in 29% yield. Chiral chromatography was employed to provide the pure compound **1** (Figure 8).

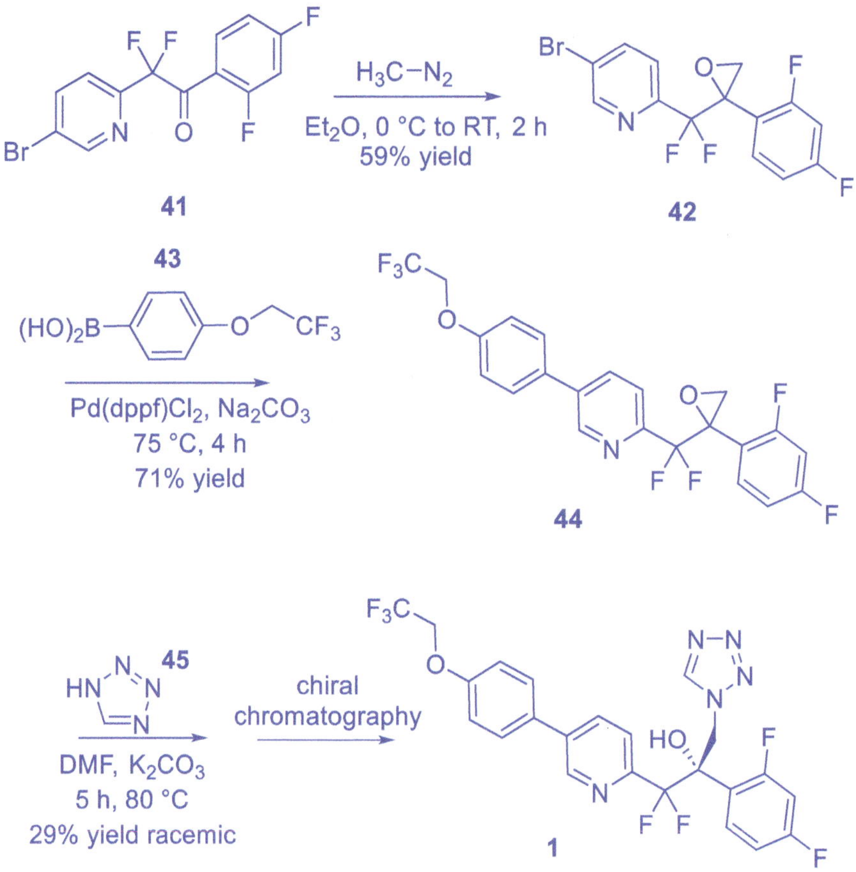

Figure 8. Bench route synthesis of oteseconazole (**1**)[49]

Oteseconazole (**1**) can also be synthesized through other routes as well. A later patent by Hoekstra et al. shows several different routes than the bench top synthesis. The tetrazole unit can be installed early in the synthetic process. The ester **39** is reacted with an ethyl ester of the tetrazole group **46**. The resulting ester group on **47** can then be removed with base and acid to provide structure **48**. The Suzuki reaction can take place here to generate **50**, which is then reacted with the 2,4-difluorophenylring that has a metal bound in the 1-position, such as a Grignard. In the presence of a chiral catalyst such as BINOL that will form the final product (Figure 9).

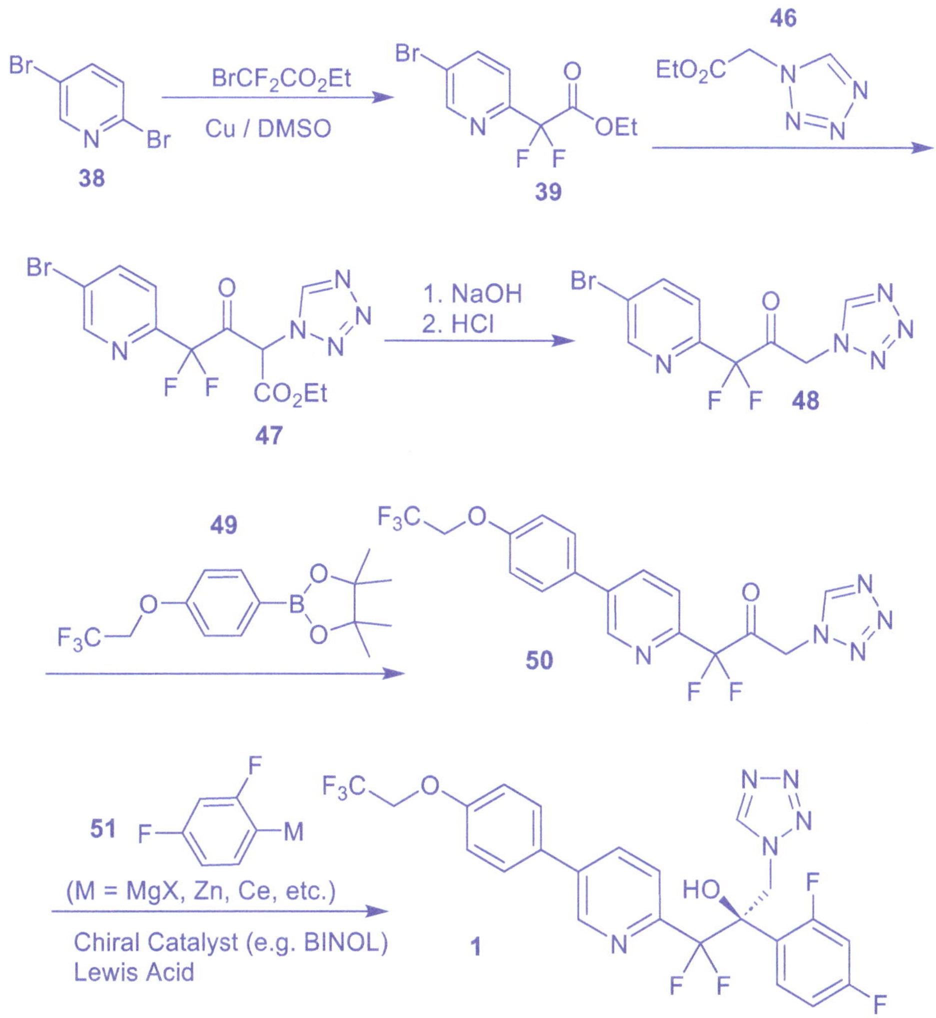

Figure 9. Installation of the tetrazole by use of a Claisen condensation reaction

Another way that oteseconazole (**1**) can be synthesized is by forming the tetrazole from a nitro group directly on the molecule by an asymmetric Henry reaction with the ketone from **41**. The resulting nitro group **52** is then reacted with sodium azide in the presence of triethyl orthoformate and acetic acid to provide the 1-tetrazole group **53**. This may be done before or after the Suzuki reaction on the pyridyl ring (Figure 10).

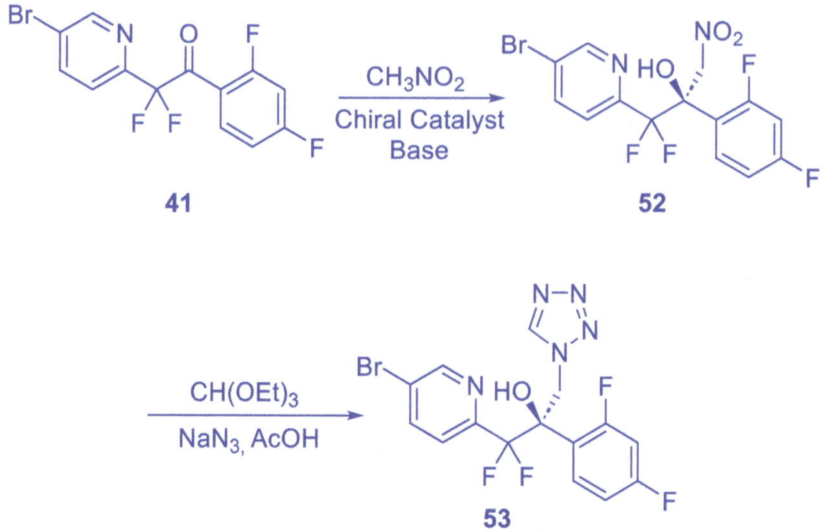

Figure 10. Asymmetric Henry reaction of (**42**)

A similar strategy may be employed by first installing a cyano group onto molecule **41** by using TMS-CN with a chiral catalyst and base. The resulting cyano group is then reduced to the corresponding amine **55** and then the tetrazole can be formed to produce **53**. This process can also be accomplished before or after the Suzuki reaction on the pyridyl ring (Figure 11).

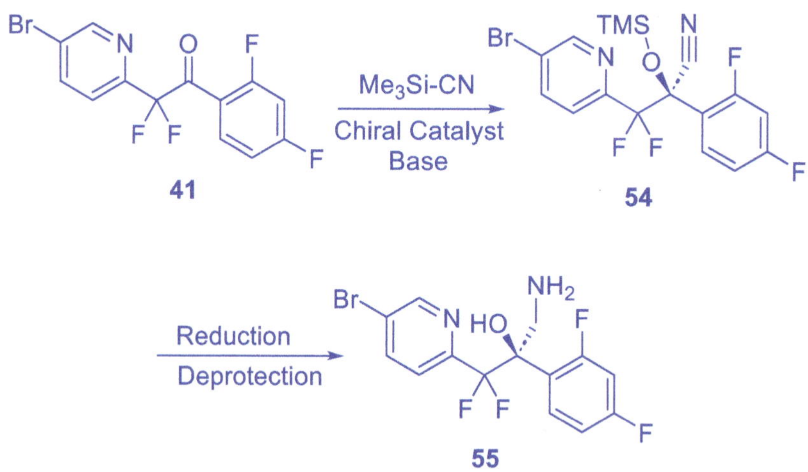

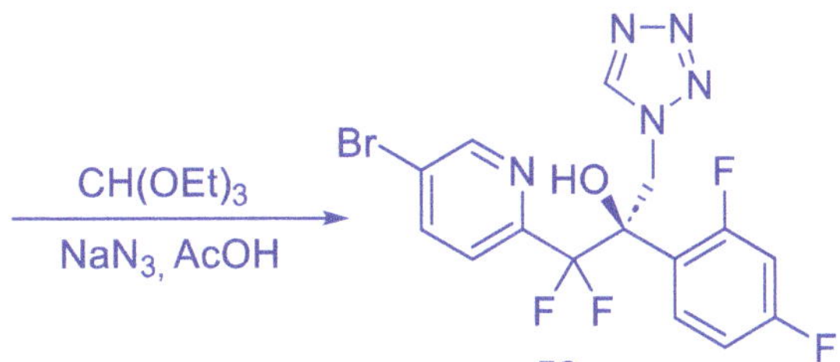

53

Figure 11. Reduction of nitrile before cyclization

7. Summary

Oteseconazole (**1**) is a new development in the treatment of *Candida* fungal infections. Its use of a tetrazole has led to an improvement in the efficacy of treatment when compared to previous medications as well as a new level of selectivity between human and fungal CYP activity. The tetrazole moiety has been shown to bind with the heme-atom in *C. albicans* CYP51 with an MIC_{50} of $\leq$ 0.001 μg/mL. It has succeeded in being an orally administered pill, while many of the previous azoles were topical creams. Oteseconazole (**1**) shows good promise in the long-term treatment of RVVC.

References

1. Sobel, J. D. Recurrent vulvovaginal candidiasis. *Am. J. Obstet. Gynecol.* **2016**, *214*, 15–21.
2. Rosati, D.; Bruno, M.; Jaeger, M.; Oever, J.; Netea, M. G. Recurrent vulvovaginal candidiasis: an immunological perspective. *Microorganisms* **2020**, *8*, 144.
3. Vivjoa Press Release. https://mycovia.com/wp-content/uploads/2022/04/FINAL-Press-Release_04.28.22.pdf (accessed 1 Apr 2023).
4. Hamilton-Miller, J. M. Chemistry and biology of the polyene macrolide antibiotics. *Bacteriol. Rev*. **1973**, 37, 166–196.
5. Oliver Lampen, J. Amphotericin B and other polyenic antifungal antibiotics. *Am. J. Clin. Pathol.* **1969**, *52*, 138–146.
6. Norman, A. W.; Spielvogel, A. M.; Wong, R. G. Polyene antibiotic–sterol interaction supported in part by United States Public Health Service Grants AM-09012 and AM-14,750. In *Advances in Lipid Research*; Elsevier, **1976**; Vol. 14, 127–170.

7. Sawaya, B. P.; Briggs, J. P.; Schnermann, J. Amphotericin B nephrotoxicity: the adverse consequences of altered membrane properties. *J. Am. Soc. Nephrol.* **1995**, *6*, 154.

8. Berney, D. Trans-*N*-cinnamyl-*N*-methyl-(1-naphthylmethyl)amine. USP4,282,251 (1981), August 4, 1981.

9. Birnbaum, J. E. Pharmacology of the allylamines. *J. Am. Acad. Dermatol.* **1990**, *23*, 782–785.

10. Favre, B.; Ryder, N. S. Characterization of squalene epoxidase activity from the dermatophyte trichophyton rubrum and its inhibition by terbinafine and other antimycotic agents. *Antimicrob. Agents Chemother.* **1996**, *40*, 443–447.

11. Gupta, A. K.; Sauder, D. N.; Shear, N. H. Antifungal agents: an overview. Part II. *J. Am. Acad. Dermatol.* **1994**, *30*, 911–933.

12. Jensen, J. C. Clinical pharmacokinetics of terbinafine (lamisil). *Clin. Exp. Dermatol.* **1989**, *14*, 110–113.

13. Elewski, B.; Tavakkol, A. Safety and tolerability of oral antifungal agents in the treatment of fungal nail disease: a proven reality. *Ther. Clin. Risk Manag.* **2005**, *1*, 299–306.

14. Lamoth, F.; Akan, H.; Andes, D.; Cruciani, M.; Marchetti, O.; Ostrosky-Zeichner, L.; Racil, Z.; Clancy, C. J. Assessment of the role of 1,3-β-d-glucan testing for the diagnosis of invasive fungal infections in adults. *Clin. Infect. Dis.* **2021**, *72*, S102–S108.

15. Free, S. J. Fungal cell wall organization and biosynthesis. In *Advances in Genetics*; Elsevier, **2013**; Vol. 81, 33–82.

16. Lesage, G.; Bussey, H. Cell wall assembly in *Saccharomyces cerevisiae*. *Microbiol. Mol. Biol. Rev.* **2006**, *70*, 317–343.

17. Wiederhold, N. P.; Lewis, R. E. The echinocandin antifungals: an overview of the pharmacology, spectrum and clinical efficacy. *Expert Opin. Investig. Drugs* **2003**, *12*, 1313–1333.

18. Sucher, A. J.; Chahine, E. B.; Balcer, H. E. Echinocandins: the newest class of antifungals. *Ann. Pharmacother.* **2009**, *43*, 1647–1657.

19. Andes, D. Pharmacokinetics and pharmacodynamics of antifungals. *Infect. Dis. Clin. North Am.* **2006**, *20*, 679–697.

20. Wiederhold, N. P.; Lewis, J. S. The echinocandin micafungin: a review of the pharmacology, spectrum of activity, clinical efficacy and safety. *Expert Opin. Pharmacother.* **2007**, *8*, 1155–1166.

21. Woolley, D. W. Some biological effects produced by benzimidazole and their reversal by purines. *J. Biol. Chem.* **1944**, 152, 225–232.

22. Fromtling, R. A. Overview of medically important antifungal azole derivatives. *Clin. Microbiol. Rev.* **1988**, *1*, 187–217.

23. Maertens, J. A. History of the development of azole derivatives. *Clin. Microbiol. Infect.* **2004**, *10*, 1–10.
24. Shafiei, M.; Peyton, L.; Hashemzadeh, M.; Foroumadi, A. History of the development of antifungal azoles: a review on structures, SAR, and mechanism of action. *Bioorganic Chem.* **2020**, *104*, 104240.
25. Allen, D.; Wilson, D.; Drew, R.; Perfect, J. Azole antifungals: 35 years of invasive fungal infection management. *Expert Rev. Anti Infect. Ther.* **2015**, *13*, 787–798.
26. Howard, K. C.; Dennis, E. K.; Watt, D. S.; Garneau-Tsodikova, S. A comprehensive overview of the medicinal chemistry of antifungal drugs: perspectives and promise. *Chem. Soc. Rev.* **2020**, *49*, 2426–2480.
27. Lee, Y.-P.; Goldman, M.; Vidt, D. G. The role of azole antifungal agents for systemic antifungal therapy. *Cleve. Clin. J. Med.* **1997**, *64*, 99–106.
28. FDA Drug Safety Communication: FDA limits usage of nizoral (ketoconazole) oral tablets due to potentially fatal liver injury and risk of drug interactions and adrenal gland problems. FDA **2019**.
29. Wood, A. *Annual Reports in Medicinal Chemistry*; Elsevier, **2006**; Vol. 41.
30. Humphrey, M. J.; Jevons, S.; Tarbit, M. H. Pharmacokinetic evaluation of UK-49,858, a metabolically stable triazole antifungal drug, in animals and humans. *Antimicrob. Agents Chemother.* **1985**, *28*, 648–653.
31. Sheehan, D. J.; Hitchcock, C. A.; Sibley, C. M. Current and emerging azole antifungal agents. *Clin. Microbiol. Rev.* **1999**, *12*, 40–79.
32. Malani, A.; Kerr, L.; Kauffman, C. Voriconazole: how to use this antifungal agent and what to expect. *Semin. Respir. Crit. Care Med.* **2015**, *36*, 786–795.
33. Herbrecht, R.; Denning, D. W.; Patterson, T. F.; Bennett, J. E.; Greene, R. E.; Oestmann, J.-W.; Kern, W. V.; Marr, K. A.; Ribaud, P.; Lortholary, O.; et al. Voriconazole versus amphotericin B for primary therapy of invasive aspergillosis. *N. Engl. J. Med.* **2002**, *347*, 408–415.
34. Kale, P.; Johnson, L.B. Second-generation azole antifungal agents. *Drugs Today* **2005**, *41*, 91.
35. Van Custem, J.; Van Gerven, F.; Janssen, P. A. J. Activity of orally, topically, and parenterally administered itraconazole in the treatment of superficial and deep mycoses: animal models. *Rev. Infect. Dis.* **1987**, *9*, S15–S32.

36. Graybill, J. R. New antifungal agents. *Eur J Clin MicrobiolInfect Dis* **1989**, *8*, 402–412.
37. Cleary, J. D.; Taylor, J. W.; Chapman, S. W. Itraconazole in antifungal therapy. *Ann. Pharmacother.* **1992**, *26*, 502–509.
38. Nomeir, A. A.; Pramanik, B. N.; Heimark, L.; Bennett, F.; Veals, J.; Bartner, P.; Hilbert, M.; Saksena, A.; McNamara, P.; Girijavallabhan, V.; et al. Posaconazole (Noxafil, SCH 56592), a new azole antifungal drug, was a discovery based on the isolation and mass spectral characterization of a circulating metabolite of an earlier lead (SCH 51048). *J. Mass Spectrom.* **2008**, *43*, 509–517.
39. Nagappan, V.; Deresinski, S. Posaconazole: a broad-spectrum triazole antifungal agent. *Clin. Infect. Dis.* **2007**, *45*, 1610–1617.
40. Wang, S.-Q.; Wang, Y.-F.; Xu, Z. Tetrazole hybrids and their antifungal activities. *Eur. J. Med. Chem.* **2019**, *170*, 225–234.
41. Yates, C. M.; Garvey, E. P.; Shaver, S. R.; Schotzinger, R. J.; Hoekstra, W. J. Design and optimization of highly-selective, broad spectrum fungal CYP51 inhibitors. *Bioorg. Med. Chem. Lett.* **2017**, *27*, 3243–3248.
42. Niwa, T.; Inoue-Yamamoto, S.; Shiraga, T.; Takagi, A. Effect of antifungal drugs on cytochrome P450 (CYP) 1A2, CYP2D6, and CYP2E1 activities in human liver microsomes. *Biol. Pharm. Bull.* **2005**, *28*, 1813–1816.
43. Niwa, T.; Shiraga, T.; Takagi, A. Effect of antifungal drugs on cytochrome P450 (CYP) 2C9, CYP2C19, and CYP3A4 activities in human liver microsomes. *Biol. Pharm. Bull.* **2005**, *28*, 1805–1808.
44. Sakaeda, T.; Iwaki, K.; Kakumoto, M.; Nishikawa, M.; Niwa, T.; Jin, J.; Nakamura, T.; Nishiguchi, K.; Okamura, N.; Okumura, K. Effect of micafungin on cytochrome P450 3A4 and multidrug resistance protein 1 activities, and its comparison with azole antifungal drugs. *J. Pharm. Pharmacol.* **2010**, *57*, 759–764.
45. Baldwin, S. J.; Bloomer, J. C.; Smith, G. J.; Ayrton, A. D.; Clarke, S. E.; Chenery, R. J. Ketoconazole and sulphaphenazole as the respective selective inhibitors of P4503A and 2C9. *Xenobiotica* **1995**, *25*, 261–270.
46. Indrayudha, P.; Rosyid As Sabiq, M. Review: diphenylphosphane derivatives of ketoconazole and oteseconazole/VT-1161, promising new azole compounds in the treatment of *Candida albicans* infections. *Proceedings of the 4th International Conference Current Breakthrough in Pharmacy. Icb-Pharma 2022 December* **2022**.

47. Lepesheva, G. I.; Waterman, M. R. Sterol 14α-demethylase cytochrome P450 (CYP51), a P450 in all biological kingdoms. *Biochim. Biophys. Acta* **2007**, *1770*, 467–477.

48. Daum, G.; Lees, N. D.; Bard, M.; Dickson, R. Biochemistry, cell biology and molecular biology of lipids of *Saccharomyces cerevisiae*. *Yeast* **1998**, 14, 1471–1510.

49. Hoekstra, W. J.; Garvey, E. P.; Moore, W. R.; Rafferty, S. W.; Yates, C. M.; Schotzinger, R. J. Design and optimization of highly-selective fungal CYP51 inhibitors. *Bioorg. Med. Chem. Lett.* **2014**, *24*, 3455–3458.

50. Vivjoa Prescribing Information. https://www.accessdata.fda.gov/drugsatfda_docs/label/2022/215888s000lbl.pdf (accessed 2023-02-28).

51. Martens, M. G.; Maximos, B.; Degenhardt, T.; Person, K.; Curelop, S.; Ghannoum, M.; Flynt, A.; Brand, S. R. Phase 3 study evaluating the safety and efficacy of oteseconazole in the treatment of recurrent vulvovaginal candidiasis and acute vulvovaginal candidiasis infections. *Am. J. Obstet. Gynecol.* **2022**, *227*, 880.e1-880.e11.

52. Brand, S. R.; Sobel, J. D.; Nyirjesy, P.; Ghannoum, M. A.; Schotzinger, R. J.; Degenhardt, T. P. A randomized phase 2 study of VT-1161 for the treatment of acute vulvovaginal candidiasis. *Clin. Infect. Dis. Off. Publ. Infect. Dis. Soc. Am.* **2020**, *73*, e1518–e1524.

Section II. ONCOLOGY DRUGS

7

Futibatinib (Lytgobi): A Selective Irreversible FGFR1–4 Inhibitor

Faridoon and
Guiping Zhang

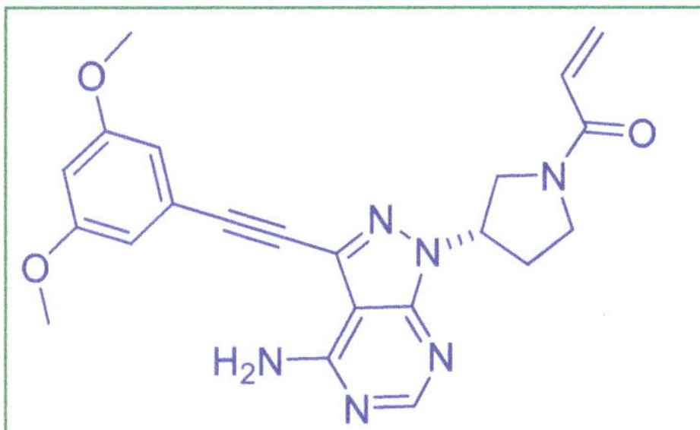

futibatinib (Lytgobi, **1**)
Taiho Oncology and
Taiho Pharmaceutical
Selective irreversible
FGFR1–4 inhibitor, 2022

1 Background

Futibatinib (Lytgobi, **1**) is an oral irreversible covalent inhibitor of fibroblast growth factor receptor (FGFR)1–4 that is being developed by Taiho Oncology and Taiho Pharmaceutical for the treatment of cancers, including cholangiocarcinoma, urothelial cancer, breast cancer, esophageal cancer, gastric cancer, and non-small-cell lung cancer. Futibatinib (**1**) got its accelerated approval in the United States on September 30, 2022, for the treatment of adult patients with previously treated, unresectable, locally advanced, or metastatic intrahepatic cholangiocarcinoma harboring FGFR2 gene fusions or other rearrangements.

Fibroblast growth factors (FGFs) and their receptors (FGFRs) play a key role in regulating a wide range of biological pathways, whereas the dysregulation of the FGFR pathway is associated with oncogenesis. Nearly 7% of all cancers harbor FGFR aberrations. Therefore, FGFR has emerged as an important therapeutic target for cancer treatment. In general, most FGFR inhibitors in development are ATP competitive,

Chemistry and Pharmacology of Drug Discovery, First Edition. Edited by Jie Jack Li.

reversible inhibitors, to which acquired resistance has been observed. Futibatinib (**1**), an irreversible FGFR1–4 inhibitor, is the most advanced covalent inhibitor in clinical development for multiple cancer types.[1]

Since the FGFR kinase domain share high sequence homology with other receptor tyrosine kinases. Therefore, the first-generation FGFR inhibitors are nonselective tyrosine kinase inhibitors (TKIs) that compete with ATP for ATP-binding site. As a result, these inhibitors not only inhibit FGFR but also inhibit a variety of other tyrosine kinases.[2] Many approved TKIs show inhibitory activity for FGFR, and some of them are being (or have been) evaluated in clinical trials for diseases where FGFR alterations are implicated.

Thanks to the rapid revolution in the drug designing strategies, numerous second generation FGFR inhibitors have been discovered with higher selectivity, potency, and safety. The three approved FGFR inhibitors, erdafitinib (JNJ-42756493, **2**), pemigatinib (**3**), and infigratinib (BGJ-398, **4**) are all selective reversible inhibitors with pan-FGFR inhibitory activity.[3] Erdafitinib (**2**) developed by Janssen was approved by FDA in April 2019 for treatment of adult patients with locally advanced or metastatic urothelial carcinoma. It is an orally available and selective pan-FGFR inhibitor[4] that inhibits FGFR1–4 with similar potency.[5] Pemigatinib (INCB054828, **3**)[6] is reversible FGFR inhibitor containing a tricyclic urea scaffold developed by Pemazyre and approved by FDA in April 2020 for the treatment of adults with previously treated, unresectable locally advanced or metastatic cholangiocarcinoma with a FGFR2 fusion or other rearrangement. Infigratinib (BGJ-398, **4**) is another ATP competitive reversible FGFR selective inhibitor developed by QED Therapeutics and Helsinn for the treatment urothelial carcinoma, cholangiocarcinoma, and other FGFR-driven conditions. Infigratinib (**4**) was granted accelerated approval by FDA in May 2021 for the treatment of previously treated, unresectable locally advanced or metastatic cholangiocarcinoma with a FGFR2 fusion or other rearrangement.[7] There are numerous other reversible pan-FGFR inhibitors in clinical trials or development. For example, AZD4547 (**5**) from AstraZeneca is a pan-FGFR inhibitor bearing the pyrazole scaffold and completed phase 2 clinical trial (NCT04439240) in multiple cancers with FGFR alterations.[3]

Covalent inhibition is the best strategy especially in the development of kinase inhibitors, which can help to improve binding affinity and selectivity. Luckily, FGFRs have the conserved cysteine in the P-loop (Cys488 in FGFR1, Cys491 in FGFR2, Cys482 in FGFR3, and Cys477 in FGFR4) and the unique Cys552 in FGFR4 in the hinge region for covalent attachment.[8] The first acrylamide-based irreversible covalent inhibitor FIIN-1 (**6**) of FGFR1–4 was discovered by Zhou et al. in 2010 which form covalent bond with a conserved cysteine (Cys486 of FGFR1) located at the rim of the P-loop.[9] After this discovery, several groups reported selective irreversible covalent inhibitors of FGFR1–4 (or more specific for FGFR2 and FGFR4). Among them, futibatinib (**1**) is the most advanced selective irreversible covalent pan-FGFR inhibitor which got accelerated FDA approval in 2022 as mentioned above.

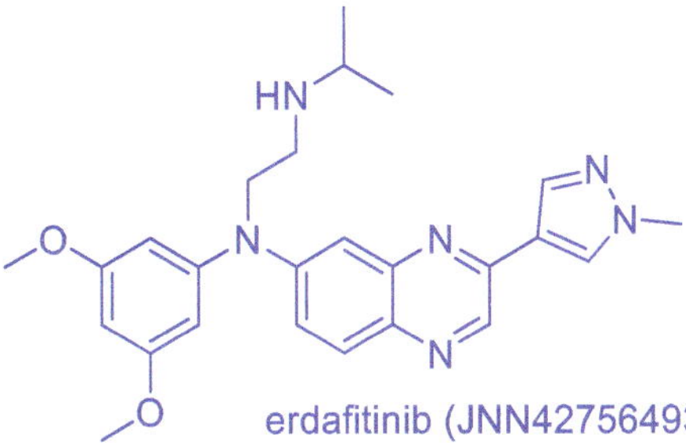

erdafitinib (JNN42756493, **2**)
Janssen, 2019
FGFR 1, IC_{50} = 1.2 nM
FGFR 2, IC_{50} = 2.5 nM
FGFR 3, IC_{50} = 3.0 nM
FGFR 4, IC_{50} = 5.7 nM

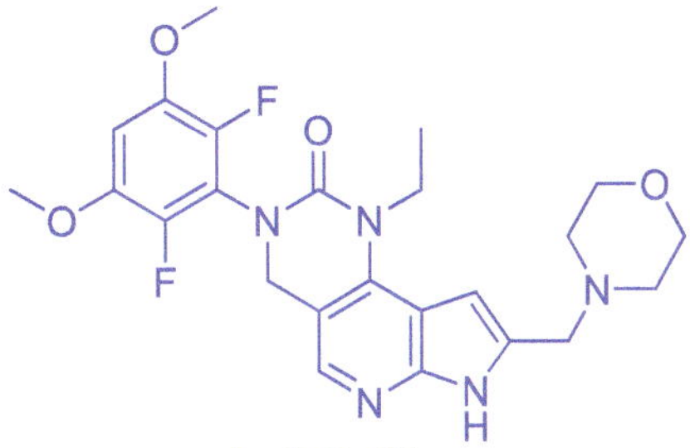

pemigatinib (**3**)
Pemazyre, 2020
FGFR 1, IC_{50} = 0.4 nM
FGFR 2, IC_{50} = 0.5 nM
FGFR 3, IC_{50} = 1.2 nM
FGFR 4, IC_{50} = 30 nM

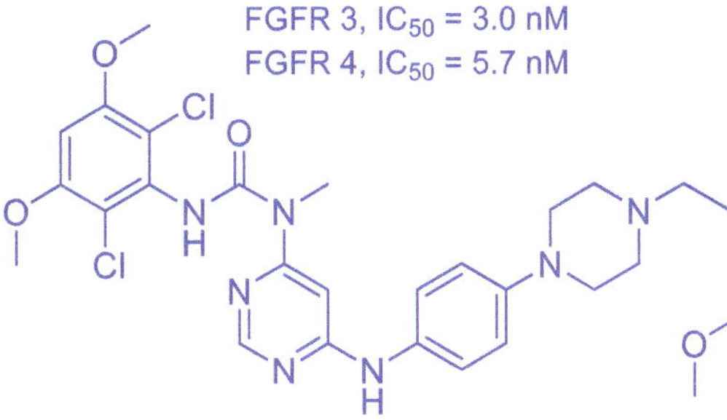

infigratinib (BGJ-398, **4**)
QED Therapeutics, 2021
FGFR 1, IC_{50} = 0.9 nM
FGFR 2, IC_{50} = 1.4 nM
FGFR 3, IC_{50} = 1.0 nM
FGFR 4, IC_{50} = 60 nM

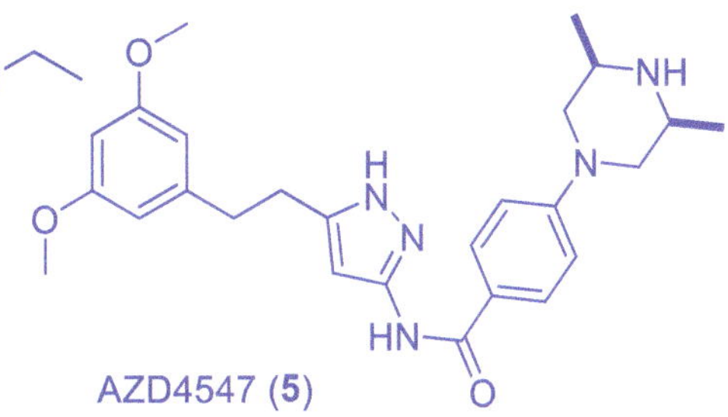

AZD4547 (**5**)
Phase II, (NCT04439240)
FGFR 1, IC_{50} = 0.2 nM
FGFR 2, IC_{50} = 2.5 nM
FGFR 3, IC_{50} = 8.0 nM

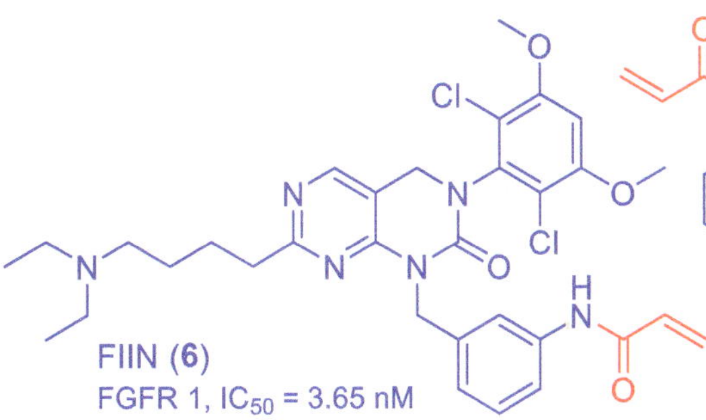

FIIN (**6**)
FGFR 1, IC_{50} = 3.65 nM
FGFR 2, IC_{50} = 2.29 nM
FGFR 3, IC_{50} = 277 nM

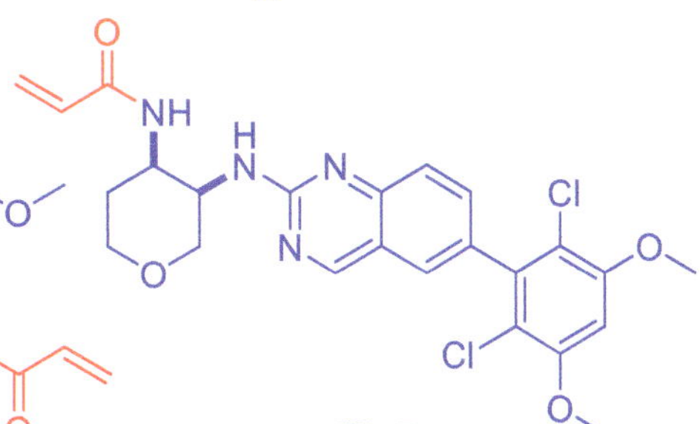

BLU554 (**7**), Phase I
FGFR 4, IC_{50} = 5 nM
FGFR 1-3, IC_{50} = > 624 nM

Beside futibatinib (**1**), several irreversible covalent-based drug candidates targeting FGFRs have entered clinical stages. For example, BLU554 (fisogatinib,7) is an orally available specific irreversible covalent inhibitor of FGFR4. It is currently in phase I clinical trial to treat patients with hepatocellular carcinoma (NCT02508467).

2 Pharmacology

The FGFRs belong to receptor tyrosine kinases (RTKs), consisting of four highly conserved genes (FGFR1–4). FGFR proteins comprise an extracellular domain with three immunoglobulin (Ig)-like domains (IgI, IgII, and IgIII), followed by a transmembrane domain and intracellular domain with kinase activity (Figure 1).

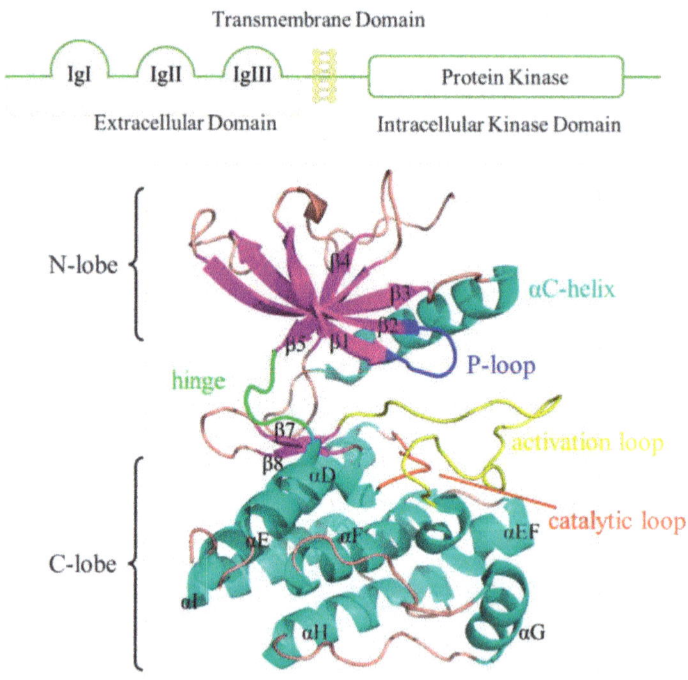

Figure 1. Schematic diagram of FGFRs and the structure of the FGFR extracellular domain (PDB: 4UXQ)

The intracellular tyrosine kinase domain is the most extensively studied region of the FGFR protein. The small N-terminal lobe (N-lobe, ~100 amino acid residues) consists of a five-stranded antiparallel β-sheet (β1–β5) and the αC-helix. Between β1 and β2, a highly flexible glycine-rich loop called P-loop is present that swings downward in the presence of ATP to create the nucleotide binding site. The larger C-terminal lobe (C-lobe, ~200 amino acid residues) is predominately composed of seven α helices (αD, αE, αEF, and αF-αI) (Figure 1). The N-lobe and C-lobe are linked by a hinge region. The activation loop, which is responsible for ATP and substrate-binding proteins and encompasses tyrosine phosphorylation sites, is located in a clef between the two lobes (Figure 1). At the beginning of the activation loop, an Asp-Phe-Gly triad forms the highly conserved DFG-motif, which works as indicator of the active/inactive states of kinase. There is another important loop (named as catalytic loop) present between αE and β7. It contains the His-Arg-Asp (HRD) motif. The Asp of HRD-motif interacts with the hydroxyl group of the substrate tyrosine and contributes to the phosphorylation.[3,10]

The FGFR pathway contains a family of 22 FGFs ligands, which primarily regulate cellular signals through four FGFRs. Typically, FGFR activation induces cell proliferation and migration. Beside this, it can also drive cell differentiation or negatively regulate proliferation. Aberrant FGFR signaling can encourage tumorigenesis, assist tumor survival, and help in making resistance to chemotherapy through anti-apoptotic signaling which render FGFR altered tumors difficult to treat.[11]

Futibatinib (**1**) is a novel small molecule inhibitor of FGFR with high selectivity and potency. It specifically targets the P-loop of the FGFR tyrosine kinase domain, forming an irreversible covalent adduct with a conserved cystine amino acid residue of the protein (Figure 2).[12–14] As this cysteine residue is conserved across all four FGFR receptors, therefore futibatinib (**1**) inhibits the kinase activities of all four FGFR isoforms. The distinct binding site and irreversible binding mode render futibatinib (**1**) less susceptible to drug resistance mutations than reversible, ATP-competitive inhibitors of FGFRs.

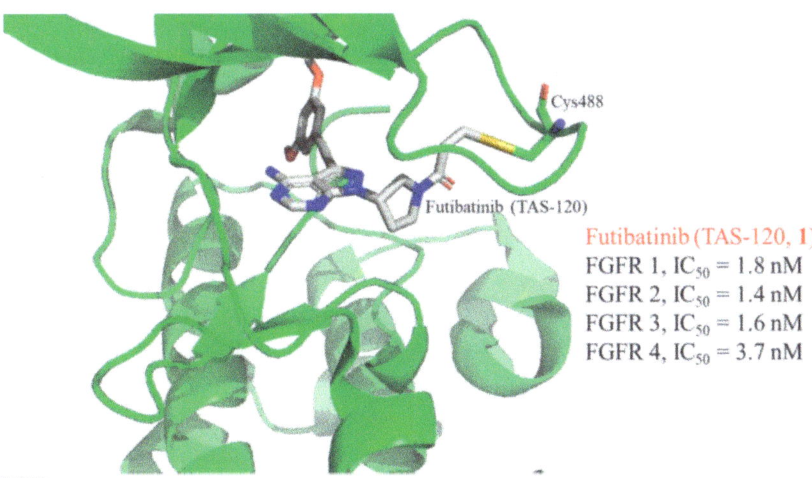

Figure 2. X-ray crystal structure of TAS-120 (**1**) in complex with FGFR1(PDB; MZW)

The inhibition of FGFR by futibatinib (**1**) inhibits FGFR phosphorylation and downstream signaling, which consequently decrease cell viability in cancer cell lines harboring FGFR alterations (fusions, rearrangements, amplifications, and mutations). Futibatinib (**1**) showed potent antitumor activity in FGFR-deregulated cancer cell lines and xenograft models. In the pharmacodynamic study of animal models, futibatinib (**1**) showed synergistic antitumor effects with PI3K pathway inhibitors, selective AKT inhibitor and cytotoxic agents.[15] Futibatinib (**1**) achieved measurable clinical benefits in patients with cholangiocarcinoma harboring FGFR2 fusion and other rearrangements. It showed enhanced antitumor activity in patients with FGFR2 fusion-positive intrahepatic cholangiocarcinoma who developed resistance to ATP-competitive FGFR inhibitors due to multiple FGFR2 mutations in the kinase domain. Furhtermore, futibatinib (**1**) also demonstrated clinical efficacy in patients with intrahepatic cholangiocarcinoma harboring FGFR2 extracellular domain in-frame deletions.[15]

3 Structure–Activity Relationship (SAR)

Table 1. Kinase inhibitory activities of 3-Quinoline analogues

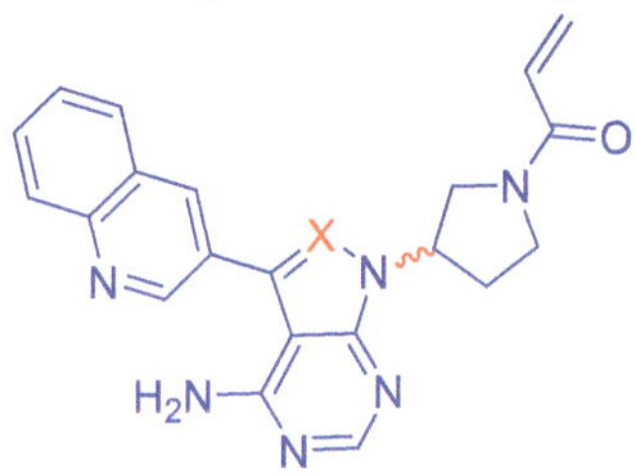

Compound	X	Chirality	Kinase inhibition IC_{50} (nM) FGFR2	EGFR del19/T790M
8	C	racemate	57	1.2
9a	C	*R*	313	0.4
9b	C	*S*	17	111
10a	N	*R*	2,260	5.4
10b	N	*S*	255	741

During systematic kinome screening of an in-house compound collection, Taiho pharmaceutical identified compound **8** as a hit with inhibitory activities of both FGFR2 and EGFR del19/T790M mutations, with IC_{50} of 57 and 1.2 nM respectively.[13] Compound **8** consists of an ATP mimicking pyrrolopyrimidine core structure, a racemic pyrrolidine moiety, and an acrylamide covalent warhead. Since acrylamide has been widely used as an electrophilic warhead to covalently target the thiol group of cysteine residues of the target protein. Therefore, compound **8** is a good starting point to develop structurally novel irreversible covalent inhibitors for both mutant EGFR and FGFR. The enantiomers **9a–b** of compound **8** and its pyrazolopyrrolidine analogues **10a–b** showed a different trend in the FGFR and EGFR inhibitory activity. EGFR or FGFR inhibitory activities of these compounds were controlled solely by stereochemistry at the 3-position of the pyrrolidine ring.

Interestingly the *R* enantiomers **9a** and **10a** showed more potent inhibitory activity against mutant EGFR than FGFR2. On the other hand, the *S* enantiomers **9b** and **10b** showed better FGFR2 inhibition (Table 1). Although **9b** (*S*) showed approximately 15-fold better inhibitory activity on FGFR2 than **10b** (*S*), **10b** (*S*) was selected as an initial lead molecule for rapid structure–activity relationship (SAR) determination, because of the easily accessible synthesis of the pyrazolopyrimidine core. The quinoline moiety in **10b** is assumed to extend to the well-known hydrophobic pocket close to the gatekeeper residue in the ATP-binding site. In the discovery of kinase inhibitors, this pocket is generally targeted to increase both potency and selectivity. In general, most of the FGFR inhibitors reported so far are bearing 3,5-dimethoxybenzene ring. Therefore, the quinoline ring in **10b** was replaced by 3,5-dimethoxybenzene via various reactions by use of commercial building blocks to identify more potent and selective lead molecules. The direct bonding of 3,5-dimethoxybenzene to the pyrazolopyrimidine core in **11** did not improve the activity while the alkyne compound **1** resulted in 200-fold increase in FGFR inhibitory activity over **10b**. When the rigid alkyne is replaced by flexible alkyl and amide analogues in **12** and **13**, respectively, the activity is lost over more than 200-fold (Table 2). The initial SAR data suggest that the rigid straight-chain alkyne could properly place the 3,5-dimethoxybenzene ring in the well-known unique hydrophobic pocket of FGFR.

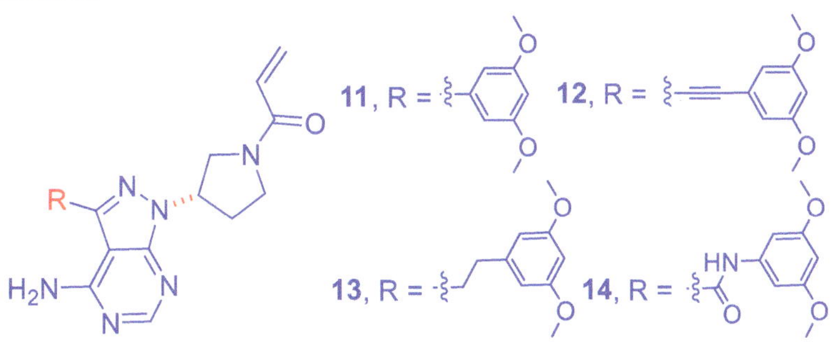

Table 2. Kinase inhibitory activities of 3,5-dimethoxybenzene analogues

Compound	Kinase inhibition IC_{50} (nM)	
	FGFR2	EGFR del19/T790M
11	270	>5000
1	1.0	>5000
12	239	>5000
13	670	>5000

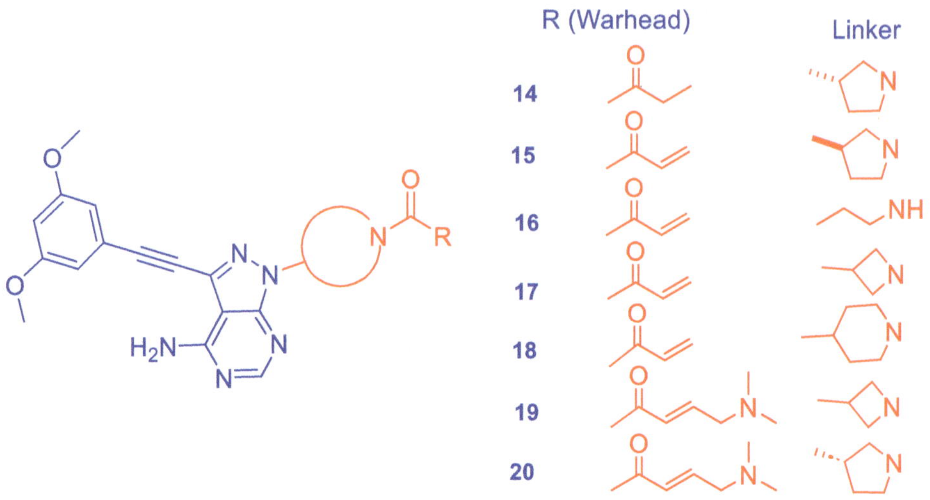

Table 3. Kinase inhibitory activities of alkyne analogues

Compound	Kinase inhibition IC_{50} (nM) FGFR2	VEGFR2
1	1.0	2358
14	244	—
15	29	1486
16	207	7010
17	1.5	3029
18	34	3377
19	3.4	3686
20	22	1206

Propionamide analogue **14** of compound **1** was synthesized to determine the effect of the acrylamide warhead on FGFR2 inhibition. It has been found that compound **14** reduced the inhibitory activity 240-fold compared with compound **1**. From this data it can be indirectly confirmed the covalent inhibitory mechanism of compound **1**. The targeted cysteine 491 (Cys491) of FGFR2 is in a P-loop that is conserved in protein kinases and known to be highly flexible. Relatively it was difficult to design linkers to achieve covalent bonding with Cys491. Thus, several commercially available amines were used to identify the optimal linker structure to capture Cys491 covalently in a rapid, parallel manner. Reversing the stereochemistry of **1** in compound **15** resulted in a 30-fold decrease in FGFR2 inhibition compared with that of the parent compound **1**. Compound **16**, with a flexible ethylene linker, showed an approximate 200-fold decrease in FGFR2 inhibitory activity. Reducing the size of the ring to four-membered azetidine in compound **17** almost retained the same activity as compound **1**, while six-membered piperidine-based compound **18** resulted in 30-fold decrease in activity. Replacing acrylamides by a 4-(dimethylamino)-but-2-enoic amide warhead in **19** and **20** resulted in several fold decrease in activity. Regarding kinase selectivity, generally first-generation multi-kinase FGFR inhibitors inhibited vascular endothelial growth factor receptor 2 (VEGFR2). Therefore, VEGFR2 was used as the initial counter kinase to explore FGFR specificity. Luckily, all analogues in Table 3 showed good selectivity over VEGFR2.

Compound with good inhibitory activity for FGFR2 were screened for cellular potency and selectivity against SNU-16 (FGFR2-amplified human gastric carcinoma that exhibits FGFR dependent cell proliferation) and SNU-1 (human gastric carcinoma without FGFR gene alteration). Generally, the antiproliferative activity of compounds tested against the SNU-16 cell line is in the nanomolar range (Table 4). On the other hand, the inhibitory activity against the SNU-1 cell line was approximately more than 5 μM. This clear difference in sensitivity observed between SNU-16 and SNU-1 cells indicates that these compounds are highly selective for FGFR.

Table 4. Cellular potency of alkyne analogues

Compound	Growth inhibition IC_{50} (nM) SNU-16	SNU-1
1	3.7	>5000
15	113	>5000
16	362	>5000
17	7.4	>5000
18	33	>5000
19	8.4	>5000
20	43	>5000

Compounds **1**, **17**, and **19** showed potent inhibition of cell proliferation in SNU-16 cells and were further evaluated for their drug metabolism and pharmacokinetics (DMPK) profiles. All the three compounds showed moderate to good microsomal stability. Only compound **1** showed good exposure on oral administrative at 50 mg/kg in mice (Table 5). The lower exposure of compound **17** and **19** may be due to increased intrinsic reactivity.

Table 5. DMPK profile of representative compounds

Compound	Microsomal stability, %age remaining, human/mouse	Mouse exposure study (50 mg/kg, oral), AUC_{0-6} h (μM h)
1	66/69	23.9
17	83/92	0.07
19	77/98	1.27

Compound **1** was selected for further characterization and named as futibatinib. The high selectivity of futibatinib (**1**) for FGFR1–4 was confirmed by testing the active-site directed competition binding assay using a KINOME*scan* panel containing 387 wild-type kinases. Among the 387 kinases other than FGFR1–4, only two kinases exhibited >50% inhibition by futibatinib (**1**) at 100 nM: Mitogen-activated protein kinase 12 (MAPK12; 69%) and insulin receptor (INSR; 55%). Then the enzyme-inhibitory activities of MAPK12 and INSR were tested in cell-free enzyme assays, and the inhibition of these two kinases by futibatinib (**1**) was limited. Thus, these results suggest that futibatinib (**1**) is a highly potent and selective covalent FGFR1−4 inhibitor.

4 Pharmacokinetics and Drug Metabolism

The pharmacokinetic properties of oral futibatinib (**1**) 20 mg qd (once daily) have been evaluated in patients with advanced solid tumors. Futibatinib (**1**) exposure increased in a dose-proportional manner over a dose range of 4–24 mg.[15] On 20 mg oral administration of futibatinib (**1**), the maximum plasma concentration is achieved at 2 h while the geometric mean apparent volume of distribution of futibatinib (**1**) is 66 L. Futibatinib (**1**) is 95% bound to human plasma protein. It bound primarily to albumin and α1-acid glycoprotein.[15] Futibatinib (**1**) is mainly metabolized by CYP3A, and to a lesser extent by CYP2C9 and CYP2D6 as well as by glutathione conjugation. The mean elimination half-life of futibatinib (**1**) is 2.9 h with mean apparent clearance of 20 L/h. After a single 20 mg oral dose of radiolabeled futibatinib (**1**) administration, nearly 91% of the total radioactivity was recovered in feces and 9% in urine. The pharmacokinetics of futibatinib (**1**) are not significantly affected by age, sex, race, or body weight.

Concomitant use of drugs that are dual P-gp and strong CYP3A inhibitors with futibatinib (**1**) may increase its exposure which may as a result increase the incidence and severity of adverse reactions. Similarly, the concomitant use of drugs that are dual P-gp and strong CYP3A inducers may decrease futibatinib (**1**) exposure, which may reduce the efficacy of futibatinib (**1**).

5 Efficacy and Safety

Efficacy of futibatinib (**1**) was evaluated in TAS-120-101 (NCT02052778), a multicenter, open-label, single-arm trial that enrolled 103 patients with previously treated, unresectable, locally advanced/metastatic intrahepatic cholangiocarcinoma harboring a FGFR2 gene fusion or other rearrangement. Patients received 20 mg of futibatinib (**1**) orally once daily until disease progression or unacceptable toxicity. The major efficacy outcome measures were overall response rate (ORR) and duration of response (DoR) as determined by an independent review committee according to RECIST v1.1. ORR was

42% (95% Confidence Interval [CI]: 32, 52) in 103 evaluable patients. In the overall population, the median DoR was 9.7 months (95% CI 7.6–17.1); DoR was ≥6 and ≥12 months in 72% and 14% of responders, respectively. Disease control rate was 82.5%, median progression-free survival (PFS) was 9.0 months and median overall survival (OS) was 21.7 months (12-month OS rate 72%).[15]

Besides the use of futibatinib (**1**) in intrahepatic cholangiocarcinoma, a phase II clinical trial (NCT04189445) is evaluating futibatinib (**1**) in tumors with specific FGFR aberrations. The phase II clinical trials in patients with urothelial cancer (NCT04601857) and breast cancer (NCT04024436; FOENIX-MBC2) are ongoing too. A phase Ib/II trial (NCT04965818) is evaluating futibatinib (**1**) in combination with binimetinib (a MEK inhibitor) in patients with KRAS mutation-positive advanced NSCLC. The phase Ib trial (JapicCTI-195063) of futibatinib (**1**) in combination with pembrolizumab in patients having advanced solid tumors, including esophageal cancer, is also ongoing.

The most common adverse reactions of futibatinib (**1**) occurring in 20% or more of patients were nail toxicity, musculoskeletal pain, constipation, diarrhea, fatigue, dry mouth, alopecia, stomatitis, abdominal pain, dry skin, arthralgia, dysgeusia, dry eye, nausea, decreased appetite, urinary tract infection, palmar-plantar erythrodysesthesia syndrome, and vomiting. The most common serious side effects of futibatinib (**1**) (which may affect more than 1% of patients), include intestinal obstruction and migraine.

6 Synthesis

Th first synthesis of futibatinib (**1**) was disclosed in a patent by Taiho pharmaceutical in 2013.[16] Recently, synthesis of futibatinib (**1**) was published as the supplemental information (SI) of Taiho's 2023 *ACS Medicinal Chemistry Letter* article.[13] Starting from pyrazolopyrimidine **21**, nucleophilic substitution reaction was applied to install the chiral moiety pyrrolidine **22** to give **23** with chiral center inversion (48% yield). In the next step, by Sonogashira coupling reaction, the arylalkyne **24** was coupled with **23** to afford **25** with excellent 96% yield. Compound **25** was then treated with 4 N HCl solution in 1,4-dioxane for the deprotection of Boc to get **26**. Compound **26** was used in the next step without further purification and was treated with acryloyl chloride in the presence of TEA and chloroform to get the target compound **1** with good yield 84% in the two steps.

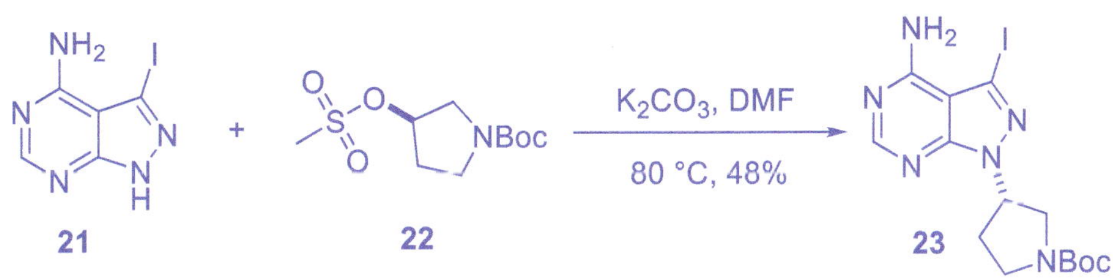

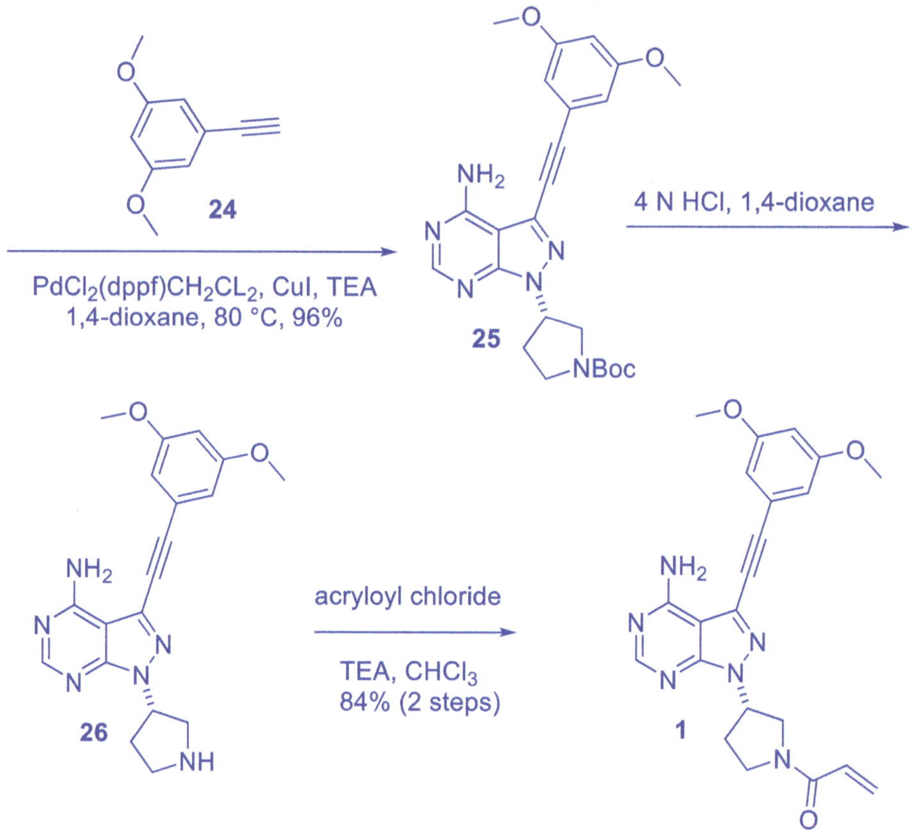

The same company reported the synthesis of futibatinib (**1**) in 2015. In this synthetic route, the sequence of Sonogashira coupling and SN2 reaction was reversed to get **25** and then followed by deprotection and acylation to get futibatinib (**1**).[17]

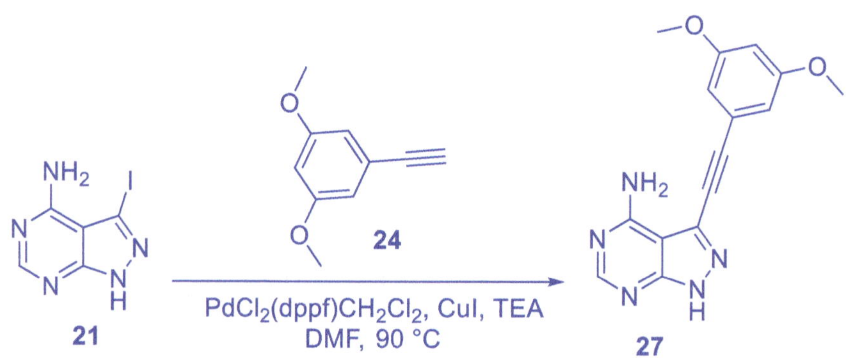

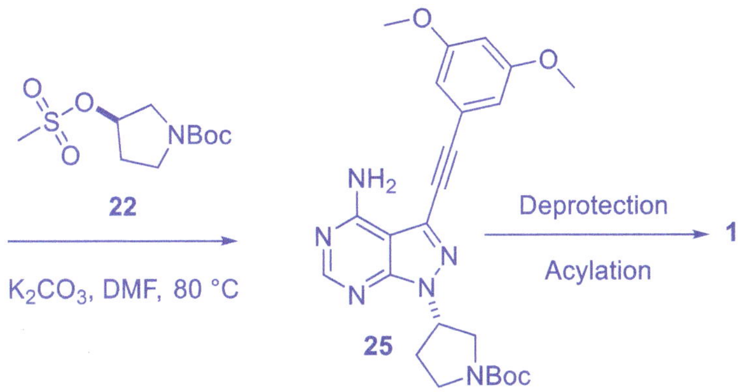

The original medicinal chemistry synthesis of futibatinib (**1**) as described above proved sufficient for the delivery of multigram quantities; however, several issues with this approach necessitated further optimization of the synthetic route to facilitate future scale-up activities. First, the original synthetic route had silica gel column chromatography purification, while chloroform was used as the reaction solvent, although it is not preferable for commercial production of API. Second, it is important to control the formation and residual of the main impurity like diamide **28**. Several other impurities are also identified that are not mentioned here. For detail structures of these impurities, please see Reference 18.

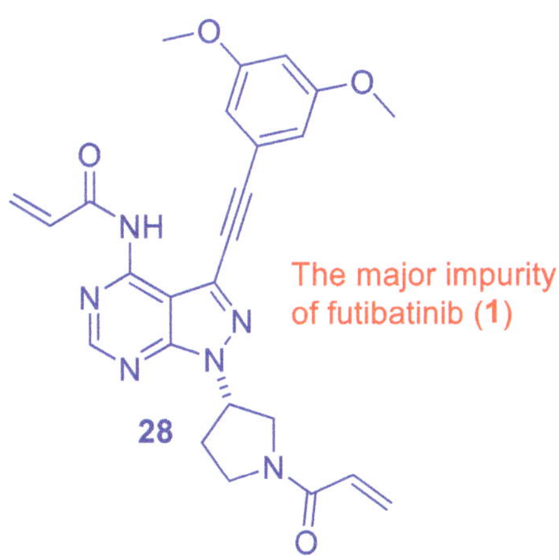

Considering the above-mentioned problems of the early discovery synthesis, an alternative approach was needed to synthesize futibatinib (**1**) on the large-scale manufacture. Initially the reaction condition using acryloyl chloride was examined, however, it was difficult to control the formation of compound **28** and a high reaction conversion rate of futibatinib (**1**). These problems were resolved by changing the acylating reagent from acryloyl chloride to 3-chloropropionyl chloride, the base was changed to an aqueous sodium hydroxide solution, and the reaction solvent was changed to a mixture of acetonitrile and water turning out to be biphasic due to the salting-out effect. So, in the final manufacturing process, the installation of acrylamide covalent warhead in futibatinib (**1**) includes a twostep reaction. In the first step compound **26** was treated with 3-chloropropionyl chloride (**29**) to obtain **30**. In the second step (elimination), compound **30** is treated with NaOH, after completion of reaction, the reaction mixture is neutralized by *aq.* HCl and then crystalized by water to afford crude futibatinib (**1**). These two steps, acylation and elimination are biphasic organic–aqueous system reactions of acetonitrile and water. In the final step of purification, the crude futibatinib (**1**) is treated with activated carbon and filtered. Then recrystallization and drying yielded pure futibatinib (**1**).

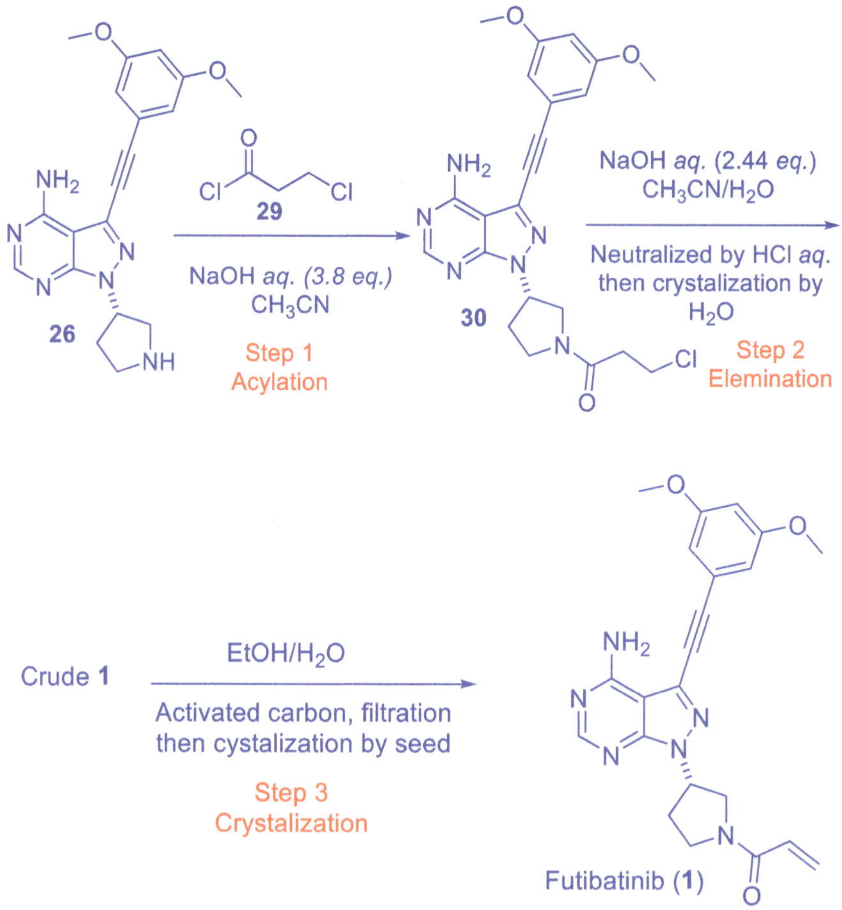

7 Summary

Although there are several FGFR inhibitors in the clinical market and several are currently in clinical development, futibatinib (**1**) has a unique mechanism of action as an irreversible covalent FGFR1–4 inhibitor with potential activity against acquired secondary FGFR kinase domain mutations. In early clinical studies, futibatinib (**1**) demonstrated excellent activity in diverse tumor types harboring various FGFR aberrations. Based on durable and encouraging responses and manageable safety in the phase II FOENIX-CCA2 study futibatinib (**1**) was approved for the treatment of patients with intrahepatic cholangiocarcinoma harboring FGFR2 fusions/rearrangements in September 2022. The excellent clinical results along with the unique irreversible mechanism of action, makes futibatinib (**1**) distinguish from other drugs approved for this target and sets it apart as a leading second-generation FGFR inhibitor. However, further studies are required to evaluate mechanisms of acquired resistance to futibatinib (**1**) and future opportunities of combination therapies with a view to provide hopes for more successful treatment in the near future for the patients with intrahepatic cholangiocarcinoma.

References

1. Javle, M.; King, G.; Spencer, K.; Borad, M. J. Futibatinib, an irreversible FGFR1-4 inhibitor for the treatment of FGFR-aberrant tumors. *The Oncol.* **2023**, *28*, 928–943.
2. Huang, L.; Jiang, S.; Shi, Y. Tyrosine kinase inhibitors for solid tumors in the past 20 years (2001–2020). *J. Hematol. Oncol.* **2020**, *13* (1), 143.
3. Zheng, J.; Zhang, W.; Li, L.; He, Y.; Wei, Y.; Dang, Y.; Nie, S.; Guo, Z. Signaling pathway and small-molecule drug discovery of FGFR: a comprehensive review. *Front. Chem.* **2022**, *10*, 860985.
4. Perera, T. P. S.; Jovcheva, E.; Mevellec, L.; Vialard, J.; DeLange, D.; Verhulst, T.; Paulussen, C.; Van De Ven, K.; King, P.; Freyne, E.; Rees, D. C. Discovery and pharmacological characterization of JNJ-42756493 (Erdafitinib), a functionally selective small-molecule FGFR family inhibitor. *Mol. Cancer Ther.* **2017**, *16* (6), 1010–1020.
5. Markham, A., Erdafitinib: first global approval. *Drugs* **2019**, *79* (9), 1017–1021.
6. Wu, L.; Zhang, C.; He, C.; Qian, D.; Lu, L.; Sun, Y.; Xu, M.; Zhuo, J.; Liu, P. C. C.; Klabe, R.; Wynn, R. Discovery of pemigatinib: a potent and selective fibroblast growth factor receptor (FGFR) inhibitor. *J. Med. Chem.* **2021**, *64* (15), 10666–10679.

7. Kang, C. Infigratinib: first approval. *Drugs* **2021**, *81* (11), 1355–1360.
8. Dai, S.; Zhou, Z.; Chen, Z.; Xu, G.; Chen, Y. Fibroblast growth factor receptors (FGFRs): structures and small molecule inhibitors. *Cells* **2019**, *8* (6), 614.
9. Zhou, W.; Hur, W.; McDermott, U.; Dutt, A.; Xian, W.; Ficarro, S. B.; Zhang, J.; Sharma, S. V.; Brugge, J.; Meyerson, M.; Settleman, J.; Gray, N. S. A structure-guided approach to creating covalent FGFR inhibitors. *Chem. Biol.* **2010**, *17* (3), 285–295.
10. Vijayan, R. S. K.; He, P.; Modi, V.; Duong-Ly, K. C.; Ma, H.; Peterson, J. R.; Dunbrack, R. L., Jr.; Levy, R. M. Conformational analysis of the DFG out kinase motif and biochemical profiling of structurally validated type II inhibitors. *J. Med. Chem.* **2015**, *58* (1), 466–479.
11. Sootome, H.; Fujita, H.; Ito, K.; Ochiiwa, H.; Fujioka, Y.; Ito, K.; Miura, A.; Sagara, T.; Ito, S.; Ohsawa, H.; Otsuki, S.; Funabashi, K.; Yashiro, M.; Matsuo, K.; Yonekura, K.; Hirai, H. Futibatinib is a novel irreversible FGFR 1-4 inhibitor that shows selective antitumor activity against FGFR-deregulated tumors. *Cancer Res.* **2020**, *80* (22), 4986–4997.
12. Kalyukina, M.; Yosaatmadja, Y.; Middleditch, M. J.; Patterson, A. V.; Smaill, J. B.; Squire, C. J. TAS-120 cancer target binding: defining reactivity and revealing the first fibroblast growth factor receptor 1 (FGFR1) irreversible structure. *ChemMedChem* **2019**, *14* (4), 494–500.
13. Ito, S.; Otsuki, S.; Ohsawa, H.; Hirano, A.; Kazuno, H.; Yamashita, S.; Egami, K.; Shibata, Y.; Yamamiya, I.; Yamashita, F.; Kodama, Y.; Funabashi, K.; Kazuno, H.; Komori, T.; Suzuki, S.; Sootome, H.; Hirai, H.; Sagara, T. Discovery of futibatinib: the first covalent FGFR kinase inhibitor in clinical use. *ACS Med. Chem. Lett.* **2023**, *14* (4), 396–404.
14. Qu, L.; Chen, X.; Wei, H.; Guo, M.; Dai, S.; Jiang, L.; Li, J.; Yue, S.; Chen, Z.; Chen, Y. Structural insights into the potency and selectivity of covalent pan-FGFR inhibitors. *Commun. Chem.* **2022**, *5* (1), 5.
15. Syed, Y. Y. Futibatinib: first approval. *Drugs* **2022**, *82* (18), 1737–1743.
16. Sagara, T.; Ito, S.; Otsuki, S.; Sootome, H. Preparation of 3,5-disubstituted alkynylbenzene compounds as FGFR inhibitors. WO2013108809 A1 **2013**.
17. Ochiiwa, H.; Hirai, H. Antitumor drug containing (3,5-disubstituted-phenylethynyl)heterocyclic compound as FGFR inhibitor for intermittent administration and treatment of cancer patients. WO2015008839 A1 **2015**.

18. Abe, Y.; Emori, K. Application of a statistical approach to process development of futibatinib by employing quality-by-design principles. Part 1: identification of critical process parameters for impurities. *Org. Process Res. Dev.* **2022**, *26* (1), 43–55.

8

Pacritinib (Vonjo): A Dual JAK2/IRAK1 Inhibitor for Treating Myelofibrosis

Faridoon and
Guiping Zhang

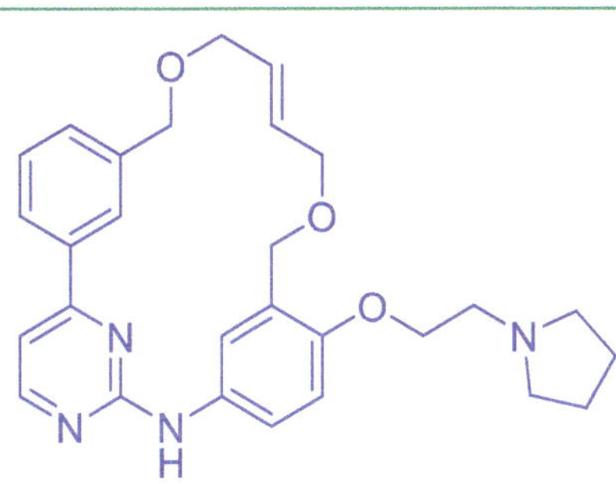

pacritinib (Vonjo, **1**)
S*BIO Pte Ltd's /CTI BioPharma
JAK2 inhibitor, 2022

1 Background

Pacritinib (Vonjo **1**), formerly known as SB1518, is an orally administered, small molecule JAK2/interleukin-1 receptor-associated kinase 1 (IRAK1) inhibitor that does not inhibit JAK1. It was developed by CTI BioPharma/S*BIO Pte. Ltd. for the treatment of conditions such as myelofibrosis and graft-vs-host disease.

Myelofibrosis is a rare type of bone marrow cancer that disrupts the body's normal production of blood cells. It causes extensive scarring in bone marrow, leading to severe anemia that can cause weakness and fatigue. Bone marrow scarring can also cause a low number of blood-clotting cells called platelets, which increases the risk of bleeding. Myelofibrosis often results in an enlarged spleen. Graft-vs-host disease is an immune-mediated condition resulting from a complex interaction between donor and recipient adaptive immunity. It occurs when T cells of donor respond to host histoincompatible

Chemistry and Pharmacology of Drug Discovery, First Edition. Edited by Jie Jack Li.

antigens on the host tissues. In April 2012, CTI BioPharma Corp. (formerly known as Cell Therapeutics Inc., CTI hereafter) and S*BIO Pte. Ltd. entered into an asset purchase agreement under which CTI would acquire worldwide rights to pacritinib from S*BIO. CTI completed the acquisition of pacritinib (**1**) in June 2012. CTI would have sole responsibility for developing and commercializing pacritinib (**1**) worldwide.

Pacritinib (**1**) received its first approval on February 28, 2022, in the United States for the treatment of adults with intermediate- or high-risk primary or secondary (post-polycythemia vera or post-essential thrombocythemia) myelofibrosis with a platelet count below 50 × 109/L. This accelerated approval is based on the demonstration of spleen volume reduction in pacritinib (**1**) recipients in phase III PERSIST-2 trial. However, the continued approval of pacritinib (**1**) is conditional upon an ongoing trial (PACIFICA) being completed to confirm the clinical benefit of this spleen volume reduction.[1] Pacritinib (**1**) is available on the market as 100 mg capsules for oral use. The recommended dosage of pacritinib (**1**) is 200 mg administered twice daily, with or without food. Pacritinib (**1**) is currently under investigation in phase II trial for prostate cancer and phase I/II in graft-vs-host disease and breast cancer. Clinical trials of pacritinib (**1**) in other indications previously started (e.g., colorectal cancer, acute myeloid leukemia, and lymphoma) has now been discontinued.

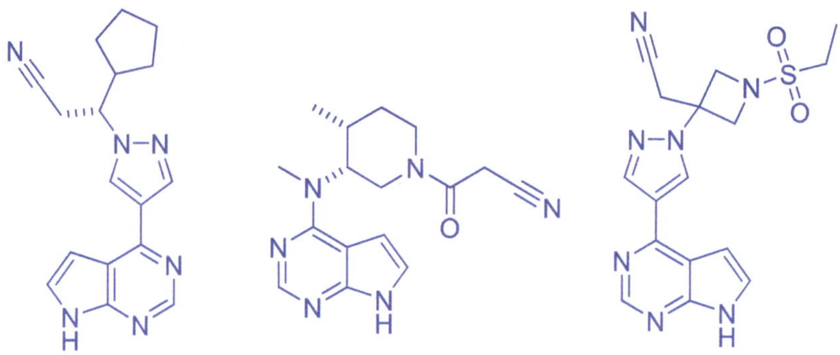

ruxolitinib (**2**)
Incyte/Novartis, 2011
JAK1,2 inhibitor

tofacitinib (**3**)
Pfizer, 2012
JAK1,2,3 inhibitor

baricitinib (**4**)
Eli Lilly, 2017
JAK1,2 inhibitor

After the pioneer work of Meydan et al.[2] in 1996 to develop the small molecular inhibitor of JAK inhibitor with antileukemic activity, great efforts have been made to develop potent and selective JAK inhibitors in the last two decades. Ruxolitinib (**2**) is a potent, orally available, selective inhibitor of both JAK1 and JAK2 of the JAK-STAT signaling pathway developed by Incyte Corp and Novartis. Ruxolitinib (**2**) was the first JAK inhibitor approved by the US Food and Drug Administration (FDA) for the treatment of patients with intermediate or high-risk myelofibrosis, including primary myelofibrosis, post-polycythemia Vera myelofibrosis, and post-essential thrombocythemia myelofibrosis in November 2011. Tofacitinib (**3**) is an orally active small molecule inhibitor of JAK1, 2, and 3 developed by Pfizer, for the treatment of immunological disorders. It was approved for the treatment of rheumatoid arthritis by FDA in 2012.[3] Baricitinib (**4**) is an orally active small-molecule inhibitor of JAK1/2 developed by Eli Lilly. It was approved by the European Medicine Agency (EMA) in 2017 for the treatment of rheumatoid arthritis. Baricitinib (**4**) was approved by the FDA for the treatment of moderate-to-severe rheumatoid arthritis in adults in June 2018.[4]

Upadacitinib (**5**) an orally administered JAK-1 inhibitor, is being developed by AbbVie for the treatment of rheumatoid arthritis was approved by FDA in 2019.[5] In the same year, FDA approved fedratinib (**6**) developed by Impact Biomedicines, Inc. for adults with intermediate-2 or high-risk primary or secondary (post-polycythemia vera or post-essential thrombocythemia) myelofibrosis.[6] Abrocitinib (**7**) is another oral small-molecule inhibitor of JAK-1 being developed by Pfizer for the treatment of moderate-to-severe atopic dermatitis (AD) got FDA approval in September 2020.[7]

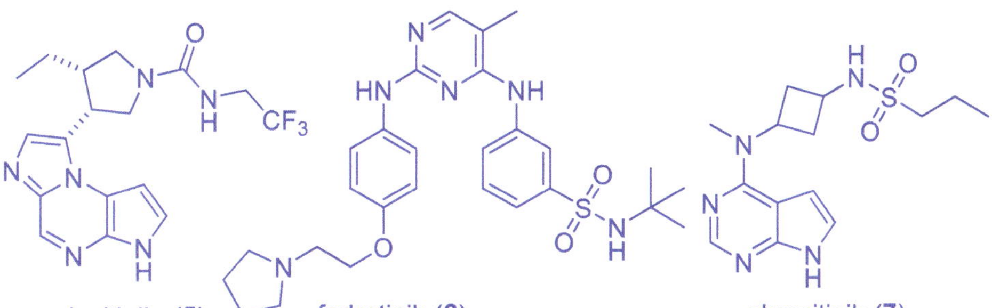

upadacitinib, (**5**)
AbbVie, 2019
JAK1 inhibitor

fedratinib (**6**)
Impact Biomedicines, 2019
JAK2 inhibitor

abrocitinib (**7**)
Pfizer, 2020
JAK1 inhibitor

2 Pharmacology

The JAK/STAT signaling pathway plays a crucial role in many cellular functions, including cell proliferation, stem cell maintenance and differentiation as well as in inflammation and autoimmune diseases.[8] In the reality JAK/STAT signaling is much more complex. The basic and simple version of molecular mechanism underlying cytokine signaling through the JAK/STAT pathway is shown in Figure 1.[9] Cytokines bind to the cytokine receptors and form activated and phosphorylated homo- or heteropolymers complex with their JAK partners. Then the cytosolic DNA-binding STAT proteins bind to the receptor-JAK complex and are themselves phosphorylated by JAK family members. Upon phosphorylation, phosphorylated STAT proteins form homo- and heteropolymers which then enter the nucleus. Phosphorylated STAT dimer transcription factors then bind specific DNA binding sites regulating gene transcription and cellular function.

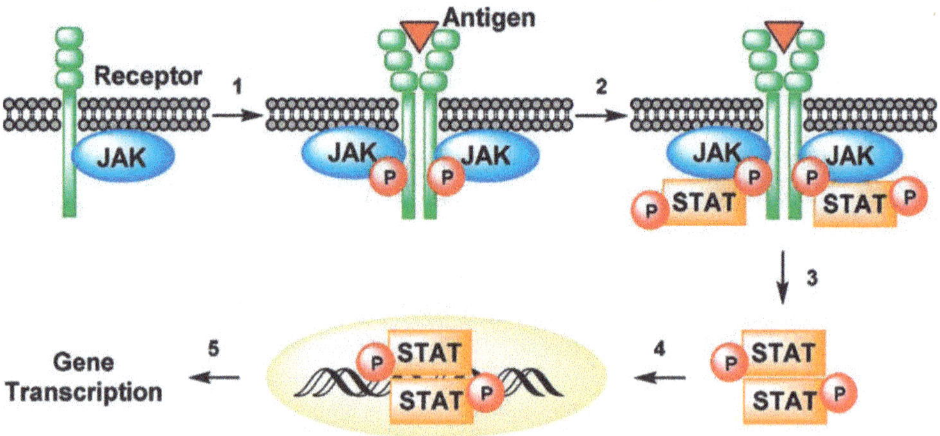

Figure 1. JAK/STAT pathway. Step 1: Cytokine binding, complex formation, activation, and phosphorylation of JAK. Step 2: Recruitment and phosphorylation of STAT. Step 3: Phosphorylated STAT (p-STAT) dimerization. Step 4: Nuclear translocation and DNA binding of p-STAT dimer. Step 5: Gene transcription. Source: Bryan and Rajapaksa [9]/with permission of American Chemical Society

The JAK family includes four members JAK1, JAK2, JAK3, and TY2K. JAKs have seven homology domains, JH1–JH7 (Figure 2). Starting from the C-terminal, JH1 is the first domain, also known as the kinase domain. JH1 has the kinase activity and phosphorylates a substrate. JH2 is a pseudokinase domain that is structurally similar to the kinase domain but has no kinase activity. It regulates the activity of the kinase domain. JH3 and one-half of JH4 constitutes the Src-homology SH2 domain, while the combination of one-half of JH4, JH5, JH6, and JH7 constitutes the FERM domain. These two domains, SH2 and FERM mainly regulate the binding of JAK and cytokine-receptor membrane-proximal box1/2 regions.[10]

Among these kinases, JAK1 plays a crucial role in allergic rhinitis, pruritic dermatitis, inflammatory bowel disease, and asthma. Small molecule inhibitors of JAK1 have proved efficacy in the treatment of these diseases.[11–13] Furthermore, several small molecules with JAK1 and JAK2 inhibitory activity have also provided therapeutic benefits in the treatment of psoriasis, rheumatoid arthritis, and pruritis. Besides this, several JAK3 selective inhibitors have been evaluated for their efficacy in the treatment of rheumatoid arthritis. Moreover, selective inhibitors of TYK2 may also be useful in the treatment of autoimmune diseases.[14] Excessive activation of JAKs has also been reported in different types of cancer. The JAK/STAT3 pathway plays an important role in the proliferation and angiogenesis of solid tumors.[15]

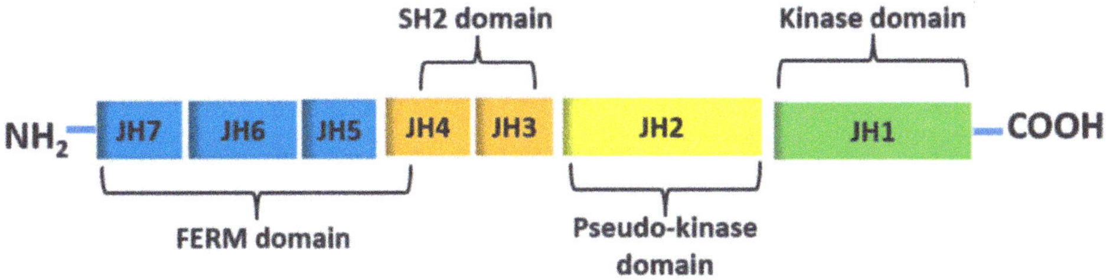

Figure 2. Domain structure of JAKs

Pacritinib (**1**) is thought to exert clinical efficacy by inhibiting two distinct pathways, JAK/STAT and TLR/Myddosome/IRAK1, leading to suppression of NFκB and downstream inflammatory cytokine cascade, reduction in splenomegaly, and myelofibrosis symptom control.[16] Pacritinib (**1**) is a potent competitive reversible kinase inhibitor with high specificity for JAK2 (IC_{50} = 23 nM), JAK2^{V617F} (IC_{50} = 19 nM). Relative to JAK2, pacritinib (**1**) is 2-fold less potent against TYK2 (IC_{50} = 50 nM), 23-fold less potent against JAK3 (IC_{50} = 520 nM) and 56-fold less potent against JAK1 (IC_{50} = 1280 nM). Pacritinib (**1**) also a potent inhibitor of FLT3 (IC_{50} = 22 nM) and its mutant FLT3^{D835Y} (IC_{50} = 6 nM), which impact in myelofibrosis is unknown (Table 1).[17] Pacritinib (**1**) is significantly more effective than best available therapy (BAT), including ruxolitinib (**2**), for reducing splenomegaly and symptoms in patients with myelofibrosis and thrombocytopenia.

Pacritinib (**1**) is orally administered and readily absorbed regardless of food intake, with a mean terminal elimination half-life of 27.7 h. It is predominantly metabolized by the CYP3A4 isozyme and biliary excretion. Pacritinib (**1**) achieves maximum plasma concentration (C_{max}) at approximately 4–5 h post dose and has an apparent volume of distribution of 229 L with 98.8% plasma protein binding (PBB). Pacritinib (**1**) is generally metabolized via CYP3A4, which will be discussed in detail later.[18]

Table 1. *In vitro* kinase spectrum of pacritinib (**1**)

Compound	Kinase	IC_{50} (nM)	Selectivity vs JAK2
	JAK1	1280	56
	JAK2	23	1.0
	JAK2^{V617F}	19	0.8
Pacritinib (**1**)	JAK3	520	23
	TYK2	50	2.2
	FLT3	22	1.0
	FLT3^{B835Y}	6	0.3

3 Structure–Activity Relationship (SAR)

From in-house library screening, S*BIO Pte Ltd's[19] identified compound **8** that showed broad kinase inhibition with reasonable activity in cell lines tested. To avoid any intellectual property contradiction, they connected the open ends of **8**, to form macrocycle **9**. Interestingly, they found that the binding mode to the kinase hinge region is not compromised. A study of various linkers with and without R1 groups demonstrated that these templates inhibited JAK2/FLT3 with selectivity against JAK1/3 is shown in Table 2. They found that the most potent JAK2 linkers were 8 atoms in length with symmetrical dibenzylic linkers, as employed in compounds **10e** and **10f**, were quite potent against JAK2 with good selectivity over JAK1 and 3. **10e** inhibit JAK2 with IC_{50} of 70 nM and is 27- and 17-fold selective against JAK1 and JAK3, respectively. They decided to focus on the progression of the most potent dibenzylic compound **10e** and **10f**. One concern of this high cLogP compounds was the very low solubility. They found the solubility of **10b** and **10d** less than 10 µg/mL in PBS buffered at pH 7.0. Therefore, they search for locations to install a solubilizing group in this template to improve solubility.

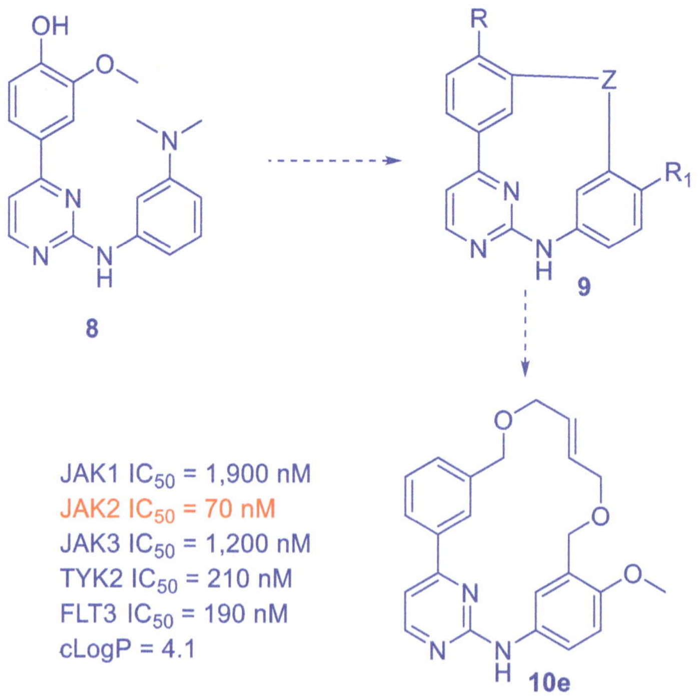

Table 2. Search for a suitable linker for selectivity toward JAK2

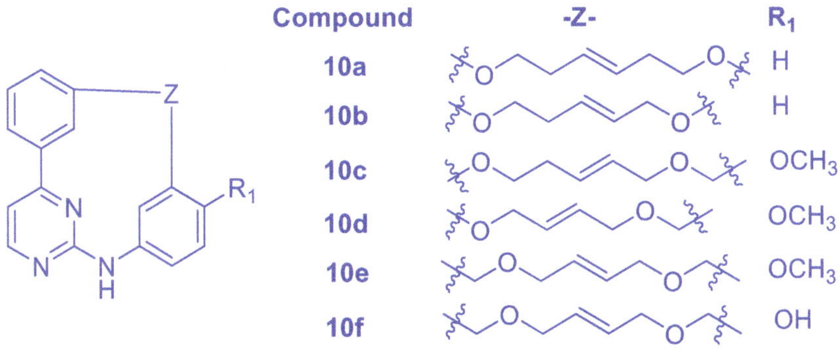

Compd	JAK2 IC50 (nM)	JAK1 IC50 (nM)	JAK3 IC50 (nM)	TYK2 IC50 (nM)	FLT3 IC50 (nM)	cLogP
10a	1,200	>10,000	>10,000	5200	1100	4.7
10b	230	>10,000	>10,000	70	880	4.3
10c	130	3000	ND	150	310	4.6
10d	260	5800	ND	120	78	4.0
10e	70	1900	1200	210	190	4.1
10f	53	400	1100	170	170	3.4

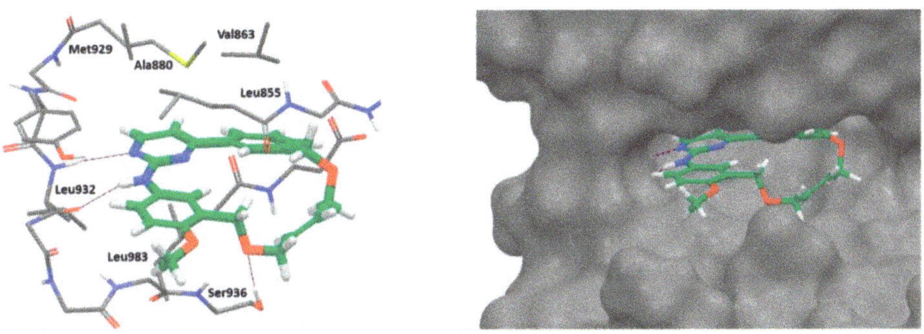

Figure 3. Compound **10e** docked into the ATP-binding site of JAK2. Source: William et al. [19]/with permission of American Chemical Society

Docking of **10e** into the JAK2 ATP binding pocket (Figure 3) showed that the macrocycle structure fills out the available space in the binding site quite well with the opportunities for building additional interactions with the protein from various sites. The docking pose of **10e** showed that the R_1 substituent points directly toward a channel out to the solvent that may offer potential for installation of a basic center to increase the solubility.

Based on their docking results, they explored both the R_1 and R_2 positions on the benzylic ring. They found that a range of N and O-substituents were very well tolerated. Though sterically large groups like morpholine and piperazines were potent and selective for JAK2, but sparingly soluble in aqueous media. They used side chain like hydroxyethyl piperazine, or less sterically open chain analogues and amide (structures not mentioned here). However, improved solubility with good potency was achieved with oxygen-linked substituents. For example, compounds **1** and **11a,b**, containing an aminoethyl ether side-chain, were found uniformly potent against JAK2 and selective against JAK1/3. The direct R_2 analogue of **1**, **11c**, was also potent against JAK2 but with sub-micromolar CDK2 activity. Morpholine analogue **11d**, however, was less CDK2 potent but suffered a 5-fold reduction in JAK2 potency over the most potent R_1 compounds (Table 3).

Table 3. SAR of optimization solubility

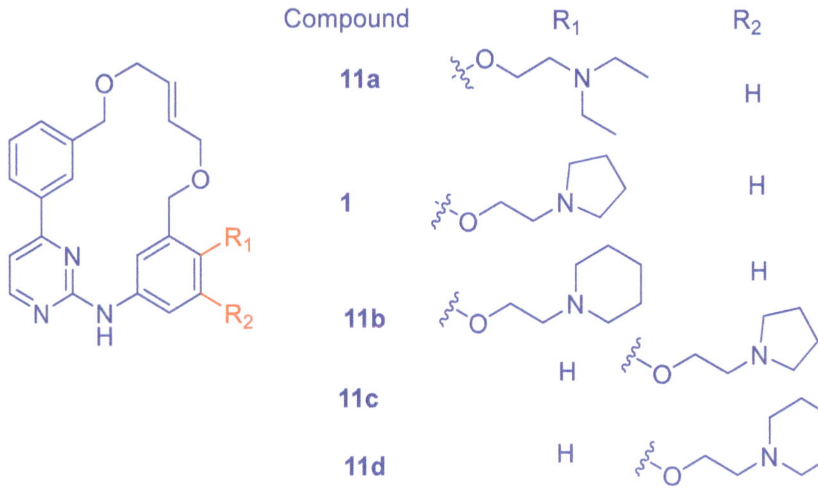

Compd	JAK2 IC_{50} (nM)	JAK1 IC_{50} (nM)	JAK3 IC_{50} (nM)	TYK2 IC_{50} (nM)	FLT3 IC_{50} (nM)	SDK2 IC_{50} (nM)	Solubility (μg/mL)
11a	24	3500	620	79	29	4700	147
1	23	1280	520	50	22	3900	>150
11b	15	950	750	84	34	2700	ND
11c	48	2100	830	80	22	570	145
11d	96	2400	3300	81	19	2700	40

In search of better drug candidate compound, they further explored the effect of small groups substitution on the aromatic rings (Table 3). Indeed 5′-methyl substituted pyrimidine **12b** exhibited 4-fold higher potency for JAK2 (IC_{50} 6 nM) as compared to **1** with good solubility. However, this compound was compromised somewhat by its selectivity against JAK3 which had decreased by nearly15-fold. When 5-methyl group was replaced by electron-withdrawing fluoro **12c**, it reduced JAK2 activity by 3-fold, which confirming that hydrophobic groups are ultimately preferred. Substitution at R_4 **12d** with methoxy is unfavorable for JAK family activity. On the other hand, substitution at R_5 **12e** with electron-withdrawing fluoro is well tolerated but reduces selectivity over JAK1 and solubility. Substituting methoxy at the R_6 position in **12f** resulted in excellent potency and selectivity for JAK2 over JAK1 (Table 4). Though the effects of adding JAK1 activity to a JAK2 inhibitor are poorly understood. Therefore, they prioritized more selective compounds in the expectation that they would have reduced off-target toxicity. Addition of an additional substituent to the right-hand side aniline ring of **1** resulted in electron rich **12a** which did not compromise JAK2 activity or selectivity and showed improved FLT3 activity as compared with **1**.

Table 4. SAR Exploration of aromatic ring substitutions with small groups

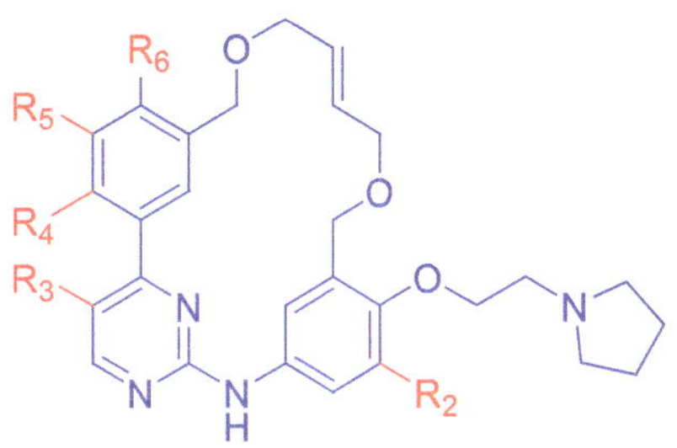

Compound	R_2	R_3	R_4	R_5	R_6
12a	OCH_3	H	H	H	H
12b	H	CH_3	H	H	H
12c	H	F	H	H	H
12d	H	H	CH_3	H	H
12e	H	H	H	F	H
12f	H	H	H	H	OCH_3
12g	H	H	H	H	F

Compd	JAK2, IC_{50} (nM)	JAK1, IC_{50} (nM)	JAK3, IC_{50} (nM)	TYK2, IC_{50} (nM)	FLT3, IC_{50} (nM)	SDK2, IC_{50} (nM)	Solubility μg/mL
12a	36	1700	ND	170	6	3400	154
12b	7	1000	89	57	19	>10	178
12c	17	830	1000	140	15	1550	ND
12d	330	4900	7200	620	12	>10	ND
12e	24	380	ND	36	8	2000	2.6
12f	19	>10	890	180	92	>10	60.8
12g	25	4600	720	ND	40	3250	ND

For further *in vitro* study, they chosed compounds **1**, **11a**, **11b**, **12b,** and **12f**. **1** inhibited the JAK2-mediated production of p-STAT5 and p-STAT3 dose dependently in Ba/F3 cells. Although, **12b** exhibiting desirable single digit nanomolar potency toward JAK2 but inhibit CYP3A4 with IC_{50} of 2.5 μM. Similarly **12b** also inhibit CYP3A4 with IC_{50} of 0.37 μM. On the other hand, **1** and **12f** were much less active against CYP3A4 and both compounds had good selectivity and microsomal stability. Although, compound **12f** has excellent enzyme selectivity, was less active against FLT3. Compound **1** shows an overall balanced profile meeting all target criteria (Table 5).[19] It was also active against the V617F mutant of JAK2 with IC_{50} = 19 nM and the D835Y mutant of FLT3 with IC_{50} = 6 nM. Its protein kinase selectivity was evaluated by testing against more than 50 other protein kinases covering all major families of the human protein kinome. Thus, they selected **1** was their drug candidate compound.

Table 5. ADME profile of pacritinib (**1**)

	Properties	Values
Pacritinib (**1**)	HLM ($t_{1/2}$, min)	>60
	DLM ($t_{1/2}$, min)	41
	RLM ($t_{1/2}$, min)	18
	MLM ($t_{1/2}$, min)	22
	Human CYP inhibition IC_{50} (μM)	>5
	PPB (%) in human	99.88
	PPB (%) in dog	99.63
	PPB (%) in mouse	99.41
	Permeability ($P_{app, A\text{-}B}$, $\times 10^{-6}$ cm/s)	16

4 Pharmacokinetics and Drug Metabolism

It has been noticed that increases in pacritinib (**1**) exposure are less than dose-proportional over a dose range of 100–600 mg in patients with advanced myeloid malignancies.[20] When pacritinib (**1**) 200 mg is administered twice daily in patients with myelofibrosis, the mean peak concentration (C_{max}) of pacritinib (**1**) at steady state is 8.4 mg/L. Generally, pacritinib (**1**)'s C_{max} is reached after ~ 4–5 h dose administration. Administration of pacritinib (**1**) with food does not significantly impact its pharmacokinetics. A steady state is achieved within a week of commencing twice-daily administration of pacritinib (**1**) and its accumulation is 386%. At steady state, the median apparent volume of distribution of pacritinib (**1**) is 229 L. As mentioned in Table 4, pacritinib (**1**) is highly bound to plasma proteins (~ 99%). Pacritinib (**1**) pharmacokinetics are not significantly affected by age, sex, race, or body weight.[1]

The metabolic study of pacritinib (**1**) revealed that it is mostly metabolized by CYP3A4. The metabolism of pacritinib (**1**), investigated by Jayaraman et al. revealed the formation of four metabolites by liver microsomes in both humans and mice (Figure 4).[18] The identified metabolites include two oxidized metabolites formed by oxidation of the pyrrole ring (M1) and pyrrole nitrogen (M3). While the third metabolite (M2) formed by O-dealkylation of the pyrrole-bearing side chain and the fourth metabolite (M4) was formed by reduction of the double bond of the linker of pacritinib (**1**). Pacritinib (**1**) has a mean apparent steady-state clearance of 2.09 L/h and a mean effective half-life of 27.7 h. When a single dose of radiolabeled pacritinib (**1**) 400 mg was orally administered in healthy adults, the radioactivity was mostly recovered in feces (87%, none as unchanged drug). While a small proportion of the drug administered was recovered in urine (6%, with 0.12% as unchanged drug).

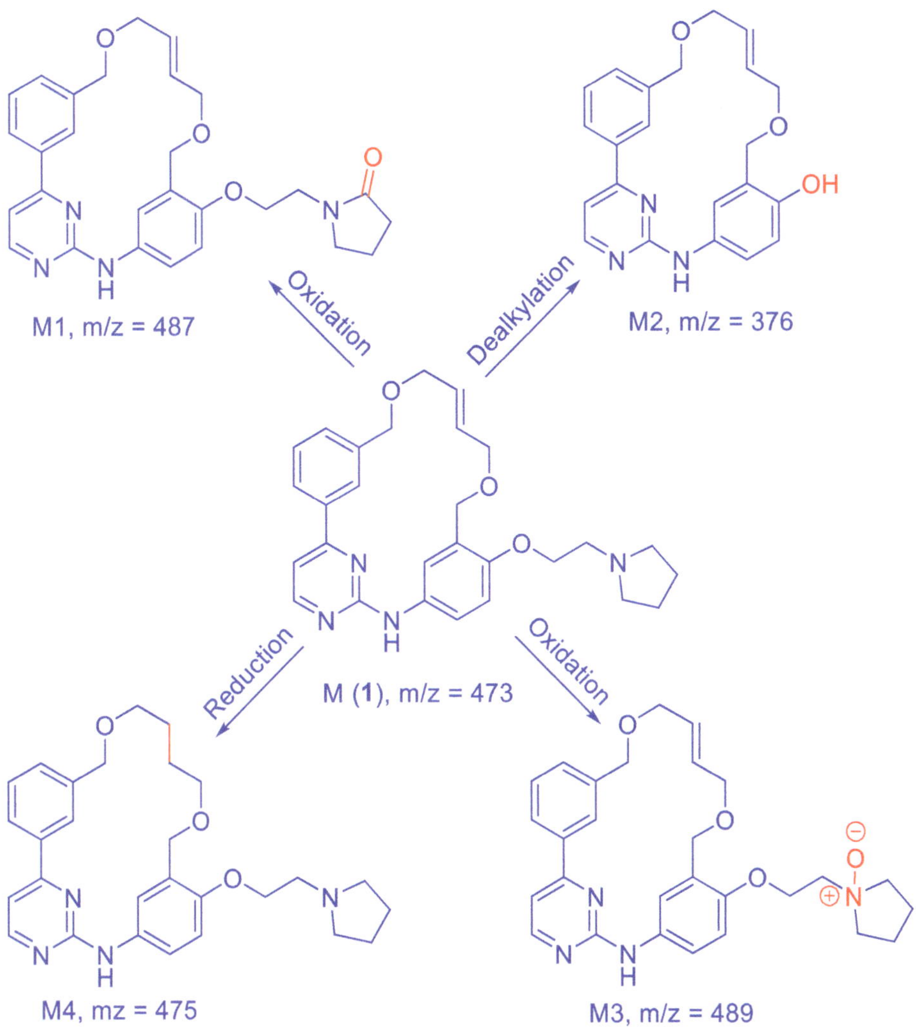

Figure 4. Schematic representation of molecular ionic structures of pacritinib (**1**) and its metabolites

5 Efficacy and Safety

In the preclinical study, pacritinib (**1**) potently blocked JAK/STAT signaling pathway, induced apoptosis, and effectively abridged neoplastic cell proliferation in a dose-dependent manner. Pacritinib (**1**) recapitulated these results in a JAK2 V617F-dependent SET-2 and BaF3 xenograft model with resolution of hepatosplenomegaly in the absence of hematologic toxicities such as anemia or thrombocytopenia or leukopenia and prolonged survival.

In phase I/II trial (NCT00719836), pacritinib (**1**) demonstrated promising clinical activity in patients with myelofibrosis and other advanced myeloid malignancies.[20] In phase I dose-escalation component, adults with advanced myeloid malignancies were treated with once-daily (qd) pacritinib (**1**) 100–600 mg ($n = 43$). Among the several dose levels tested, there was a less than proportionate increase in systemic exposure at doses from 100 to 400 mg, with negligible increase in exposure >400 mg daily. Therefore, 400 mg daily (*qd*) was identified as the recommended phase II dose for further evaluation.[21]

In the phase I study, the majority of the patients (86%) experienced clinical benefit. In patients with myelofibrosis, the clinical benefit rate was 94% (vs 43% in patients with acute myeloid leukemia). In the phase II study pacritinib (**1**) showed promising result: 23.5% of evaluable patients achieved a spleen volume reduction of ≥ 35% by MRI from baseline to week 24.[20] However, phase III studies of pacritinib (**1**), PERSIST-1 and PERSIST-2, were placed on hold by the FDA in February 2016 due to due to concerns over interim survival results, bleeding, and cardiovascular events.[22] This hold was removed by FDA on January 5, 2017, upon review of the final phase III PERSIST-1 data, final data from the phase III PERSIST-2 study, and planned dose comparison protocol in patients with failure of prior JAK2-directed therapy.[23]

Phase III PERSIST-2 randomized 311 patients with baseline thrombocytopenia (platelet count ≤ 100 × 109/L) to pacritinib (**1**) 200 mg twice daily, 400 mg daily, and BAT (45% received ruxolitinib). Combined analysis of patients on both pacritinib (**1**) arms found improved rates of ≥35% spleen reduction compared to BAT (18% vs 3%, $p = 0.001$) based on computed tomography (CT) or magnetic resonance imaging (MRI). Pacritinib (**1**) 200 mg twice daily was most effective at improving hemoglobin levels and reducing the transfusion burden. Plans are currently underway to launch the randomized phase III PACIFICA trial in patients with severe thrombocytopenia (platelet count ≤ 50 × 109/L) comparing pacritinib (**1**) 200 mg daily vs physician's choice.

The most common adverse reactions of pacritinib (**1**) in the clinical trials were diarrhea, thrombocytopenia, anemia, nausea, and peripheral enema. The most frequent serious adverse reactions included pneumonia, cardiac failure, disease progression, pyrexia, and squamous cell carcinoma of skin.

6 Synthesis

The synthesis of pacritinib (**1**) was first disclosed in the patent published in 2007.[24] Later in 2011, the modified synthesis of pacritinib (**1**) was published in the *J. Med. Chem.* by William et al. from the same company.[19] According to this synthetic route, fragment **16** was synthesized by Suzuki coupling of commercially available 2,4-dichloropyrimidine **13** and boronic acid **14** affording the corresponding biaryl alcohol **15** which was then subjected to the reaction with allyl bromide in the presence of tetrabutylammonium hydrogensulfate (TBAHSO$_4$) to give the diether **16**.

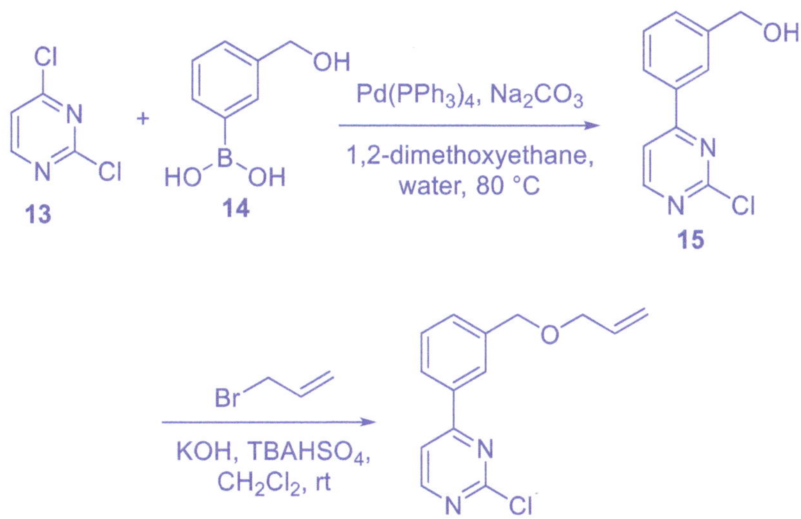

13 **14** **15** **16**

The second fragment **21** was synthesized starts with the nucleophilic attack of 2-hydroxy-5-nitrobenzaldehyde **17** to 1-bromo-2-chloroethane in the presence of potassium carbonate to afford the aromatic ether **18**. The reduction of the aldehydic group of compound **18** afforded the benzylic alcohol **19** which was then subjected to the reaction with allyl bromide in the presence of tetrabutylammonium hydrogen sulfate (TBAHSO$_4$) to give the diether **20**. The nitro group of diether **20** was reduced to amine under iron/ammonium chloride conditions to produce the amine **21**.

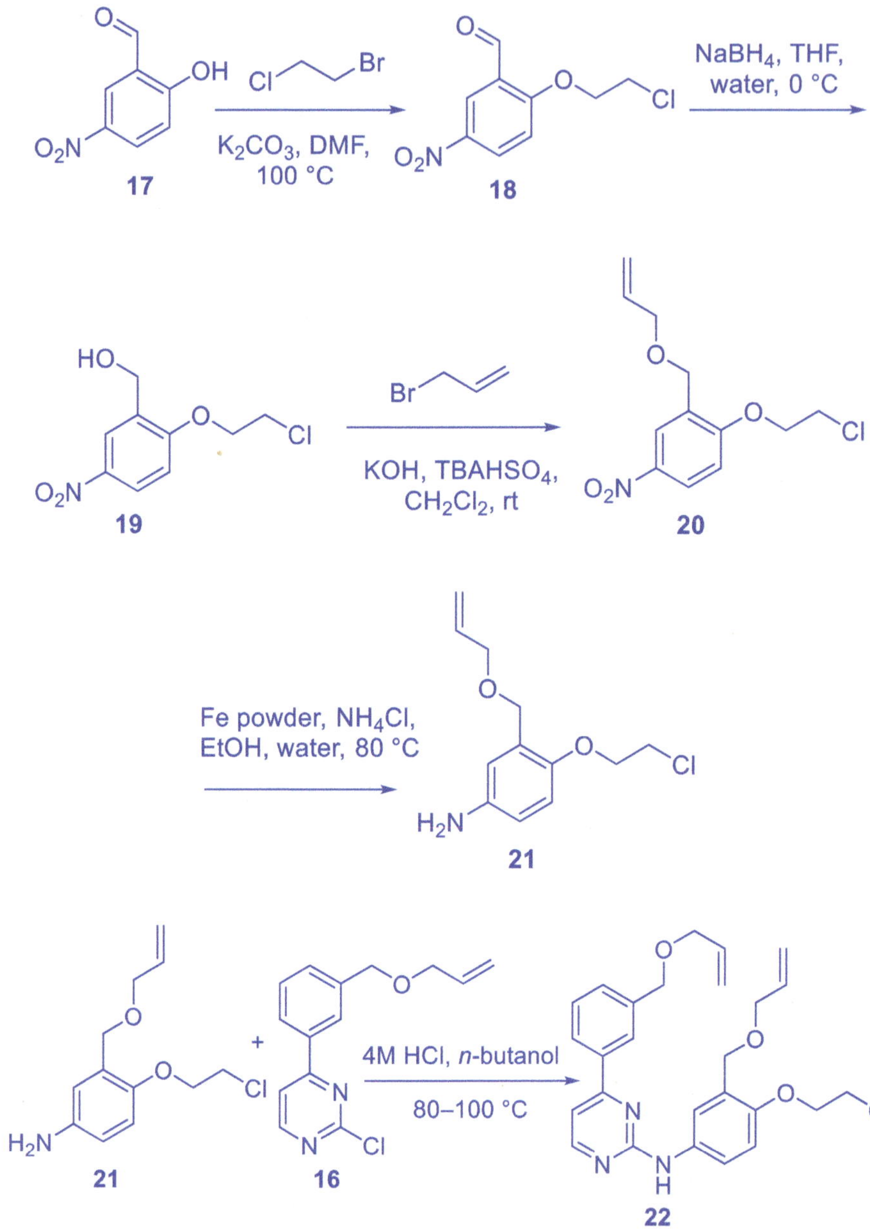

After the nucleophilic aromatic substitution of compound **21** with compound **16** to get compound **22**, the ring-closing metathesis (RCM) afforded the macrocycle **23** via ruthenium-based approach using the Zhan catalyst 1B. The macrocycle **23** was obtained as inseparable mixtures of approximately 85:15 *E*/*Z* geometry. Compound **24** was finally obtained via nucleophilic addition of pyrrolidine to the chloride of macrocycle **23** in excellent yield and used directly for biological study as reported in the journal of medicinal chemistry.[19] In 2018, a patent[25] claimed a method of separating the isomers applying an ion exchange medium comprising silver ions to give pure pacritinib (**1**).

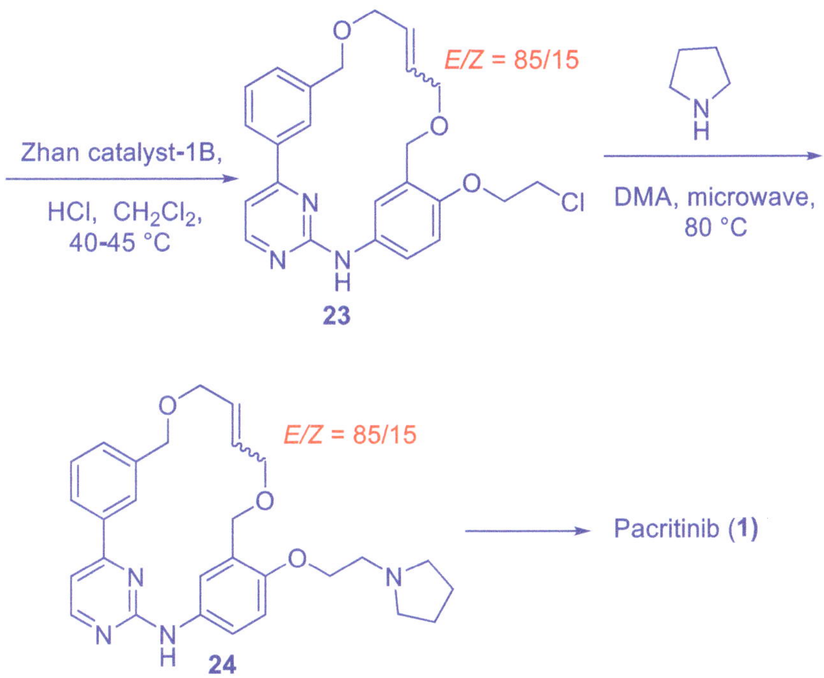

Later in 2017, the selectivity of the RCM of diene **22** was further improved by Shen et al.[26] using their molybdenum monoaryloxide pyrrolide complex catalyst (Mo-1). They initially examined the RCM of boryl-diene **25**. Subjection of **25** to 10 mol% **Mo-1** at ambient temperature for 12 h resulted in an inefficient reaction (25% conversion; 17% product **26**) with higher *E* selectivity (93/7 *E*/*Z*). When the same reaction was performed with an equivalent of $B(C_6F_5)_3$ to avoid catalyst deactivation, product **26** could be isolated in 60% yield and 95/5 *E*/*Z* selectivity.

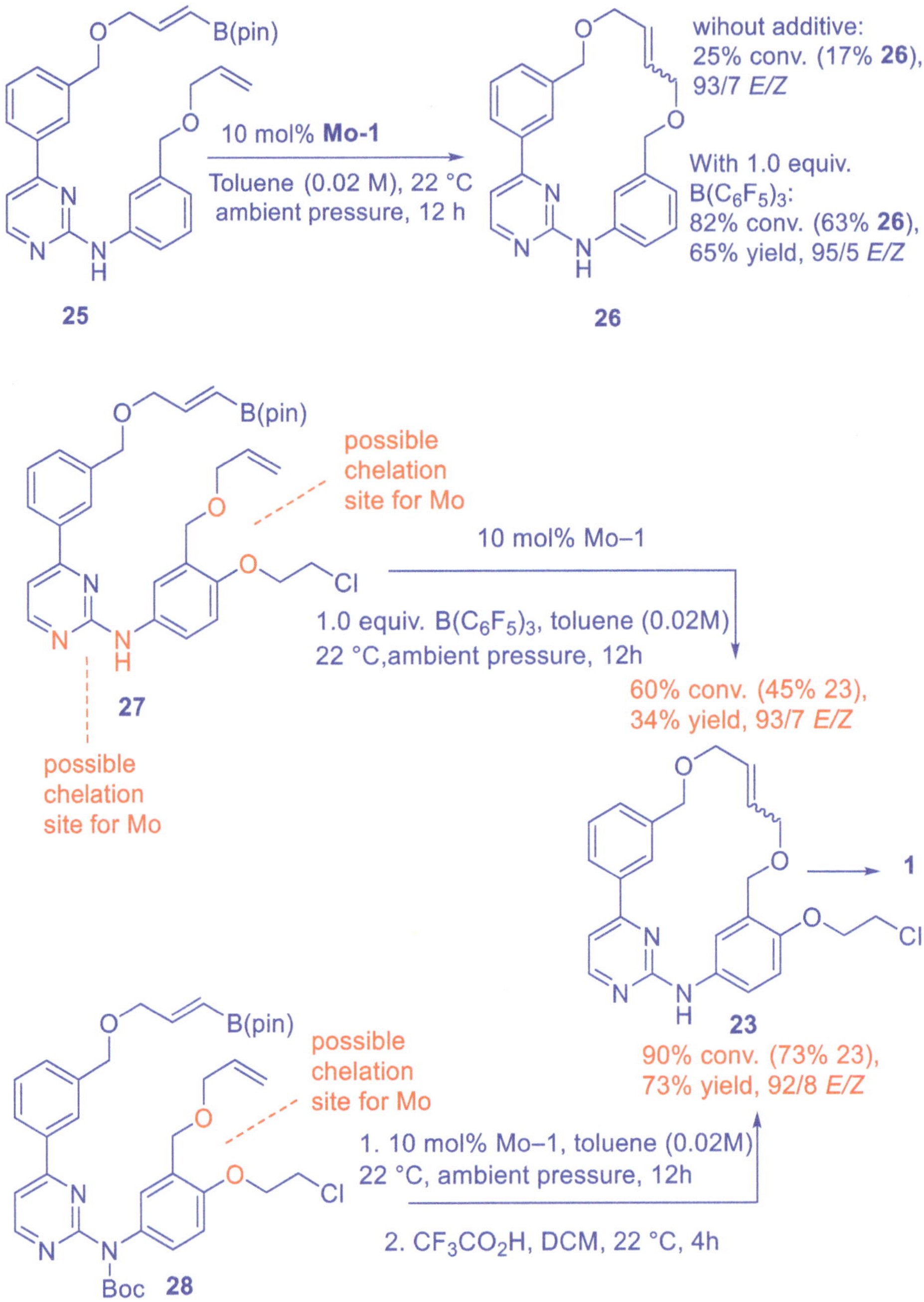

The same procedure with triether **27** is highly stereoselective, but the yield is lower (34%). Triether **27** contains an additional di-ether fragment that might coordinate to the Lewis acidic molybdenum center to cause reduced catalyst activity; this was a concern because the basicity of the new ether oxygen might be enhanced by the *para* amino group. Therefore, they investigated the Boc-protected variant **28** assuming that this modification would more firmly diminish the Lewis basicity of the pyrimidine and the bis(ether) moiety. Thus subjecting **28** to the same reaction conditions but without $B(C_6F_5)_3$, and then removing the protecting group, resulted in product **23** at an overall yield of 73% and 92/8 *E*/*Z* selectivity.

7 Summary

Though the phase III studies of pacritinib (**1**) were placed on hold by the FDA in February 2016 due to concerns over interim survival results, bleeding, and cardiovascular events. Its developers did not lost courage and came with more detailed and comprehensive data to satisfy FDA to remove the hold on its clinical trial and then pacritinib (**1**) received its first accelerated conditional approval on February 28, 2022, in the United States for the treatment of thrombocythemia and myelofibrosis. This accelerated approval of pacritinib (**1**) will facilitate robust collection of real-world data that would further help to understand the fullest extent of efficacy and safety of pacritinib (**1**) in patients with myelofibrosis and severe thrombocytopenia. In patients with myelofibrosis and thrombocytopenia, including those with prior anti-JAK therapy, pacritinib (**1**) twice daily is more effective than BAT, including ruxolitinib (**2**), for reducing splenomegaly and symptoms.

References

1. Lamb, Y. N. Pacritinib: first approval. *Drugs* **2022**, *82*, 831–838.
2. Meydan, N.; Grunberger, T.; Dadi, H.; Shahar, M.; Arpaia, E.; Lapidot, Z.; Leeder, J. S.; Freedman, M.; Cohen, A.; Gazit, A.; et al. Inhibition of acute lymphoblastic leukaemia by a JAK-2 inhibitor. *Nature* **1996**, *379*, 645–648.
3. Coricello, A.; Mesiti, F.; Lupia, A.; Maruca, A.; Alcaro, S. Inside perspective of the synthetic and computational toolbox of JAK inhibitors: Recent updates. *Molecules* **2020**, *25*, 3321.
4. Markham, A. Baricitinib: first global approval. *Drugs* **2017**, *77*, 697–704.
5. Duggan, S.; Keam, S. J. Upadacitinib: first approval. *Drugs* **2019**, *79*, 1819–1828.
6. Blair, H. A. Fedratinib: frst approval. *Drugs* **2019**, *79*, 1719–1725.

7. Deeks, E. D.; Duggan, S. Abrocitinib: first approval. *Drugs* **2021**, *81*, 2149–2157.

8. Hin Tang, J. J.; Hao Thng, D. K.; Lim, J. J.; Toh, T. B. JAK/STAT signaling in hepatocellular carcinoma. *Hepatol. Oncol.* **2020**, *7*, HEP18.

9. Bryan M. C.; Rajapaksa N. S. Kinase inhibitors for the treatment of immunological disorders: recent advances. *J. Med. Chem.* **2018**, *61*, 9030–9058.

10. Hu, X.; Li. J.; Fu, M.; Zhao, X.; Wang, W. The JAK/STAT signaling pathway: from bench to clinic. *Signal Transduction Targeted Ther.* **2021** *6*, 402.

11. Wills-Karp, M.; Luyimbazi, J.; Xu, X.; Schofield, B.; Neben, T. Y.; Karp, C.L.; Donaldson, D.D. Interleukin-13: central mediator of allergic asthma. *Science* **1998**, *282*, 2258–2261.

12. Virtanen, A.T.; Haikarainen, T.; Raivola, J.; Silvennoinen, O. Selective JAKinibs: prospects in inflammatory and autoimmune diseases. *BioDrugs* **2019**, *33*, 15–32.

13. Zak, M.; Hanan, E. J.; Lupardus, P.; Brown, D. G.; Robinson, C.; Siu, M.; Lyssikatos, J. P.; Romero, F. A.; Zhao, G.; Kellar, T.; et al. Discovery of a class of highly potent Janus Kinase 1/2 (JAK1/2) inhibitors demonstrating effective cell-based blockade of IL-13 signaling. *Bioorg. Med. Chem. Lett.* **2019**, *29*, 1522–1531.

14. Schwartz, D. M.; Kanno, Y.; Villarino, A.; Ward, M.; Gadina, M.; O'Shea, J. J. JAK inhibition as a therapeutic strategy for immune and inflammatory diseases. *Nat. Rev. Drug Discovery* **2017**, *16*, 843–862.

15. Buchert, M.; Burns, C. J.; Ernst, M. Targeting JAK kinase in solid tumors: emerging opportunities and challenges. *Oncogene* **2016**, *35*, 939–951.

16. Mascarenhas J. Pacritinib for the treatment of patients with myelofibrosis and thrombocytopenia. *Expert Rev. Hematol.* **2022**, *15*, 671–684.

17. Hart S.; Goh, K. C.; Novotny-Diermayr, V.; Hu, C. Y.; Hentze, H.; Tan, Y. C.; Madan, B.; Amalini, C.; Loh, Y. K.; Ong, L. C.; William, A. D.; Lee, A.; Poulsen, A.; Jayaraman, R.; Ong, K. H.; Ethirajulu, K.; Dymock, B. W.; Wood, J. W. SB1518, a novel macrocyclic pyrimidine-based JAK2 inhibitor for the treatment of myeloid and lymphoid malignancies. *Leukemia* **2011**, *25*, 1751–1759.

18. Jayaraman, R.; Pasha, M. K.; Williams, A.; Goh, K. C.; Ethirajulu, K. Metabolism and disposition of pacritinib (SB1518), an orally active

Janus kinase 2 inhibitor in preclinical species and humans. *Drug Metab. Lett.* **2015**, *9*, 28–47.

19. William, A. D.; Lee, A. C.; Blanchard, S.; Poulsen, A.; Teo, E. L.; Nagaraj, H.; Tan, E.; Chen, D.; Williams, M.; Sun, E. T.; Goh, K. C.; Ong, W. C.; Goh, S. K.; Hart, S.; Jayaraman, R.; Pasha, M. K.; Ethirajulu, K.; Wood, J. M.; Dymock, B. W. Discovery of the macrocycle 11-(2-pyrrolidin-1-yl-ethoxy)-14,19-dioxa-5,7,26-triaza-tetracyclo[19.3.1.1(2,6).1(8,12)]heptacosa-1(25),2(26),3,5,8,10,12(27),16,21,23-decaene (SB1518), a potent Janus kinase 2/fms-like tyrosine kinase-3 (JAK2/FLT3) inhibitor for the treatment of myelofibrosis and lymphoma. *J. Med. Chem.* **2011**, *54*, 4638–4658.

20. Verstovsek, S.; Odenike, O.; Singer, J.W.; Granston, T.; Al-Fayoumi, S.; Deeg, H. J. Phase 1/2 study of pacritinib, a next generation JAK2/FLT3 inhibitor, in myelofibrosis or other myeloid malignancies. *J. Hematol. Oncol.* **2016**, *9*, 137.

21. Verstovsek, S.; Odenike, O.; Scott, B.; Estrov, Z.; Cortes, J.; Thomas, D. A.; Wood, J.; Ethirajulu, K.; Lowe, A.; Zhu, H. J.; Kantarjian, H.; Deeg, H. J., Phase I dose-escalation trial of SB1518, a novel JAK2/FLT3 inhibitor, in acute and chronic myeloid diseases, including primary or post-essential thrombocythemia/polycythemia vera myelofibrosis. *Blood* **2009**, *114*, 3905–3905.

22. Shantzer, L.; Berger, K.; Pu, J. J., Primary myelofibrosis and its targeted therapy. *Ann. Hematol.* **2017,** *96*, 531–535.

23. Garmezy, B.; Schaefer, J. K.; Mercer, J.; Talpaz, M., A provider's guide to primary myelofibrosis: pathophysiology, diagnosis, and management. *Blood Rev.* **2021,** *45*, 100691.

24. Blanchard, S.; Lee, C. H. A.; Nagaraj, H. K. M.; Poulsen, A. S.; Eric, T.; Tan, Y. L. E.; William, A. D. Preparation of oxygen linked pyrimidine macrocyclic derivatives as antiproliferative agents. WO 2007058627 A1, **2007**.

25. Hagooly, Y.; Laskavy, A.; Yosef G. T.; Ben-Ari, E. A process for obtaining purified unsaturated macrocyclic compounds. WO 2018104856 A1, **2018**.

26. Shen, X.; Nguyen, T. T.; Koh, M. J.; Xu, D.; Speed, A. W. H.; Speed, A. W., Schrock, R. R. and Hoveyda, A. H. Kinetically E-selective macrocyclic ring-closing metathesis. *Nature* **2017**, *541*, 380–385.

9

Tucatinib (Tukysa): An Oral, Selective HER2 Inhibitor for the Treatment of HER2-Positive Solid Tumors

Fengtao Zhou and
Ke Ding

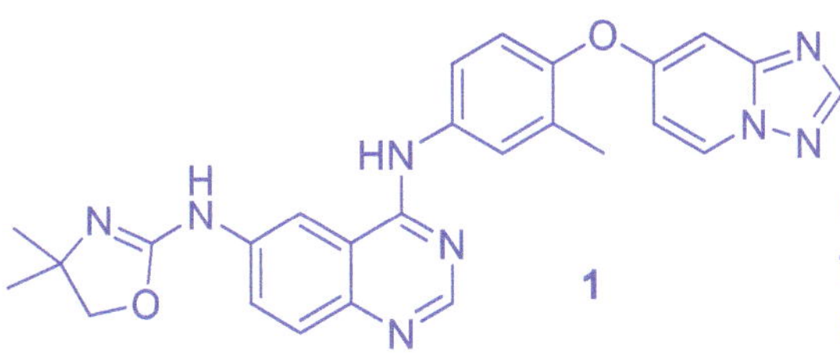

1

USAN: tucatinib
Trade name: Tukysa
Selective Her-2 Inhibitor
Launched: 2020

Tucatinib (**1**, Tukysa, ONT-380) was an oral highly selective human epidermal growth-factor receptor-2 (HER-2) tyrosine kinase inhibitor, developed by Seattle Genetics (SeaGen). Tucatinib (**1**) was approved in April 2020 for the treatment of unresectable or metastatic HER2-positive breast cancer in combination with trastuzumab and capecitabine.

Before the approval of tucatinib (**1**), the dual tyrosine kinase inhibitors lapatinib (**2**) and covalent inhibitor neratinib (**3**) have been clinically used to treat patients with HER2-positive metastatic breast cancer (Figure 1). However, both lapatinib (**2**) and neratinib (**3**) displayed off-target adverse effects, including diarrhea and skin rashes, due to the no selective blockage of both HER2 and EGFR. Tucatinib (**1**) is the first highly selective HER2 inhibitor with approximately 500-fold more potent activities against HER2 than EGFR, which exhibits fewer side effects resulting from the inhibition of EGFR.[1] Furthermore, tucatinib (**1**) displays great potential to be used for the treatment of HER-2 positive breast cancer harboring brain metastases since it could cross the blood-

Chemistry and Pharmacology of Drug Discovery, First Edition. Edited by Jie Jack Li.

brain barrier more easily compared to monoclonal antibodies. Thus, tucatinib (**1**) offers a promising therapeutic strategy for the treatment of HER-2-positive breast cancer with improved treatment efficacy and safety in combination with trastuzumab and capecitabine.

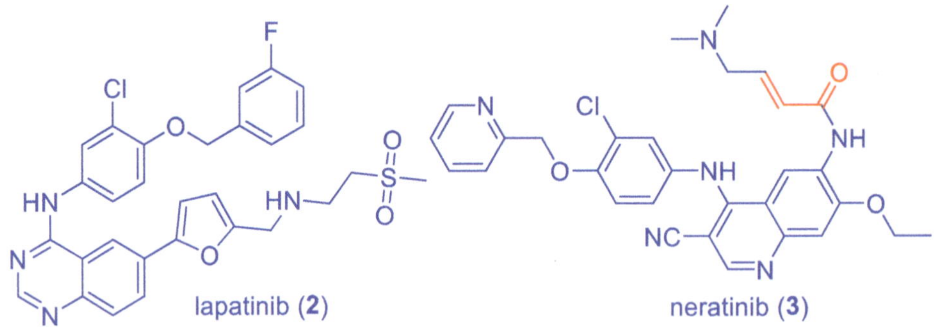

Figure 1. Chemical structure of HER2 tyrosine kinase inhibitors lapatinib (**2**) and neratinib (**3**)

1 Background

Breast cancer is a second major cause of cancer-related mortality in women and HER2-positive accounts for about 15–20% of all breast cancer patients.[2] In contrast to HER2-negative breast cancer, HER2-overexpressing breast cancer has been related to aggressive disease and low survival rates.[3] HER2 overexpression is also detected in other types of human cancers, including bladder, colon, ovary, esophagus, stomach, lung, uterus, and prostate cancer.[3,4] Thus, HER2 is an effective therapeutic drug target in patients with breast cancer possessing HER2 overexpression and/or amplification.[3]

During the last few decades, significant progress has been made in the treatment of HER2-positive disease, especially in HER2-positive breast cancer. A series of novel HER2-targeted agents have been developed that have remarkably improved the prognosis of patients harboring HER2-positive breast cancer and extended their lives. Currently, approved drugs for patients with HER2-positive breast cancer include the following three classes:[5] (1) Monoclonal bodies, such as trastuzumab and pertuzumab. (2) Antibody–drug conjugates (ADCs), such as trastuzumab deruxtecan (DS-8201) and trastuzumab emtansine (T-DM1). (3) Small-molecule tyrosine kinase inhibitors (TKIs) lapatinib[6] and neratinib.[7] However, several clinic unmet needs exist in patients with HER2-positive metastatic breast cancer. First, although the treatment of the patient with trastuzumab could improve prognosis in patients with HER2-positive metastatic breast cancer, primary and acquired resistance remains a significant clinical challenge in most patients with HER2-amplified metastatic breast cancer. Second, adverse effect profiles for small-molecule tyrosine kinase inhibitors, such as lapatinib (**2**) and neratinib (**3**), display roughly equipotent inhibition activities against EGFR and HER2, which possesses off-target adverse effects, including diarrhea and skin rashes, resulting from EGFR inhibition.[8] Thus, the urgent clinic needs to develop novel tyrosine kinase inhibitors that block HER2 activity with no inhibition activity against other EGFR subfamily proteins to improve efficacy and safety.

Recently, tucatinib (**1**), developed by Seattle Genetics, received its first approval in April 2020 for the treatment of unresectable or metastatic HER-2 positive breast cancer in combination with trastuzumab and capecitabine. It was the first oral HER2 tyrosine kinase inhibitor with a highly selective enzymatic activity against HER2 and EGFR (HER2 IC_{50} = 6.9 nM; EGFR IC_{50} = 449 nM).[1] This selectivity may improve tolerability and efficacy compared to earlier HER2 inhibitors such as lapatinib (**2**) and neratinib (**3**) which harbor common EGFR-associated toxicities (Figure 2).[9]

2 Pharmacology

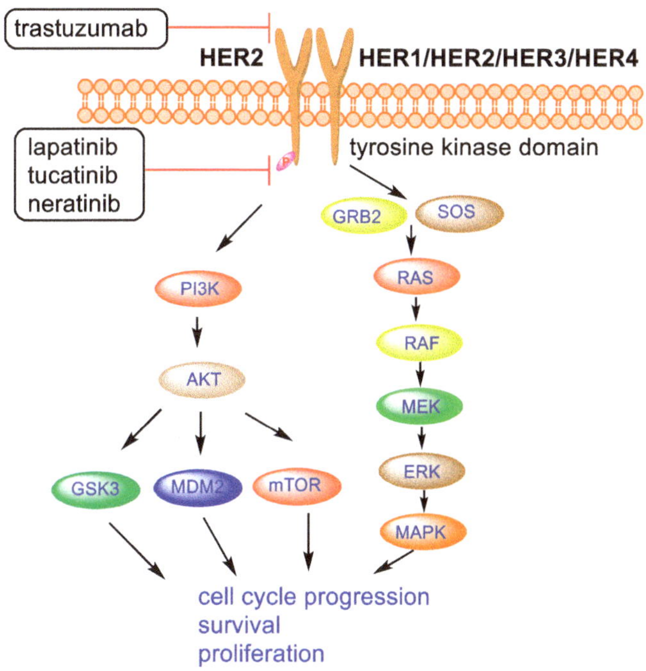

Figure 2. The HER2 signaling pathway

HER2 belongs to the HER family comprising four members, namely EGFR, HER2, HER3, and HER4. These proteins structurally have an extracellular ligand-binding domain, a single transmembrane domain, and a cytoplasmic catalytic tyrosine kinase domain.[10] The binding of HER proteins with their ligands triggers the formation of homodimers and heterodimers, which upregulate downstream signaling pathways. Unlike the other HER proteins, HER2 lacks a specific ligand and is the preferred partner for the homodimerization and heterodimerization with itself or other HER proteins.[11] Subsequently, the homodimerization and heterodimerization of HER2 protein triggers the activation of the kinase activity of the HER2, resulting in stimulation of downstream signaling pathways, including activation of the Ras/MAPK and PI3K/AKT/mTOR signaling pathways.[12]

HER2-directed kinase inhibitors can compete with ATP, resulting in the blockade of phosphorylation and preventing activation of downstream signaling pathways. Tucatinib is the first high-selective HER2 kinase inhibitor, which reversibly binds to the cytoplasmic catalytic kinase domain of the HER2 and competes with ATP, preventing tyrosine phosphorylation and inhibiting downstream MAPK and AKT signaling pathways. The recent data indicates the combination of tucatinib and trastuzumab exhibited synergized improved anti-tumor activity *in vitro* and *in vivo*.[1]

3 Pharmacokinetics and Drug Metabolism

Tucatinib tablets are available as 50 and 150 mg film-coated tablets, which can be taken every 12 h with or without food in combination with trastuzumab. Tucatinib tablets were recommended to be taken orally at a dosage of 300 mg until disease progression or unacceptable toxicity appears. It should be taken regularly at the scheduled time, even if a dose of tucatinib is vomited or missed. Furthermore, for the patient with severe hepatic impairment, the dosage should be reduced to 200 mg orally twice daily.[13]

When tucatinib was taken by oral administration, its t_{max} was approximately 2 h. After continuously using tucatinib 300 mg twice daily for 14 days, a steady state is reached in approximately 4 days. As a result, the AUC increases by 1.7-fold while the C_{max} increases by 1.5-fold, respectively. With 97% plasma protein binding, the apparent volume of distribution is 1670 L. The half-life ($T_{1/2}$) of tucatinib is approximately 8.5 h. After a 300 mg dose of tucatinib, approximately 86% of tucatinib is excreted in feces (16% of tucatinib is unmetabolized) and 4% of it is excreted in urine. Although the AUC_{∞} modestly rises 1.5-fold and the t_{max} is postponed from 1.5 to 4 h following a high-fat meal, these changes are not clinically useful. Additionally, participants with severe hepatic impairment had higher exposure to tucatinib than volunteers with mild impairment (1.61- and 1.15-fold geometric mean ratio $AUC_{0\to\infty}$ increases, respectively). Volunteers with modest hepatic impairment and those with normal hepatic function had comparable tucatinib plasma concentrations.[13]

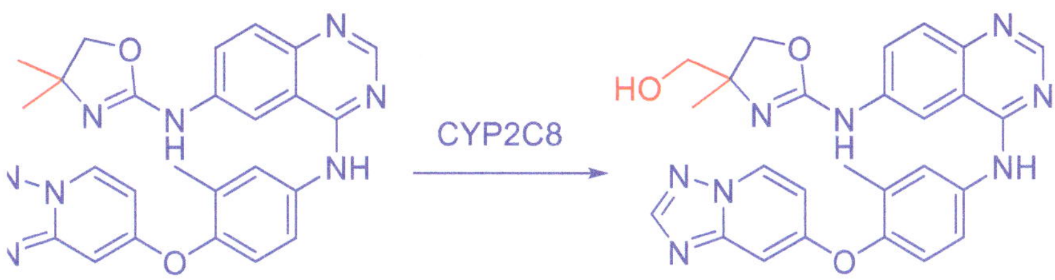

Figure 3. Predominant metabolic pathway of tucatinib (**1**)

A phase I clinic trial (NCT03758339) demonstrated that tucatinib (**1**) is metabolized mainly by CYP2C8 and to a lesser degree by CYP3A. It is found that the chief metabolite of tucatinib is ONT-993 (**4**), generated by CYP2C8-mediated hydroxylation (Figure 3). It is shown that the cytotoxicity of ONT-993 (**4**) is 2- to 3-fold less than that of tucatinib and its potency-adjusted exposure accounts for less than 10% of the overall pharmacological activity. Therefore, ONT-993 (**4**) is not anticipated to improve tucatinib's safety or efficacy remarkably.[14]

According to research on drug–drug interactions, patients are advised against using potent CYP2C8 inhibitors along with tucatinib concurrently. Tucatinib (**1**) is advised to be taken 100 mg twice daily if unavoidable using a strong CYP2C8 inhibitor concurrently.[15]

4 Efficacy and Safety

The safety of tucatinib (**1**) in combination with trastuzumab and capecitabine was investigated in a randomized (2:1), double-blind, placebo-controlled clinical trial. The patients with HER2-positive unresectable locally advanced or metastatic breast cancer were selected for this clinical trial.[16] The study treatment consisted of oral tucatinib 300 mg or placebo twice a day, regular dosages of trastuzumab either intravenously or subcutaneously, and on days 1–14 of each 21-day cycle, oral capecitabine 1000 mg/m^2.

A total of 612 individuals were randomly assigned to receive treatment with trastuzumab and capecitabine (410 to tucatinib and 202 to placebo). The primary outcome for the first 480 randomized patients was PFS as determined by a blinded independent review committee. Tucatinib greatly improves the primary endpoint of progression-free survival (PFS). For patients treated with tucatinib (**1**), trastuzumab, and capecitabine, the median PFS was 7.8 months compared with 5.6 months for those receiving placebo. Among the 612 patients in the trial, the group receiving tucatinib, trastuzumab, and capecitabine had a median overall survival (OS) of 21.9 months, while the group receiving placebo, trastuzumab, and capecitabine had a median OS of 17.4 months.[8]

The safety of tucatinib (**1**) in combination with trastuzumab and capecitabine was assessed in 601 patients with HER2-positive metastatic breast cancer who received at least one dosage of study treatment on HER2CLIMB. The most common (incidence > 50%) all-grade adverse reactions include diarrhea (81%), palmar-plantar erythrodysesthesia syndrome (63%) and nausea (58%). The most frequent grade 3–4 adverse reactions (incidence > 5%) include palmar-plantar erythrodysesthesia (13%), diarrhea (13%), and hepatotoxicity (9%).[17]

5 Synthesis

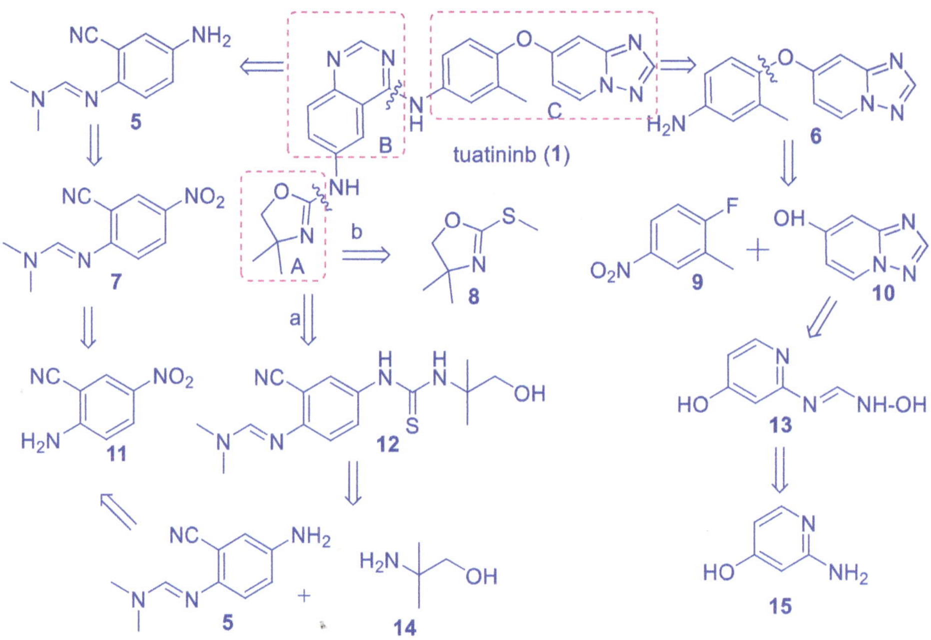

Scheme 1. Retro-synthetic analysis of tucatinib (**1**)

The retro-synthetic analysis of tucatinib (**1**) is shown in Scheme 1. Tucatinib (**1**) is assembled from fragments A, B, and C. Fragments A and B are connected by carbon-nitrogen bonds via aromatic nucleophilic substitution (S_NA). A cyclization connects fragments **5** and **6**. The fragment **5** could be synthesized from commercially available aniline **11** by a condensation followed by a reduction reaction. On the other hand, fragment **6** is connected by the S_NAr reaction of **9** with **10**. Fragments **10** is assembled from starting materials **15** by condensation followed by cyclization. There are two connection methods to construct fragments A and B. One is to directly install fragment **8** into fragment B by S_NAr reaction. The other is made by coupling **5** and **14** followed by cyclization.

The first synthesis route for tucatinib (**1**) was reported by Array Biopharma in 2007.[18] First, the treatment of aniline **11** with *N,N*-dimethylformamide dimethyl acetal (DMF–DMA) delivered formamidine **7** in 87% yield. It was followed by catalytic hydrogenation reduction of the nitro group to give intermediate **5** in 90% yield. Subsequently, the intermediate **5** was treated with the 1,1′-thiocarbonyl diimidazole (TCDI) and 2-amino-2-methyl-1-propanol **14** at –10 °C for 16 h, leading to thiourea **12** in 34% isolated yield, which then performed an intermolecular [5 + 1] cycloaddition with fragment **5** to give aminoquinazoline derivative **17** in 62% yield. Finally, tucatinib (**1**) was obtained in 68% yield through an intramolecular cyclization by the treatment with *p*-toluenesulfonyl chloride (*p*-TsCl) and NaOH at room temperature. In summary, tucatinib (**1**) was prepared in 16.3% yields in five steps from starting materials **5** (Scheme 2).

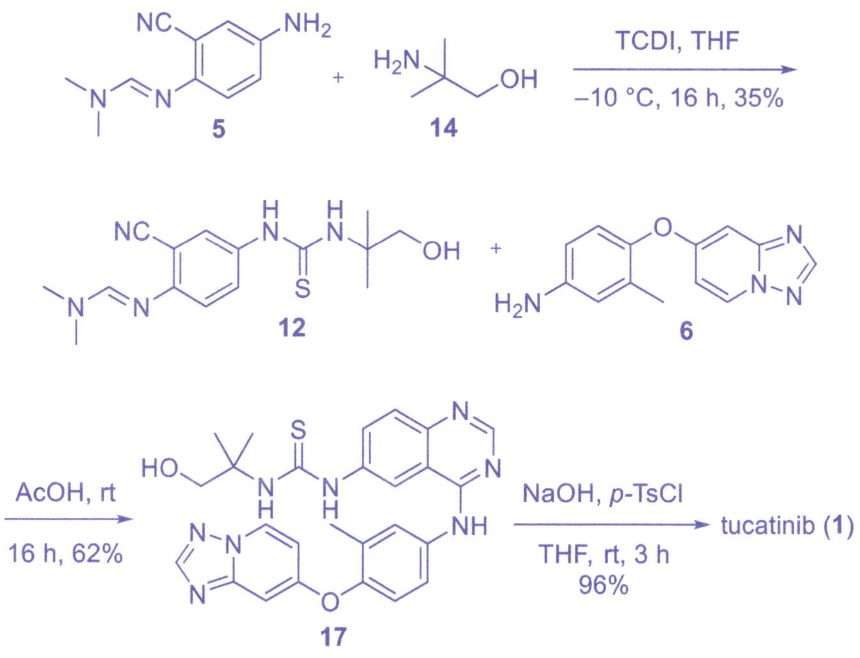

Scheme 2. Array BioPharma's synthesis route of tucatinib (**1**)

Mao and coworkers recently designed a convergent synthetic route for synthesizing tucatinib (**1**, Scheme 3).[19] At first, the S_NAr nucleophilic substitution of 2-methyl-4-nitrophenol **9** with 4-chloropyridin-2-amine **18** to give diaryl ether **19** in the presence of DIPEA as a base at 150 °C for 72 h in 64% yield, followed by the reaction of DMF–DMA and hydroxylamine hydrochloride to afford the formamidine **20**.

Subsequently, the cyclization of formamidine **20** in the presence of trifluoroacetic acid anhydride (TFFA) to deliver the triazolopyridine **16**. Next, the catalytic hydrogenation reduction of the nitro group resulted in the intermediate **6** in 68% yield, followed by the condensation with the *N,N*-dimethylphenylformimidamide **7** to give intermediate **21** in 83% yield. The catalytic hydrogenation reduction of the nitro group gave intermediate **22**, which underwent the substitution with dimethyl-2-(methylthio)-4,5-dihydrooxazole **8** to give **1** in 76% yield. In summary, it takes seven steps to furnish the final product **1** with an overall 19.0% yield.

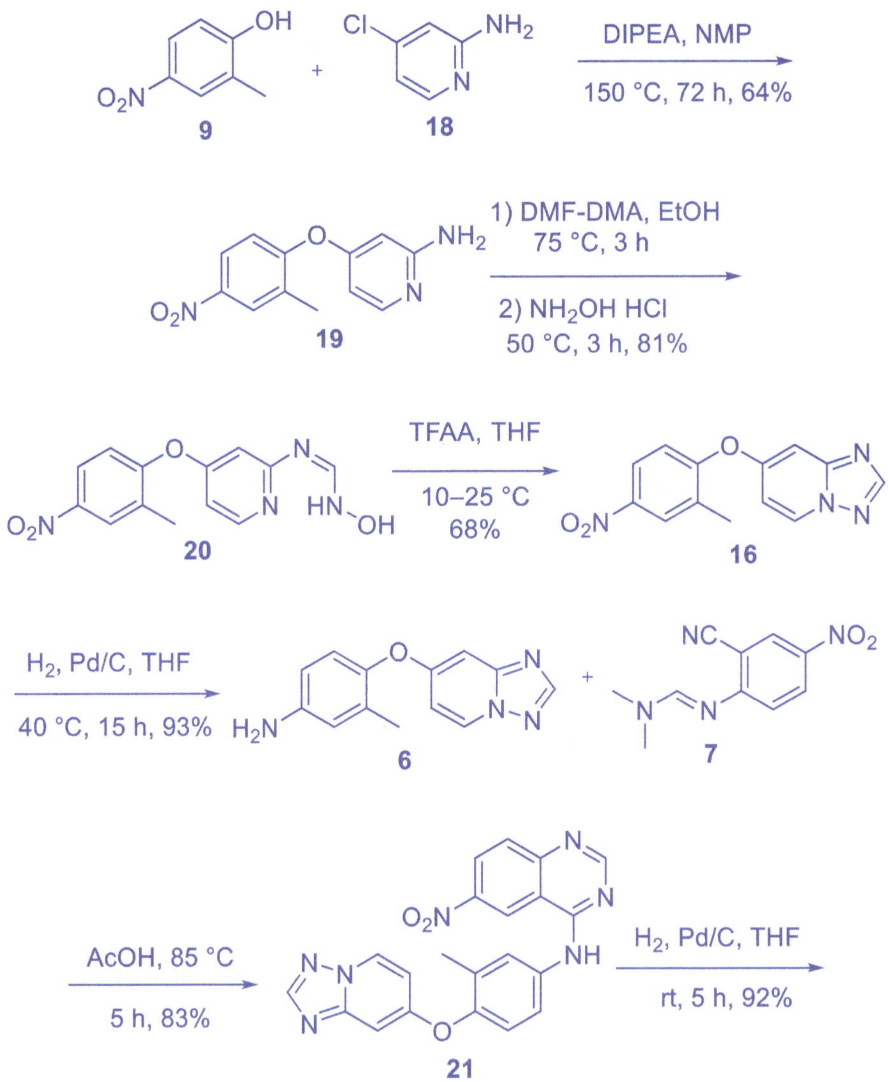

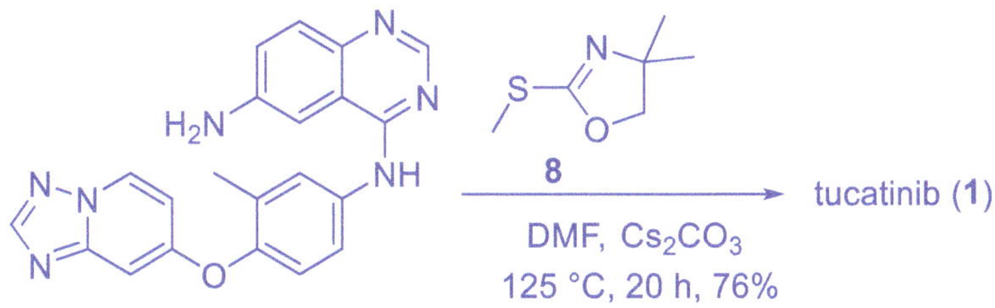

22

Scheme 3. Mao's synthetic route to tucatinib (**1**)

Mao and coworkers recently developed a practical process for preparing tucatinib (**1**, Scheme 4).[20] Shortly, the condensation of 4-chloropyridin-2-amine (**18**) with DMF–DMA, followed by the treatment with hydroxylamine hydrochloride, leading to the *N*-hydroxy-formimidamide **23** in 89% yield. Subsequently, compound **23** was subjected to TFAA, giving the triazolo[1,5-*a*]pyridine derivative **24** in 71% yield. Next, the nucleophilic substitution of 4-amino-2-methylphenol (**25**) with compound **24** in DMF/K_2CO_3 at 130–140 °C for 16 h, to deliver intermediate **6** in 73% yield. The subsequent cyclization reaction was conducted by heating **6** and **11** with acetic acid to furnish compound **21**, followed by catalytic hydrogenation to afford aniline **22**. Subsequently, compound **22** reacted sequentially with TCDI and 2-amino-2-methyl-1-propanol **14** in DMF for 4 h, and compound **17** was obtained in 81% yield. Finally, **17** was treated with NaOH and *p*-TsCl in THF at 50–60 °C for 3 h to give tucatinib (**1**) in 77% yield.

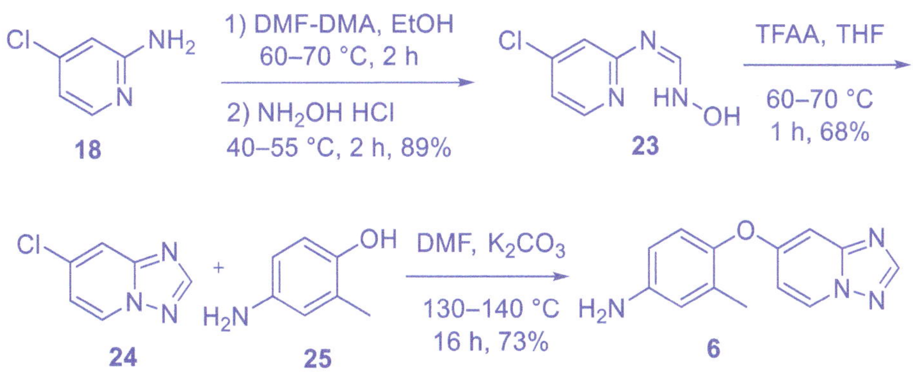

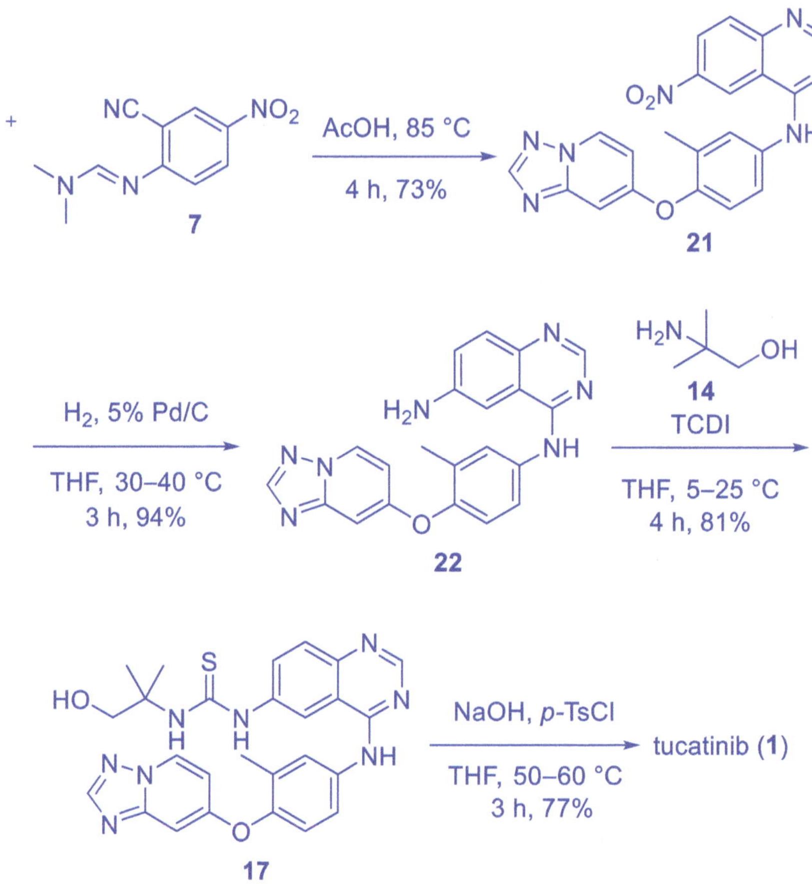

Scheme 4. Mao's synthetic route to tucatinib (**1**)

6 Summary

In summary, tucatinib (**1**) is an orally bioavailable, selective HER2 inhibitor, developed by Seattle Genetics and received its approval in April 2020 for the treatment of HER2 positive breast cancers. Compared to lapatinib and neratinib associated with gastrointestinal and dermatologic adverse effects, tucatinib displays high selectivity against HER2, with improved tolerability and efficacy. However, the clinical applicability of reversible HER2 TKIs is limited due to clinical resistance caused by the HER2 mutation.[21] Therefore, targeted protein degradation may emerge as an alternative strategy for the novel treatment of HER2-positive breast cancers in the future.

References

1. Lee, A. Tucatinib: first approval. *Drugs* **2020**, *80*, 1033–1038.
2. Loibl, S.; Gianni, L. HER2-positive breast cancer. *Lancet* **2017**, *389*, 2415–2429.
3. Oh, D.-Y.; Bang, Y.-J. HER2-targeted therapies–a role beyond breast cancer. *Nat. Rev. Clin. Oncol.* **2020**, *17*, 33–48.
4. Yan, M.; Schwaederle, M.; Arguello, D.; Millis, S. Z.; Gatalica, Z.; Kurzrock, R. HER2 expression status in diverse cancers: review of results from 37992 patients. *Cancer Metastasis Rev.* **2015**, *34*, 157–164.
5. Swain, S. M.; Shastry, M.; Hamilton, E. Targeting HER2-positive breast cancer: advances and future directions. *Nat. Rev. Drug Discovery* **2023**, *22*, 101–126.
6. Tsang, R. Y.; Sadeghi, S.; Finn, R. S. Lapatinib, a dual-targeted small molecule inhibitor of EGFR and HER2, in HER2-amplified breast cancer: from bench to bedside. *Clin. Med. Insights Ther.* **2011**, *3*, 1–13
7. Tesch, M. E.; Gelmon, K. A. Targeting HER2 in breast cancer: latest developments on treatment sequencing and the introduction of biosimilars. *Drugs* **2020**, *80*, 1811–1830.
8. Moulder, S. L.; Borges, V. F.; Baetz, T.; McSpadden, T.; Fernetich, G.; Murthy, R. K.; Chavira, R.; Guthrie, K.; Barrett, E.; Chia, S. K. Phase I study of ONT-380, a HER2 inhibitor, in patients with HER2(+)-advanced solid tumors, with an expansion cohort in HER2(+) metastatic breast cancer (MBC). *Clin. Cancer Res.* **2017**, *23*, 3529–3536.
9. Shah, M.; Wedam, S.; Cheng, J.; Fiero, M. H.; Xia, H.; Li, F.; Fan, J.; Zhang, X.; Yu, J.; Song, P.; et al. FDA approval summary: tucatinib for the treatment of patients with advanced or metastatic HER2-positive breast cancer. *Clin. Cancer Res.* **2021**, *27*, 1220–1226.
10. Kulukian, A.; Lee, P.; Taylor, J.; Rosler, R.; de Vries, P.; Watson, D.; Forero-Torres, A.; Peterson, S. Preclinical activity of HER2-selective tyrosine kinase inhibitor tucatinib as a single agent or in combination with trastuzumab or docetaxel in solid tumor models. *Mol. Cancer Ther.* **2020**, *19*, 976–987.
11. Li, D.; Tu, Y.; Jin, K.; Duan, L.; Hong, Y.; Xu, J.; Chen, N.; Zhang, Z.; Zuo, H.; Gong, W.; et al. Discovery of SPH5030, a selective, potent, and irreversible tyrosine kinase inhibitor for HER2-amplified and HER2-mutant cancer treatment. *J. Med. Chem.* **2022**, *65*, 5334–5354.

12. O'Brien, N. A.; Huang, H. K. T.; McDermott, M. S. J.; Madrid, A. M.; Luo, T.; Ayala, R.; Issakhanian, S.; Gong, K. W.; Lu, M.; Zhang, J.; et al. Tucatinib has selective activity in HER2-positive cancers and significant combined activity with approved and novel breast cancer–targeted therapies. *Mol. Cancer Ther.* **2022**, *21*, 751–761.

13. Food and Drug Administration. Drug Approval Package: Tukysa. Available from:https://www.accessdata.fda.gov/drugsatfda_docs/nda/2020/213411Orig1s000MultidisciplineR.pdf

14. Topletz-Erickson, A.; Lee, A.; Rustia, E. L.; Sun, H.; Mayor, J. G.; Abdulrasool, L. I.; Walker, L.; Endres, C. J. Evaluation of safety and clinically relevant drug–drug interactions with tucatinib in healthy volunteer's. *Clin. Pharmacokinet.* **2022**, *61*, 1417–1426.

15. Sun, H.; Cardinal, K. A.; Wienkers, L.; Chin, A.; Kumar, V.; Neace, C.; Henderson, C.; Endres, C. J.; Topletz-Erickson, A.; Regal, K.; et al. Elimination of tucatinib, a small molecule kinase inhibitor of HER2, is primarily governed by CYP2C8 enantioselective oxidation of gem-dimethyl. *Cancer Chemother. Pharmacol.* **2022**, *89*, 737–750.

16. Le Du, F.; Dieras, V.; Curigliano, G. The role of tyrosine kinase inhibitors in the treatment of $HER2^+$ metastatic breast cancer. *Eur. J. Cancer* **2021**, *154*, 175–189.

17. Murthy, R. K.; Loi, S.; Okines, A.; Paplomata, E.; Hamilton, E.; Hurvitz, S. A.; Lin, N. U.; Borges, V.; Abramson, V.; Anders, C.; et al. Tucatinib, trastuzumab, and capecitabine for HER2-positive metastatic breast cancer. *N. Engl. J. Med.* **2020**, *382*, 597–609.

18. Lyssikatos, J. P.; Marmsater, F. P.; Zhao, Q.; Greschuk, J. M. *N*-4-Phenyl-quinazline-4-amine derivatives and related compounds as Erbb type I receptor tyrosine kinase inhibitors for the treatment of hyperproliferative diseases. WO 2007059257A2 (2007).

19. Yin, L. F.; Mao, Y. J.; Liu, Y. W.; Bu, L. H.; Zhang, L.; Chen, W. X. New synthetic route to tucatinib. *Synthesis* **2019**, *51*, 2660–2664.

20. Lyu, Y. D.; Huang, L. L.; Zhu, X. L.; Lu, S.; Mao, Y. J. A practical alternate synthesis of tucatinib. *Org. Prep. Proced. Int.* **2021**, *53*, 554–561.

21. Hu, M.; Li, Y.; Li, J.; Zhou, H.; Liu, C.; Liu, Z.; Gong, Y.; Ying, B.; Xie, Y. Discovery of potent and selective HER2 PROTAC degrader based tucatinib with improved efficacy against HER2 positive cancers. *Eur. J. Med. Chem.* **2022**, *244*, 114775.

10

Tazemetostat (Tazverik): An EZH2 Inhibitor for Treatment of Epithelioid Sarcoma and Follicular Lymphoma

Ruheng Zhao and
Timothy A. Cernak

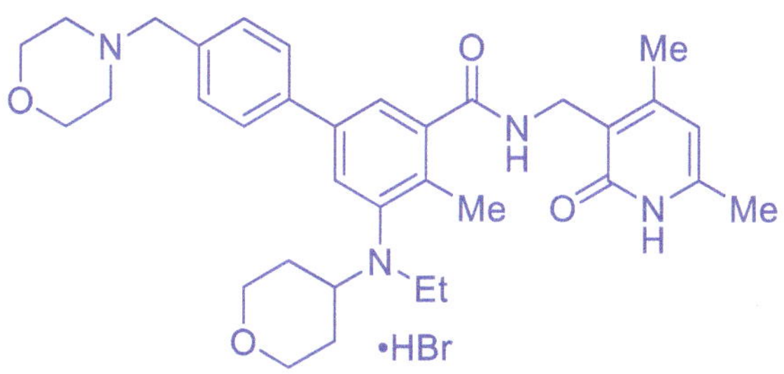

Tazemetostat
(Tazverik, **1**)
Epizyme, 2020
EZH2 inhibitor

1. Background

Enhancer of zeste homolog 2 (EZH2) is a core component of the polycomb repressive complex 2 (PRC2) that imparts posttranslational modifications (PTMs) to histone proteins, thereby regulating gene expression. Its critical roles in cellular processes such as proliferation, apoptosis, and senescence have been well established.[1] Beyond these basic cellular processes, overexpression, or mutations in EZH2 have been implicated in various cancer-associated phenotypes, including tumorigenesis, metastasis, altered metabolism, drug resistance, and immune evasion.[2]

Due to the vast implications of EZH2 in cancer progression and maintenance, pharmaceutical interventions targeting EZH2 emerged as a promising therapeutic strategy. However, the development of specific and potent inhibitors has been

Chemistry and Pharmacology of Drug Discovery, First Edition. Edited by Jie Jack Li.

challenging, presenting obstacles such as achieving target specificity, managing off-target effects, and optimizing pharmacokinetics.

Tazemetostat (**1**), developed by Epizyme (later acquired by Ipsen), represents a paradigm shift in the field of EZH2 targeting. Unlike its predecessor, DZNep (**2**), which inhibits EZH2 in a non-direct manner, tazemetostat (**1**) emerged as the first direct and selective inhibitor of EZH2. Its development not only highlighted the importance of methylation in PTMs but also is a beautiful art adopting "Magic Methyl" effect in drug discovery. This phenomenon suggests that the strategic incorporation of a small methyl group can substantially enhance drug properties such as potency, selectivity, and pharmacokinetics by inducing conformational changes in the target protein.[3, 4]

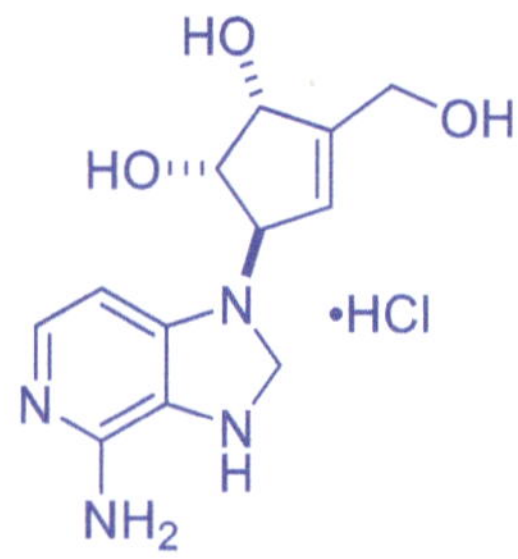

DZNeP (**2**)

In 2020, tazemetostat (**1**), marketed as Tazverik, received accelerated approval for two clinical indications:

1. Treatment of metastatic or locally advanced epithelioid sarcoma, a rare and aggressive soft tissue malignancy with limited treatment options.[5]
2. Therapy for relapsed or refractory follicular lymphoma in patients harboring EZH2 mutations who had undergone at least two prior systemic treatments.[5]

While the approval of tazemetostat (**1**) represents a significant milestone in EZH2-targeted therapies, a deeper understanding of the drug discovery process and its underlying principles is essential for advancing the field of medicinal chemistry.

2. Pharmacology

2.1. Epigenetic Regulation in Gene Expression

Epigenetic regulation is foundational to the hallmarks of cancer. Disruptions in epigenetic patterns or chromatin configurations can induce oncogenic properties or potentiate gene repression, both propelling tumorigenesis. Central to this is the realm of PTMs, which modulate gene expression without amending the DNA sequence. Identified PTMs include acetylation, methylation, ubiquitination, phosphorylation, and glycosylation. These modifications are being targeted in drug development efforts to deliver more precise and effective therapies.

Lysine methylation, one of the most common histone PTMs, plays a key role in regulating gene expression.[6] The impact of lysine methylation depends on both the specific lysine residue being modified and the state of methylation. Methylation states involve the addition of one (me1), two (me2), or three (me3) methyl groups to the amine tails of lysine residues in histones. The different methylation states instruct different chromatin structure and the accessibility of transcription factors and other regulatory proteins to DNA, ultimately influencing gene expression.

2.2. Posttranslational Modifications of EZH2 and Cancer Progression

The nuanced world of methylation requires a series of enzymes for reading, writing, and erasing the methyl code. A critical player here is the polycomb repressive complex 2 (PRC2), a multi-protein epigenetic regulatory complex, orchestrating transcription regulation. EZH2, the catalytic engine of PRC2 and a representative of the Protein Lysine Methyltransferase family, steers cellular processes like cell cycle progression, DNA repair, and autophagy. Depending on context, EZH2 can suppress or co-activate transcription either in a PRC2-dependent manner or through a PRC2-independent pathway. A prime example is the methylation of Lys-27 on histone 3 (H3K27). A hotspot mutation in the SET domain of EZH2 is Y641X (X = F, N, S, C, or H). The SET domain responsible for transferring methyl groups to H3K27, and this mutation alters its catalytic function.[7] This mutation, found in cancers such as DLBCL and melanoma, recalibrates EZH2's specificity, escalating H3K27 di-methylation to tri-methylation.[8] The resulting H3K27me3, associated with gene repression, plays an instrumental role in cellular decisions, often culminating in oncogenic outcomes. The mutation's molecular genesis possibly arises from the loss of tyrosine, reshaping the enzyme's active site and facilitating enhanced methylation.

2.3. Progress and Mechanisms of Action for EZH2 Inhibitors

EZH2's SET domain functions by transferring methyl groups from *S*-adenosyl-methionine (SAM) to target lysines. DZNep (**2**), the pioneer EZH2 inhibitor, indirectly inhibits EZH2 by increasing levels of SAH, a byproduct that represses SAM-dependent activity.

Over recent years, potent SAM-competitive inhibitors, in addition to tazemetostat (**1**), like GSK126 (**3**), PF-06821497 (**4**), and CPI-1205 (**5**), been developed.[9–11] These drugs, by binding to the SET domain, occupy the SAM binding pocket, aiming to reduce H3K27 trimethylation and reverse the gene repression associated with EZH2 mutation. New-age drugs also seek to perturb the PRC2 complex's integrity or triggering EZH2 degradation. However, to date, only tazemetostat (**1**) has been approved approval for specific cancer types.

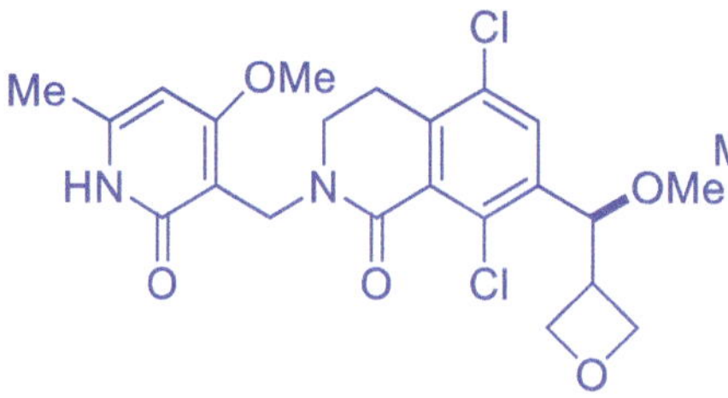

PF-06821497 (**3**)

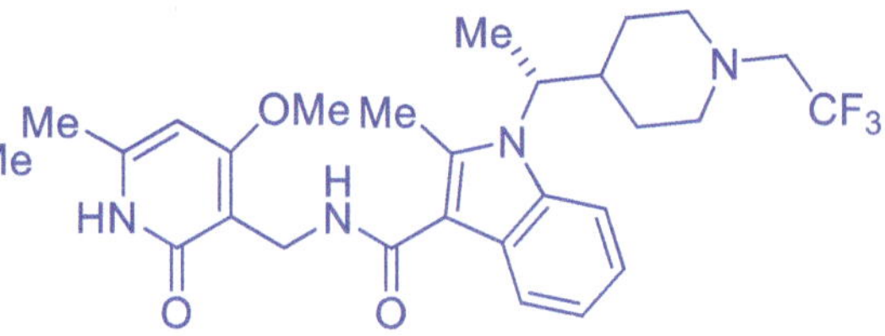

CPI-1205 (**4**)

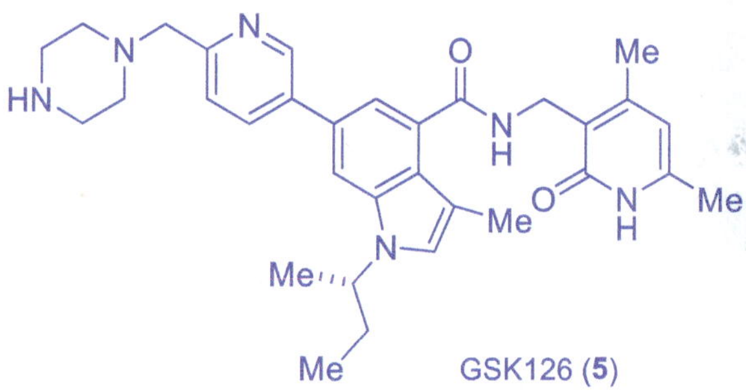

GSK126 (**5**)

3. Structure–Activity Relationship (SAR)

3.1. The Discovery of Initial Hit Using High-Throughput Screening

The role of epigenetic regulator EZH2 in cancer progression has garnered significant attention, highlighting the urgent need for targeted therapies. To address this, Epizyme embarked on the discovery of potent EZH2 inhibitors. Utilizing high-throughput screening (HTS) on a 175,000-compound library targeting the wild-type PRC2 complex, they pinpointed an initial hit.[1] Subsequent structural similarity assessments led to additional screenings and a pyridone-containing compound **6** emerged as a potential inhibitor. Despite its potency, further attempts to use this compound in cellular assays was limited by poor solubility and low oral bioavailability.

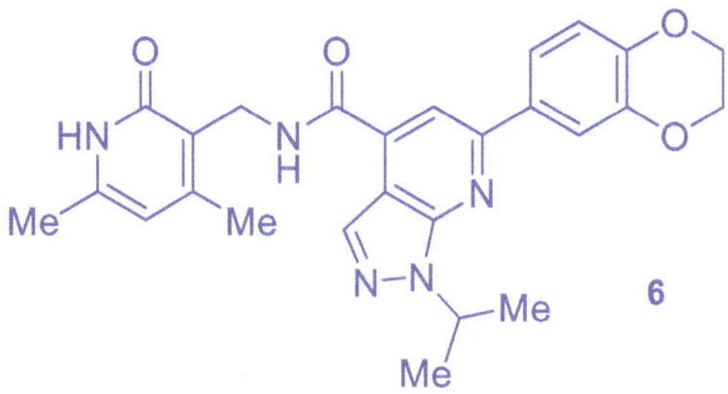

6

3.2. Lead Optimization

The solubility issue was tackled first, by introducing polarity to the molecule. Polar substitutions at the 6-position of the 7-azaindazole can be tolerated. Modifications were made to the corresponding position, leading to a series of compounds with improved solubility, oral bioavailability, and cellular activity. From the numerous synthesized variants, it was found that the basic amine and small nonpolar group can be well tolerated at this position without significant loss of potency, and the amine can link directly or via an aryl spacer. Ultimately, a morpholine at the 4-position of the phenyl ring was chosen as the candidate **7** because it significantly improved solubility, oral bioavailability, and ensuring moderate clearance and modestly increase cellular activity. While considering the ease of synthesis and the modest potency shift with this morpholine substitutions, hydrogen or halogen was employed for subsequent SAR studies.[12]

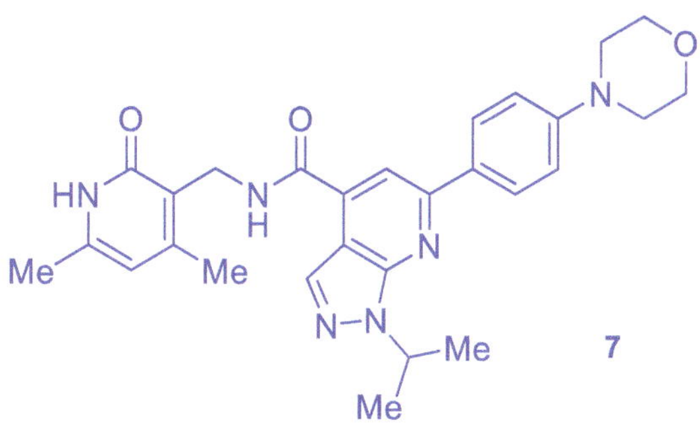

7

Considering the structure variations between the initial hits and compound **6** generated from the extended screening post-clustering, the central difference lies in the core scaffold. Thus, the core pyrazolopyridine scaffold underwent an in-depth evaluation (Table 1). Different heterocycle that bearing different electron densities and distinct target engagement activities was evaluated. A shift to indazole (**9**) enhanced potency, while benzimidazole (**10**) and triazole (**11**) caused significant potency loss, highlighting the loss of potency may attribute to the formation of intramodular hydrogen bond between the amide and ortho N atom. Furthermore, the optimal substitution occurs at the 4-position when the amide is ortho substituted to the fused ring system. Taking both the intramodular hydrogen bond and substitution position into account, conformation offers the most plausible explanation for compounds performance. Specifically, a planer conformation of amide relative to the ring adversely affects activity.

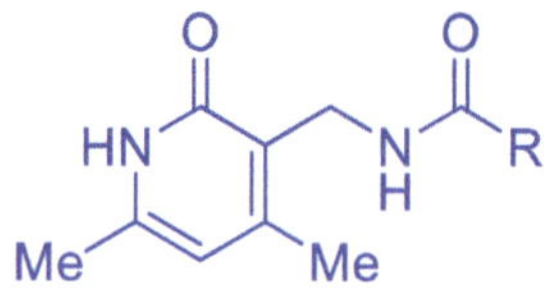

Compound	R	EZH2 IC_{50} (μM)
8	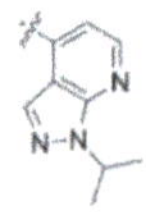	3.1 ± 1.0
9	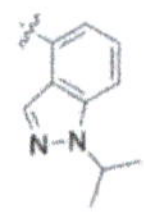	1.1 ± 0.3
10	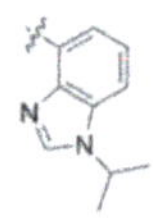	>50
11	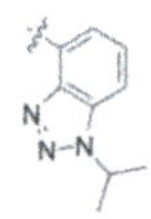	>50
12	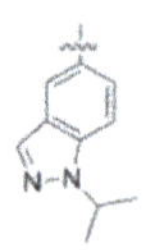	22 ± 6

Table 1. Optimization of 5,6-bicylic core

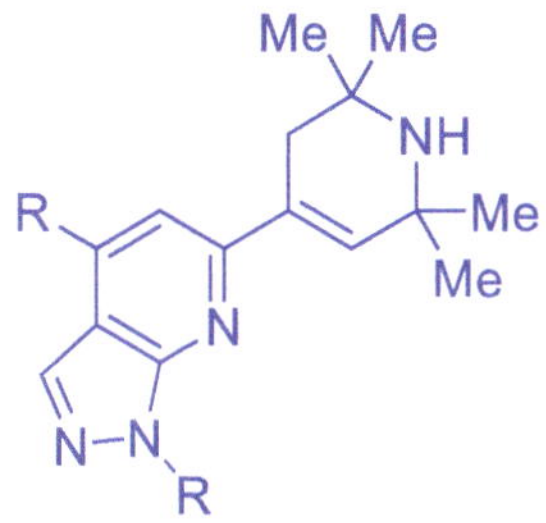

13: R = Me, IC_{50} = 39 ± 13 μM
14: R = Et, IC_{50} = 5 ± 3 μM
15: R = *i*-Pr, IC_{50} = 0.3 ± 0.2 μM

The N1-substitution on the core scaffold also underwent evaluation, when adopting pyrazolopyridine as the core motif, a bulkier, hydrophobic isopropyl group on **13** outperformed 10-fold than ethyl analog **14** and 100-fold than methyl counterpart **15**, showcasing the importance of substituent size. Furthermore, non-polar substituents were generally more favorable than polar groups. However, when assessing more broader range of synthesized analogs, further improvements were only marginal, particularly when weighing the balance between physicochemical and drug-like attributes.

Since progress had stalled on the optimization of the nonpolar bulky substitution on the indazole, the iterative optimization strategy was used, which returning to the earlier phase to revisit and reevaluate the core scaffold.

The SAR data previously showed that planarity restricted by hydrogen bond or lack of ortho substitution between the amide and the core rings diminished potency. Recognizing this, the researchers opted for a less planar core scaffold, which could potentially diversify the binding profile, and offering new avenues for functional group additions. Accordingly, the disubstituted aniline structure was evaluated, achieved by opening the 5-membered pyrazole ring (Table 2).

The concept of the "magic methyl" effect is fundamental to understanding the optimizations in the lead compound. When a methyl group was introduced at the R_1 position, there was a significant enhancement in potency—a 1000-fold improvement compared to the compounds **16** and **17**.

Further understanding of the "magic methyl" effect's mechanism, garnered from reputable databases such as the Protein Data Bank and the Cambridge Structural Database, elucidated its role beyond simple pocket occupation. The introduction of a methyl group at the R_1 position instigated a conformational shift in the secondary amide, twisting it away from a planar disposition and introducing a torsion angle in the range of 60°–140° or 220°–300°. Intriguingly, the influence of this methylation was not confined to the amide. Computational studies highlighted that the ortho methylation induced significant conformational changes in the disubstituted aniline. In this context, the methyl group, precisely positioned ortho, serves a dual purpose to enhancing molecular potency: it acts as a conformational lock to avoid planar structures for both amide and aniline. In the presence of the methyl group, the molecule adopted a twisted conformation, contrasting with the planar orientation that was preferred in its absence.

Further refinements focused on the other substituents of the disubstituted aniline (Table 2). The R_2 substituent was eventually settled on ethyl derivative **17**, which presented a modest advantage over the methyl group (**18**), resulting in a 10-fold increase in potency compared to monosubstituted aniline (**20**). The exploration of the R_3 substituent took cues from previous studies on N1-substituents of the indazole ring. While the preference for bulkier groups was also observed, the removal of the rigid indazole ring widened the possibilities, making polar groups an attractive alternative with performances comparable to their nonpolar counterparts. Ultimately, the THP moiety (**17**) was chosen for the R_3 position, given its favorable influence on other physicochemical properties, notably reducing logD and decreasing metabolic clearance.

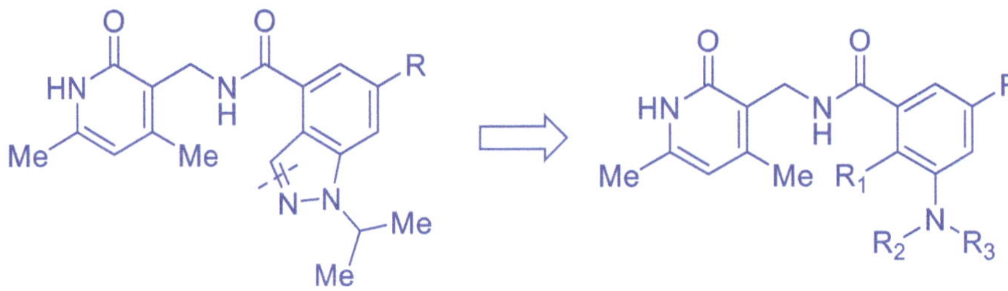

Compound	R_1	R_2	R_3	R_4	EZH2 IC_{50}
16	H	Et	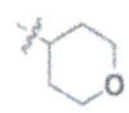	Cl	14 ± 2
17	Me	Et	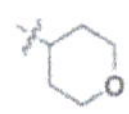	Cl	0.01 ± 0.01
18	Me	Me	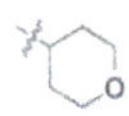	Br	0.03 ± 0.02
19	Me	Me	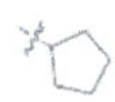	Br	0.03 ± 0.02
20	Me	H	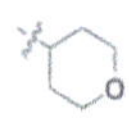	Br	0.5 ± 0.2
21	Me	Me	H	Cl	2.7 ± 1.2

Table 2. Optimization of substituted benzene core

Despite the enhanced potency achieved through extensive optimization, the molecule still faced challenges regarding bioavailability. Insights from HTS brought the 4,6-dimethyl pyridone into the molecule, which was then assessed to identify further optimization sites (Table 3). During this exploration, it became clear that preserving the integrity of the molecule's "warhead" was crucial. Any modifications, such as the addition of a benzylic carbon (**23**) or the methylation of the amide **22**, drastically reduced potency, with reductions ranging from 10- to 100-fold per alteration. Maintaining the 4,6-dimethyl substitutions was essential; removing a single methyl group (**24**, **25**) diminished potency 10-fold, and removing both resulted in a significant 200-fold drop (**26**).

To balance the physicochemical properties, benzyl morpholine was finally incorporated, leading to the creation of the compound now recognized as tazemetostat (**1**). This optimized molecule not only showcases superior potency, minimal clearance, and impressive bioavailability but also demonstrated its effectiveness in subsequent clinical trials.

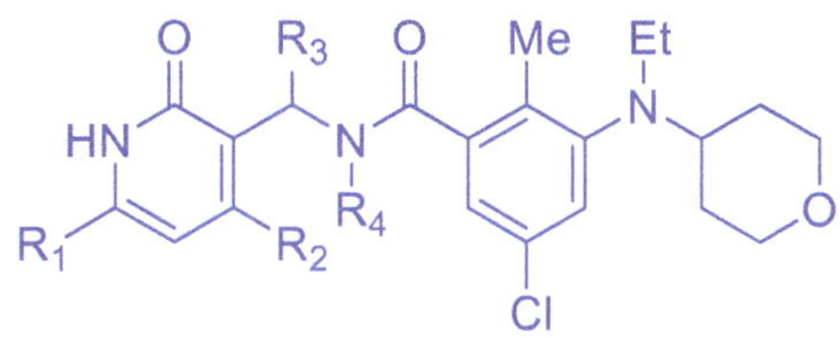

Compound	R_1	R_2	R_3	R_4	EZH2 IC_{50}
17	Me	Me	H	H	0.01 ± 0.01
22	Me	Me	H	Me	0.2 ± 0.1
23	Me	Me	Me	H	3 ± 1
24	Me	H	H	H	0.2 ± 0.01
25	H	Me	H	H	0.1 ± 0.04
26	H	H	H	H	3.3 ± 1.2
27	Me	CF_3	H	H	0.03 ± 0.01

Table 3. Optimization of pyridone warhead

4. Pharmacokinetics and Drug Metabolism

Epizyme marshaled a multidisciplinary team of scientists, dedicating many years of focused research to develop tazemetostat (**1**). The challenge lay not just in achieving potent inhibitory action but also in ensuring optimal physicochemical properties for oral bioavailability.

Tazemetostat (**1**) is administered in its HBr salt form. The recommended oral dose is 800 mg, taken twice daily. This dosing regimen was established based on rigorous pharmacokinetics and pharmacodynamics studies, alongside safety and efficacy evaluations from dose escalation studies.

The drug's mean absolute oral bioavailability stands at 33%. Post-administration, tazemetostat (**1**)'s C_{max} (maximum plasma concentration) is achieved within a 1–2-h window, and its average half-life spans 3–4 h. Its primary metabolic pathway involves N-dealkylation via CYP3A, leading to the formation of metabolites EPZ-6930 (**28**, resulting from the loss of tetrahydropyran) and EPZ-6931 (**29**, through de-ethylation). In some instances, the drug loses both N-alkyl functionalities. These metabolites, however, are considerably less potent, translating to negligible pharmacological effects *in vivo*.[13, 14]

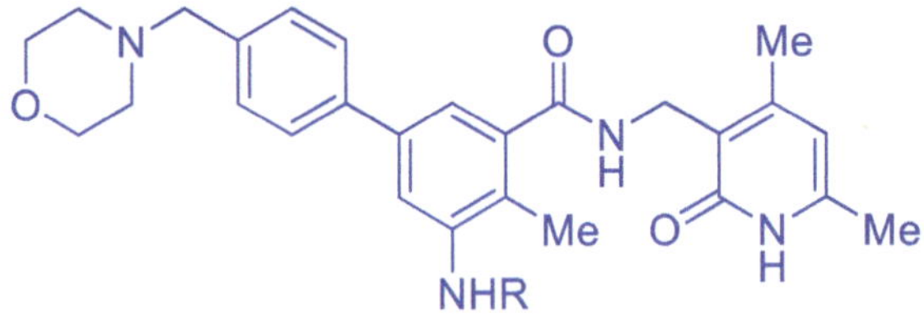

EPZ6930 (**28**) : R = Et
EPZ6931 (**29**) : R = THP

Elimination predominantly occurs through fecal excretion (79%) and, to a lesser extent, via urine (15%) over a 12-day period. Tazemetostat (**1**) has a substantial apparent volume of distribution (V_d) at 1230 L, indicating extensive distribution into body tissues. *In vitro* studies have shown that the drug is 88% bound to human plasma proteins. As a significant substrate of CYP3A, tazemetostat (**1**) has potential drug–drug interaction (DDI) risks when co-administered with other CYP3A inhibitors or inducers. Additionally, it acts both as a substrate and an inhibitor of P-gp; hence, the concurrent use of tazemetostat (**1**) with P-gp inhibitors is not recommended. Notably, the drug's exposure remains largely unaffected by a high-fat meal.[15]

5. Efficacy and Safety

In preclinical evaluations, the in vivo efficacy of tazemetostat (**1**) was consistently demonstrated across multiple tumor xenograft models, highlighting its potential therapeutic application. Tazemetostat (**1**) showcased excellent selectivity over other enzyme, especially over 30-fold selectivity for EZH1, underscoring its targeted mechanism of action.

Transitioning to clinical investigations, tazemetostat's (**1**) promising preclinical efficacy transferred well in human subjects. An initial phase I study assessed both the safety and efficacy of tazemetostat (**1**) as a standalone treatment in patients with relapsed or refractory (*R*/*R*) non-Hodgkin lymphoma and those at high-risk for solid tumors. During the dose-escalation phase, tazemetostat (**1**) dosages ranged between 100 and 1600 mg, administered twice daily. Based on an assessment of the overall response and safety data, a subsequent expansion cohort received an optimized dose of 800 mg, again twice daily ($n = 64$). Importantly, tazemetostat (**1**) was well-tolerated: the predefined maximum tolerated dose threshold was not reached, emphasizing its safety profile[13].

Following this, a phase II trial was initiated, involving 99 *R*/*R* Follicular Lymphoma patients, each having undergone at least two prior therapeutic regimens. Distinct response rates were observed: in the EZH2^{mut} cohort ($n = 45$), a significant 69% objective response rate (ORR), while in the EZH2^{wild} group ($n = 54$), the ORR stood at 35%, which suggests the adoption of this biomarker to determine whether use this drug. The median response duration reached 10.9 months in the EZH2 mutant cohort, with several patients progressing from a partial to a complete response upon extended treatment.[16] Such compelling outcomes underline the potential of EZH2 inhibition as a strategy for managing challenging cases of follicular lymphoma.

It is noteworthy that the combination of tazemetostat (**1**)'s minimal treatment-related adverse events and the robust efficacy data from this trial persuaded the FDA to grant marketing approval in 2020—even before the initiation of a pivotal phase III clinical trial. As of the latest updates, tazemetostat (**1**)'s efficacy is under evaluation across a spectrum of cancers, including epithelioid sarcoma (ES), metastatic castration-resistant prostate cancer (mCRPC), and diffuse large B-cell lymphoma (DLBCL). While approvals have been secured for specific indications, others remain under rigorous clinical scrutiny.[14, 17, 18]

6. Synthesis

The synthesis of tazemetostat (**1**) employs a convergent approach, involving the preparation and subsequent coupling of three distinct molecular fragments through amide coupling and Suzuki coupling. This synthetic route not only is pivotal in the medicinal chemistry development of tazemetostat (**1**) but also plays a critical role in its manufacturing process.[12, 19]

The synthesis begins with the commercially available 2-methyl-3-nitrobenzoic acid (**30**). After bromination with dibromatin, the resulting compound **31** is then subjected to methylation to protect the carboxylic acid. This step involves treating the carboxyl group with Na_2CO_3 and CH_3I, resulting in a methyl benzoate derivative **32**. Then the nitro group of the methyl benzoate derivative is reduced to an amine using Fe and NH_4Cl. This reduction yields the amine compound **33**.

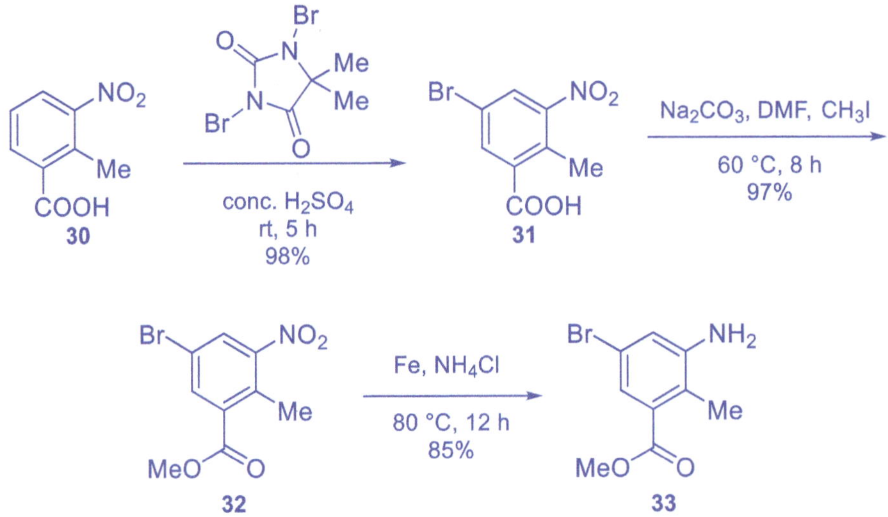

The formation of the tertiary amine core of tazemetostat (**1**) is constructed through two consecutive reductive amination reactions. These two steps conducted under standard reductive amination reaction condition employing $NaBH(OCOCH_3)_3$ as the reducing agent to achieving **35**. Following the dual reductive amination strategies, a base-promoted ester hydrolysis was employed, leading to the formation of the first fragment **36**, which encompasses both bromide and carboxylic acid functional groups for later coupling.

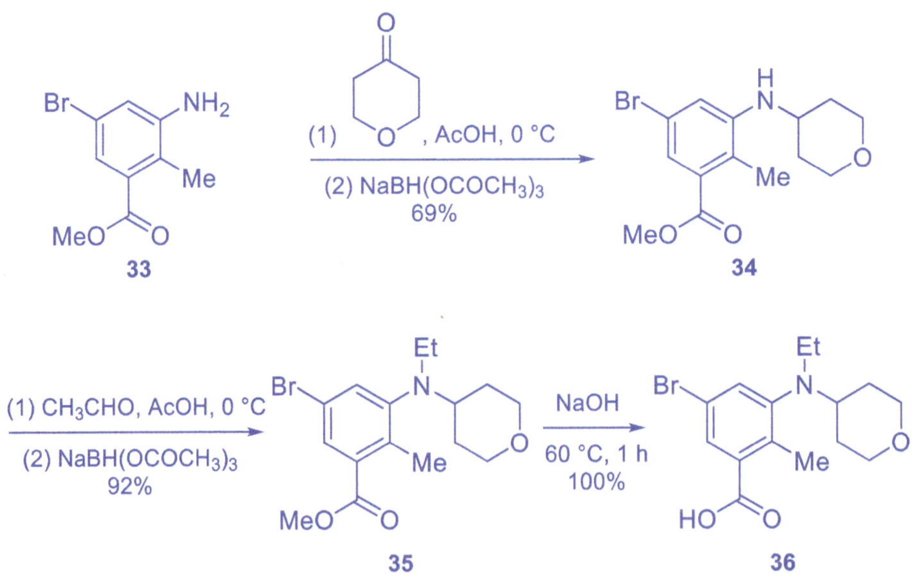

The second fragment **37**, which bears an amine moiety is then coupled to the benzoic acid core through the amide coupling reaction, which install the warhead toward EZH2. This step achieves a 74% yield when using PYBOP as the coupling reagent.

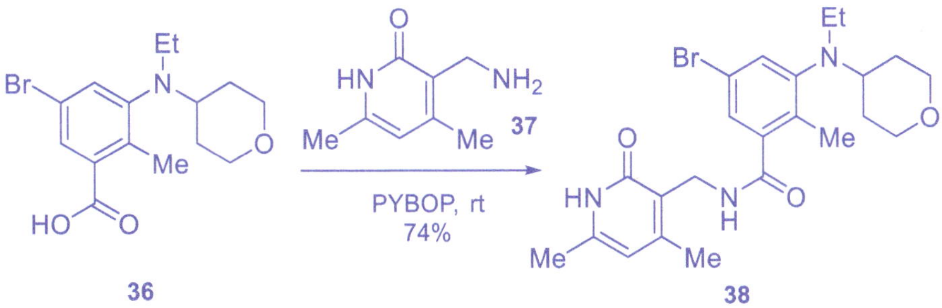

The final step in the synthesis of tazemetostat (**1**) is marked by a Suzuki coupling reaction. This reaction couples compound **39** and corresponding boronic ester **40** using $Pd(PPh_3)_4$. The successful execution of this reaction results in the formation of tazemetostat (**1**) in 71% yield.

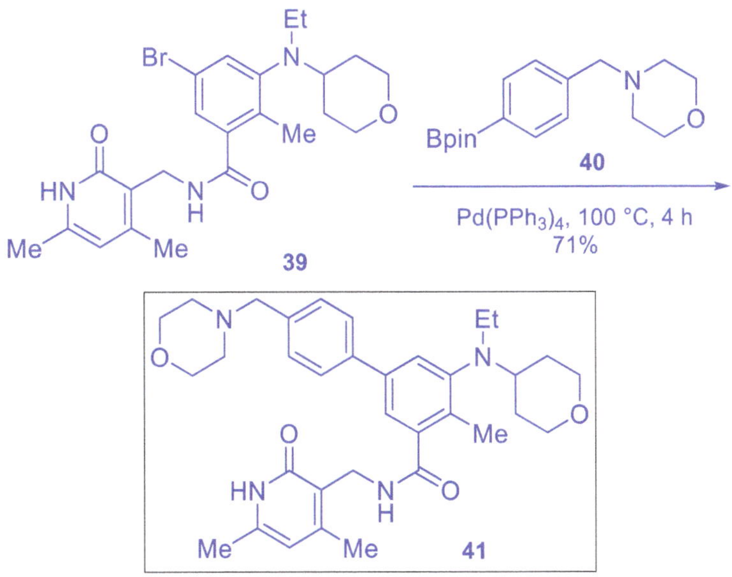

In summary, the medicinal chemistry route for tazemetostat (**1**) is a multistep, convergent process that efficiently assembles the drug through strategic fragment couplings and a series of chemical transformations. This route not only demonstrates the intricacies of organic synthesis but also underscores the importance of each step in achieving the final therapeutic molecule with high purity and yield.

7. Summary

The development of tazemetostat (**1**), a pioneering EZH2 inhibitor, highlights the enduring importance of molecular conformation in drug discovery. This principle is crucial not only for tazemetostat (**1**) but also forms a cornerstone in the creation of numerous effective therapeutic agents. The strategic manipulation of molecular conformation plays a key role in enhancing drug properties such as potency and selectivity.

While the concept of molecular conformation has been known for decades, recent advancements in computing power have provided an additional tool, making the analysis and prediction of molecular shapes more accessible. This enhancement in computational capabilities allows for more sophisticated modeling of drug–target interactions, offering valuable insights that guide the drug development process. However, it is important to note that this is just one aspect of a much broader and multifaceted approach.

The journey of tazemetostat (**1**), particularly through the "magic methyl" effect, demonstrates how subtle structural modifications can significantly alter a drug's efficacy profile. In conclusion, molecular conformation remains a fundamental aspect of drug discovery, essential for the development of new and improved therapies. The modern era, with its advanced computational tools, offers exciting opportunities to deepen this understanding, and the core principles of medicinal chemistry continue to guide this ever-evolving field.

References

1. Knutson, S. K.; Wigle, T. J.; Warholic, N. M.; Sneeringer, C. J.; Allain, C. J.; Klaus, C. R.; Sacks, J. D.; Raimondi, A.; Majer, C. R.; Song, J.; et al. A selective inhibitor of EZH2 blocks H3K27 methylation and kills mutant lymphoma cells. *Nat. Chem. Biol.* **2012**, *8*, 890–896.
2. Kim, K.; Roberts, C. Targeting EZH2 in cancer. *Nat. Med.* **2016**, *22*, 128–134.
3. Barreiro, E. J.; Kümmerle, A. E.; Fraga, C. A. M. The methylation effect in medicinal chemistry. *Chem. Rev.* **2011**, *111*, 5215–5246.

4. Schönherr, H.; Cernak, T. Profound methyl effects in drug discovery and a call for new C–H methylation reactions. *Angew. Chem. Int. Ed.* **2013**, *52*, 12256–12267.

5. Hoy, S. Tazemetostat: first approval. *Drugs* **2020**, *80*, 513–521.

6. Martin, C.; Zhang, Y. The diverse functions of histone lysine methylation. *Nat. Rev. Mol. Cell Biol.* **2005**, *6*, 838–849.

7. Sneeringer, C. J.; Scott, M. P.; Kuntz, K. W.; Knutson, S. K.; Pollock, R. M.; Richon, V. M.; Copeland, R. A. Coordinated activities of wild-type plus mutant EZH2 drive tumor-associated hypertrimethylation of lysine 27 on histone H3 (H3K27) in human B-cell lymphomas. *PNAS* **2010**, *107*, 20980–20985.

8. Souroullas, G. P.; Jeck, W. R.; Parker, J. S.; Simon, J. M.; Liu, J.; Paulk, J.; Xiong, J.; Clark, K. S.; Fedoriw, Y.; Qi, J.; et al. An oncogenic Ezh2 mutation cooperates with particular genetic alterations to induce tumors in mice and redistributes H3K27 trimethylation throughout the genome. *Nat. Med.* **2016**, *22*, 632–640.

9. McCabe, M. T.; Ott, H. M.; Ganji, G.; Korenchuk, S.; Thompson, C.; Van Aller, G. S.; Liu, Y.; Graves, A. P.; Della Pietra, A., 3rd; Diaz, E.; et al. EZH2 inhibition as a therapeutic strategy for lymphoma with EZH2-activating mutations. *Nature* **2012**, *492*, 108–112.

10. Kung, P. P.; Bingham, P.; Brooun, A.; Collins, M.; Deng, Y. L.; Dinh, D.; Fan, C.; Gajiwala, K. S.; Grantner, R.; Gukasyan, H. J.; et al. Optimization of orally bioavailable enhancer of zeste homolog 2 (EZH2) inhibitors using ligand and property-based design strategies: identification of development candidate (*R*)-5,8-dichloro-7-(methoxy(oxetan-3-yl)methyl)-2-((4-methoxy-6-methyl-2-oxo-1,2-dihydropyridin-3-yl)methyl)-3,4-dihydroisoquinolin-1(2*H*)-one (PF-06821497). *J. Med. Chem.* **2018**, *61*, 650–665.

11. Vaswani, R. G.; Gehling, V. S.; Dakin, L. A.; Cook, A. S.; Nasveschuk, C. G.; Duplessis, M.; Iyer, P.; Balasubramanian, S.; Zhao, F.; Good, A. C. Identification of (R)-*N*-((4-methoxy-6-methyl-2-oxo-1,2-dihydropyridin-3-yl)methyl)-2-methyl-1-(1-(1-(2,2,2-trifluoroethyl)piperidin-4-yl)ethyl)-1*H*-indole-3-carboxamide (CPI-1205), a potent and selective inhibitor of histone methyltransferase EZH2, suitable for phase I clinical trials for B-cell lymphomas. *J. Med. Chem.* **2016**, *59*, 9928–9941.

12. Kuntz, K. W.; Campbell, J. E.; Keilhack, H.; Pollock, R. M.; Knutson, S. K.; Porter-Scott, M.; Richon, V. M.; Sneeringer, C. J.; Wigle, T. J.; Allain, C. J.; et al. The importance of being me: magic methyls, methyltransferase inhibitors, and the discovery of tazemetostat. *J. Med. Chem.* **2016**, *59*, 1556–1564.

13. Italiano, A.; Soria, J. C.; Toulmonde, M.; Michot, J. M.; Lucchesi, C.; Varga, A.; Coindre, J. M.; Blakemore, S. J.; Clawson, A.; Suttle, B.; et al. Tazemetostat, an EZH2 inhibitor, in relapsed or refractory B-cell non-hodgkin lymphoma and advanced solid tumours: a first-in-human, open-label, phase 1 study. *Lancet Oncol.* **2018**, *19*, 649–659.

14. Gounder, M.; Schöffski, P.; Jones, R. L.; Agulnik, M.; Cote, G. M.; Villalobos, V. M.; Attia, S.; Chugh, R.; Chen, T. W.-W.; Jahan, T.; et al. Tazemetostat in advanced epithelioid sarcoma with loss of INI1/SMARCB1: an international, open-label, phase 2 basket study. *Lancet Oncol.* **2020**, *21*, 1423–1432.

15. Package insert, available at: https://dailymed.nlm.nih.gov/dailymed/drugInfo.cfm?setid=7db07b5f-4e22-467c-9c0a-f830b08dbb1d

16. Morschhauser, F.; Tilly, H.; Chaidos, A.; McKay, P.; Phillips, T.; Assouline, S.; Batlevi, C. L.; Campbell, P.; Ribrag, V.; Damaj, G. L.; et al. Tazemetostat for patients with relapsed or refractory follicular lymphoma: an open-label, single-arm, multicentre, phase 2 trial. *Lancet Oncol.* **2020**, *21*, 1433–1442.

17. Choudhury, A. D.; Xie, W.; Tewari, A.; Miyamoto, D. T.; Kochupurakkal, B.; Ellis, L.; Bandel, M.; Leisner, C.; Shapiro, G.; D'Andrea, A. D.; et al. A phase Ia/Ib study of talazoparib in combination with tazemetostat in metastatic castration-resistant prostate cancer (mCRPC). *J. Clin. Oncol.* **2022**, *40*, TPS5098.

18. Ribrag, V.; Morschhauser, F.; McKay, P.; Salles, G. A.; Batlevi, C. L.; Schmitt, A.; Tilly, H.; Cartron, G.; Thieblemont, C.; Fruchart, C.; et al. Interim results from an ongoing phase 2 multicenter study of tazemetostat, an EZH2 inhibitor, in patients with relapsed or refractory (R/R) diffuse large B-cell lymphoma (DLBCL). *Blood* **2018**, *132*, 4196.

19. Wang, L.; Li, R.; Song, C.; Chen, Y.; Long, H.; Yang, L. Small-molecule anti-cancer drugs from 2016 to 2020: synthesis and clinical application. *Nat. Prod. Commun.* **2021**, *16,* 1–42.

Section III. CNS DRUGS

11

Ozanimod (Zeposia): An S1P Receptor Modulator for Treating Multiple Sclerosis and Inflammatory Bowel Diseases

Shaohui Yu and
Xi Wang

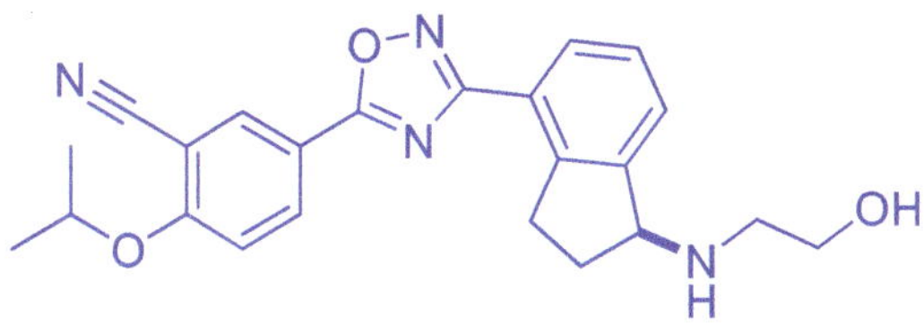

ozanimod (Zeposia, **1**)
Celgene/BMS, 2020
S1PRM

1. Background

1.1. MS and IBD

One of the most remarkable properties of the immune system is its ability to distinguish between self-cells and foreign cells. In some individuals, there's a flaw in the process and autoimmune disease occurs because of the inappropriate self-attack.[1]

Multiple sclerosis (MS) is a chronic autoimmune disease of the central nervous system (CNS). In MS, myelin which is the fatty tissue that protects nerve fibers, is attacked upon immune cell activation forming scar tissues called sclerosis.[2,3] As a result, the nerves are damaged and lose the ability to conduct electrical impulses to and from the brain, triggering progressive neurodegeneration of the brain and spinal cord. MS is typically present in young adults with the average onset age between 20 and 30 years.[4] Common symptoms of MS include fatigue, depression, bowel and bladder dysfunction,

Chemistry and Pharmacology of Drug Discovery, First Edition. Edited by Jie Jack Li.

pain, impaired mobility, cognitive problems that often interfere with personal relationships, and impact quality of life (QOL).[5] Over the last decade, the prevalence of MS has been increasing worldwide and ranges from 5 to 300 per 100,000 people.

Relapsing-remitting multiple sclerosis (RRMS) is the most common type of MS, marked by discrete, day-to-week attacks (relapses), followed by a week-to-month symptom-free interval (remissions). Approximately 85% of people with MS are initially diagnosed with RRMS. The exact etiology and pathogenesis of MS are still uncertain, but it is commonly believed that both genetic and environmental factors contribute to the aberrant immune responses leading to disease progression. Currently there's no cure for MS patients and the most common treatment has evolved from immunosuppressant in mid-1990s to disease-modifying therapies (DMTs) today. The first disease-modifying drug for RRMS, interferon beta-1 (IFNβ-1) was the key breakthrough for the treatment of MS.[6,7] It can dramatically reduce the number of attacks and disease progression in the early relapsing phases of MS, but not in the advanced phases.[8]

Inflammatory bowel diseases (IBDs) are chronic relapsing inflammatory disorders of the gastrointestinal tract characterized by two major clinically defined forms: ulcerative colitis (UC) and Crohn's disease (CD). Like MS, the etiology of IBD is not fully understood with immunologic abnormalities, genetics, environmental factors, and the microbiota all playing a role.[2,9] UC involves localized superficial inflammation that affects the colon and only invades the inner lining of bowel tissue. CD manifests as transmural inflammation that can spread anywhere in the gastrointestinal (GI) tract and penetrates the intestinal layers from inner to outer lining.[2,10] The global prevalence of IBD is high, with approximately 4.9 million cases in 2019.[11,12] Conventional treatments for UC include aminosalicylates, corticosteroids, and thiopurines.[8] In the late 1990s, the advent of anti-tumor necrosis factors (TNFs) made a major impact on the therapeutic algorithms and until now they remain the frontline treatment of IBD.[13] However, many patients still require alternative therapies due to non-response, loss of response, or intolerance of anti-TNFs.[13]

At first glance, MS and IBD are clinically distinct diseases. However, for decades, mounting evidence have prompted researchers to suspect about a possible link between the two.[14–16] Both MS and IBD are diseases of a relapsing and remitting nature and are mediated by the impairment of the immune system; they are also prevalent in young people in the developed countries.[17] Many patients with IBD (particularly UC) exhibit abnormal T2 signals in the white matter, accompanied or not by clinical symptoms that mimic MS.[18,19] Other research indicates an increased risk of IBD among patients with MS compared to the general population.[20,21] A systematic review showed a 50% increased risk of comorbidities between IBD and MS without any significant difference for this association between the UC and CD.[22]

Besides the evidence that commensal gut bacteria play a role in both IBD and MS, scientists have revealed that bidirectional communication of inflammatory signals through the gut-brain axis is important to relay the health status of the host and stimulate regulatory responses.[23,24] The migration of intestinal immune cells to CNS may contribute to the pathogeneses of neurological and neurovegetative diseases such as MS.[24] More recent studies at molecular level have demonstrated in both MS and IBD, Th17 cell play important roles. Th17 cells promote inflammation primarily through the secretion of cytokines such as IL-17 and TNF-α.[25,26] Either alone or in conjunction with TNF-α, IL-17 promote the secretion of inflammatory mediators, chemokines and proteases. These factors induce inflammation and promote the enlisting, activation and movement of neutrophil to target tissues.[27,28] In addition, evidence has shown that in both MS and IBD, the overproduction of TNF contributes to various pathological hallmarks such as demyelination and transmural inflammation.[13,25,26]

The mysterious relationship between IBD and MS continues when taking the therapies into picture. If TNF overproduction plays an important role in both IBD and MS, then why anti-TNF drugs such as Infliximab and Adalimumab can significantly increase MS incidence in IBD patients?[29,30] To add to the riddle, why INF-β therapies, that worked miracle in some MS patients, was reported to exacerbate IBD symptoms?[31] Considering the common immunological and pathophysiological overlaps between IBD and MS, the discrepancies in their responses to the same treatment suggest additional mechanism might exist that scientists yet to find out.

1.2. Treatment That "Kills Two Birds with One Stone"

Natalizumab is one of the successful treatments for both MS and CD developed around 2000s. As a humanized monoclonal IgG4κ antibody, Natalizumab can selectively bind to α4-integrin and inhibit the interaction of α4β1 with VCAM-1 and α4β7 with MAdCAM-1.[32] Both VCAM-1 and MAdCAM-1 are found on endothelial cells and interact with α4β1 and α4β7 on leukocytes for firm adherence before extravasation into inflamed tissues.[33] As a result, natalizumab prevents migration of autoreactive leukocytes out of blood vessels into target organs.[34] Despite its success, evidence has emerged to show that some integrins such as α4β1, αLβ2, also play important roles in other immune phenomena such as the formation of immune synapse and the differentiation of Th1 lymphocytes. Therefore, long-term administration of blocking agents such as Natalizumab may result in adverse effects.[35]

Fingolimod was discovered in 1995–1996 following a chemical derivatization program based on the fungal metabolite myriocin (also known as ISP-1).[36,37] Researchers found the binding of fingolimod to a new molecular target directly located in the lymph node (LN) and not at the vascular levels that led to the unveiling of the metabolisms behind S1P and its receptors.[38] Since then, this class of drug known as S1P receptor modulators has gained great interest and the second-generation compounds such as ponesimod (ACT-128800), siponimod (BAF312), ceralifimod (ONO-4641), ozanimod

(**1**, RPC-1063), and amiselimod (MT-1303) with higher receptor selectivity were developed.[39] Comparing with anti-integrin therapies, which requires invasive route of administration, S1PR modulators are orally administered with other inherent advantages such as low manufacturing cost and the absence of immunogenicity (Figure 1).

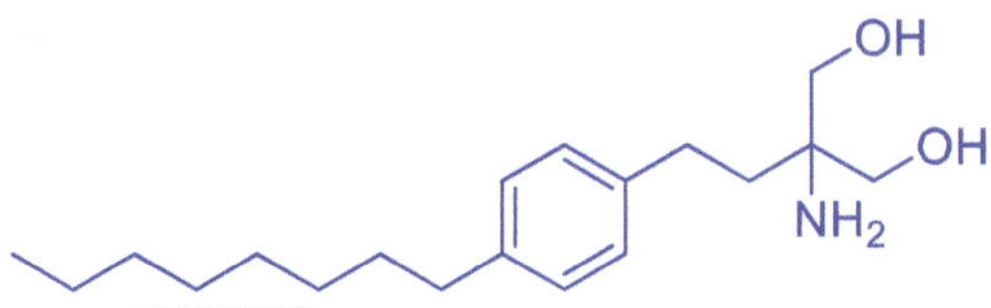

Figure 1. The structure of fingolimod

Ozanimod (**1**) is the first S1P modulator that has the potential to become a combinatory treatment for both MS and IBD. It was approved by the US Food and Drug Administration (FDA) for the management of MS in 2020 and UC in 2021. The ongoing clinical trial YELLOWSTONE (Phase 3) is designed to test the safety and efficacy of ozanimod (**1**) in patients with CD.[40]

Ozanimod (**1**) is also the first drug tested in human subjects after the establishment of the Molecular Libraries Program (MLP) funded by National Institute of Health (NIH).[41] The initial discovery of ozanimod (**1**, RPC-1063) was reported by Rosen, Roberts and their teams at Scripps Research Institute in a series of papers from 2002 to 2008.[42–45] In 2009, Scripps Research licensed ozanimod (**1**) to a biotechnology start-up Receptos while MLP undertook the assay development, high-throughput screening (HTS) and medicinal chemistry.[46] Within 18 months, the company optimized ozanimod (**1**) began its phase 1 study.[47] Receptos was subsequently purchased by Celgene in 2015 for US$ 7.3 billion, which was later acquired by Bristol-Myers Squibb in 2019 (Figure 2).[48]

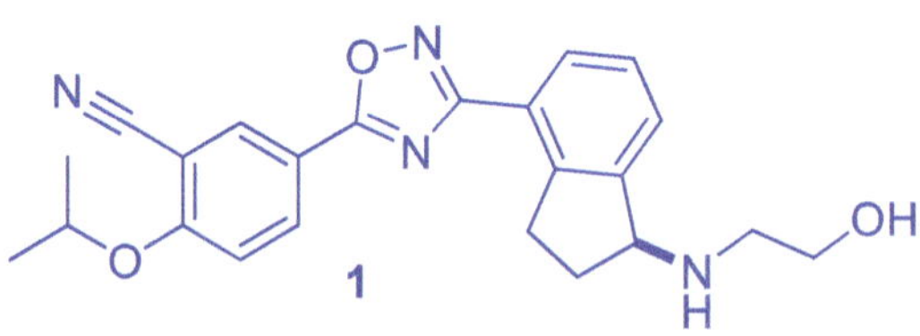

1

Figure 2. The structure of ozanimod (**1**)

2. Pharmacology

2.1. S1P and Its Receptors

Sphingosine-1-phosphate (S1P) is lipid molecule with bioactive properties involved in numerous cellular processes such as cell growth, movement, programmed cell death, self-degradation, cell specialization, aging, and immune system reactions.[49] With sphingoid bases as building blocks, all cells make S1P intracellularly as an essential element in membrane structure (Figure 3).[38,50]

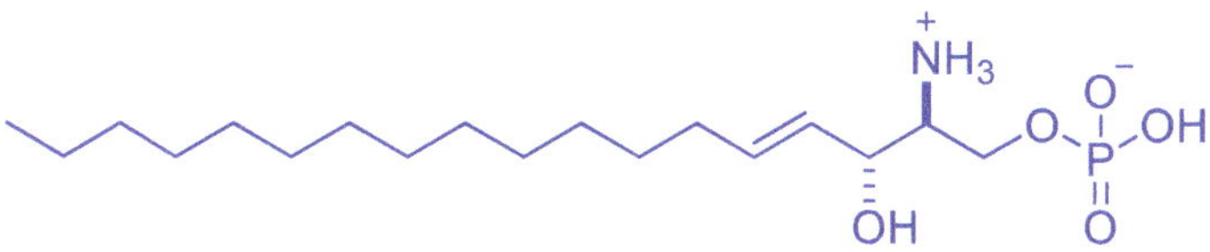

Figure 3. The structure of sphingosine-1-phosphate

S1P is also secreted. Extracellular S1P acts as a signaling molecule and is involved in many physiological processes.[10,51] In immune cells, S1P plays an important role in supporting T-cell survival.[52] At sites of inflammation, S1P is involved in the recruitment of immune cells, which could further exacerbate the inflammatory process.[53] It is important to note that S1P signaling can be essential in an inflammatory response regardless of its concentration gradients (more details below).[54] S1P also holds an important function in the regulation of vascular integrity as a suppressor of angiogenesis.[49]

Extracellular S1P work as a ligand of high affinity G-protein-coupled receptors (GPCRs) on cell surfaces.[55] These receptors comprise five distinct subtypes which are linked to diverse signaling pathways and are regulated by distinct mechanisms. T and B lymphocytes, as well as endothelial cells, all express distinctive profiles of S1P receptors, which are major regulators of development, recirculation, tissue homing patterns and chemotactic responses to chemokines of B and T cells.[56]

S1P receptor subtype 1 (S1PR1) is expressed by lymphocytes, dendritic cells, cardiomyocytes and vascular endothelial cells and is involved in the regulation of chronic inflammation via mediation of lymphocyte egress from secondary lymphoid organs.[57,58] When S1PR1's function is blocked, lymphocytes are sequestered in the LNs and thymus.[59] S1PR1 is also expressed on endothelium where it is involved in dendritic cell recruitment and vascular permeability.[59] During inflammation, S1P-S1PR1 may play a role in lymphocyte retention within inflamed tissues.[56] The S1P-S1PR1 axis also participates in naïve B cell trafficking from bone marrow to blood, egress from secondary lymphoid organs, and marginal B cell localization in spleen.[56] Stimulating S1PR1 leads to the activation of G protein and subsequently Ras, phosphatidylinositol 3-kinase (PI3K), phospholipase C (PLC), signal transducers, activators of transcription 3 (STAT3), and nuclear factor kappa B (NF-κB). RAS and PI3K, in return, activate extracellular regulated protein kinase (ERK), Rac and protein kinase B (Akt).[60–62] All these events result in T cells egress from LNs, negative inotropy, cardio-protection and endothelial function.[63] S1PR2 receptors might have opposite functions to S1PR1 with a pro-inflammatory role.[56] It is also involved in mast cell degranulation, histamine secretion, bronchial smooth muscle contraction and hair cell survival in the inner ear.[64] Together with S1PR1, S1PR3 is important in the regulation of heart rate in human.[65] In addition, S1PR3 modulation was found linking to heart rate effects and the risk for type I and II atrioventricular block.[66] S1PR4 is expressed predominantly on immune cells but also on airway smooth muscle cells, which regulates cytokine production by T cells.[67] S1PR4 stimulation contributes to the activation of dendritic cells.[68] S1PR5 is expressed primarily in CNS white matter tracts, predominantly by oligodendrocytes, the survival of which is promoted by S1P activation.[69] In addition, S1PR5 also regulates natural killer cell development and trafficking from bone marrow.[59] S1PR1, S1PR2, and S1PR3 are broadly expressed in most of organs, while S1PR4 and S1PR5 are primarily expressed in lymphoid tissues and the lung and brain, leukocytes and spleen respectively.[70]

Currently, there are four S1PR modulators: fingolimod, siponimod, ozanimod (**1**), and ponesimod that obtained regulatory approvals. Fingolimod (FTY720) is a prodrug that requires phosphorylation into their bioactive form whereas others are direct acting. As a first generation S1P receptor modulator, fingolimod targets four of the five receptors (S1PR1, 3, 4, and 5) and Phosphorylated FTY720 (pFTY720) binds both S1PR1 and S1PR5 with high affinity.[42,71] In the case of S1PR1, FTY720 has mixed agonist–antagonist effects, where it initially induces strong activation of the receptor and likely causes the acute cardiac side effects in patients.[72] However, pFTY720 also shifts S1PR1 into a conformation that results in preferential ubiquitination and degradation, and S1PR1 is ultimately lost from the cell surface after FTY720 treatment.[73] For immune cell trafficking, it is not essential to resolve whether FTY720 acts primarily as an agonist or antagonist, as the cells would not follow S1P gradients because it is seeing an S1P receptor signal either everywhere (if FTY 720 is an agonist) or nowhere (if FTY720 is an antagonist).[54] Following the success of FTY720, at least a dozen additional drugs targeting S1P signaling have entered clinical trials.[74] Siponimod and ozanimod (**1**) are S1PR1 and S1PR5 modulators, while Ponesimod is specific for S1PR1.

2.2. S1P Concentration Gradient and Lymphocytes Egress

Through a series of balanced enzymatic pathways, synthetic and degradative enzymes are responsible for the synthesis of S1P and the maintenance of its concentration in various tissue types. S1P is generated intracellularly through phosphorylation of sphingosine by sphingosine kinase 1 (SphK1) and sphingosine kinase 2 (SphK2) and can be reversibly dephosphorylated to sphingosine by sphingosine phosphatases 1, 2 and irreversibly degraded by S1P lase.[75] S1P concentration is much higher in blood and lymph than interstitial spaces including immune organs such as spleen, thymus and LNs. This differential in S1P concentrations across various tissue types is known as the S1P gradient and plays a key role in the trafficking of lymphocytes and other leukocytes from immune organs to sites of inflammation.[76] There are many theories on how S1P gradients are created and maintained, and many reviews have in depth discussions regarding this process.[77–80]

S1PR1 on lymphocytes is very sensitive to S1P exposure and gets completely internalized after incubation with as little as 1nM S1P for 20 min *ex vivo.* The high concentration of S1P (~1 μM) in blood guides immune cells out of the thymus, spleen, and bone marrow.[81] In response to the abundant blood S1P, T cells in blood fully internalize their surface S1PR1.[82] If S1PR1 internalization is blocked, T cells enter LNs from blood inefficiently, presumably because they do not lose their attraction to blood.[82] Plasma S1P turns over rapidly with a half-life of only 15 mins.[54,83]

Lymph S1P is supplied by lymphatic endothelial cells with a high concentration of ~0.1 μM.[81] Immune cells are guided out of LNs, Peyer patches, and nonlymphoid tissues by lymph S1P and then travels into circulation. Plasma S1P entering tissues are destroyed to maintain the S1P gradients between tissues and circulation, but it remains a mystery how it is achieved.[54]

S1P signaling plays a critical role in mature T-cell exit from the thymus and a supporting role in immune cell exit from the bone marrow.[54] The S1P gradient between the thymic parenchyma and blood is tightly controlled and T-cell exit is very sensitive to disruption of the gradient.[54] T cells leave the thymus via specialized vessels at the corticomedullary junction.[84] The process ensures only mature T cells that express abundant S1PR1 to navigate selection and leave.[54] Comparing with the exit from the thymus into blood, exit from bone marrow into blood is only weekly dependent on S1P gradients.[54] It was found that loss of any of the receptors known to mediate exit from bone marrow (S1P receptors, CCR2 for monocytes, or CXCR2 for neutrophils), leads to a much less profound block in bone marrow exit than loss of S1PR1 for thymic exit.[85] It is possible that the highly porous bone marrow endothelium, which permits passive export of red blood cells, can permit passive export of white blood cells in the absence of strong retention signals.[85] Lymphocytes exit from the spleen has been more difficult to study than those exiting from LNs, because both entry and exit are via blood. However strong evidence suggests that splenic egress is driven by S1P gradients between spleen and blood.[54] Follicular B cell exit from the spleen into blood in an S1PR1-dependent manner that has been confirmed by intravital two photon microscopy: S1PR1-deficient B cells exited the B follicle into the marginal zone at a rate at least 10-fold lower than that of wild-type B cells.[86]

Mature lymphocytes circulate among secondary lymphoid organs including LNs, spleen, and Peyer patches.[51] Lymphocytes from blood enter LNs and mucosa associated lymphoid tissues via high endothelial venules (HEVs) located in the T cell zones.[54] When a naïve T cell enters a lymphoid node where S1P concentration is low, it gradually recovers the surface expression of S1PR1 and regains the ability to migrate out from lymphoid nodes toward the higher S1P concentrations in blood and lymph.[87,88] This re-expression takes several hours (~4–24 h), which allows a naïve T cell to interact with antigen-presenting cells.[89] If a naïve T cell is activated by antigen in the lymph node, the re-expression of S1PR1 is then suppressed for several days through antigenic stimulation and interferon binding, which renders the antigen-activated T cells transiently insensitive to the S1P gradients.[89] After several days of clonal expansion and differentiation into effector cells, S1PR1 is re-expressed which makes the cells responsive to S1P concentration and exit the lymph node.[89] In most cases, activated lymphocytes exit LNs to travel to sites of infection. However, LNs egress also enables tissue damage in autoimmune diseases and in some cases, also contributes to disease dissemination.[54]

In the case where a naïve T cell is not activated by antigen in the lymph node, the cells re-express S1PR1 after several hours, exit lymph nodes and eventually return to the blood stream which completes one circulation loop.[89] Thus, lymphocytes utilize this S1P gradient between lymphoid organs and the circulation as a cue for the egress process. Disruption of the S1P gradient either by inhibition of sphingosine kinase or by inhibition of S1P lyase results in lymphopenia due to defect in the lymphocyte egress.[87] In addition, S1P production is also triggered by the pro-inflammatory cytokines such as interleukin-1 (IL-1), TNF-α, and vascular endothelial growth factor (VEGF), increasing S1P concentration in efferent lymphatic vessels.[90] This increase of the S1P gradient between the LNs and the efferent lymphatic vessel during inflammation causes lymphocytes to exit the LNs, re-enter blood stream, be mobilized to the site of inflammation and perpetuate the inflammatory response.[91]

The function of S1P receptors to guide immune cells from the low S1P environment of the LNs to the high S1P environment of the lymph has triggered intense research activities. However, one should understand it can be extremely challenging to measure extracellular signaling-available S1P in LNs. As S1P is a lipid, there is no mRNA to track, and there is also no single protein can serve as a proxy.[54] Therefore, many data to test the S1P signaling can be more elusive than one might expect.[54] In addition, S1P signaling contributes to multiple aspects of an immune response and it has been difficult to construct a holistic picture of how S1P signaling affect diseases. Therefore, it seems likely that the efficacy of S1P modulating drugs is attributable to multiple processes.

2.3. Ozanimod Mechanism of Action

In vitro studies suggested ozanimod (**1**) showed its greatest affinity for S1PR1 and S1PR5 specifically. Its affinity for S1PR1 was found to be 27-times greater than for S1PR5 and greater than 10,000-fold over S1PR2,3,4.[10] There are several proposed mechanisms of action for ozanimod (**1**).

It is proposed that the primary mechanism of action for ozanimod (**1**) is its effects as an S1PR modulator on the trafficking of immune cells. And the model of "functional antagonism" as a mode of action for S1PR1 agonists such as ozanimod (**1**) is now well established.[92] By artificially binding S1P1R on lymphocytes, ozanimod (**1**) first exerts an agonistic effect.[93–95] Binding to S1PR1 leads to agonist recruitment of beta-arrestins to the receptor complex,[96] which in turn promotes receptor internalisation.[97,98] As a result, cells no longer respond to the gradient and remain within the LNs, resulting in pharmacodynamic effects characterized by a significant but rapidly reversible reduction in circulating lymphocyte subsets.[97] Downregulation of the lymphocyte expressed S1PR1 prevents their ability to respond to the S1P gradient and egress from peripheral lymphoid tissues. This results in a "functional antagonism" and reduces circulating lymphocytes, including auto-reactive lymphocytes, preventing trafficking to sites of inflammation and contribution to disease.[10] The effect of ozanimod (**1**) is primarily observed on naïve and central memory B and T lymphocytes expressing the chemokine receptor CCR7.[99]

In clinical studies, ozanimod (**1**) was found to produce manageable chronotropic cardiac effects at initiation followed by expected reductions in peripheral blood lymphocyte count.[100] The pharmacodynamic reduction in heart rate can be circumvented by implementation of drug titration for the first 4–5 days of treatment.[101,102] This profile is consistent with transient agonism and subsequent functional antagonism of S1PR1.[98] It is worth mentioning that all agents that engage S1PR1 will have some form of cardiac effect.[98] The possible involvement of S1PR3 in humans remains unclear and new modulators such as siponimod and ozanimod (**1**) with minimal S1PR3 binding does not prevent first-dose cardiac effects.[103,104]

Although the internalization of S1PR1 in response to ligand binding that leads to functional antagonism is well established, evidence suggests that the binding of ozanimod (**1**) to S1PR5 may not lead to receptor down-modulation or halt signaling in the same manner.[98] Ozanimod (**1**) may act as an agonist for S1PR5, although inhibition by non-internalization mechanism remains a possibility.

Ozanimod (**1**) may also act through a completely different mechanism. Multiple S1P receptors are found on endothelial surfaces and play a role in regulating endothelial function and stabilization.[76] Specifically, stimulation of S1PR1 is believed to enhance cellular junctions.[90] As a S1PR1 agonists, ozanimod (**1**) induced "tightening" of cell junctions, inhibiting trans-endothelial migration and blocking lymphocyte egress from the lymph node.[42,105] Consequently, lymphocytes are trapped, decreasing circulating lymphocyte counts. Research showed that upon withdrawal of the agonist agent, the permeability is restored and lymphocyte trafficking resumes.[106]

In addition to the effect of ozanimod (**1**) on circulating lymphocytes, another mode of action has been proposed in the applicability of the drug as a neuroprotective agent. This observation was based on a study in which ozanimod (**1**) showed the exertion of a beneficial effect on experimental autoimmune encephalomyelitis (EAE) mice even when absolute lymphocyte count (ALC) was normal.[107] Ozanimod (**1**) can readily cross the blood–brain barrier (BBB) and thus potentially exerts beneficial effects by directly establishing interactions with brain cells, possibly via S1PR5 signaling. Studies in mice have shown that ozanimod (**1**) reduces axonal damage, thereby preserving CNS tissue morphology after induced demyelination.[108] The anti-inflammatory and neuroprotective effects of Ozanimod were tested in *ex vivo* studies in EAE brains.[98] Ozanimod (**1**) treatment ameliorated EAE-driven striatal glutamatergic synapse alterations in an S1PR1-dependent manner and lowered the expression level of pro-inflammatory markers in EAE brain slices and microglial cell cultures.[107]

In fact, the ability to cross the blood–brain barrier for all S1P receptor modulators is probably one of the most important features that makes them a potential combinatory treatment for both MS and IBD. The blood–brain barrier that resists drugs from penetrating might be the true "barrier" for relating IBD and MS together.[109] The questions we raised early in this chapter regarding TNF and IFN-β treatment might find an answer by bearing the "barrier" hypothesis in mind. It is possible that the efficacy discordance of TNF blockade in MS compared with IBD is the inability of TNF antagonists to pass through the BBB to neutralize and prevent TNF mediated injury.[110] The case for why IFN-β works for MS but seem to exacerbate IBD patients is more complicated. Researchers have found IFN-β has proinflammatory functions and aggravates the pathogenesis of Th17-mediated inflammatory disease with only exception, MS.[109,111] It is hypothesized that IFN-β only has an indirect effect in MS by tightening the blood–brain barrier and resisting inflammatory cells or cytokines across rather than treating MS itself directly.[109,111,112] While there are no direct data on the effect of ozanimod (**1**) on BBB integrity in MS, another S1PR modulator siponimod has been shown to have beneficial effects on occluding and zonula occludens-1 in a mouse model of traumatic brain injury.[113] In addition, cellular and animal studies of fingolimod and siponimod have demonstrated a range of effects including actions on astrocytes and microglia and involvement in remyelination and repair.[98,114–116] Together these evidence suggested that positive effects on tight junction (TJ) and BBB integrity as well as neuroprotection may be a class effect of S1PR modulators.[98]

3. Drug Metabolism and Pharmacokinetics

Ozanimod (**1**) does not require *in vivo* phosphorylation for activation.[108] It is metabolized by multiple enzymes to form two major active plasma metabolites [CC112273 (**14**) and CC1084037 (**5**)], a major inactive plasma metabolite (RP101124) and several minor active plasma metabolites (RP101075, RP101988, RP101442, RP112289, and RP112509).[58] All metabolites show similar selectivity for S1PR1 and S1PR5. The FDA approval of ozanimod (**1**) was initially delayed for 2 years by insufficient preclinical characterization of the active metabolite CC112273 (**14**), causing billions of dollars of lost sales. The ozanimod (**1**) example illustrate the need for conducting timely radiolabeled human metabolites studies and assessment of exposure coverage during drug development (Figure 4).[58]

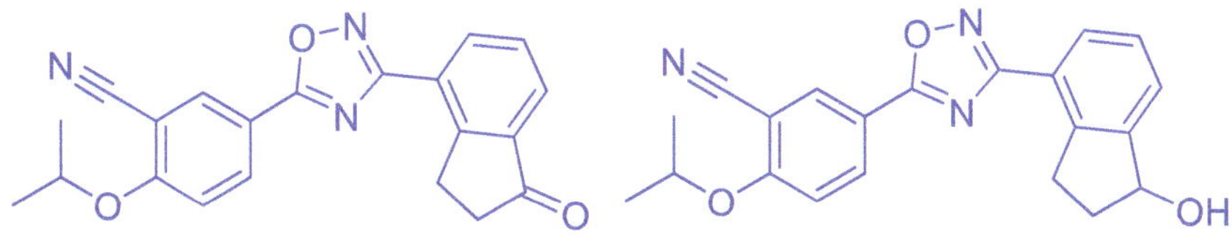

CC112273 (**14**) CC1084037 (**5**)

Figure 4. Structures of ozanimod's major metabolites CC112273 (**14**) and CC1084037 (**5**)

Ozanimod (**1**) causes a dose-dependent and reversible reduction in the ALC. In a 7-day multiple ascending dose regimen, the median reduction in ALC was 14%, 49% and 68% respectively, for doses of 0.3 mg, 1 mg, and 2 mg. ALC returned to normal ranges within 6 days of stopping. Though patients experiencing excess S1P modulatory effects can experience lymphopenia,[100] with a short half-life of 19 h, ozanimod (**1**) allows for rapid lymphocyte recovery.[77] Careful review of events in clinical trials further indicated that severe lymphopenia is not a risk, and is manageable with dose adjustment or interruption, and is not associated with serious adverse events.[117–119]

Ozanimod (**1**) is metabolized by three primary pathways, including aldehyde dehydrogenase and alcohol dehydrogenase, cytochrome P450 isoforms 3A4 and 1A1, and reductive metabolism by gut microflora. The primary metabolite RP101075 is further metabolized to form major active metabolite CC112273 (**14**) by monoamine oxidase B, which further undergoes reduction by carbonyl reductases to form CC1084037 (**5**) or CYP2C8-mediated oxidation to form RP101509. CC1084037 (**5**) is oxidized rapidly to form CC112273 (**14**) by aldo-keto reductase 1C1/1C2 and or 3β-and 11β-hydroxysteroid dehydrogenase and this reversible oxidoreduction between two active metabolites favors CC112273.[58]

Preclinical studies have shown that the pharmacokinetic properties of the minor active metabolites (RP101988 and RP101075) are similar to those of ozanimod (**1**) with similar affinities for S1PR1 and 5.[98,120] The two major metabolites are CC112273 and CC1084307. Approximately 94% of total circulating active drug is made up of Ozanimod (6%), CC112273 (73%) and CC1084037 (15%). The areas under the concentration-time curves (AUCs) of CC112273 and CC1084037 are 13-fold and 2.5-fold greater than that of ozanimod (**1**).[58] CC112273 is several folds less potent than Ozanimod for S1PR1 and 5 and is thought to account for the majority of ozanimod (**1**) activity in humans owing to its relative abundance.[58] The less abundant major active metabolite CC1084307 is about 10 times as potent as CC112273.[58] CC112273 and CC1084037 were bound 98.2%, 99.8% and 99.3% to human plasma protein respectively. The times to reach steady state for ozanimod (**1**) and CC112273 in healthy individuals are 102 h and 45 days and the time to reach maximum plasma concentration T_{max} is 6–8 h for Ozanimod and 10 h for CC112273. Food intake regardless of high or low fat and caloric content in healthy individuals did not affect the exposure of ozanimod (**1**) or its active metabolites.[121]

Ozanimod (**1**) metabolism is complex and a metabolic pathway in human was only recently proposed.[58] Two of the key enzymes involved are cytochrome CYP3A4 and MAO-B.[58] Ozanimod (**1**) is eliminated through the urinary tract and has a mean single oral dose excretion of 0.03% and 0.06%, a seven-day regimen mean excretion of 0.04% to 0.09%, and a 28-day regimen mean excretion of 0.03% to 0.06%. Ozanimod was shown to have a single dose renal clearance range from 0.116 to 0.287 L/h, seven-day dosing regimen renal clearance range from 0.189 to 0.435 L/h and a 28-day regimen renal clearance range from 0.229 to 0.291 L/h.[122]

4. Structure–Activity Relationship (SAR)

S1PR1 consists of an extracellular N terminus with an alpha-helix structure that folds over the top of the receptor, preventing ligand accessibility to the inner binding pocket.[123] In total S1PR1 includes seven transmembrane domains, hydrophilic extracellular and intracellular loops. S1PR1 modulators were classified into class I and II agents.[63] Class I modulators possess a polar head group with/without an acyl chain. Examples include ML056, SEW2871, and ASP4058 (Table 1). The polar head group of ML056 is composed of a phosphonate group and a primary amine, protonated at the physiological conditions, while its tail consists of a phenyl ring with a meta-substitution of an alkyl chain.[124] The phosphonate and protonated amine enable ML056 to effectively interact with charged residues such as Tyr29, Lys34, Asn101, Arg120, and Glu121 at the binding pocket of S1PR1 (Figure 5).[124] The phenyl acyl tail of ML056 fills the hydrophobic part of the S1PR1 binding site, establishing selective hydrophobic interactions with the aromatic and short aliphatic residues.[124] ML056 is an antagonist of S1PR1. However, change in its structure such as lengthening its acyl tail and moving phenyl substituent to a para position result in the conversion of antagonism to full agonism.[124] Fingolimod phosphate, AFD(R), and amiselimod phosphate are also examples where lengthening the alkyl chain and positioning of phenyl substituents led to the conversion of antagonism to full agonistic activities on S1PR1.[71,125] Although class I S1PR1 modulators always need a polar head group for interaction with the receptor, the presence of an acyl tail is not necessary for effective binding such as SEW2871 and ASP4058.[45,126] A trifluoromethyl moiety on phenyl ring provides the polar head group for SEW2871. In ASP4058, the trifluoromethyl and a trifluoropropan-2-oxy group act as the polar head and establish the ion–dipole interactions with Arg120 and Glu121 in S1PR1.[126,127] Though SEW2871 and

ASP4048 are devoid of acyl tails, they possess bulky aromatic rings which enable the interactions with aromatic residues at the S1PR1 binding pocket.

Table 1. S1PR1 modulators, chemical structures, pEC_{50} values, indication and clinical trial phases

S1PR1 modulators	Chemical structure	pEC_{50}	Indication/clinical trial
ML056	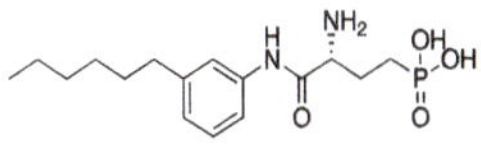		
Fingolimod	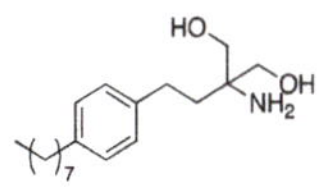	6.1	RRMS/approved Rett syndrome/I, II Chemotherapy-induced peripheral neuropathy/I cognition, brain volume loss/IV Stroke inflammation/II Amyotrophic lateral sclerosis/II Schizophrenia/II Chronic inflammatory demyelinating Polyradiculoneuropathy/II
AFD (R)	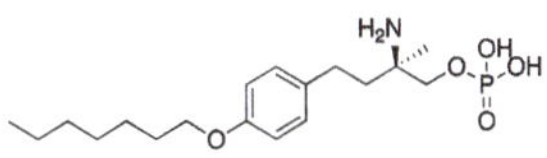	8.4–8.8	
Amiselimod	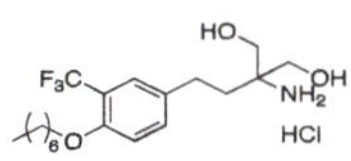	10.1–10.9	RRMS/II CD/II UC/II Plaque psoriasis/II
SEW2871	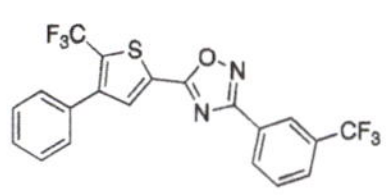	5.5–7.7	
ASP4058	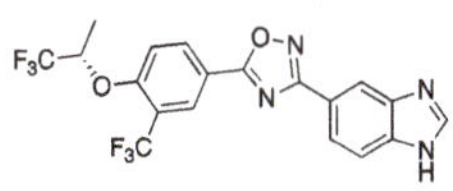	8.1	RRMS/preclinical Intracranial aneurysm/preclinical
Ozanimod (**1**)	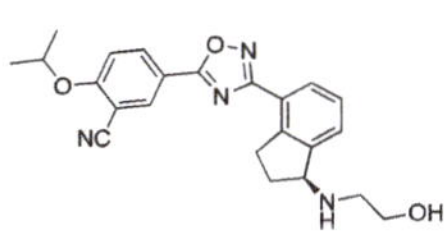	9.4–9.8	RRMS/approved Liver disease/I digestive system disease/III CD/approved UC
CYM-5442	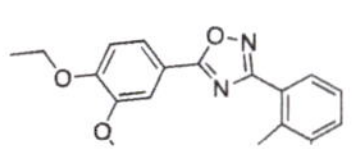	8.9	Chronic graft-vs-host

			disease/preclinical
RP-001	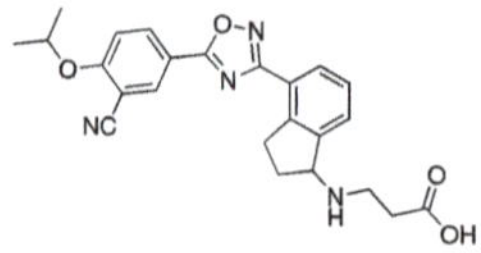	11.1	RRMS/discontinued
Siponimod	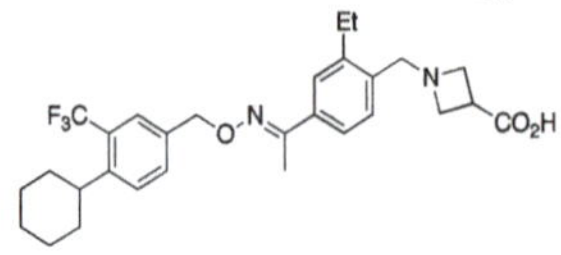	9.4–10.1	SPMS/approved Active dermatomyositis/II Polymyositis/II
Etrasimod	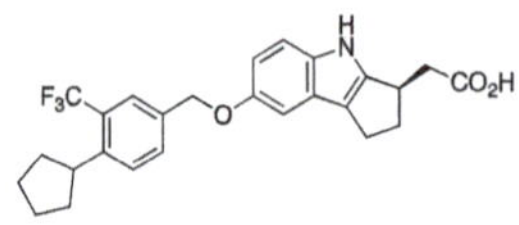	8.2–9.2	UC/III eosinophilic esophagitis/II Alopecia areata/II CD/II, III atopic dermatitis/II Primary biliary cholangitis/II
Ponesimod	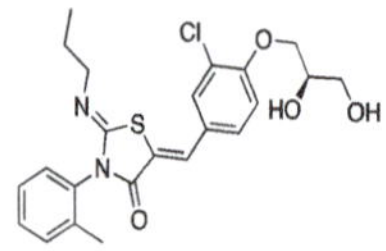	8	RRMS/approved Chronic graft-vs-host disease/II Plaque psoriasis/II
Cenerimod	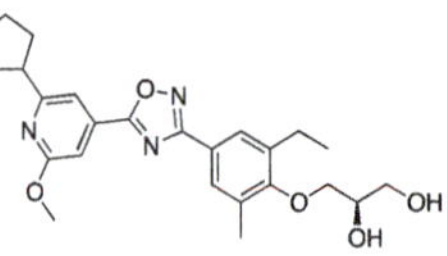	9	Systemic lupus erythematosus/II
Compound 26 (PMID:16190743)	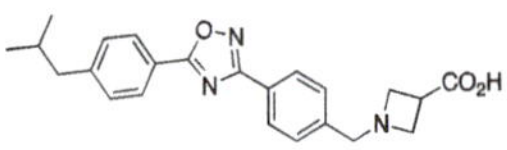	9.2	Preclinical
Compound 43 (PMID:26751273)	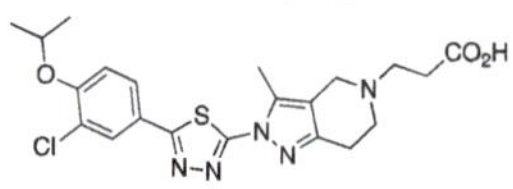	8.8	
GSK2018682	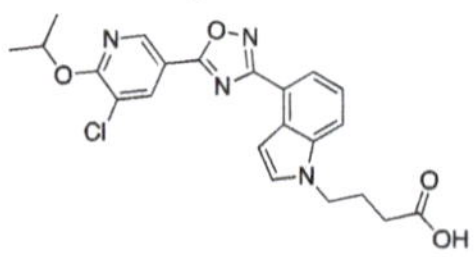	7.7	RRMS/discontinued
SAR247799	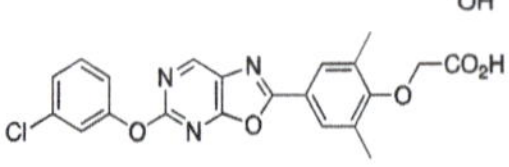	6.3–7.9	Microvascular coronary artery disease/I
BMS986104 Derivative 12	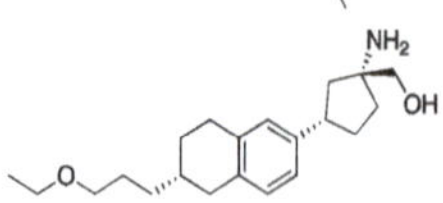	8.7	Rheumatoid arthritis/I
RP-101075	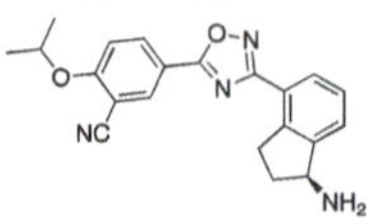	9.6	Microvascular thrombosis/preclinical

BMS986104 Derivative 24	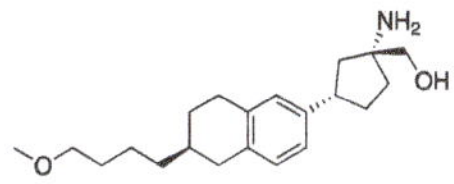	8.7	Autoimmune disease/preclinical

Source: Kandjani et al. [63]/with permission of Elsevier

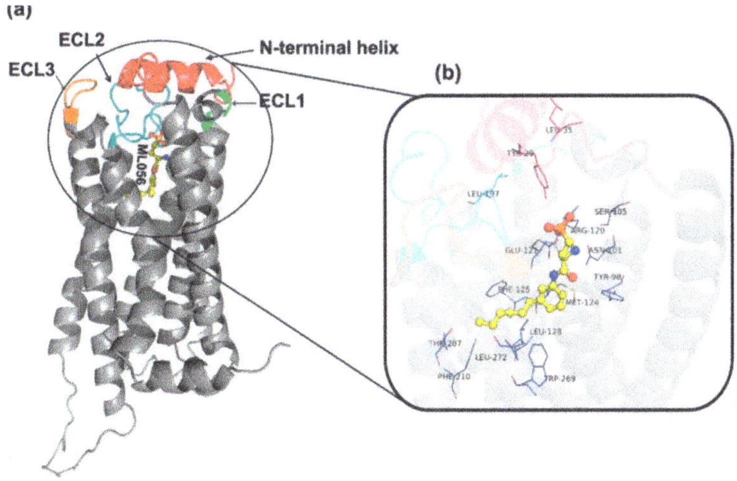

Figure 5. S1PR1 in complex with ML056. Source: Adapted from Hanson et al. [124]. (a) S1PR1 consists of an extracellular N terminus with an alpha-helix structure that folds over the top of the ligand-binding pocket,[124] seven transmembrane domains, and hydrophilic extracellular and intracellular loops.[44] (b) ML056 consists of a head group with a polar phosphonate and primary amine and a phenyl acyl tail which reflects the highly amphipathic nature of the S1PR1 binding pocket. The residues involved in the ligand binding have been highlighted. Source: Kandjani et al. [63]/with permission of Elsevier

Class II S1PR1 modulators do not require a polar head group to interact with the receptor. Instead of an acyl tail, the presence of a phenyl ring enables them to establish specific aromatic interactions with Phe210, Phe265, and Trp269.[63] CYM-5442 is a small-molecule S1PR1 agonist that does not require interactions with either Arg120 or Glu121 yet is fully active *in vivo* to induce lymphopenia.[45] The aromatic interaction between diethoxyphenyl moiety of CYM-5442 with the S1PR1 aromatic residues, Phe210, Phe265 and Trp269 is necessary for ligand binding and activity, while the aromaticity of the binding site is not necessary for binding of S1P to the receptor.[63] The diethoxyphenyl group of CYM-5442 also induces further selectivity as this moiety has steric interactions with Phe263 of S1PR3, but not with its equivalent resident Leu276 in S1PR1.[45]

Most second generation S1PR1 modulators including RP-001, siponimod, etrasimod, ozanimod (**1**), ponesimod, cenerimod, and GSK-2018682 are class II S1PR1 modulators.[10,128–133] Cryo-electron microscopy structure of siponimod in complex with

S1PR1 showed that in addition to Phe263, Siponimod also has steric clashes with I284 in S1PR3 binding pocket which further explains its high S1PR1 over S1PR3 selectivity.[123] One of the common feature of class II modulators is the disubstituted phenyl ring with a bulky substituent at the para position.[63] The relocation of the bulky substituent from para to meta position reduces the activities.[63]

Using cryo-EM data, structures of S1PR1 bound to siponimod, cenerimod, ozanimod (**1**), and SEW 2871 were reconstructed by Yuan et al.[123] The polar headgroups from the three agonists siponimod, cenerimod, and ozanimod (**1**) form direct contacts with $K34^{N\text{-}ter}$ from the N-terminal capping helix.[123] In addition, upon aligning four agonists in the S1PR1 structures, scientists found that the hydrophobic portion of the ligands was located in a nearly identical narrow hydrophobic pocket of S1PR1. A sub-pocket of the orthosteric site was also identified to accommodate the subgroups of agonists, such as the trifluoromethyl moiety of siponimod and SEW2871, the methoxy moiety of cenerimod and the cyano moiety of ozanimod (**1**). These moieties form hydrophobic interactions with the surrounding residues $C206^{5.43}$, $T207^{5.44}$, $F210^{5.47}$, $L272^{6.51}$, and $F273^{6.52}$ from TM5 and TM6. When alanine replacement was generated in the sub-pocket, the results of the signaling assay indicated that the activation potency of all receptor variants decreased when sensing different agonists.

SAR247799, BMS986104 and compounds 12 and 24 are new S1PR1 modulators called S1PR3-sparing biased agonists.[134,135] SAR247799 activates S1PR1 on endothelium without imposing receptor desensitization effect leading to the activation of protection pathways in human endothelial cells with no lymphopenia.[63] This compound possesses a polar head group of hydroxyacetic acid to interact with S1PR1 polar residues at the binding site and the chlorophenoxy moiety at the other end to establish aromatic interactions with the S1PR1 binding site residues.[63] BMS-986104 derivatives possess an alkyl tail attached to a tetrahydronaphthalene and a polar head group.[63]

5. Efficacy and Safety

The safety of ozanimod (**1**) in humans was first evaluated in a phase I, randomized double-blind, placebo-controlled trial. The study found that ozanimod (**1**) was well tolerated up to 3 mg as a single dose. Treatment emergent adverse events (TEAEs) were mild to moderate and there were no severe TEAEs or dose-limiting toxicities. No cardiovascular or pulmonary events were considered of clinical concern, and there were no hepatic or ophthalmologic effects identified.[100] Patients showed a dose-dependent heart rate reduction which leads to the dose-escalation protocol in all subsequent clinical trials.[122] This study also determined that ozanimod (**1**) did not prolong the QTc interval at therapeutic and supratherapeutic doses and did not raise any new safety concerns.[108,122]

5.1. Ozanimod to Treat Multiple Sclerosis

The phase II trial RADIANCE was a 24-week, double blind, placebo-controlled clinical trial in RRMS patients, with the primary endpoint of the cumulative number of total gadolinium-enhancing MRI lesions at weeks 12–24. The mean cumulative number of gadolinium-enhancing lesions at weeks 12–24 was 11.1 (SD 29.9) with placebo compared with 1.5 (3.7) with ozanimod hydrochloride 0.5 mg and 1.5 (3.4) with ozanimod hydrochloride 1 mg.[136] Both doses of ozanimod were well tolerated, with nasopharyngitis and headache being the most common TEAEs. The maximum reduction in mean heart rate by Holter monitoring during the first 6 h in ozanimod-treated participants were less than 2 beats per min (bpm) compared with baseline, with no patient having a minimum hourly heart rate less than 45 bpm. Electrocardiograms and 24-h Holter monitoring showed no increased incidence of atrioventricular block or sinus pause with ozanimod (**1**).[136,137] A 2-year extension trial followed where ozanimod-treated patients continued the treatment, and the placebo group were randomly switched to ozanimod hydrochloride 0.5 or 1 mg/d. The efficacy results were confirmed with a low number of new T2 lesions and GdE lesions on brain scans in all groups on ozanimod (**1**) during the study period.[138]

RADIANCE B phase III trial was a multicenter, randomized, double-blinded doubly-dummy, parallel group, active-controlled study with relapsing MS at 147 medical centers and clinical practices in 21 countries.[102] Patients were randomized (1.1:1) to once-daily ozanimod 0.46 mg (equivalent to ozanimod hydrochloride 0.5 mg), 0.92 mg (equivalent to ozanimod hydrochloride 1.0 mg) or once weekly intramuscular IFN β-1a 30 μg for 24 months. The primary endpoint was the annualized relapse rate (ARR) in comparison to IFN β-1a 30 μg at 24 months. Adjusted ARRs were 0.17 with Ozanimod hydrochloride 1.0 mg, 0.22 with Ozanimod hydrochloride 0.5 mg, and 0.28 with IFN β-1a, with rate ratios vs IFN β-1a 0.62 for ozanimod hydrochloride 1.0 mg and 0.79 for ozanimod hydrochloride 0.5 mg.[102,137] The study also concluded that ozanimod 0.92 mg has greater efficacy than Ozanimod 0.46 mg and both doses were at least as effective as IFN β-1a.[102,122]

RADIANCE B study showed treatment with ozanimod (**1**) resulted in less brain volume loss, which suggested that ozanimod might protect against structural changes associated with disease progression. During dose escalation, no clinically meaningful cardiac findings were reported including second-degree or third-degree atrioventricular block or symptomatic bradycardia.[102] Treatment with ozanimod for 24 months resulted in greater efficacy on clinically meaningful measures of disease activity (relapses and MRI lesions) than treatment with IFN β-1a.[102]

A similar 12-month, multicenter, double-blind, phase III trial SUNBEAM was conducted between December 2014 and November 2015.[101] The study was adequately powered with a sufficient sample size to assess the primary efficacy endpoint. The results were similar to those of RADIANCE B. Both RADIANCE B and SUNBEAM trials reported a reduction in the loss of whole-brain volume, cortical grey matter, and thalamic volume with ozanimod (**1**) when compared to IFN β-1a. These results suggest a possible role of ozanimod in protecting against structural changes associated with disease progression over time. SUNBEAM also confirmed that both low and high-dose ozanimod was as effective to IFN β-1a in reducing active disease in relapsing MS, with ozanimod 0.92 mg demonstrating numerically greater efficacy than Ozanimod 0.46 mg.[124] The study found ozanimod was more effective than IFN β-1a in lowering the ARR and lowering the number of new T2 lesions and GdE lesions over 12 months.

A post hoc analysis comparing the effects of ozanimod and IFN β-1a on cognitive processing speed (CPS) in SUNBEAM study showed ozanimod had modestly beneficial effects on CPS in relapsing MS participants. It improved Symbol Digit Modalities Test (SDMT) scores at months 6 and 12. A more significant percentage of Ozanimod-treated participants had clinically meaningful improvements in SDMT scores vs IFN β-1a: 30.0% vs 22.2% at month 6 and 35.6%. vs 27.9% at month 12.

SUNBEAM study also reported TEAEs by 59.8% of patients on ozanimod 0.92 mg/d, 57.2% of patients on ozanimod 0.46 mg/d, and 75.5% patients on IFN β-1a. The TEAEs that occurred in more than 5% of patients treated with ozanimod were nasopharyngitis, upper respiratory tract infections and headache. No clinically significant bradycardia or second or third-degree atrioventricular block was observed. No serious opportunistic infections have been reported to date.[108]

The long-term safety and efficacy of ozanimod in relapsing MS was studied in the DAYBREAK open-label extension (OLE) trial. Patients with relapsing MS who completed a phase 1–3 ozanimod trial were eligible for the open-label study of ozanimod 0.92 mg/d. The study characterizes safety and efficacy of ozanimod for up to 5 years and up to 8 continuous years of exposure in patients with RMS. The overall safety and tolerability profile of ozanimod was consistent with the phase 3 trials. MS clinical and radiologic disease activity remained low in patients who received ozanimod 0.92 mg continuously since the parent trials. Patients who switched to ozanimod 0.92 mg from either IFN β-1a or ozanimod 0.46 mg experienced reductions in ARR and brain MRI lesion counts. The study found that the risks of oral herpes and herpes zoster didn't increase with longer ozanimod usage. COVID-19 cases during DAYBREAK were largely nonserious and a vast majority of patients recovered from COVID-19 without sequelae. Incidence of relapse after ozanimod discontinuation was low in DAYBREAK, and none of the relapses resulted in persistent, severe disability.[139]

Another phase III clinical trial ENLIGHTEN is currently ongoing to document any changes in cognitive processing speed over 3 years as measured by the Symbol Digit Modalities Test (SDMT) in adult participants with RMS who receive treatment with ozanimod (**1**).[140]

5.2. Ozanimod to Treat Ulcerative Colitis

TOUCHSTONE was a randomized, double-blind, placebo-controlled phase II trial that included a 5-week screening, 9-week induction (including an initial 7-day dose-escalation period), 24-week maintenance and optional OLE period in patients with moderately to severely active UC.[141] Clinical remission was achieved by greater proportions of patients who received oral ozanimod 0.46 mg and ozanimod 0.92 mg once daily compared with placebo after 8 weeks of induction therapy (14%, 16% vs 6% respectively) and an additional 24 weeks of maintenance therapy (26%, 21% vs 6% respectively). Statistical significance ($P < 0.05$) was observed for only ozanimod 0.92 mg during induction and for both doses during maintenance. OLE study showed that for OLE patients who received ozanimod 0.92 mg once daily, the observed Mayo scores decreased within 4 to 8 weeks and those were maintained up to 200 weeks.[141,142]

TRUENORTH was a 52-week, phase III, randomized, double-blind, placebo-controlled trial of ozanimod as induction and maintenance therapy in patients with moderately to severely active UC. It is designed to further evaluate ozanimod 0.92 mg for UC treatment. In the 10-week induction period, patients were assigned to receive ozanimod or placebo. At 10 weeks, patients with a clinical response to ozanimod underwent randomization to receive double-blind ozanimod or placebo for the maintenance period (through week 52). Clinical remission was achieved by significantly more patients who received ozanimod compared with placebo at weeks 10 (18.4% vs 6%, $P < 0.001$) and 52 (37% vs 18.5%, $P < 0.001$). The results of TRUENORTH phase III trial demonstrated that a once-daily oral formulation of ozanimod provided clinical efficacy in patients with moderately to severely active UC. Treatment with ozanimod led to significant improvements as compared with placebo, in the incidence of clinical remission and in all key secondary clinical endoscopic and histologic end points at week 10 and 52. These results were observed in patients with active disease that had been inadequately controlled by conventional agents.[143]

Cancer, opportunistic infection, and macular edema were observed in patients who received ozanimod, but the incidences were low. Non-serious infections were more common with ozanimod than with placebo during the maintenance phase of the trial. The incidences of elevated alanine aminotransferase levels were higher among patients who received ozanimod than among those who received placebo. Liver events were mostly mild or moderate in severity. Overall, the results were consistent with safety findings that have previously been reported regarding ozanimod therapy in phase III trials involving patients with MS.[40,143]

5.3. Ozanimod to Treat Crohn's Disease

STEPSTONE was a 12-week, phase II, uncontrolled, multicenter trial in adults with moderately to severely active CD. All patients began treatment with a 7-day dose escalation of ozanimod hydrochloride (4 days on 0.25 mg daily followed by 3 days at 0.5 mg daily), followed by a daily dosage of ozanimod hydrochloride 1.0 mg for 11 weeks.[144] A reduction from baseline in the mean Simple Endoscopic Score for Crohn's Disease (SES-CD) was reported at Week 12, and endoscopic response and remission, assessed by a blinded central reader, were achieved at estimates substantially higher than anticipated based on results of a recent analysis following placebo treatment.[145] Additional objective measures of disease activity (Global Histologic Disease Activity Score [GHAS] and Robarts Histopathology Index [RHI]) were also reduced substantially from baseline at Week 12.[146] Other endpoints were also consistent with a treatment benefit. The overall mean decrease from baseline in (Crohn's Disease Activity Index) CDAI score was −130.4.

In general, the study concluded that ozanimod was well tolerated, with a similar safety profile to that in UC, with no new safety concerns identified. No clinically important changes in heart rate were observed at treatment initiation, consistent with previously reported findings in MS.[144]

Based on the positive phase II results, a phase III program (YELLOWSTONE) for ozanimod to treat CD was initiated in 2019 and is currently ongoing. The YELLOWSTONE clinical trial was designed to assess the safety and efficacy of ozanimod treatment for adults with moderately to severely active CD with inadequate response or intolerance to other treatments.[40]

6. Synthesis

In their patents, Rosen, Roberts, and their teams at Scripps reported a racemic synthesis of ozanimod (**1**).[147,148] The route was concise, involving only three steps (Scheme 1). Amidoxime **3** was prepared from a benzonitrile substrate **2**. The oxadiazole core was constructed through the condensation of a benzoic acid derivative **4** with the amidoxime substrate **3**. This method served as the foundation for several subsequent approaches. In the final step, the C–N bond was formed via an S_N2 reaction using CC1084037 (**5**) as the starting material, yielding a racemic product. The enantiomer ozanimod (**1**) was isolated by chiral HPLC separation to achieve 99% purity by weight.

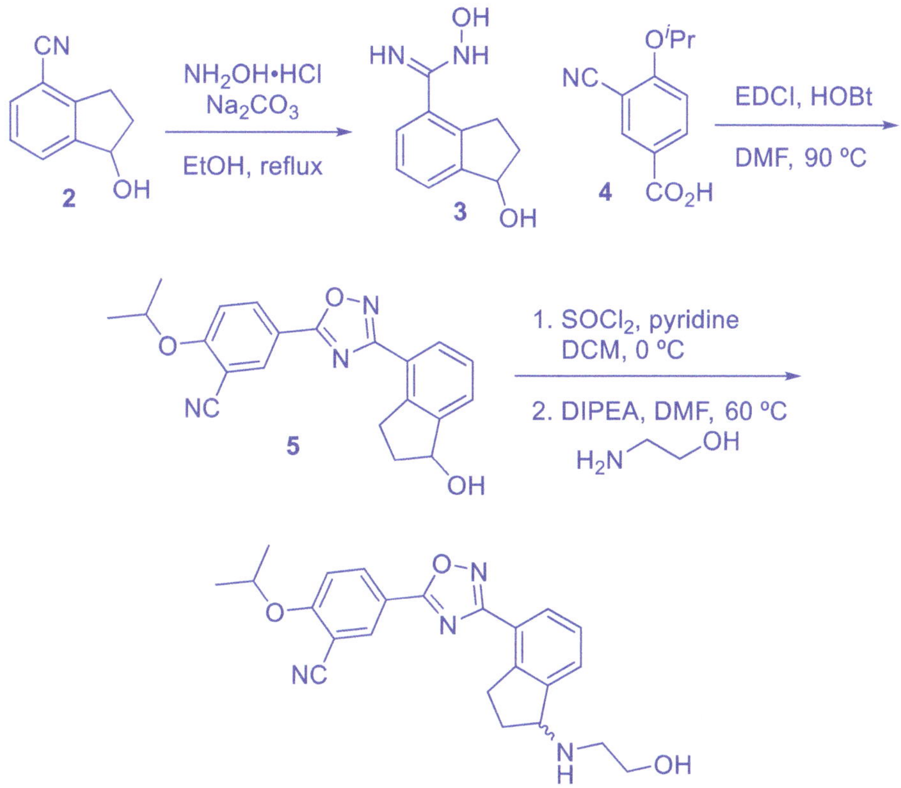

Scheme 1. Racemic synthesis of ozanimod (**1**)

A Hungarian company, Egis Pharmaceuticals, patented the process for the production of a single enantiomer of ozanimod (**1**) in 2018 (Scheme 2).[149] The synthetic route takes advantage of enantiopure sulfonamide precursor, which is key to the formation of C–N bond with desired configuration in the diastereoselective imine reduction step. The sulfonamide also assists in the subsequent amine alkylation step. In the last step, hydrolysis in the acid liberates both the free amine and hydroxy group to afford the final product **1**.

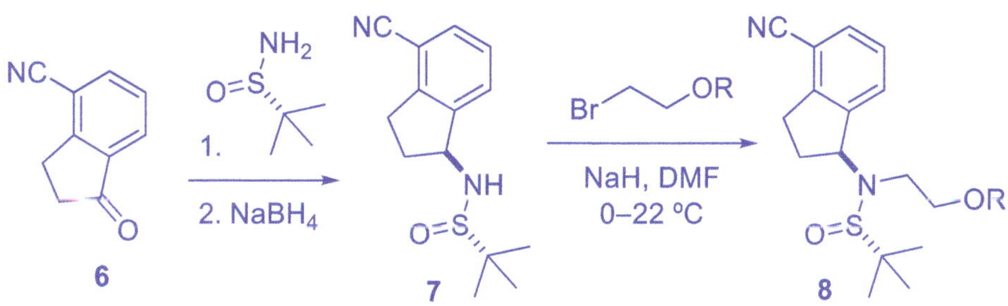

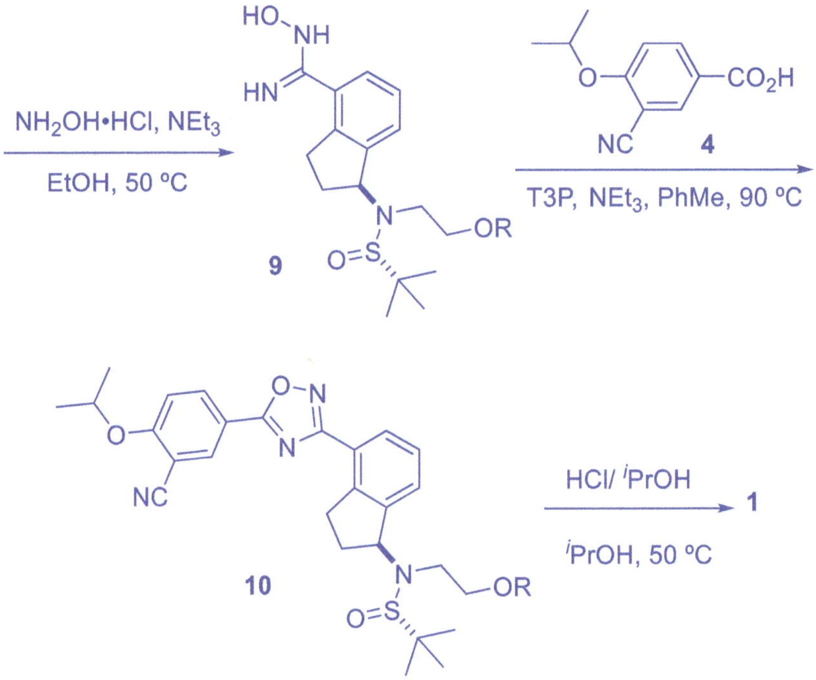

Scheme 2. Synthesis of enantiopure ozanimod (**1**) using a chiral sulfonamide auxiliary

Another synthesis was published by Taddei and co-workers in 2021, which included an enantioselective reaction to afford product **1** (Scheme 3).[150] In the final step, a ruthenium-catalyzed asymmetrical hydrogenation of the *in situ* formed imine afforded the amine product **1** with 99% enantiomeric excess (*ee*).

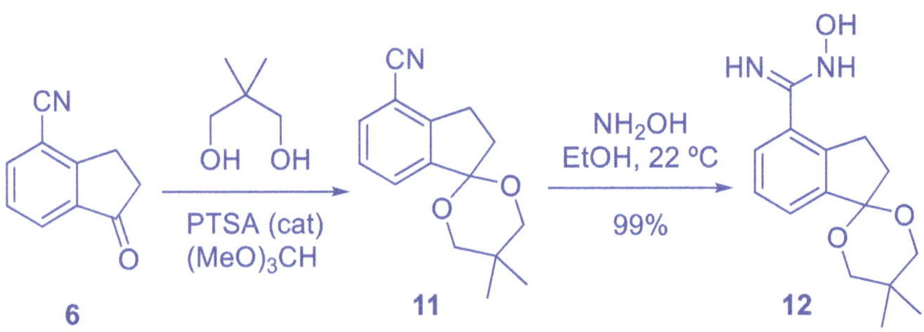

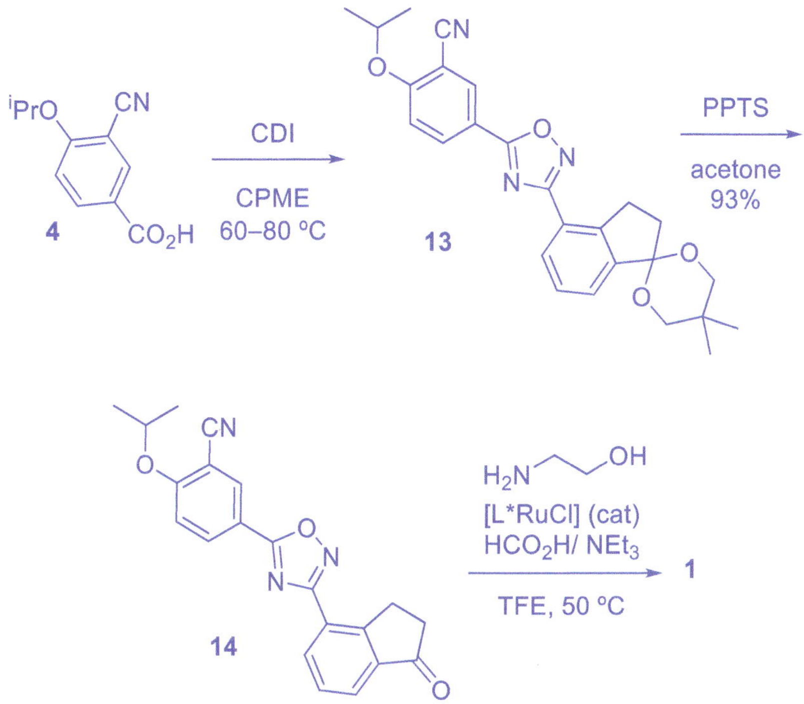

Scheme 3. Enantioselective synthesis of ozanimod (**1**)

7. Summary

Ozanimod (**1**) is the first S1P modulators approved by FDA for the treatment of MS and UC. It demonstrated efficacy and a favorable side effect profile, providing a unique option among the currently available therapies. The once daily oral administration is convenient for the patients and adverse events that emerged during the phase II/III studies suggest a favorable profile. The commonly accepted mechanism of action is the selective modulation of S1PR1 and 5, which prevents autoreactive lymphocytes from exiting the LNs.

Ozanimod (**1**) was approved in 2020 by FDA and EMA for the treatment of clinically isolated syndrome, relapsing-remitting and secondary progressive forms of MS after the conclusion of the RADIANCE phase II and III studies and the SUNBEAM phase III study that showed that the drug was effective and well tolerated. In addition, the two phase III trials have shown statistical superiority of ozanimod (**1**) over intramuscular IFN β-1a regarding reduction in ARR and MRI outcomes.[137] Furthermore, ozanimod (**1**) is currently the only US FDA approved S1PR modulator that does not require first-dose observation or genetic testing.[108,151]

In 2021, ozanimod (**1**) was approved by FDA as the first S1P modulator for the treatment of adults with moderately to severely active UC.[121,151] It is also likely that Ozanimod's distinct feature and favorable side effect profile would lend itself to clinical trials as part of combination therapy with other IBD treatments.[142] The oral route of small molecule therapy may prove to be cost saving, especially compared to infusion. Additional studies are required to evaluate its efficacy when compared with other available therapies.

References

1. Cantorna, M. T. *Prog. Biophys. Mol. Biol.* **2006**, *92* (1), 60–64.
2. What are Crohns and Colitis? https://crohnsandcolitis.ca/About-Crohn-s-Colitis/What-are-Crohns-and-Colitis (accessed May 16, 2023).
3. Relapsing-Remitting Multiple Sclerosis (RRMS) https://www.nationalmssociety.org/What-is-MS/Types-of-MS/Relapsing-remitting-MS (accessed May 18, 2023).
4. McGinley, M. P.; Goldschmidt, C. H.; Rae-Grant, A. D. *JAMA* **2021**, *325* (8), 765.
5. Tullman, M. J. *Am. J. Manag. Care* **2013**, *19* (2), 20–28.
6. Kappos, L.; Polman, C. H.; Freedman, M. S.; Edan, G.; Hartung, H. P.; Miller, D. H.; Montalban, X.; Barkhof, F.; Bauer, L.; Jakobs, P.; Pohl, C.; Sandbrink, R. *Neurol. Ther.* **2006**, *67* (7), 1242–1249.
7. Kappos, L.; Freedman, M. S.; Polman, C. H.; Edan, G.; Hartung, H.-P.; Miller, D. H.; Montalbán, X.; Barkhof, F.; Radü, E.-W.; Bauer, L.; Dahms, S.; Lanius, V.; Pohl, C.; Sandbrink, R. *Lancet* **2007**, *370* (9585), 389–397.
8. Inglese, M.; Petracca, M. *Schizophr. Res.* **2015**, *161* (1), 94–101.
9. Danese, S.; Panés, J. *Gastroenterology* **2014**, *147* (5), 981–989.
10. Scott, F. L.; Clemons, B.; Brooks, J.; Brahmachary, E.; Powell, R.; Dedman, H.; Desale, H. G.; Timony, G. A.; Martinborough, E.; Rosen, H.; Roberts, E.; Boehm, M. F.; Peach, R. J. *Br. J. Pharmacol.* **2016**, *173* (11), 1778–1792.
11. Gohil, K.; Carramusa, B. *Pharm. Therap.* **2014**, *39* (8), 576–577.
12. Wang, R.; Li, Z.; Liu, S.; Zhang, D. *BMJ Open* **2023**, *13* (3), e065186.
13. Orfanoudaki, E.; Foteinogiannopoulou, K.; Theodoraki, E.; Koutroubakis, I. E. *J. Clin. Med.* **2023**, *12* (7), 2452.
14. Healey, L. A.; Backes, M. B. *N. Engl. J. Med.* **1989**, *321* (11), 762–764.
15. Pokorny, C. S.; Beran, R. G.; Pokorny, M. J. *Intern. Med. J.* **2007**, *37* (10), 721–724.

16. Kimura, K. *Mayo Clin. Proc.* **2000**, *75* (8), 802–806.
17. Shahmohammadi, S.; Sahraian, M. A.; Shahmohammadi, A.; Doosti, R.; Zare-Mirzaie, A.; Naser Moghadasi, A. *Mult. Scler. Relat. Disord.* **2018**, *22*, 22–26.
18. Alkhawajah, M. M.; Caminero, A. B.; Freeman, H. J.; Oger, J. J. F. *Mult. Scler. J.* **2013**, *19* (3), 259–265.
19. Katsanos, A. H.; Katsanos, K. H. *Expert Rev. Clin. Immunol.* **2014**, *10* (3), 363–373.
20. Marrie, R. A.; Reider, N.; Cohen, J.; Stuve, O.; Sorensen, P. S.; Cutter, G.; Reingold, S. C.; Trojano, M. *Mult. Scler. J.* **2015**, *21* (3), 282–293.
21. Castelo-Branco, A.; Chiesa, F.; Bengtsson, C. E.; Lee, S.; Minton, N. N.; Niemcryk, S.; Lindholm, A.; Rosenlund, M.; Piehl, F.; Montgomery, S. *Mult. Scler. J. - Exp. Transl. Clin.* **2020**, *6* (3), doi:10.1177/2055217320947761.
22. Kosmidou, M.; Katsanos, A. H.; Katsanos, K. H.; Kyritsis, A. P.; Tsivgoulis, G.; Christodoulou, D.; Giannopoulos, S. *J. Neurol.* **2017**, *264* (2), 254–259.
23. Ochoa-Repáraz, J.; Mielcarz, D. W.; Ditrio, L. E.; Burroughs, A. R.; Foureau, D. M.; Haque-Begum, S.; Kasper, L. H. *J. Immunol.* **2009**, *183* (10), 6041–6050.
24. Agirman, G.; Yu, K. B.; Hsiao, E. Y. *Science* **2021**, *374* (6571), 1087–1092.
25. Fresegna, D.; Bullitta, S.; Musella, A.; Rizzo, F. R.; De Vito, F.; Guadalupi, L.; Caioli, S.; Balletta, S.; Sanna, K.; Dolcetti, E.; Vanni, V.; Bruno, A.; Buttari, F.; Stampanoni Bassi, M.; Mandolesi, G.; Centonze, D.; Gentile, A. *Cells* **2020**, *9* (10), 2290.
26. Kollias, G.; Douni, E.; Kassiotis, G.; Kontoyiannis, D. *Ann. Rheum. Dis.* **1999**, *58* (Supplement 1), i32–i39.
27. Chen, L.; Ruan, G.; Cheng, Y.; Yi, A.; Chen, D.; Wei, Y. *Front. Immunol.* **2022**, *13*, 1055914.
28. Prajeeth, C. K.; Kronisch, J.; Khorooshi, R.; Knier, B.; Toft-Hansen, H.; Gudi, V.; Floess, S.; Huehn, J.; Owens, T.; Korn, T.; Stangel, M. *J. Neuro-Oncol.* **2017**, *14* (1), 204.
29. Avasarala, J.; Guduru, Z.; McLouth, C. J.; Wilburn, A.; Talbert, J.; Sutton, P.; Sokola, B. S. *Mult. Scler. Relat. Disord.* **2021**, *51*, 102942.
30. Li, L.; Aviña-Zubieta, J. A.; Bernstein, C. N.; Kaplan, G. G.; Tremlett, H.; Xie, H.; Peña-Sánchez, J.-N.; Marrie, R. A.; Etminan, M. *Neurol. Ther.* **2023**, *100* (6), E558–E567.
31. Rodrigues, S.; Magro, F.; Soares, J.; Nunes, A. C. R.; Lopes, S.;

Marques, M.; Rio, E.; Macedo, G. *Inflamm. Bowel Dis.* **2010**, *16* (12), 2001–2003.

32. Hutchinson, M. *Ther. Clin. Risk Manage.* **2007**, *3* (2), 259–268.
33. Davenport, R.; Munday, J. *Drug Discovery Today* **2007**, *12* (13–14), 569–576.
34. Selewski, D. T.; Shah, G. V.; Segal, B. M.; Rajdev, P. A.; Mukherji, S. K. *Am. J. Neuroradiol.* **2010**, *31* (9), 1588–1590.
35. González-Amaro, R.; Mittelbrunn, M.; Sánchez-Madrid, F. *Immunology* **2005**, *116* (3), 289–296.
36. Suzuki, S.; Li, X. -K.; Enosawa, S.; Shinomiya, T. *Immunology* **1996**, *89* (4), 518–523.
37. Adachi, K.; Kohara, T.; Nakao, N.; Afita, M.; Chiba, K.; Mishina, T.; Sasaki, S.; Fujita, T. *Bioorg. Med. Chem. Lett.* **1995**, *5* (8), 853–856.
38. Danese, S.; Furfaro, F.; Vetrano, S. *J. Crohn's Colitis* **2018**, *12* (suppl_2), S678–S686.
39. Juif, P.-E.; Kraehenbuehl, S.; Dingemanse, J. *Expert Opin. Drug Metab. Toxicol.* **2016**, *12* (8), 879–895.
40. Feagan, B. G.; Schreiber, S.; Afzali, A.; Rieder, F.; Hyams, J.; Kollengode, K.; Pearlman, J.; Son, V.; Marta, C.; Wolf, D. C.; D'Haens, G. G. *Contemp. Clin. Trials* **2022**, *122*, 106958.
41. Baillargeon, P.; Fernandez-Vega, V.; Sridharan, B. P.; Brown, S.; Griffin, P. R.; Rosen, H.; Cravatt, B.; Scampavia, L.; Spicer, T. P. *SLAS Discovery* **2019**, *24* (3), 386–397.
42. Mandala, S.; Hajdu, R.; Bergstrom, J.; Quackenbush, E.; Xie, J.; Milligan, J.; Thornton, R.; Shei, G.-J.; Card, D.; Keohane, C.; Rosenbach, M.; Hale, J.; Lynch, C. L.; Rupprecht, K.; Parsons, W.; Rosen, H. *Science* **2002**, *296* (5566), 346–349.
43. Goetzl, E. J.; Rosen, H. *J. Clin. Invest.* **2004**, *114* (11), 1531–1537.
44. Rosen, H.; Stevens, R. C.; Hanson, M.; Roberts, E.; Oldstone, M. B. A. *Annu. Rev. Biochem.* **2013**, *82* (1), 637–662.
45. Gonzalez-Cabrera, P. J.; Jo, E.; Sanna, M. G.; Brown, S.; Leaf, N.; Marsolais, D.; Schaeffer, M. T.; Chapman, J.; Cameron, M.; Guerrero, M.; Roberts, E.; Rosen, H. *Mol. Pharmacol.* **2008**, *74* (5), 1308–1318.
46. Highlights 2015 | NIH Common Fund https://commonfund.nih.gov/highlights2015 (accessed Dec 20, 2023).
47. Bristol Myers Squibb - U.S. Food and Drug Administration Approves Bristol Myers Squibb's Zeposia® (ozanimod), an Oral Treatment for Adults with Moderately to Severely Active Ulcerative Colitis https://news.bms.com/news/details/2021/U.S.-Food-and-Drug-Administration-Approves-Bristol-Myers-Squibbs-Zeposia-ozanimod-

an-Oral-Treatment-for-Adults-with-Moderately-to-Severely-Active-Ulcerative-Colitis1/default.aspx (accessed Dec 20, 2023).

48. Wadman, M. *Nat. Biotechnol.* **2012**, *30* (4), 309–312.

49. Xiao, S.; Peng, K.; Li, C.; Long, Y.; Yu, Q. *Cell Death Discovery* **2023**, *9* (1), 380.

50. Hannun, Y. A.; Obeid, L. M. *Nat. Rev. Mol. Cell Biol.* **2018**, *19* (3), 175–191.

51. Cyster, J. G.; Schwab, S. R. *Annu. Rev. Immunol.* **2012**, *30* (1), 69–94.

52. Mendoza, A.; Fang, V.; Chen, C.; Serasinghe, M.; Verma, A.; Muller, J.; Chaluvadi, V. S.; Dustin, M. L.; Hla, T.; Elemento, O.; Chipuk, J. E.; Schwab, S. R. *Nature* **2017**, *546* (7656), 158–161.

53. Sands, B. E.; Schreiber, S.; Blumenstein, I.; Chiorean, M. V.; Ungaro, R. C.; Rubin, D. T. *J. Crohns. Colitis* **2023**, *17* (12), 2012–2025.

54. Baeyens, A. A. L.; Schwab, S. R. *Annu. Rev. Immunol.* **2020**, *38* (1), 759–784.

55. Pérez-Jeldres, T.; Alvarez-Lobos, M.; Rivera-Nieves, J. *Drugs* **2021**, *81* (9), 985–1002.

56. Aoki, M.; Aoki, H.; Ramanathan, R.; Hait, N. C.; Takabe, K. *Mediators Inflammation* **2016**, *2016*, 1–11.

57. Schwab, S. R.; Cyster, J. G. *Nat. Immunol.* **2007**, *8* (12), 1295–1301.

58. Surapaneni, S.; Yerramilli, U.; Bai, A.; Dalvie, D.; Brooks, J.; Wang, X.; Selkirk, J. V.; Yan, Y. G.; Zhang, P.; Hargreaves, R.; Kumar, G.; Palmisano, M.; Tran, J. Q. *Drug Metab. Dispos.* **2021**, *49* (5), 405–419.

59. McGinley, M. P.; Cohen, J. A. *Lancet* **2021**, *398* (10306), 1184–1194.

60. Sugimoto, N.; Takuwa, N.; Okamoto, H.; Sakurada, S.; Takuwa, Y. *Mol. Cell. Biol.* **2003**, *23* (5), 1534–1545.

61. Ryu, Y.; Takuwa, N.; Sugimoto, N.; Sakurada, S.; Usui, S.; Okamoto, H.; Matsui, O.; Takuwa, Y. *Circ. Res.* **2002**, *90* (3), 325–332.

62. Lee, M.-J.; Thangada, S.; Paik, J.-H.; Sapkota, G. P.; Ancellin, N.; Chae, S.-S.; Wu, M.; Morales-Ruiz, M.; Sessa, W. C.; Alessi, D. R.; Hla, T. *Mol. Cell* **2001**, *8* (3), 693–704.

63. Kandjani, O. J.; Yaqoubi, S.; Vahdati, S. S.; Borhannejad, B.; Dastmalchi, S.; Alizadeh, A. A. *Eur. J. Med. Chem.* **2023**, *250*, 115182.

64. Herr, D. R.; Reolo, M. J. Y.; Peh, Y. X.; Wang, W.; Lee, C.-W.; Rivera, R.; Paterson, I. C.; Chun, J. *Sci. Rep.* **2016**, *6* (1), 24541.

65. Li, Q.; Li, Y.; Lei, C.; Tan, Y.; Yi, G. *Clin. Chim. Acta* **2021**, *519*, 32–39.

66. Germana Sanna, M.; Vincent, K. P.; Repetto, E.; Nguyen, N.; Brown, S. J.; Abgaryan, L.; Riley, S. W.; Leaf, N. B.; Cahalan, S. M.; Kiosses, W. B.; Kohno, Y.; Brown, J. H.; McCulloch, A. D.; Rosen, H.; Gonzalez-Cabrera, P. *J. Mol. Pharmacol.* **2016**, *89* (1), 176–186.

67. Ohotski, J.; Long, J. S.; Orange, C.; Elsberger, B.; Mallon, E.; Doughty, J.; Pyne, S.; Pyne, N. J.; Edwards, J. *Br. J. Cancer* **2012**, *106* (8), 1453–1459.

68. Dillmann, C.; Ringel, C.; Ringleb, J.; Mora, J.; Olesch, C.; Fink, A. F.; Roberts, E.; Brüne, B.; Weigert, A. *J. Immunol.* **2016**, *196* (4), 1579–1590.

69. Doorn, R.; Lopes Pinheiro, M. A.; Kooij, G.; Lakeman, K.; Hof, B.; Pol, S. M.; Geerts, D.; Horssen, J Van Der Valk, P.; Van Der Kam, E.; Ronken, E.; Reijerkerk, A.; de Vries, H. E. *J. Neuro-Oncol.* **2012**, *9* (1), 133.

70. Takuwa, Y.; Okamoto, Y.; Yoshioka, K.; Takuwa, N. *BioFactors* **2012**, *38* (5), 329–337.

71. Brinkmann, V.; Davis, M. D.; Heise, C. E.; Albert, R.; Cottens, S.; Hof, R.; Bruns, C.; Prieschl, E.; Baumruker, T.; Hiestand, P.; Foster, C. A.; Zollinger, M.; Lynch, K. R. *J. Biolumin. Chemilumin.* **2002**, *277* (24), 21453–21457.

72. Camm, J.; Hla, T.; Bakshi, R.; Brinkmann, V. *Am. Heart J.* **2014**, *168* (5), 632–644.

73. Myat, L. O.; Thangada, S.; Wu, M. T.; Liu, C. H.; Macdonald, T. L.; Lynch, K. R.; Lin, C. Y.; Hla, T. *J. Biolumin. Chemilumin.* **2007**, *282* (12), 9082–9089.

74. Stepanovska, B.; Huwiler, A. *Pharmacol. Res.* **2020**, *154*, 104170.

75. Pérez-Jeldres, T.; Tyler, C. J.; Boyer, J. D.; Karuppuchamy, T.; Bamias, G.; Dulai, P. S.; Boland, B. S.; Sandborn, W. J.; Patel, D. R.; Rivera-Nieves, J. *Inflamm. Bowel Dis.* **2019**, *25* (2), 270–282.

76. Becher, N.; Swaminath, A.; Sultan, K. *Ther. Clin. Risk Manage.* **2022**, *18*, 913–927.

77. Chaudhry, B. Z.; Cohen, J. A.; Conway, D. S. *Neurotherapeutics* **2017**, *14* (4), 859–873.

78. Hampton, H. R.; Chtanova, T. *Front. Immunol.* **2019**, *10*, 1168.

79. Pappu, R.; Schwab, S. R.; Cornelissen, I.; Pereira, J. P.; Regard, J. B.; Xu, Y.; Camerer, E.; Zheng, Y.-W.; Huang, Y.; Cyster, J. G.; Coughlin, S. R. *Science* **2007**, *316* (5822), 295–298.

80. Serra, M.; Saba, J. D. *Adv. Enzyme Regul.* **2010**, *50* (1), 349–362.

81. Yanagida, K.; Hla, T. *Annu. Rev. Physiol.* **2017**, *79* (1), 67–91.

82. Janssen, A.; Van Der Burg, M.; Szuhai, K.; Kops, G. J. P. L.; Medema, R. H. *Science* **2011**, *333* (6051), 1895–1898.

83. Kharel, Y.; Morris, E. A.; Congdon, M. D.; Thorpe, S. B.; Tomsig, J. L.; Santos, W. L.; Lynch, K. R. *J. Pharmacol. Exp. Ther.* **2015**, *355* (1), 23–31.

84. Zachariah, M. A.; Cyster, J. G. *Science* **2010**, *328* (5982), 1129–1135.
85. Lim, V. Y.; Zehentmeier, S.; Fistonich, C.; Pereira, J. P. *Adv. Immunol.* **2017**, *134*, 47–88.
86. Arnon, T. I.; Horton, R. M.; Grigorova, I. L.; Cyster, J. G. *Nature* **2013**, *493* (7434), 684–688.
87. Obinata, H.; Hla, T. *Int. Immunol.* **2019**, *31* (9), 617–625.
88. Walzer, T.; Chiossone, L.; Chaix, J.; Calver, A.; Carozzo, C.; Garrigue-Antar, L.; Jacques, Y.; Baratin, M.; Tomasello, E.; Vivier, E. *Nat. Immunol.* **2007**, *8* (12), 1337–1344.
89. Abbas, A. K.; Lichtman, A. H.; Pillai, S.. *Cellular and Molecular Immunology*, 10th ed.; Elsevier, **2021**.
90. Brinkmann, V. *Pharmacol. Ther.* **2007**, *115* (1), 84–105.
91. Woodland, D. L.; Kohlmeier, J. E. *Nat. Rev. Immunol.* **2009**, *9* (3), 153–161.
92. Bigaud, M.; Guerini, D.; Billich, A.; Bassilana, F.; Brinkmann, V. *Biochim. Biophys. Acta, Mol. Cell. Biol. Lipids* **2014**, *1841* (5), 745–758.
93. Chipuk, J. E.; Bouchier-Hayes, L.; Kuwana, T.; Newmeyer, D. D.; Green, D. R. *Science* **2005**, *309* (5741), 1732–1735.
94. Matloubian, M.; Lo, C. G.; Cinamon, G.; Lesneski, M. J.; Xu, Y.; Brinkmann, V.; Allende, M. L.; Proia, R. L.; Cyster, J. G. *Nature* **2004**, *427* (6972), 355–360.
95. Brinkmann, V.; Cyster, J. G.; Hla, T. *Am. J. Transplant.* **2004**, *4* (7), 1019–1025.
96. Chen, H.; Qin, Y.; Chou, M.; Cyster, J. G.; Li, X. *Elife* **2023**, *12*, e88204.
97. Pyne, N.; Pyne, S. *Mol. Ther.* **2017**, *22* (3), 344.
98. Chun, J.; Giovannoni, G.; Hunter, S. F. *Drugs* **2021**, *81* (2), 207–231.
99. Pham, T. H. M.; Okada, T.; Matloubian, M.; Lo, C. G.; Cyster, J. G. *Immunity* **2008**, *28* (1), 122–133.
100. Tran, J. Q.; Hartung, J. P.; Peach, R. J.; Boehm, M. F.; Rosen, H.; Smith, H.; Brooks, J. L.; Timony, G. A.; Olson, A. D.; Gujrathi, S.; Frohna, P. A. *J. Clin. Pharmacol.* **2017**, *57* (8), 988–996.
101. Comi, G.; Kappos, L.; Selmaj, K. W.; Bar-Or, A.; Arnold, D. L.; Steinman, L.; Hartung, H. P.; Montalban, X.; Kubala Havrdová, E.; Cree, B. A. C.; Sheffield, J. K.; Minton, N.; Raghupathi, K.; Ding, N.; Cohen, J. A. *Lancet Neurol.* **2019**, *18* (11), 1009–1020.
102. Cohen, J. A.; Comi, G.; Selmaj, K. W.; Bar-Or, A.; Arnold, D. L.; Steinman, L.; Hartung, H.-P.; Montalban, X.; Kubala Havrdová, E.;

Cree, B. A. C.; Sheffield, J. K.; Minton, N.; Raghupathi, K.; Huang, V.; Kappos, L. *Lancet Neurol.* **2019**, *18* (11), 1021–1033.

103. Urbano, M.; Guerrero, M.; Rosen, H.; Roberts, E. *Bioorg. Med. Chem. Lett.* **2013**, *23* (23), 6377–6389.

104. Xu, J.; Gray, F.; Henderson, A.; Hicks, K.; Yang, J.; Thompson, P.; Oliver, J. *Clin. Pharmacol. Drug Dev.* **2014**, *3* (3), 170–178.

105. Singer, I. I.; Tian, M.; Wickham, L. A.; Lin, J.; Matheravidathu, S. S.; Forrest, M. J.; Mandala, S.; Quackenbush, E. J. *J. Immunol.* **2005**, *175* (11), 7151–7161.

106. Harris, S.; Tran, J. Q.; Southworth, H.; Spencer, C. M.; Cree, B. A. C.; Zamvil, S. S. *Neurol. Neuroimmunol. Neuroinflamm.* **2020**, *7* (5), e839.

107. Musella, A.; Gentile, A.; Guadalupi, L.; Rizzo, F. R.; De Vito, F.; Fresegna, D.; Bruno, A.; Dolcetti, E.; Vanni, V.; Vitiello, L.; Bullitta, S.; Sanna, K.; Caioli, S.; Balletta, S.; Nencini, M.; Buttari, F.; Stampanoni Bassi, M.; Centonze, D.; Mandolesi, G. *Cells* **2020**, *9* (5), 1290.

108. Fronza, M.; Lorefice, L.; Frau, J.; Cocco, E. *Drug Des. Devel. Ther.* **2021**, *15*, 1993–2004.

109. Lin, C. H.; Kadakia, S.; Frieri, M. *Autoimmun. Rev.* **2014**, *13* (2), 114–116.

110. Robinson, W. H.; Genovese, M. C.; Moreland, L. W. *Arthritis Rheum.* **2001**, *44* (9), 1977–1983.

111. Kraus, J.; Ling, A. K.; Hamm, S.; Voigt, K.; Oschmann, P.; Engelhardt, B. *Ann. Neurol.* **2004**, *56* (2), 192–205.

112. Harzheim, M.; Stepien-Mering, M.; Schröder, R.; Schmidt, S. *J. Interf. Cytokine Res.* **2004**, *24* (12), 711–716.

113. Cuzzocrea, S.; Doyle, T.; Campolo, M.; Paterniti, I.; Esposito, E.; Farr, S. A.; Salvemini, D. *J. Neuro-Oncol.* **2018**, *35* (13), 1452–1466.

114. Hunter, S. F.; Bowen, J. D.; Reder, A. T. *CNS Drugs* **2016**, *30* (2), 135–147.

115. Grassi, S.; Mauri, L.; Prioni, S.; Cabitta, L.; Sonnino, S.; Prinetti, A.; Giussani, P. *Front. Pharmacol.* **2019**, *10*, 807.

116. Groves, A.; Kihara, Y.; Chun, J. *J. Neurol. Sci.* **2013**, *328* (1–2), 9–18.

117. Comi, G.; Hartung, H.-P.; Bakshi, R.; Williams, I. M.; Wiendl, H. *Drugs* **2017**, *77* (16), 1755–1768.

118. Fox, E. J.; Lublin, F. D.; Wolinsky, J. S.; Cohen, J. A.; Williams, I. M.; Meng, X.; Ziehn, M.; Kolodny, S.; Cree, B. A. C. *Neurol. Neuroimmunol. Neuroinflamm.* **2019**, *6* (6), e614.

119. Fox, E.; Edwards, K.; Burch, G.; Wynn, D. R.; Laganke, C.; Crayton, H.; Hunter, S. F.; Huffman, C.; Kim, E.; Pestreich, L.; McCague, K.; Barbato, L. *Mult. Scler. Relat. Disord.* **2014**, *3* (5), 607–619.

120. Tran, J. Q.; Hartung, J. P.; Olson, A. D.; Mendzelevski, B.; Timony, G. A.; Boehm, M. F.; Peach, R. J.; Gujrathi, S.; Frohna, P. A. *Clin. Pharmacol. Drug Dev.* **2018**, *7* (3), 263–276.

121. Paik, J. *Drugs* **2022**, *82* (12), 1303–1313.

122. Lassiter, G.; Melancon, C.; Rooney, T.; Murat, A.-M.; Kaye, J. S.; Kaye, A. M.; Kaye, R. J.; Cornett, E. M.; Kaye, A. D.; Shah, R. J.; Viswanath, O.; Urits, I. *Neurol. Int.* **2020**, *12* (3), 89–108.

123. Yuan, Y.; Jia, G.; Wu, C.; Wang, W.; Cheng, L.; Li, Q.; Li, Z.; Luo, K.; Yang, S.; Yan, W.; Su, Z.; Shao, Z. *Cell Res.* **2021**, *31* (12), 1263–1274.

124. Hanson, M. A..; Roth, C. B..; Jo, E.; Griffith, M. T..; Scott, F. L.; Reinhart, G.; Desale, H.; Clemons, B.; Cahalan, S. M.; Schuerer, S. C..; Sanna, M. G.; Han, G. W.; Kuhn, P.; Rosen, H.; Stevens, R. C. *Science* **2012**, *335* (6070), 848–851.

125. Davis, M. D.; Clemens, J. J.; Macdonald, T. L.; Lynch, K. R. *J. Biolumin. Chemilumin.* **2005**, *280* (11), 9833–9841.

126. Yamamoto, R.; Okada, Y.; Hirose, J.; Koshika, T.; Kawato, Y.; Maeda, M.; Saito, R.; Hattori, K.; Harada, H.; Nagasaka, Y.; Morokata, T. *PLoS One* **2014**, *9* (10), e110819.

127. Sugahara, K.; Maeda, Y.; Shimano, K.; Mogami, A.; Kataoka, H.; Ogawa, K.; Hikida, K.; Kumagai, H.; Asayama, M.; Yamamoto, T.; Harada, T.; Ni, P.; Inoue, S.; Kawaguchi, A. *Br. J. Pharmacol.* **2017**, *174* (1), 15–27.

128. Cahalan, S. M.; Gonzalez-Cabrera, P. J.; Sarkisyan, G.; Nguyen, N.; Schaeffer, M. T.; Huang, L.; Yeager, A.; Clemons, B.; Scott, F.; Rosen, H. *Nat. Chem. Biol.* **2011**, *7* (5), 254–256.

129. Piali, L.; Birker-Robaczewska, M.; Lescop, C.; Froidevaux, S.; Schmitz, N.; Morrison, K.; Kohl, C.; Rey, M.; Studer, R.; Vezzali, E.; Hess, P.; Clozel, M.; Steiner, B.; Bolli, M. H.; Nayler, O. *Pharmacol. Res. Perspect.* **2017**, *5* (6), e00370.

130. Demont, E. H.; Bailey, J. M.; Bit, R. A.; Brown, J. A.; Campbell, C. A.; Deeks, N.; Dowell, S. J.; Eldred, C.; Gaskin, P.; Gray, J. R. J.; et al. *J. Med. Chem.* **2016**, *59* (3), 1003–1020.

131. Pan, S.; Gray, N. S.; Gao, W.; Mi, Y.; Fan, Y.; Wang, X.; Tuntland, T.; Che, J.; Lefebvre, S.; Chen, Y.; Chu, A.; Hinterding, K.; Gardin, A.; End, P.; Heining, P.; Bruns, C.; Cooke, N. G.; Nuesslein-Hildesheim, B. *ACS Med. Chem. Lett.* **2013**, *4* (3), 333–337.

132. Buzard, D. J.; Kim, S. H.; Lopez, L.; Kawasaki, A.; Zhu, X.; Moody, J.; Thoresen, L.; Calderon, I.; Ullman, B.; Han, S.; et al. *ACS Med. Chem. Lett.* **2014**, *5* (12), 1313–1317.

133. Bolli, M. H.; Abele, S.; Binkert, C.; Bravo, R.; Buchmann, S.; Bur, D.; Gatfield, J.; Hess, P.; Kohl, C.; Mangold, C.; et al. *J. Med. Chem.* **2010**, *53* (10), 4198–4211.
134. Gilmore, J. L.; Xiao, H. Y.; Dhar, T. G. M.; Yang, M.; Xiao, Z.; Yang, X.; Taylor, T. L.; McIntyre, K. W.; Warrack, B. M.; Shi, H.; et al. *J. Med. Chem.* **2021**, *64* (3), 1454–1480.
135. Poirier, B.; Briand, V.; Kadereit, D.; Schäfer, M.; Wohlfart, P.; Philippo, M.-C.; Caillaud, D.; Gouraud, L.; Grailhe, P.; Bidouard, J.-P.; Trellu, M.; Muslin, A. J.; Janiak, P.; Parkar, A. A. *Sci. Signaling* **2020**, *13* (634), 8050.
136. Cohen, J. A.; Arnold, D. L.; Comi, G.; Bar-Or, A.; Gujrathi, S.; Hartung, J. P.; Cravets, M.; Olson, A.; Frohna, P. A.; Selmaj, K. W. *Lancet Neurol.* **2016**, *15* (4), 373–381.
137. Rasche, L.; Paul, F. *Expert Opin. Pharmacother.* **2018**, *19* (18), 2073–2086.
138. Cohen, J. A.; Comi, G.; Arnold, D. L.; Bar-Or, A.; Selmaj, K. W.; Steinman, L.; Havrdová, E. K.; Cree, B. A. C.; Montalbán, X.; Hartung, H. P.; Huang, V.; Frohna, P.; Skolnick, B. E.; Kappos, L. *Mult. Scler. J.* **2019**, *25* (9), 1255–1262.
139. Cree, B. A. C.; Selmaj, K. W.; Steinman, L.; Comi, G.; Bar-Or, A.; Arnold, D. L.; Hartung, H. P.; Montalbán, X.; Havrdová, E. K.; Sheffield, J. K.; Minton, N.; Cheng, C. Y.; Silva, D.; Kappos, L.; Cohen, J. A. *Mult. Scler. J.* **2022**, *28* (12), 1944–1962.
140. Study Details | Study Describing Cognitive Processing Speed Changes in Relapsing Multiple Sclerosis Subjects Treated With Ozanimod (RPC-1063) | ClinicalTrials.gov https://clinicaltrials.gov/study/NCT04140305 (accessed Dec 20, 2023).
141. Sandborn, W. J.; Feagan, B. G.; Hanauer, S.; Vermeire, S.; Ghosh, S.; Liu, W. J.; Petersen, A. K.; Charles, L.; Huang, V.; Usiskin, K.; Wolf, D. C.; D'Haens, G. *J. Crohn's Colitis* **2021**, *15* (7), 1120–1129.
142. Sandborn, W. J.; Feagan, B. G.; Wolf, D. C.; D'Haens, G.; Vermeire, S.; Hanauer, S. B.; Ghosh, S.; Smith, H.; Cravets, M.; Frohna, P. A.; Aranda, R.; Gujrathi, S.; Olson, A. *N. Engl. J. Med.* **2016**, *374* (18), 1754–1762.
143. Sandborn, W. J.; Feagan, B. G.; D'Haens, G.; Wolf, D. C.; Jovanovic, I.; Hanauer, S. B.; Ghosh, S.; Petersen, A.; Hua, S. Y.; Lee, J. H.; Charles, L.; Chitkara, D.; Usiskin, K.; Colombel, J.-F.; Laine, L.; Danese, S. *N. Engl. J. Med.* **2021**, *385* (14), 1280–1291.
144. Feagan, B. G.; Sandborn, W. J.; Danese, S.; Wolf, D. C.; Liu, W. J.;

Hua, S. Y.; Minton, N.; Olson, A.; D'Haens, G. *Lancet Gastroenterol. Hepatol.* **2020**, *5* (9), 819–828.

145. Duijvestein, M.; Jeyarajah, J.; Guizzetti, L.; Zou, G.; Parker, C. E.; van Viegen, T.; VandeCasteele, N.; Khanna, R.; Van der Aa, A.; Sandborn, W. J.; Feagan, B. G.; D'Haens, G. R.; Jairath, V. *Clin. Gastroenterol. Hepatol.* **2020**, *18* (5), 1121–1132.e2.

146. Novak, G.; Parker, C. E.; Pai, R. K.; MacDonald, J. K.; Feagan, B. G.; Sandborn, W. J.; D'Haens, G.; Jairath, V.; Khanna, R. *Cochrane Database Syst. Rev.* **2017**, *2017* (7), 1–69.

147. Roberts, E.; Rosen, H.; Brown, S.; Morales, M. A.; Peng, X.; Poddutoori, R. NOVEL MODULATORS OF SPHINGOSINE PHOSPHATE RECEPTORS Filing date: 14 May 2009, Publication date: 17 December 2009.

148. Roberts, E.; Rosen, H.; Brown, S.; Guerrero, M. A.; Peng, X.; Poddutoori, R. MODULATORS OF SPHINGOSINE PHOSPHATE RECEPTORS Filed date: Jun. 23, 2014, Publication date: July 5, 2016.

149. Porcs-Makkay, M.; Simig, G.; Molnar, E.; Petho, J.; Volk, B.; Szlavik, L.; Szabo, É.; Halasz, J. PROCESS FOR THE PRODUCTION OF OZANIMOD Filed date: May 18, 2018, Publication date: Nov 29, 2018..

150. Cianferotti, C.; Barreca, G.; Bollabathini, V.; Carcone, L.; Grainger, D.; Staniland, S.; Taddei, M. *Eur. J. Org. Chem.* **2021**, *2021* (12), 1924–1930.

151. Choi, D.; Stewart, A. P.; Bhat, S. *Ann. Pharmacother.* **2022**, *56* (5), 592–599.

12

Ciprofol (Cipepofol): A γ-Aminobutyric Acid Receptor Agonist for Induction of Anesthesia

Ji Zhang and
Dao-Qian Chen

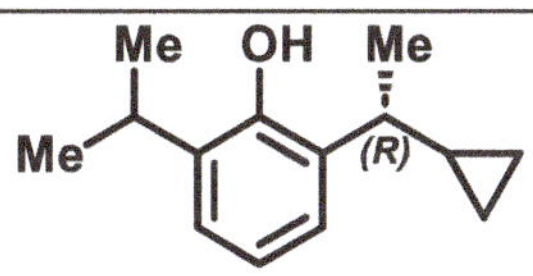

Ciprofol (**1**)

Trade name: Sishuning
Formerly Name: HSK3486/Cipepofol
Haisco Pharmaceutical Group
Launched: 2020 (China)

Today's anesthetic agents are mainly divided into inhalational agents and intravenous agents based in drug administration,[1] the former one including nitrous oxide, halothane and flurane (isoflurane/sevoflurane/desflurane) that are volatile, and the later one including propofol, barbiturates (thiopental was banned), benzodiazepines (midazolam and diazepam), and etomidate.

Based in China National Medical Products Administration, ciprofol (**1**, SiShuning, Cipepofol, formerly HSK3486) was approved to indications including sedation in gastrointestinal endoscopy, followed by anesthesia during the surgical/procedure of nontracheal intubation, induction and maintenance of general anesthesia, and sedation during intensive care by China in 2020.[2] Ciprofol (**1**) is independently developed and manufactured by Haisco Pharmaceutical Group Co., Ltd. (Chengdu, China), then commercialized in China now, which is still an investigational drug in United States as of drafting time. Ciprofol (**1**), a novel propofol (**2**) analog, is emerging as a captivating alternative in future, drawing increasing interest within the medical community with less dose and less injection pain and better cardiovascular stability.

Chemistry and Pharmacology of Drug Discovery, First Edition. Edited by Jie Jack Li.

1. Background

Undoubtedly, propofol (**2**) does build up the start of a new era in anesthesiology and expire the patent, which is widely used in clinics. Propofol (**2**), marketed as Diprivan, also known as 2,6-diisopropylphenol, is the most popular drug for anesthesia since approved by FDA in 1989. It appears to be an ideally rapidly acting short-acting agent that results in a decreased level of consciousness and a lack of memory for events, while pain on injection, lipid accumulation and bacterial contamination still bring some concerns.[1] Occasionally, propofol (**2**) causes excitation.[3] Since its discovery in 1973,[4] propofol (**2**) was meticulously developed in the United Kingdom by Imperial Chemical Industries under the designation ICI 35868, patented by John Glen and Roger James.[5] Interestingly, the initial propofol (**2**) formulation, a 1% preparation formulated in Cremophor EL as the vehicle, which was introduced in 1977, faced withdrawal due to the occurrence of anaphylactic reactions associated with the Cremophor EL. Consequently, it was reformulated as an emulsion comprising a mixture of soybean oil and propofol (**2**) in water, and subsequently relaunched in 1986 in Europe and was approved by the FDA in 1989. Current formulations comprises 1% or 2% (w/v) propofol (**2**), 10% soya bean oil, 1.2% egg phosphatide, and 2.25% glycerol. Also, 0.005% disodium edetate (EDTA) or metabisulfite is allowed to delay bacterial and fungal growth in this formulation. Propofol (**2**) serves as a reliable agent for surgical anesthesia, as well as procedural sedation and intensive care unit (ICU) sedation over the past 30 years. In 2016, FDA issued a warning regarding the high risk of brain injury associated with anesthesia in infants.[6] The prolonged exposure to anesthesia exceeding three hours, as well as the repeated use of general anesthetics and sedative drugs in infants or women during mid-pregnancy, can have a detrimental impact on the development of children's brains. In a rhesus monkey study, neurotoxicity was demonstrated to be associated with the application of propofol (**2**).[7] Together, propofol (**2**) has demonstrated its success for its rapid onset of anesthesia, rapid recovery in clinical practice, while adverse drug reactions (ADRs) including local pain, cardiovascular and respiratory depression, and drop in blood pressure, are reported synchronously.[2]

Similar research about this kind of phenol analogs, like PF0713 (**3**),[8] and fospropofol (**4**, trade name Lusedra used by Eisai Inc, one prodrug),[9] are developed to clinical phase and approved, respectively.

PF0713 (**3**), (*R,R*)-2,6-di-sec-butylphenol, is a single diastereomer containing two defined chiral carbons of the *R*-configuration, and improves the potency with 10-fold more than propofol (**2**) (0.38 vs 4.1 μg/mL), inducing brief propofol (**2**)-like anesthesia without pain on injection.[10]

Fospropofol (**4**),[10] is a water-soluble phosphate pro-drug of propofol (**2**), which is converted to propofol (**2**) within a few minutes of i.v. injection as moderate sedation without pain on injection. However, it was refused for more profound sedation. The most frequently recorded adverse reaction is paresthesia (including burning, tingling, and stinging) and/or pruritus, usually manifested in the perineal region within 5 minutes after administration. FDA requires that it should be administered only by persons trained in the administration of general anesthesia and not involved in the conduct of the diagnostic or therapeutic procedure.

Herein, ciprofol (**1**) is designed and expected to improve potency of previous propofol (**2**) with lower side effects (Figure 1).

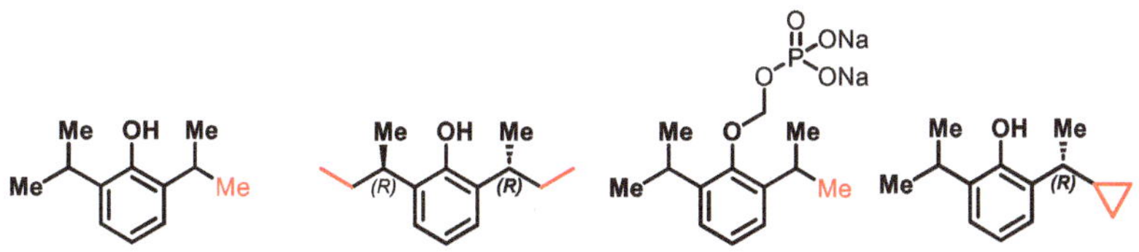

Propofol (**2**) Launched in 1977 | PF0713 (**3**) | Fospropofol (**4**) Launched in 2008 | Ciprofol (**1**) Launched in 2020

Figure 1. 2,6-Disubstituted phenol anesthesia-propofol (**2**) analogs

Table 1. Approved 2,6-substituted phenol for anesthesia

Propofol (2) (Imperial Chemical Industries)
Indications:
Induction of General Anesthesia for patients greater than or equal to 3 years of age;
Maintenance of General Anesthesia for patients greater than or equal to 2 months of age;
Initiation and maintenance of Monitored Anesthesia Care (MAC) sedation in Adult Patients;
Sedation for Adult patients in combination with Regional Anesthesia;
Intensive Care Unit (ICU) sedation of intubated, Mechanically ventilated adult patients.

Fospropol (4) (Eisai Inc.)
Indications: Monitored Anesthesia Care (MAC) sedation in adult patients.

Ciprofol (1) (Haisco Pharmaceutical Group Co., Ltd)
Indications:
Sedation in gastrointestinal endoscopy;
Anesthesia during the surgical/procedure of nontracheal intubation;
Induction and maintenance of General Anesthesia;
Sedation during intensive care.

Deserved of mentioning, propofol (**2**) has been approved to treat five indications (Table 1), followed by fospropofol (**4**) and ciprofol (**1**). Fospropofol (**4**) has been only approved to one indication (Monitored anesthesia care [MAC] sedation in adult patients) without more profound sedation so far. Subsequently, ciprofol (**1**) has four indications now. According to the GABA-A receptor binding assay,[11] ciprofol (**1**) exhibits much higher potency than propofol (**2**) and PF0713 (**3**). In a phase II clinical trial including 39 Chinese ICU patients,[12] ciprofol (**1**) shows good tolerance and efficacy for sedation with 5 times dosage less than propofol (**2**). Related treatment emergent adverse events (TEAEs) are hypotension (7.7% vs 23.1%, $P = 0.310$) and sinus bradycardia (3.8% vs 7.7%, $P = 1.000$) in the ciprofol (**1**) and propofol (**2**) groups, respectively. Ciprofol (**1**) is rapidly metabolized and accumulates at the low concentrations after continuous infusion in a stable circulatory system compared to propofol (**2**).

Many clinical trials about the efficacy of ciprofol (**1**) have been carried out in different indications in different Phase in Australia and China, mostly in head-to-head evaluation with propofol (**2**) (Table 2).[2] More and more clinic data shows ciprofol (**1**) has a comparable efficacy and safe profile in 5 times less dosage with propofol (**2**), bringing patients to lower injection pain.

Table 2. Clinical trials of ciprofol (**1**)

Clinicial Trails	Administration		Result
Anesthesia in patients undergoing kidney transplantation[13]	Ciprofol (N = 52) induction:0.4 mg/kg maintain:0.8-2.4 mg/kg/h	Propofol (N = 53) induction:2.0 mg/kg maintain:4-12 mg/kg/h	Shorter eyelash reflex disapperance (p < 0.001); Longer Awakening time (p < 0.001); Lower injection pain (1.9% vs 60.4%) Intraoperative hypotension benifit
Anesthesia/sedation in patients undergoing colonoscopy[14]	Ciprofol (N = 31/32) induction:0.4/0.5 mg/kg	Propofol (N = 31) induction:2.0 mg/kg	Rate of AEs (16.1% vs 21.9% vs 25.8%) Lower injection pain (12.9% vs 3.6% vs 45.2%)
Induction of deep sedation during gastroscopy and colonoscopy procedure[15]	Ciprofol (15 gastroscopy and 129 andcolonoscopy) induction:0.4 mg/kg	Propofol (15 gastroscopy and 129 andcolonoscopy) induction:1.5 mg/kg	Success rate of colonoscopy (100% vs 99.2%) Lower injection pain (4.9% vs 52.4%) Lower AEs: (31.3% vs 62.8%)
Anesthesia/sedation in patients undergoing fiberoptic bronchoscopy[16-17]	Ciprofol (N = 46/134) induction: 0.3/0.4 mg/kg	Propofol (N = 46/133) induction: 1.2/:2.0 mg/kg	Non-inferior to propofol-remifentanil Good safety profile/lower injection pain (6.52% vs 36.96%)/(4.4% vs 39.4%)
Induction/maintenance of general anesthesia in elective surgery[18-19]	Ciprofol (N = 88)0.4 mg/kg	Propofol (N = 88)2.0 mg/kg	Lower injection pain 6.8% vs 20.5%
Sedation in ICU patients with mechanical ventilation[20]	Ciprofol (N = 26) induction: 0.1–0.2 mg/kg maintain: 0.3 mg/kg/h	Propofol (N = 13) induction:0.5-1.0 mg/kg maintain: 1.5 mg/kg/h	Comparable tolerance and efficacy Hypotension (7.7% vs 23.1%) Sinus bradycardia (3.8% vs 7.7%)
Elderly and Non-elderly Healthy Chinese[21]	Ciprofol (4*N = 8) Elderly : 0.2/0.3/0.4mg/kg; Non-elderly : 0.4 mg/kg		Slightly lower AEs and similar PK (non-elderly than elderly); 0.3 mg/kg dose is recommended for elderly patients.
Induction of short-term general anesthesia[22]	Ciprofol 0.15 mg/kg, n = 2/0.4 mg/kg, n = 10 0.6 mg/kg, n = 6/0.9 mg/kg, n = 6		Well-tolerated 0.4 to 0.9 mg/kg Rapid onset and fast recovery; >0.4 mg/kg recommended for phase 2
Healthy Subjects[23]	Ciprofol (N = 8) induction:0.75 mg/kg maintain:0.5 mg/kg/h	Propofol (N = 8) induction:3.0 mg/kg maintain:2.0 mg/kg/h	Lower injection pain and respiratory depression similar PK/PD
Induction and maintenance of general anaesthesia[24]	Ciprofol (N = 86) induction:0.4 mg/kg maintain:0.8-2.4 mg/kg/h	Propofol (N = 42) induction:2.0 mg/kg maintain:4-12 mg/kg/h	Noninferior efficacy profile Lower injection pain (8.1% vs 21.4%) Rash (1.2% vs 0%)
General anesthesia in patients undergoing gynecological surgery[25]	Ciprofol (N = 60) induction:0.4 mg/kg	Propofol (N = 60) induction:2.0 mg/kg	Lower injection pain (16.7% vs 58.3%) Lower AEs: (20% vs 48.33%)
Sedation for Same-Day Bidirectional Endoscopy[26]	Ciprofol (N =45)+ esketamine Ciprofol (N =45)+ placebo	Propofol (N =45)+ esketamine Propofol (N =45)+ placebo	Improve cardiorespiratory stability
Patients under painless gastroenteroscopy anesthesia[27]	Ciprofol (N = 47) induction:0.4 mg/kg	Propofol (N = 49) induction:2.0 mg/kg	Equally safe and effective No early-cognitive dysfunction Lower injection pain (2.1% vs 71.4%)
Patients with hepatic impairment[28]	Ciprofol (N=24) induction: 0.4 mg/kg for 1 min/maintenance: 0.4 mg/kg/h		No clinically relevant differences in exposure/PD AEs: Moderate hepatic impairment is highest
Induction of anesthesia in adult patients[29]	Ciprofol induction:0.3 (N = 31)/0.4 (N = 8)/0.5 mg/kg (N = 29)	Propofol induction: 2.0 (N =31)/2.5 mg/kg (N =10)	Lower injection pain (6.4%/0%/13.8 vs 22.6%/10%) Similar efficacy and safety (ciprofol-0.5 mg/kg vs propofol-2.0 mg/kg)
Procedural sedation and anesthesia in non-operating room settings[30]	Ciprofol induction:6 (N =69)/8 mg/kg/h (N =69) maintain:1–2.5 mg/kg/h	Propofol (N = 69) induction:40 mg/kg/h maintain: 5–12.5 mg/kg/h	No pain on injection in ciprofol groups Similar safety profile/neurological outcomes/inflammatory parameters
Anesthesia for one 72-year-old man in cardiac surgery with CPB[31]	Ciprofol (N = 1) induction: 0.4 mg/kg; maintain: 0.8 mg/kg/h		Safe and effective agent in CPB surgery
Anesthesia in gynecological ambulatory surgery[32]	Ciprofol (N = 64) induction:0.5 mg/kg maintain:1.0 mg/kg/h	Propofol (N = 64) induction:2.0 mg/kg maintain:5.0 mg/kg/h	Lower injection pain (1.6% vs 76.6%) Lower AEs: (56.2% vs 92.2%) Time of loss of consciousness (1.6 min vs 1.4 min).

AEs: adverse events

In a phase 3, multicenter, randomized, double-blind, comparative study,[19] both ciprofol (**1**, 0.4 mg/kg) and propofol (**2**, 2.0 mg/kg) groups demonstrated anesthesia induction success rates of 100.0%. The lower limit of the 95% confidence interval (−4.18% difference) suggests that ciprofol (**1**) is non-inferior to propofol (**2**), substantiating its comparable effectiveness in achieving successful anesthesia induction. Furthermore, the ciprofol (**1**) group exhibits a significantly lower incidence of injection pain compared to the propofol (**2**) group, with rates of 6.8% and 20.5% respectively ($p < 0.05$). This finding highlights the favorable profile of ciprofol (**1**) in terms of reduced discomfort upon administration when compared to propofol (**2**). Additionally, fewer cases were observed where the Bispectral Index (BIS) exceeded 60 within 15 minutes of intravenous administration, suggesting that ciprofol (**1**) had the potential to achieve a more optimal sedation level during the post-induction period, even under an equivalent dosage regimen to propofol (**2**). This study highlights the advantages of ciprofol (**1**) in terms of cardiovascular stability and maintaining the desired sedation state.

2. Pharmacology

Provided by previous research about propofol (**2**), ciprofol (**1**) shares similar mechanism of action with propofol (**2**), targeting the intricate γ-aminobutyric acid (GABA) receptor system preferentially, whose activation rapidly increases Cl^- conductance and hyperpolarization of the postsynaptic membrane GABA receptor. GABA is an inhibitory neurotransmitter,[33] which acts a key role in the balanced inhibitory/excitatory neuronal network in the central nervous system. A series of distinct classes of drugs (benzodiazepines and benzodiazepine-like compounds, beta-carbolines steroids, barbiturates, alcohols, picrotoxin, tertbutylbicyclophosphorothionate (TBPS) exert their effects by interacting with specific modulatory sites on the GABA receptor.[34]

About mechanism of action for ciprofol (**1**), competitive binding assays on the α1β2γ2 subtype of γ-aminobutyric acid type A (GABAA) receptors demonstrated that ciprofol (**1**) triggers GABA-evoked chloride currents at lower concentration, and enhances GABA-induced activation at higher concentrations.[35] Besides, in whole-cell patch-clamp experiments, the affinity of ciprofol (**1**) with GABAA receptors was 4-5 times higher than that of propofol (**2**). Ciprofol (**1**) approximately has high affinity for the picrotoxin-binding site or has an allosteric inhibition of TBPS/TBOB for binding to GABAA receptor, while there is no affinity for benzodiazepine and GABA sites.

Considering unclear mechanism of action at atomic resolution, more and more evidence is found with structural biology methods advancing. Generally, the formation of an intermolecular hydrogen bond involving hydroxyl group of propofol (**2**) plays a dominant role in its molecular recognition with receptors that lead to hypnosis. Atomically, X-ray structure of GLIC (a homopentameric member of the pLGIC family)-pentameric ligand-gated ion channels with propofol (**2**) clarifies that it binds at the entrance of the cavity and is sandwiched between M1 and M3 and interacts mainly with T255 and Y254 via van der Waals contacts, in which the propofol (**2**) hydroxyl group could form a hydrogen bond with Y254.[36] Mutations of GLIC (V242M and T255A I202A) lining the binding site profoundly affect its general-anesthetic pharmacology and show a significant mobility of propofol (**2**) within the cavity. However, no X-ray structure of ciprofol (**1**) was reported to disclose its atomic mechanism of action.

Yuki et al. demonstrated that propofol (**2**) bound to the lovastatin site in leukocyte function–associated antigen-1 (LFA-1) and inhibited the production of interleukin-2 via LFA-1 in a dependent manner.[37] It is worth noting that this lovastatin site is also the binding site for anesthetics isoflurane and sevoflurane.

In electrophysiological assays,[38] propofol (**2**) allosterically enhanced the actions of GABA at the GABA-A receptor. Propofol (**2**) significantly shifted the dose–response curve of GABA-activated current towards the left without affecting the maximum of the GABA response. It profoundly slows down the desensitization process of GABA-A receptors, which plays a pivotal role during rapid and repetitive activation of inhibitory synapses (Bai et al. 1999). Early neurochemical studies[39] demonstrated that propofol (**2**) markedly enhanced [^{3}H]-GABA binding in the rat cerebral cortex and dose-dependently inhibits the binding of [^{35}S]-TBPS. Furthermore, propofol (**2**) also influenced presynaptic mechanisms of GABAergic transmission such as GABA uptake and [^{3}H]-GABA uptake into purified striatal synaptosomes was inhibited in a dose dependent and reversible manner by propofol (**2**) (IC_{50} = 46 μM) GABA release. In 1988, Collins reported propofol (**2**) (20–50 μM) increased GABA-mediated inhibitory transmission in rat olfactory cortex slices and influenced positively both pre- and post-synaptic GABA-dependent neuronal inhibition.[40]

According to Eckenhoff's study, fluorine-substituted analogue (fropropofol, see Figure 2) demonstrated no hypnotic activity, but rather weak excitatory activity. Furthermore, hydrogen bonding is an essential molecular characteristic for propofol (**2**) protein binding sites inducing hypnosis.[41] Interestingly, in a rigorous three-layer ONIOM (M06-2X/6-31 + G*:PM6: AMBER) study, fropropofol can bind to the protein with lower binding affinity molecular binding via a halogen-bonding and benzene interaction.[42]

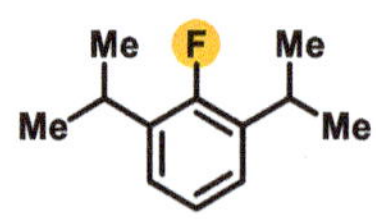

Fropropofol

Figure 2. Fluorine-substituted analogue

3. Structure–Activity Relationship (SAR)

Here we describe the structure–activity relationship about ciprofol (**1**) mainly based in James's (1980),[43] Trapani's (1998)[44] and Qin's works[11] (Figure 3). Their potencies will be largely affected by steric hindrance of the other five positions (R^2-R^6/R^3-R^5/R^4). Hydroxyl group is a key characteristic for hypnosis and the ortho substitution will modify the hydrogen bond binding strength. Most structures are alkyl group substitution that is prone to be hydrophobic domain.

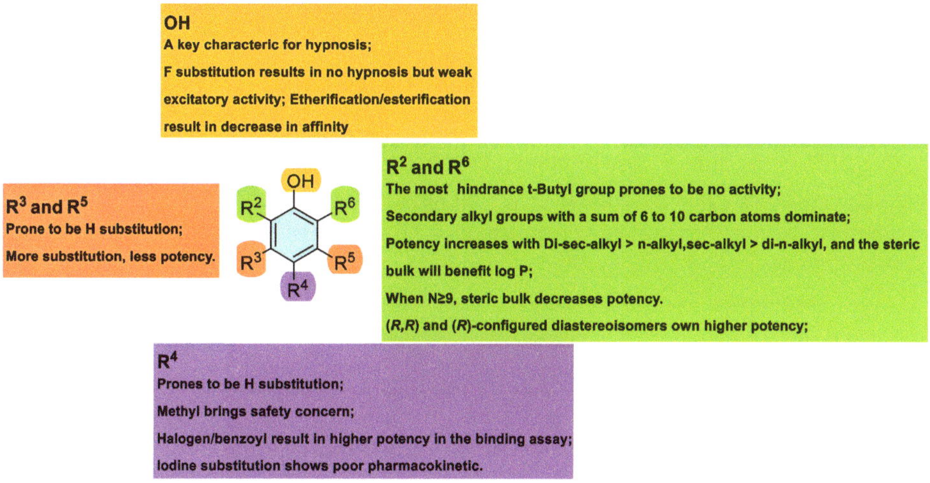

Figure 3. Structure–activity relationships (SAR) of alkylphenols as anesthetic agents

For free OH group, F substitution shows poor hypnosis but a weak excitatory activity.[41] Etherification or esterification of the phenolic hydroxyl group was demonstrated to decrease in binding affinity.[44]

For R^2 and R^6 position, secondary alkyl groups with a sum of 6–10 carbon atoms show an optimal anesthetic activity apart from the most hindrance *t*-butyl group. Their potencies increase with di-*sec*-alkyl > *n*-alkyl, *sec*-alkyl > di-*n*-alkyl, in which steric bulk will benefit logP, while the steric bulk decreases potency with the carbon number of chain ($N \geq 9$). As for diastereoisomers, (*R,R*) and (*R*)-configured one are prone to own higher potency than the other isomers.

For R^3 and R^5 position, substitutions in these two positions almost show poor potencies, as well as R^4 position. Furthermore, methyl-substitution in R^4 will cause a delayed death, which may result from a *p*-quinonemethide intermediate via a different metabolism mode.[43] Interestingly, halogens (Cl, Br, and I) or benzoyl substituents[44] benefit in inhibiting [^{35}S]-TBPS binding, in which iodo-substitution[45] exhibits poorer PK than propofol (**2**) via intraperitoneal injection, resulting in failure of inducing anesthesia in rodents without detected in mouse serum or brain. Also, I-substituent shows approximately even six-fold less than that of propofol (**2**) by a regular intravenous injection.

In summary, previous SAR around R^3, R^4, and R^5 position were mainly developed by two papers[43,44] to provide a good understanding to modification of these three positions,

R^2 and R^6 as well. Noteworthy, this chapter focuses on renewing the structure-activity relationships (SAR) related to ciprofol (**1**) by Qin' work.

In 2014 and 2017, Qin and coworkers patented and published the design and evaluation of ciprofol (**1**) as a promising anesthetic for head-to-head evaluation with propofol (**2**) in detail, allowing us to directly clarify the effect of different groups in R^2 and R^6. As shown in Figure 4, a validated animal model of general anesthesia[46] is measured by loss of righting reflex (LORR) experiment, which is recorded in three periods (induction, duration of anesthesia and recovery time), respectively. Using this LORR data, ED_{50} (mg/kg, 50% of the mice to lose righting reflex, being like the HD_{50} of James' work), LD_{50} (mg/kg, median lethal dose), and $HD_{10\ min}$ (mg/kg, the dose required to produce 10 min of anesthesia) are gained to evaluate the potency of 2,6-disubstituted phenol derivatives.

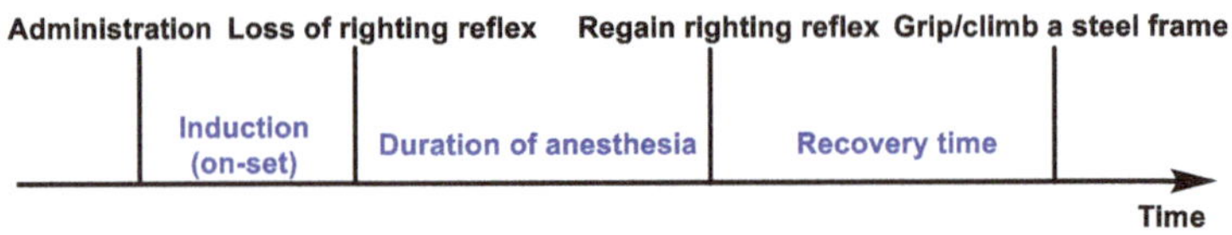

Figure 4. Loss of righting reflex (LORR) experiment

As shown in Table 3, 2-isopropyl was pre-introduced in one side chain of phenol as a model class based in propofol (**2**).[11] The displacement of cyclopropyl will significantly increase the potency (ED_{50} = 3.7 mg/kg) with *N* (the carbon number of both side chains) = 8 and (*R*)-configured enantiomer ciprofol (**1**) exhibits the highest potency with the least HD_{10min} (4.8 mg/kg). Compound **6** containing a cyclopropyl group evidences a similar trend that (*R*)-enantiomer performs higher potency than the other isomers. Together, configuration obviously has an influence on the ED_{50} and LD_{50}. Interestingly, more steric hindrance including [**5** (*N* = 9), **6** (*N* = 9), **7** (*N* = 10), **8** (*N* = 11)] decreases potency with increasing the carbon number of chain, which is opposite to the trend of $N \leq 8$.[43]

The SAR of the other side chain with cyclopropylethyl are shown in Table 4.[11] No substitution (compound **9**) brings a deep anesthesia but with long recovery time, which means a great risk to surgical operations. Simple substitutions (Me-(**10**)/Et-(**11**)//OMe-(**12**)) result in better potencies with similarly low ED_{50} and shorter recovery time. Additionally, compound **12**, OMe results in reduced recovery time compared with compounds **10** and **11**. The introductions of more steric hindrance (**1**, **14–19**) in the other side chain seemly benefit in reduced ED_{50} compared with propofol (**2**). However, most of them cannot exhibit a quick recovery except (**1**, (*R,R*)-**16** and (*R,S*)-**16**). Even though both ciprofol (**1**) and (*R,R*)-**16** exhibit lower ED_{50} and HD_{50} compared with PF0713 (**3**), data of duration and recovery time seemly are not improved. Besides, (*R,R*)-**16** shows a better profile than ciprofol (**1**), as well as $GABA_A$ receptor binding assay. Why is ciprofol (**1**) the lucky one to be developed?

Table 3. *In vivo* studies on sedation and SAR study of 2,6-disubstituted phenol derivatives

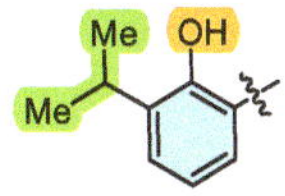

2-Isopropylphenols

	Side chain	On-set (sec)	Duration (sec)	Recovery time (sec)	ED_{50} (mg/kg)	LD_{50} (mg/kg)	LD_{50}/ED_{50}	HD_{10min} (mg/kg)	*c*Log P
2 Propofol	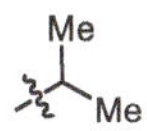	<10	303.2 ± 97.82	50.00 ± 10.37	11.70	31.3	2.7	25.0	3.929
1 Ciprofol	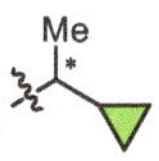	<10	518.50 ± 48.47	78.00 ± 37.60	3.7	22.7	6.1	10.0	4.373
		21.86 ± 9.04	313.43 ± 43.65	64.80 ± 14.57	1.5	9.9	6.6	4.8	*(R)*
		5.71 ± 1.38	344.00 ± 68.96	168.33 ± 71.71	7.9	50.0	6.3	30.0	*(S)*
5	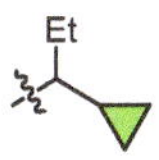	13.60 ± 8.26	523.57 ± 23.16	50.38 ± 42.52	6.2	53.2	8.6	—	4.902
6	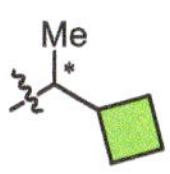	43.50 ±25.56	364.00 ± 21.44	46.00 ± 21.53	7.1	40.0	5.6	—	4.932
		<10	443.50 ± 62.70	89.11 ± 59.96	6.6	38.2	5.8	—	*(R)*
		15.63 ± 2.67	450.43 ± 50.32	191.80 ± 4.38	46.1	107.0	2.3	100.0	*(S)*
7	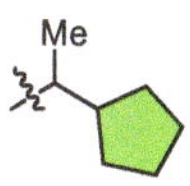	5.00 ± 0.71	737.33 ± 182.65	441.60 ± 88.05	19.9	115	5.8	40.0	5.491
8	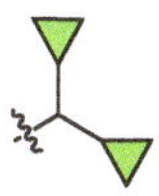	<10	2616.4 ± 311.81	220.17 ± 38.91	30.1	>100.0	3.3	50.0	4.871

Source: Adapted from Qin et al. [11]

In vivo studies (Table 5), tested compounds are administered to fasted rat at 1 mg/kg intravenously ($n = 3$).[11] Both ciprofol (**1**) and (*R,R*)-**16** show higher C_{max} (maximum plasma concentration = 834, 713 ng/mL), lower CL (100, 95 mL/kg min) and shorter $T_{1/2}$ (28, 31 min) than propofol (**2**) (C_{max} = 384 ng/mL; CL = 204 mL/kg min; $T_{1/2}$ = 42 min). Highest V_{dss} of propofol (**2**) demonstrates its high distribution into body tissues among them. Together, both ciprofol (**1**) and (*R,R*)-**16** elicit superiority to systemic exposure than propofol (**2**) according to Qin's study (Table 5). More importantly, no significant differences are reported between ciprofol (**1**) and (*R,R*)-**16**. However, based on the development of CMC (Chemical, Manufacturing, and Control), (*R,R*)-**16** containing two chiral carbon center appears to be the most difficult to develop, which may answer why ciprofol (**1**) is the lucky one.

Table 4. *In vitro* activity on sedation and SAR study of 2,6-disubstituted phenol derivatives

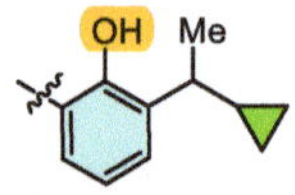

2-Cyclopropylethylphenols

	Side chain	On-set (sec)	Duration (sec)	Recovery time (sec)	ED_{50} (mg/kg)	LD_{50} (mg/kg)	LD_{50}/ED_{50}	HD_{10min} (mg/kg)	cLog P
2	**Propofol**	<10	303.2 ± 97.82	50.00 ± 10.37	11.70	31.3	2.7	25.0	3.929
9	H	<10	598.00 ± 165.64	329.60 ± 43.83	14.0	64.8	4.6	30.0	3.146
	Me	<10	198.33 ± 53.50	122.00 ± 58.17	7.4	54.9	7.4	—	3.595
11	Et	<10	409.57 ± 33.86	226.57 ± 81.01	7.7	57.1	7.4	—	3.974
12	MeO	<10	364.14 ± 27.84	79.00 ± 17.40	7.4	67.0	9.1	—	2.995
1 **Ciprofol**	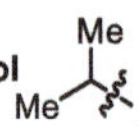	<10	518.50 ± 48.47	78.00 ± 37.60	3.7	22.7	6.1	10.0	4.373
		21.86 ± 9.04	313.43 ± 43.65	64.80 ± 14.57	1.5	9.9	6.6	4.8	*(R)*
	10	5.71 ± 1.38	344.00 ± 68.96	168.33 ± 71.71	7.9	50.0	6.3	30.0	*(S)*
13		<10	500.67 ± 46.89	193.63 ± 144.59	14.6	80.0	5.7	—	3.889
14		9.50 ± 2.07	365.75 ± 124.40	306.13 ± 97.81	10.5	40.0	3.8	45.0	4.308
15	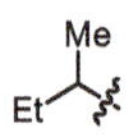	<10	418.14 ± 63.77	130.71 ± 32.37	4.5	38.1	8.5	20.0	4.902
		<10	471.88 ± 109.66	391.29 ± 65.05	2.0	14.3	8.5	7.5	*(R,R)*
		<10	545.00 ± 79.68	363.67 ± 91.20	10.1	65.4	6.5	20.0	*(R,S)*
		<10	418.00 ± 65.27	192.33 ± 41.81	1.3	8.3	6.4	4.0	*(S,R)*
		<10	643.40 ± 100.94	224.00 ± 55.96	5.3	36.8	6.9	10.0	*(S,S)*
3	**PF0713**	18 ± 3.46	459.81 ± 71.04	39.83 ± 18.84	2.69	23.8	8.8	10.0	4.987
16	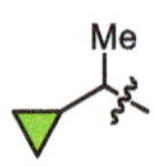	<10	563.63 ± 127.92	233.83 ± 37.98	3.6	>15.0	4.2	-	4.817
		<10	318.43 ± 56.00	61.67 ± 17.42	1.5	~6.3	4.2	5.0	*(R,R)*
		<10	492.17 ± 49.81	24.50 ± 7.42	5.9	43.6	7.4	20.0	*(R,S)*
		7.00 ± 1.41	567.33 ± 69.46	150.80 ± 37.51	19.5	149.0	7.6	40.0	*(S,S)*
17	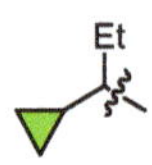	7.50 ± 1.05	670.13 ± 185.18	205.8 ± 65.30	9.0	74.7	8.3	20.0	5.346
18	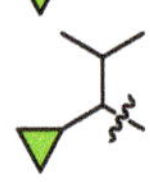	—	—	—	—	—	—	—	5.745
19	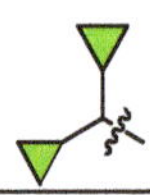	< 10	761.33 ± 189.10	326.50 ± 73.22	9.8	98.2	10	20	5.261

Source: Adapted from Qin et al. [11]

Table 5. Pharmacokinetic parameters of propofol (**2**), ciprofol (**1**) and (*R*, *R*)-**16**

Compound	Structure	AUC(0-t) (ng h/mL)	$T_{1/2}$ min	Vdss L/kg	CL mL/kg min	C_{max} mg/mL	Challenge for CMC
Propofol (2)	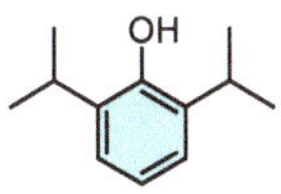	76.9	42	5.23	204	384	Increase
Ciprofol (1)	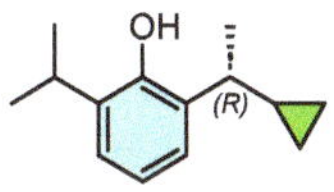	164.0	28	2.02	100	834	
***(R,R)*-16**	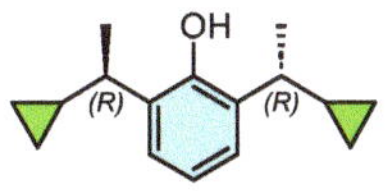	170.0	31	1.92	95	713	

CMC: Chemical, Manufacturing and Control

Source: Adapted from Qin et al. [11]

4. Pharmacokinetics and drug metabolism

In vivo study,[35] ciprofol (**1**) demonstrates a dose-dependent increase (1, 2, or 4 mg/kg) in exposure after intravenous injection (see Table 6). It is easily distributed in the tissues with V_{ss} = 7.79 L/kg (rats) and 6.06 L/kg (dogs), higher than total fluid volumes. In rats, the concentrations in the adrenal gland, fat, skin, ovary, and kidneys are 5 times higher than plasma concentrations. It also exhibits a good penetration ability in the blood–brain barrier, detected to 3.2 times in brain tissues higher than that in plasma. The residual concentration of ciprofol (**1**) is less than 10% of the peak concentration (C_{max}) after 240 min of administration, except in the fat, skin, bladder, and uterus. Interestingly, ciprofol (**1**) is demonstrated to high binding rate to plasma proteins in rat, dog, and human species, ranging from 85% to 96% within a concentration range of 80–1200 ng/mL.

Table 6. Pharmacokinetic parameters of ciprofol (**1**) after a single injection in male and female rats/dogs

PK parameters	Rats			Dogs		
Dose (mg/kg)	1	2	4	1	2	4
$AUC_{0\text{-}t}$ (h ng/mL)	55.8	130	286	165	342	716
$AUC_{0\text{-}\infty}$ (h ng/mL)	57.3	135	295	179	387	790
C_{max} (ng/mL)	236	459	1046	564	1214	2796
CL (L/h/kg)	18.2	15.1	13.8	6.67	5.32	5.25
$t_{1/2}$ (h)	0.46	0.95	0.75	1.28	1.66	1.37
V_{ss} (mL/kg)	6.84	8.38	8.16	7.36	5.75	5.06

Source: Adapted from Liao et al. [35]

In a Phase I study, a single dose of 0.4 mg/kg [^{14}C] ciprofol (**1**) is administered to six healthy subjects.[47] Being similar with propofol (88% in urine within 5 days and less than 2% in feces),[48] ciprofol (**1**) is primarily excreted through the urine (84.6% in 10 days) and feces (2.65%). Besides, propofol was observed to be expelled through exhalation in roughly equivalent to one billionth, indicating that ciprofol (**1**) could also be eliminated via the same pathway.[48] It identifies twenty metabolites and the main metabolic pathways (Figure 5). The metabolic pathways of ciprofol (**1**) *in vivo* are hypothesized to involve oxidation, followed by glucuronidation and sulfation. The primary metabolite identified is a conjugate of ciprofol (**1**) with glucuronic acid (M4) in both plasma and urine, which is consistent with propofol (**2**).[47] The predominant metabolites found are M4, M2-4, M3, and M5, which collectively represented 79.3% (plasma), 4.36% (plasma), 2.28% (urine), and 3.93% (plasma) of the total plasma and urine radioactivity, respectively. Additionally, a small amount of M7, accounting for 0.18% of the overall plasma radioactive exposure, plays an intermediate toward M2–4, M3, and M5. Provided by propofol (**2**), four similar urinary metabolites were identified as the glucuronic acid conjugate of propofol (**2**, 53%) and the glucuronic acid (18% & 13%) and sulphate (9%) conjugates of 2,6-diisopropyl-1,4-quinol.[48] Typically, the metabolite formed through glucuronidation is widely regarded as nonhypnotic and nontoxic.

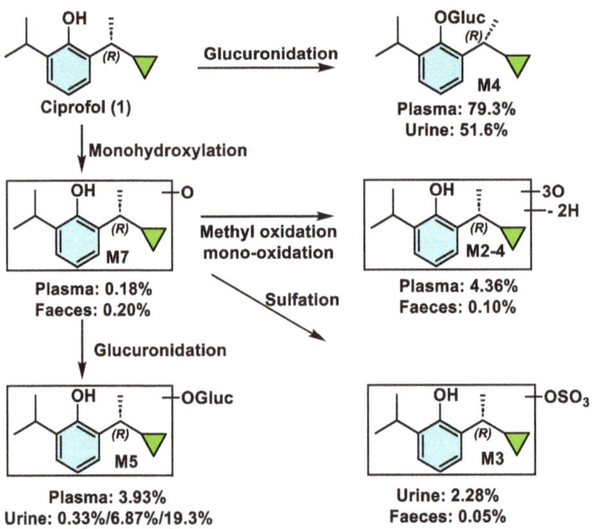

Figure 5. Proposed main metabolic pathways of ciprofol (**1**) in humans. Source: Adapted from Bian et al.[47]

Another study investigating pharmacokinetic (PK) parameters of ciprofol (**1**) in healthy subjects compared different administration models, including single-dose administration, sequential maintenance administration after the initial dose, and maintenance dose after the loading dose.[22] The results revealed the following PK parameters at a single 0.4 mg/kg dose of ciprofol (**1**): C_{max} (maximum concentration): 1330.0 ng/mL; T_{max} (time to reach C_{max}): 2.0 min; Elimination half-life ($T_{1/2}$): 2.09 h; Clearance (CL): 1.47 L/h/kg; Area under the curve ($AUC_{0-\infty}$): 271.67 ng h/mL; Apparent volume of distribution (V_d): 4.3 L/kg; These PK parameters indicate a rapid onset of action and a short peak time after a single dose of ciprofol (**1**). Additionally, it exhibited a high clearance and a small volume of distribution, aligning with the anticipated characteristics of short-term intravenous anesthesia.

In contrast, when continuous infusions of ciprofol (**1**, 0.4–0.5 mg/kg/h) for 4 and 12 h are performed with propofol (**2**, 2.0 mg/kg/h) as the positive control in another study, both T_{max} and $T_{1/2}$ are significantly prolonged, accompanied by an increased volume of distribution.[23] The plasma concentration-time profiles of ciprofol (**1**) exhibit similarity to those of propofol (**2**), albeit with lower concentrations. Following the initial dose, both ciprofol (**1**) and propofol (**2**) show a rapid increase in plasma concentration. After the maintenance infusion dose, there is an initial decrease in plasma concentration for both drugs, followed by a gradual increase over time. However, differences of pharmacokinetic (PK) parameters between ciprofol (**1**) and propofol (**2**) are complex in different studies. In a Phase II study,[18] drug exposure of propofol (**2**) is approximately 4 to 5 times higher than that of ciprofol (**1**) at a low dosage. Ciprofol (**1**) exhibits slightly lower values for half-life ($T_{1/2}$) and volume of distribution compared to propofol (**2**), indicating a relatively shorter duration of action and a more confined distribution in the body. T_{max} and CL values of ciprofol (**1**) are very similar to those of propofol (**2**), suggesting comparable rates of absorption and elimination between the two drugs.

In human liver microsome incubations,[35] CYP2B6 accounts for the major metabolism isoform for ciprofol (**1**), followed by CYP1A2 and CYP2C19.

5. Efficacy and Safety

Ciprofol (**1**) is generally well tolerated in clinical trials.[2] No evidence of reproductive toxicity associated with ciprofol (**1**) has been observed in rat fertility and early embryonic development toxicity tests, embryo-fetal developmental toxicity tests, and perinatal developmental toxicity tests. Besides, all the results of Ames with Salmonella typhimurium test, chromosome aberration test and mouse bone marrow micronucleus support negative genotoxicity.

In beagle dogs,[35] transient tachycardia can be monitored in all the ciprofol (**1**) (1, 2, or 4 mg/kg) and propofol (**2**) groups, especially with 169% increase reported in 4 mg/kg of ciprofol (**1**), while there is no significant change in respiratory rate or tidal volume after ciprofol (**1**) administration. The corrected QTcF intervals and body temperature will be obviously affected, as well as blood pressure. No post-effects are observed after recovery. about 10 times lower dosage was administrated in clinical studies,

Multiple clinical trials present the strong evidence that low dosage ciprofol (**1**) benefits patients from injection pain and incidence of AEs compared to 4–5 times higher dosage of propofol (**2**). Interestingly, anesthesia-related seizures are common when induction or relatively low concentration of an anesthetic drug.[49] Generally, increased γ-amino-butyric acid GABA-ergic inhibition, which exhibits totally opposite action of mechanism of ciprofol (**1**) and propofol (**2**), can sensitize the cortex so that seizures occur by a small amount of excitation. However, muscle fasciculation, as one of abnormal limb movements, which may not be indicative of true seizures, was reported in the ciprofol (**1**) group in a phase I clinical trial (33.3%–0.4 mg/kg, 33.3%–0.6 mg/kg and 83.3%–0.9 mg/kg). In a phase II a/b study, ciprofol (**1**) shows a lower incidence of muscle fasciculation (4.5%, 2 patients of 42), without observed in the propofol (**2**).[14] One case of myoclonus only in the ciprofol (**1**) group has been reported in a phase III trial of general anesthesia induction in patients for elective surgery.[19] The incidence of epilepsy only in the ciprofol (**1**) group was 3.8% (1 patients of 26) in a phase II trial of sedation in ICU patients with mechanical ventilation, resulting in withdrawal from the trial.[12] The incidence of body movement (patient has no conscious movement of the limbs) was higher in the ciprofol (**1**) group than in the propofol (**2**) group (9.4% vs 0%) in anesthesia in gynecological day surgery.[32] Additionally, a study for general anesthesia in patients undergoing gynecological surgery reported two cases of body movement in propofol (**2**) group (3.33%), which was ascribed to injection pain.[50] The occurrence of seizures during anesthesia might be observed at a rate of approximately 1 in every 172,592 administered anesthetics according to studies of propofol (**2**), which may be influenced by factors such as errors in administering anesthetic drugs, withdrawal of anti-epileptic medications, or a lack of oxygen supply to the brain (cerebral anoxia) or an excessive buildup of carbon dioxide (hypercarbia).[49] Since there is currently insufficient reliable evidence regarding the relationship between ciprofol (**1**) and muscle fasciculation, it is essential to conduct further research in the future.

There were no substantial indications of a drug-drug interaction between ciprofol (**1**) and mefenamic acid.[51] Mefenamic acid, which is known as an UGT1A9 inhibitor, was demonstrated to have no significant effect on the pharmacokinetics and pharmacodynamics of ciprofol (**1**) in healthy individuals when administered with mefenamic acid.

6. Synthesis

Sichuan Haisco Pharmaceutical Co. Ltd. developed and manufactured chemical process and product development of ciprofol (**1**). As to date, the strategy to ciprofol (**1**) devised by Haisco has diverted from drug discovery to drug development since the original disclosure was published in 2017.[11] Here in this chapter, we disclosed four iterative routes reported in two literatures.

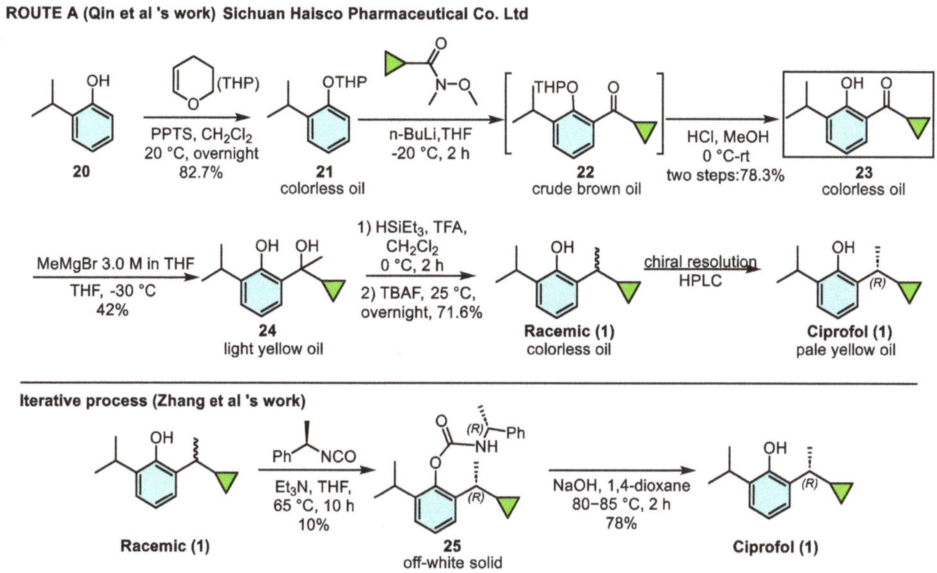

Figure 6. Haisco synthesis route to ciprofol (**1**)

Developed by Qin's work,[11] this initial route (**A**) in the drug discovery phase, starts with the THP protection of phenol using pyridinium 4-toluenesulfonate (PPTS) to afford **21** in 82.7% yield (Figure 6). A key ketone intermediate **23**, a colorless oil, was accomplished by treating with *n*-BuLi and Weinreb amide in –20 °C then a direct hydrolysis of crude oil and purified by flash chromatography in 78.3% yield. Noteworthy, 1.1 kg crude oil **23** was carried out in a telescopic reaction from 1.5 kg of **20** with general work-up, which brings concerns about purification.[52] When treated with Grignard reagent, ketone **23** was functionalized to alcohol **24** in 42% yield. A racemic (**1**) was obtained via a two-step sequence of $HSiEt_3$ in CF_3CO_2H at −30 °C, followed by TBAF. A general chiral resolution was carried out by chiral HPLC to afford R isomers, as known as ciprofol (**1**), which was difficulty to achieve scale-up. Alternatively, an iterative process was reported by Zhang et al.[55] from Haisco. Herein, the chiral resolution of ciprofol (**1**) was ultimately achieved by recrystallizing the carbamate intermediate **25** which was obtained from the reaction between (*R*)-(+)-1-phenylethyl isocyanate and racemic (**1**), followed by hydrolysis. Earlier study showed (*R*)-(+)-1-phenylethyl was helpful to isolate PF0713 (**3**).[8] However, the overall yield of the desired product was only 1–2% due to several challenges in this route including the formation of complex impurities, which needs column chromatographic purification or limited solid, and a low yield in 10% during the late-stage chiral resolution step, the utilization of *n*-BuLi as well. Additionally, there were apprehensions regarding the toxicity associated with the usage of PPTS for synthesizing the THP ether **21**.

The route **B** is an improved process for intermediate **23** (see Figure 7). Zhang et al. reported an iterative process for key intermediate **23** at the 100-gram scale.[52] Despite having one fewer synthetic step, this route still posed a challenge in terms of purification and lower yield. It necessitated a laborious column chromatographic purification process to obtain the key intermediate **23**. This purification step could potentially present difficulties when attempting to scale up the process.

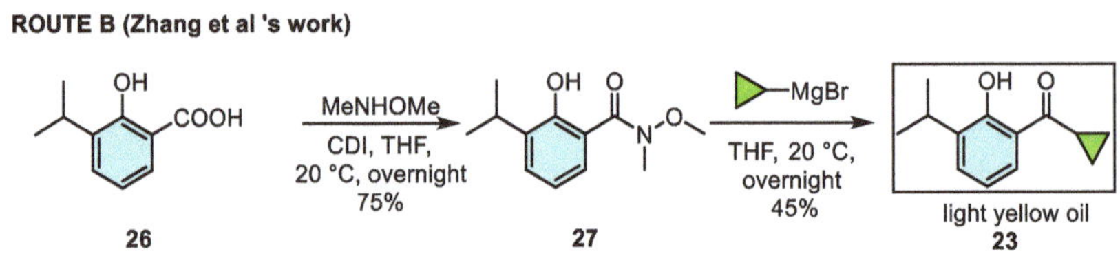

Figure 7. Haisco synthesis route B to ciprofol (**1**)

As shown in Figure 8, route C by Zhang et al.[52] described a more practical process using the same starting material 2-Isopropylphenol **20** on the kilogram scale. Under the strongly basic conditions (NaOH) at 10–15 °C, the desired ether **28** was isolated in 95% yield, which was directly used to the next step without purification. Undergoing σ-[3,3]-Claisen rearrangement reaction, **28** was converted to **29** in 53% yield in the presence of K_2CO_3. A similar strategy using Claisen rearrangement has been reported by Jenkins in 2009 before.[53] Subsequently, a classic Simmons–Smith cyclopropanation was performed with $AlEt_3$ (3 equiv)/CH_2I_2 (4 equiv) at 4 kg scale. The key racemic (**1**) was distilled to get 3.5 kg in 80% yield (96% purity), even though this cyclopropanation was slow and the quenching conditions were severe and time-consuming. After racemic (**1**) was treated with (*R*)-(+)-1-phenylethyl isocyanate/Et_3N, three times recrystallizations from heptane afforded carbamate **25** in 25% yield (97% purity, >99.5% *de*). However, despite monitoring the water content in the reaction solution below 1.0%, a side reaction between the isocyanate and water led to the formation of approximately 1–3% of urea.

ROUTE C (Zhang et al 's work)

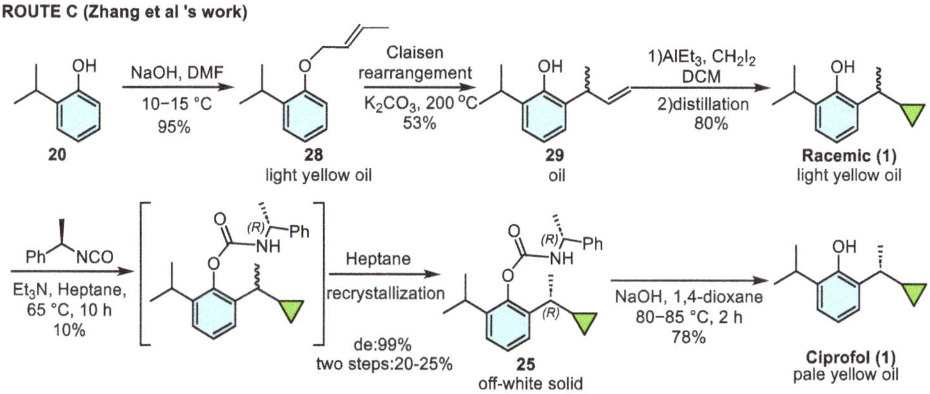

Figure 8. Haisco synthesis route C to ciprofol (**1**)

ROUTE D-endgame synthesis (Zhang et al 's work)

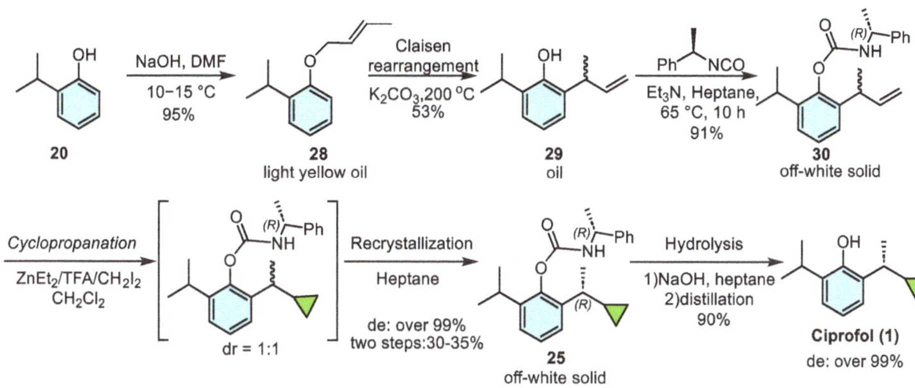

Figure 9. Haisco kilogram-scale route for clinical ciprofol (**1**)

Finally, an endgame routine (Figure 9) was carried out by adjusting the priority of carbamate reaction to increase one more solid **30** as an intermediate than route C, developed by the same group.[52] Typically, much more intermediate solids will benefit the whole process because of easily purification by recrystallizations and impurity control. During the pilot plant synthesis, **29** was obtained followed with same condition of route C. Then, an off-white solid **30** was easily afforded to 13.47 kg in 91% yield with 98% purity when treated with (*R*)-(+)-1-phenylethyl isocyanate/Et_3N in heptane. Subsequently, the cyclopropanation was completely accomplished to get racemic **25** in the presence of $ZnEt_2/CH_2I_2/CF_3CO_2H$ at room temperature at 11 kg scale, in which less loading of CF_3CO_2H than $ZnEt_2$ could avoid the hydrolysis of **25** to racemic (**1**). Recrystallization afforded the optically pure **25** in 30–35% yield (99.4% de, and 99% HPLC purity). The hydrolysis of **25** under the condition of NaOH/n-heptane will afford the desired ciprofol (**1**) in 90% yield with over 99% *de*. The overall yield of this endgame route was 12–14%.

7. Summary

Dating back to the discovery of propofol (**2**) in 1977, the patent of propofol (**2**) has expired for many years. When considering something in retrospect as medicine chemists, we must think about why ciprofol (**1**) is still discovered and developed? Firstly, drug is developed to meet the clinical needs of patients as the highest priority. Secondly, patentability is a magic weapon for drugs. Ciprofol (**1**) is currently approved to indications including sedation in gastrointestinal endoscopy, followed by anesthesia during the surgical/procedure of nontracheal intubation, induction and maintenance of general anesthesia, and sedation during intensive care by China in 2020. Numerous clinical trials investigating ciprofol (**1**) have consistently demonstrated its benefits,[2] including enhanced tolerance, higher sedation satisfaction score, and reduced occurrence of adverse reactions (respiratory depression and decrease blood pressure). Particularly noteworthy is its remarkable effectiveness in significantly decreasing the frequency of injection-related pain. There is a scarcity of long-term medication experience and comprehensive clinical data validating its efficacy in broader clinical scenarios due to a short period on the market. Still, a lack of clinical research data about ciprofol (**1**) in patients under 18 years old and pregnant and lactating women will not be recommended for their treatments. The familiarity and proficiency of anesthesiologists with propofol (**2**) make it challenging to envision a full-scale transition to ciprofol (**1**), as well as medicinal cost for patients. Compared to propofol (**2**), the complexity of chemical process is still more challenging due to a chiral center of one side chain. With further investigation and understanding, ciprofol (**1**) holds the potential to offer additional benefits to patients, ultimately enhancing the field of anesthesia.

References

1. Sneyd, J. R. Thiopental to desflurane-an anaesthetic journey. Where are we going next? *Br. J. Anaesth.* **2017**, *119*, 44–52.
2. Lu M; Liu J; Wu X.; Zhang Z. Ciprofol: a novel alternative to propofol in clinical intravenous anesthesia? *Biomed Res. Int.* **2023**, *2023*, 1–12.
3. Sneyd J. R. Excitatory events associated with Propofol anaesthesia: a review. *J. Royal Soc. Med.* **1992**, 85, 288–291.
4. Thompson, K. A.; Goodale, D. B. The recent development of Propofol (DIPRIVAN®). *Intensive Care Med.* **2000**, *26*, S400.
5. Glen J.; James R. 2,6-Diisopropylphenol as an anaesthetic agent. US4056635A **1997**.

6. Andropoulos, D. B.; Greene, M. F. Anesthesia and developing brains – implications of the FDA warning. *N. Engl. J. Med.* 2017, *376*, 905–907.

7. Creeley, C.; Dikranian, K.; Dissen, G.; Martin, L.; Olney, J.; Brambrink, A. Propofol-induced apoptosis of neurones and oligodendrocytes in fetal and neonatal rhesus macaque brain. *Br. J. Anaesth.* **2013**, *110*, 29–38.

8. Jenkins T. E. Stereoisomers propofol therapeutic compounds. US8242315B2 (2008).

9. Boules, R.; Szkiladz, A.; Nogid, A. Fospropofol disodium (lusedra) injection for anesthesia-care sedation: a clinical review. *Pharm. Ther.*, **2012**, *37*, 395.

10. Sneyd, J. R.; Rigby-Jones, A. E. New drugs and technologies, intravenous anaesthesia is on the move (again). *Br. J. Anaesth.* **2010**, *105*, 246–254.

11. Qin, L.; Ren, L.; Wan, S.; Liu, G.; Luo, X.; Liu, Z.; Li, F.;, Yu, Y.; Liu, J.; Wei, Y. Design, synthesis, and evaluation of novel 2, 6-disubstituted phenol derivatives as general anesthetics. *J. Med. Chem.* **2017**, *60*, 3606–3617.

12. Liu, Y.; Yu, X.; Zhu, D.; Zeng, J.; Lin, Q.; Zang, B.; Chen, C.; Liu, N.; Liu, X.; Gao, W.; Guan, X. Safety and efficacy of ciprofol vs. propofol for sedation in intensive care unit patients with mechanical ventilation: a multi-center, open label, randomized, phase 2 trial. *Chin. Med. J.* **2022**, *135*, 1043–1051.

13. Qin, K.; Qin, W.; Ming, S.; Ma, X.; Du, X. Effect of ciprofol on induction and maintenance of general anesthesia in patients undergoing kidney transplantation. *Eur. Rev. Med. Pharmacol. Sci.* **2022**, *26*, 5063–5071.

14. Teng, Y.; Ou, M.; Wang, X.; Zhang, W.; Liu, X.; Liang, Y.; Li, K.; Wang, Y.; Ouyang, W.; Weng, H.; Li, J. Efficacy and safety of ciprofol for the sedation/anesthesia in patients undergoing colonoscopy: phase IIa and IIb multi-center clinical trials, *Eur. J. Pharm. Sci.* **2021**, *164*, 105904.

15. Li, J.; Wang, X.; Liu, J.; Wang, X.; Li, X.; Wang, Y.; Ouyang, W.; Li, J.; Yao, S.; Zhu, Z.; Guo, Q. Comparison of ciprofol (HSK3486) versus propofol for the induction of deep sedation during gastroscopy and colonoscopy procedures: a multi-centre, non-inferiority, randomized, controlled phase 3 clinical trial. *Basic Clin. Physiol. Pharmacol.* **2022**, *131*, 138–148.

16. Wu, B.; Zhu, W.; Wang, Q.; Ren, C.; Wang, L.; Xie, G. Efficacy and safety of ciprofol-remifentanil versus propofol-remifentanil during fiberoptic bronchoscopy: a prospective, randomized, double-blind, non-

inferiority trial. *Front. Pharmacol.* **2022**, *13*, 1091579.

17. Luo, Z.; Tu, H.; Zhang, X.; Wang, X.; Ouyang, W.; Wei, X.; Zou, X.; Zhu, Z.; Li, Y.; Shangguan, W.; Wu, H. Efficacy and safety of HSK3486 for anesthesia/sedation in patients undergoing fiberoptic bronchoscopy: a multicenter, double-blind, propofol-controlled, randomized, phase 3 study. *CNS Drugs* **2022**, *36*, 301–313.
18. Zeng, Y.; Wang, D.; Lin, Z.; Liu, J.; Wei, X. C.; Deng, J.; Liu, Y. F.; Ma, E. L.; Yang, M. C.; Zheng, H.; Yu, X. D. Efficacy and safety of HSK3486 for the induction and maintenance of general anesthesia in elective surgical patients: a multicenter, randomized, open-label, propofol-controlled phase 2 clinical trial. *Eur. Rev. Med. Pharmacol. Sci.* **2022**, *26*, 1114–1124.
19. Wang, X.; Liu, J.; Zuo, Y.; Zhu, Q. M.; Wei, X. C.; Zou, X. H.; Luo, A. L.; Zhang, F. X.; Li, Y. L.; Zheng, H.; Li, H. Effects of ciprofol for the induction of general anesthesia in patients scheduled for elective surgery compared to propofol: a phase 3, multicenter, randomized, double-blind, comparative study. *Eur. Rev. Med. Pharmacol. Sci.* **2022**, *26*, 1607–1617.
20. Liu, Y.; Chen, C.; Liu, N.; Tong, L.; Nie, Y.; Wu, J.; Liu, X.; Gao, W.; Tang, L.; Guan, X. Efficacy and safety of ciprofol sedation in ICU patients with mechanical ventilation: a clinical trial study protocol. *Adv. Ther.* **2021**, *38*, 5412–5423.
21. Li, X.; Yang, D.; Li, Q.; Wang, H.; Wang, M.; Yan, P.; Wu, N.; Li, F.; Ma, S.; Ding, Y.; Liu, J. Safety, pharmacokinetics, and pharmacodynamics of a single bolus of the γ-aminobutyric acid (GABA) receptor potentiator HSK3486 in healthy Chinese elderly and non-elderly. *Front. Pharmacol.* **2021**, *12*, 735700.
22. Teng, Y.; Ou, M.; Wang, X.; Zhang, W. S.; Liu, X.; Liang, Y.; Zuo, Y. X.; Zhu, T.; Liu, B.; Liu, J. Pharmacokinetic and pharmacodynamic properties of ciprofol emulsion in Chinese subjects: a single center, open-label, single-arm dose-escalation phase 1 study. *Am. J. Transl. Res.* **2021**, *13*, 13791–13802.
23. Hu, C.; Ou, X.; Teng, Y.; Shu, S.; Wang, Y.; Zhu, X.; Kang, Y.; Miao, J. Sedation effects produced by a ciprofol initial infusion or bolus dose followed by continuous maintenance infusion in healthy subjects: a phase 1 trial. *Adv. Ther.* **2021**, *38*, 5484–5500.
24. Liang, P.; Dai, M.; Wang, X.; Wang, D.; Yang, M.; Lin, X.; Zou, X.; Jiang, K.; Li, Y.; Wang, L.; Shangguan, W. Efficacy and safety of ciprofol vs. propofol for the induction and maintenance of general anaesthesia: a multicentre, single-blind, randomised, parallel-group,

phase 3 clinical trial. *Eur. J. Anaesthesiol.* **2023**, *40*, 399–406.

25. Chen, B.; Yin, X.; Jiang, L.; Liu, J. H.; Shi, Y. Y.; Yuan, B. Y. The efficacy and safety of ciprofol use for the induction of general anesthesia in patients undergoing gynecological surgery: a prospective randomized controlled study. *BMC Anaesthesiol.* **2022**, *22*, 245.
26. Long, Y.; Feng, C.; Ding, Y.; Feng, X. M.; Liu, H.; Ji, F. H.; Peng, K. Esketamine as an adjuvant to ciprofol or propofol sedation for same-day bidirectional endoscopy: protocol for a randomized, double-blind, controlled trial with factorial design. *Front. Pharmacol.* **2022**, *13*, 821691.
27. Chen, X.; Guo, P.; Yang, L.; Liu, Z.; Yu, D. Comparison and clinical value of ciprofol and propofol in intraoperative adverse reactions, operation, resuscitation, and satisfaction of patients under painless gastroenteroscopy anesthesia. *Contrast Media Mol. Imaging*, **2022**, *2022*, 9541060.
28. Hu, Y.; Li, X.; Liu, J..; Chen, H.; Zheng, W.; Zhang, H.; Wu, M.; Li, C.; Zhu, X.; Lou, J.; Yan, P. Safety, pharmacokinetics and pharmacodynamics of a novel γ-aminobutyric acid (GABA) receptor potentiator, HSK3486, in Chinese patients with hepatic impairment. *Ann. Med.*, **2022**, *54*, 2757–2768.
29. Zhu, Q.; Luo, Z.; Wang, X.; Wang, D.; Li, J.; Wei, X.; Tang, J.; Yao, S.; Ouyang, W.; Zhang, W.; Zuo, Y. Efficacy and safety of ciprofol versus propofol for the induction of anesthesia in adult patients: a multicenter phase 2a clinical trial. *Int. J. Clin. Pharm.* **2023**, *45*, 473.
30. Zhong, J.; Zhang, J.; Fan, Y.; Zhu, M.; Zhao, X.; Zuo, Z.; Zhou, X.; Miao, C. Efficacy and safety of ciprofol for procedural sedation and anesthesia in non-operating room settings. *J. Clin. Anesth.* **2023**, *85*, 111047.
31. Yu, L.; Bischof, E.; Lu, H. Anesthesia with ciprofol in cardiac surgery with cardiopulmonary bypass: a case report. *World J. Clin. Cases* **2023**, *11*, 157–163.
32. Man, Y.; Xiao, H.; Zhu, T.; Ji, F. Study on the effectiveness and safety of ciprofol in anesthesia in gynecological day surgery: a randomized double-blind controlled study. *BMC Anesthesiol.* **2023**, *23*, 92.
33. Silverman, R. B. Design and mechanism of GABA aminotransferase inactivators. Treatments for epilepsies and addictions. *Chem. Rev.* **2018**, *118*, 4037–4070.
34. Trapani, G.; Altomare, C.; Sanna, E.; Biggio, G.; Liso, G. Propofol in anesthesia. Mechanism of action, structure-activity relationships, and drug delivery. *Curr. Med. Chem.* **2000**, *7,* 249-271.

35. Liao, J.; Li, M.; Huang, C.; Yu, Y.; Chen, Y.; Gan, J.; Xiao, J.; Xiang, G.; Ding, X.; Jiang, R.; Li, P. Pharmacodynamics and pharmacokinetics of HSK3486, a novel 2, 6-disubstituted phenol derivative as a general anesthetic, *Front. Pharmacol.* **2022**, *13*, 830791.

36. Nury, H.; Renterghem, C.; Weng, Y.; Tran, A.; Baaden, M.; Dufresne, V.; Changeux, J. P.; Sonner, J. M.; Delarue, M.; Corringer, P. J. X-ray structures of general anaesthetics bound to a pentameric ligand-gated ion channel. *Nature* **2011**, *469*, 428–431.

37. Koichi, Y.; Bu, W.; Xi, J.; Shimaoka, M.; Eckenhoff, R. Propofol shares the binding site with isoflurane and sevoflurane on leukocyte function-associated antigen-1. *Anaesth. Analg.* **2013**, *117,* 803–811.

38. Orser, B.; Wang, L.; Pennefather, P.; MacDonald, J. Propofol modulates activation and desensitization of GABAA receptors in cultured murine hippocampal neurons. *J. Neurosci.* **1994**, *14*, 7747–7760.

39. a) Concas, A.; Santoro, G.; Mascia, M. P.; Serra, M.; Sanna, E.; Biggio, G. The general anesthetic propofol enhances the function of γ-aminobutyric acid-coupled chloride channel in the rat cerebral cortex. *J. Neurochem.,* **1990**, *55*, 2135; b) Concas, A.; Santoro, G.; Serra, M.; Sanna, E.; Biggio, G. Neurochemical action of the general anaesthetic propofol on the chloride ion channel coupled with GABAA receptors. *Brain Res.* **1991**, *542*, 225.

40. Collins, G. G. Effects of the anaesthetic 2,6-diisopropylphenol on synaptic transmission in the rat olfactory cortex slice. *Br. J. Pharmacol.* **1988**, *95*, 939.

41. Woll, K.; Weiser, B.; Liang, Q.; Meng, T.; McKinstry-Wu, A.; Pinch, B.; Dailey, W. P.; Gao, W. D.; Covarrubias, M.; Eckenhoff, R. G. Role for the propofol hydroxyl in anesthetic protein target molecular recognition. *ACS Chem. Neurosci.* **2015**, *6*, 927–935.

42. Qiu, L.; Lin, J.; Liu, Q.; Wang, S.; Lv, G.; Li, K.; Shi, H.; Huang, Z.; Bertaccini, E. J. The role of the hydroxyl group in propofol–protein target recognition: insights from ONIOM studies. *J. Phys. Chem. B* **2017**, *121*, 5883–5896.

43. James, R.; Glen, J. B. Synthesis, biological evaluation, and preliminary structure-activity considerations of a series of alkylphenols as intravenous anesthetic agents. *J. Med. Chem.* **1980**, *23*, 1350–1357.

44. Trapani, G.; Latrofa, A.; Franco, M.; Altomare, C.; Sanna, E.; Usala, M.; Biggio, G.; Liso, G. Propofol analogues. Synthesis, relationships between structure and affinity at GABAA receptor in rat brain, and differential electrophysiological profile at recombinant human GABAA receptors. *J. Med. Chem.* **1998**, *41*, 1846–1854.

45. Lingamaneni, R.; Krasowski, M.; Jenkins, A.; Truong, T.; Giunta, A. L.; Blackbeer, J.; MacIver, M. B.; Harrison, N. L.; Hemmings, H. C. Anesthetic properties of 4-iodopropofol: implications for mechanisms of anesthesia. *J. Am. Soc. Anesth.* **2001**, *94*, 1050-1057.

46. a) Hill-Venning, C.; Peters, J. A.; Callachan, H.; Lambert, J. J.; Gemmell, D. K.; Anderson, A.; Byford, A.; Hamilton, N.; Hill, D. R.; Marshall, R. J.; Campbell, A. C. The anesthetic action and modulation of GABAA receptor activity by the novel water-soluble aminosteroid Org 20599. *Neuropharmacology* **1996**, *35*, 1209–1222; b) Lingamaneni, R.; Krasowski, M.; Jenkins, A.; Truong, T.; Giunta, A. L.; Blackbeer, J.; MacIver, M. B.; Harrison, N. L.; Hemmings, H. C. Anesthetic properties of 4-iodopropofol: implications for mechanisms of anesthesia. *J. Am. Soc. Anesth.* **2001**, *94*, 1050–1057.

47. Bian, Y.; Zhang, H.; Ma, S.; Jiao, Y.; Yan, P.; Liu, X.; Ma, S.; Xiong, Y.; Gu, Z.; Yu, Z.; Huang, C. Mass balance, pharmacokinetics and pharmacodynamics of intravenous HSK3486, a novel anaesthetic, administered to healthy subjects. *Br. J. Clin. Pharmacol.* **2021**, *87*, 93–105.

48. a) Simons, P. J.; Cockshott, I. D.; Douglas, E. J.; Gordon, E. A., Hopkins, K.; Rowland, M. Disposition in male volunteers of a subanaesthetic intravenous dose of an oil in water emulsion of ^{14}C-propofol. *Xenobiotica* **1988**, *18*, 429–440; b) Sahinovic, M.; Struys, M.; Absalom, A. Clinical pharmacokinetics and pharmacodynamics of propofol. *Clin. Pharmacokinet.* **2018**, *57*, 1539–1558.

49. Hu, C.; Ou, X.; Teng, Y.; Shu, S.; Wang, Y.; Zhu, X.; Kang, Y.; Miao, J. Sedation effects produced by a ciprofol initial infusion or bolus dose followed by continuous maintenance infusion in healthy subjects: a phase 1 trial. *Adv. Ther.* **2021**, *38*, 5484–5500.

50. Voss, L. J.; Sleigh, J. W.; Barnard, J. P.; Kirsch, H. E. The howling cortex: seizures and general anesthetic drugs. *Anesth. Analg.* **2008**, *107*, 1689–1703.

51. Chen, B.; Yin, X.; Jiang, L.; Liu, J.; Shi, Y. Y.; Yuan, B. Y. The efficacy and safety of ciprofol use for the induction of general anesthesia in patients undergoing gynecological surgery: a prospective randomized controlled study. *BMC Anesthesiol.* **2022**, *22*, 1–7.

52. Yang, D.; Hu, Y.; Ruan, Z.; Jiang, B.; Wang, H.; Xu, Y.; Hu, M.; Yan, M.; Lou, H. Drug–drug interaction of ciprofol injectable emulsion with mefenamic acid capsules in healthy subjects. *Br. J. Clin. Pharmacol.* **2023**, *89*, 3165–3174.

53. Zhang, X.; Yu, S.; Liu, Z.; Long, Y.; Zhao, J.; Xu, W.; Zhang, H.;

Zhang, H. Development of a kilogram-scale route for clinical sample production of the intravenous anesthetic cipepofol. *Org. Process Res. Dev.* **2022**, *26*, 1054–1062.

54. Jenkins, T. E. Analogs of propofol, preparation thereof and use as anesthetics, WO2009140275A1 (2009).

13

Rimegepant (Nurtec ODT): A CGRP Receptor Antagonist as a Treatment of Episodic Migraine

Yuqi Lavender Zha and
Guanglin Luo

FDA's 2020 approval rimegepant (Nurtec ODT, **1**), a calcitonin gene-related peptide (CGRP) receptor antagonist as a treatment of episodic migraine in adults, is good news for millions of patients suffering from episodic migraine.

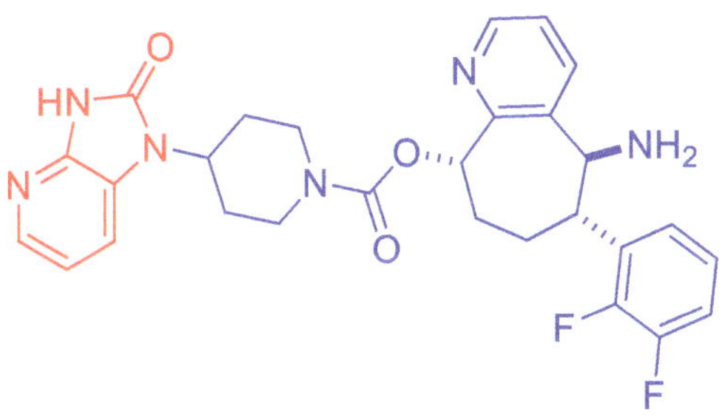

rimegepant
(Nurtec ODT, **1**)
Biohaven/Pfizer, 2020
CGRP antagonist

Rimegepant (**1**) is the second CGRP small-molecule antagonist on the market. The other two small-molecule CGRP receptor antagonists, ubrogepant (Ubrelvy, **2**) and atogepant (Qulipta, **3**) were approved in 2019 and 2021, respectively.[1]

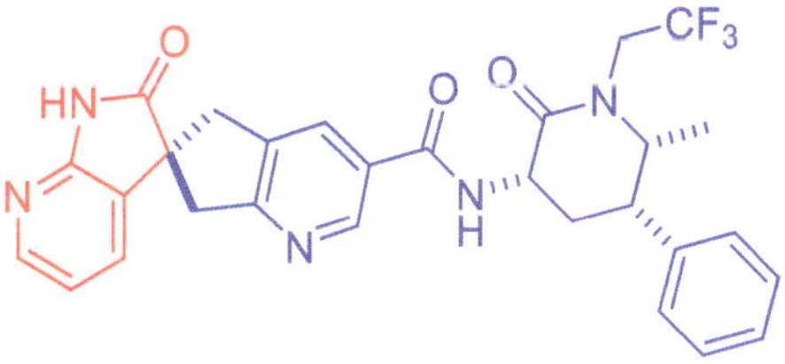

ubrogepant (Ubrelvy, **2**)
Allergan, 2019
CGRP antagonist

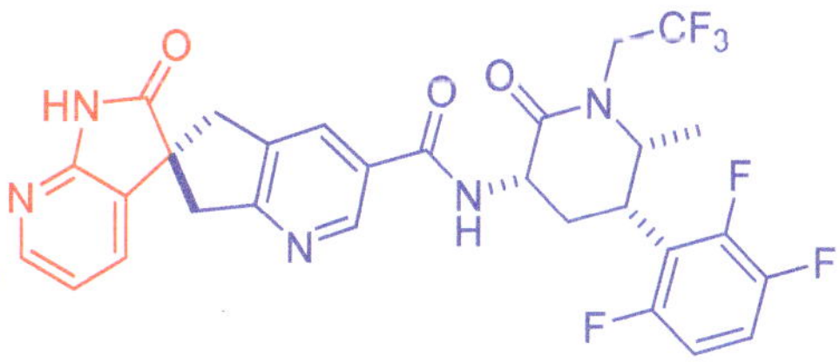

atogepant (Qulipta, **3**)
Allergan/Abbvie, 2021
CGRP antagonist

Chemistry and Pharmacology of Drug Discovery, First Edition. Edited by Jie Jack Li.

1. Background

Migraine headache is commonly characterized by recurrent headaches lasting 4–72 h with moderate to severe, pulsating pain. Associated symptoms may include nausea and/or vomiting, and increased sensitivity to light and sound (photophobia and phonophobia, respectively).[2] The influence of migraine is often underestimated by non-migraineurs. Although it is not a life-threatening disease, due to the severity of the headache, its long duration, and high prevalence, migraine is a very disabling disorder, which has long become a significant public health issue in both sexes and all age groups, especially young and middle-aged women. According to the Global Burden of Disease Study, from 1990 to 2019 (the latest one), migraine remained second among the world's causes of disability (both sexes, all ages) for 30 years, but topped the list in young women in the global burden disease of 2019.[3] Migraine was responsible for 41.1 million years lived with disability (YLDs) globally, accounting for 4.76% of all YLDs caused by 396 diseases.[4]

In the past, without understanding of the mechanism behind migraine and availability of migraine-specific drugs, nonsteroidal anti-inflammatory drugs (NSAIDs) such as, aspirin, acetaminophen, ibuprofen, and naproxen were used to treat mild migraine pain, but do not work in the majority of migraine patients.[1] Since research efforts accumulated in this area, the pathophysiology of migraine started to reveal itself, but contention remains, especially around the initiation of migraine. There are two dominant hypotheses about the mechanism behind migraine: the vascularly focused hypothesis and the neuroanatomically focused hypothesis (Figure 1).

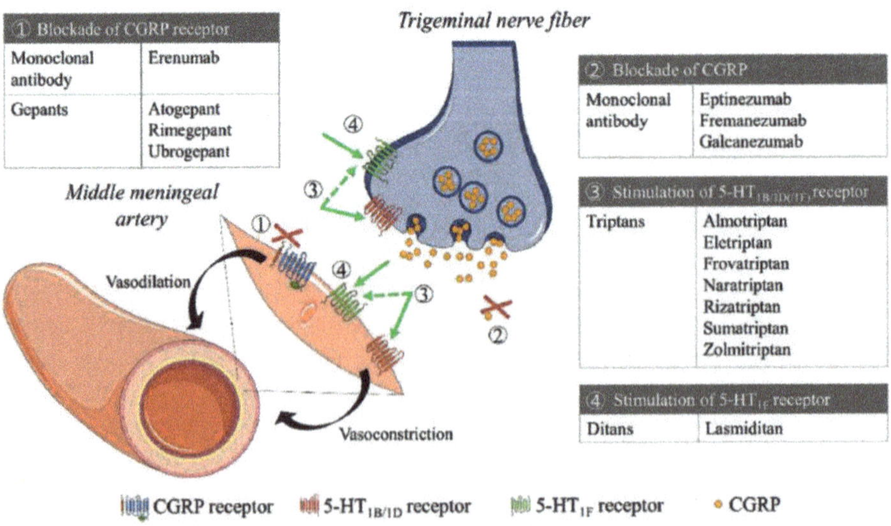

Figure 1. Overview of migraine-specific medications and their possible targets. Source: de Vries et al.[5]/with permission of Elsevier

The vascularly focused hypothesis centered around the perivascular CGRPergic transmission of the trigemino-vascular system, which proposed a three-step model of migraine initiation: (1) release of several neuropeptides including CGRP from the perivascular trigeminal nerve branches; (2) induction of vasodilation by CGRP; (3) Activation of nociceptors on the trigeminal nerve by vasodilation, which is eventually followed by pain perception.[6] This hypothesis supports the use of 5-$HT_{1B/1D}$ agonists, namely triptans for acute treatment of migraine. It is believed that 5-$HT_{1B/1D}$ triptans elicit their antimigraine action by vasoconstriction of excessively dilated intracranial, extracerebral arteries and/or inhibiting the release of inflammatory neuropeptides from perivascular trigeminal sensory neurons.[7] Triptans **4–10** are now all available as generic drugs.[8] However, because 5HT receptors are universally expressed in many organs and tissues, one of the major concerns with triptans is their potential adverse effect on the cardiovascular system, which led to restricted use of triptans in patients with cardiovascular diseases. It has been suggested that 5-HT_{1B} receptor activation results in vasoconstriction, while inhibition of neuropeptide release is mediated via the 5-HT_{1D} receptor. However, a randomized clinical trial of the 5-HT_{1D} selective agonist PNU-142633 failed to demonstrate significant efficacy over placebo.[9] 5-HT_{1F} receptor has become the next target, which is also believed to inhibit pre-junctional release of CGRP in the trigeminal ganglion and does not play a role in vasoconstriction. One brain-penetrant 5-HT_{1F} receptor agonist, Lasmiditan has been approved by the FDA for the acute treatment of migraine with improved cardiovascular safety.[5]

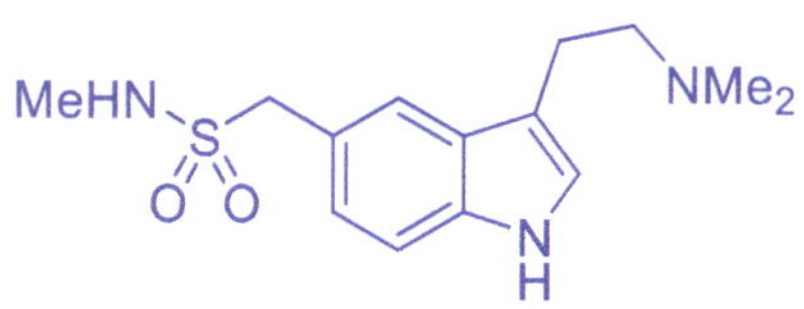

Sumatriptan (Imitrex, **4**)
GlaxoSmithKline, 1995

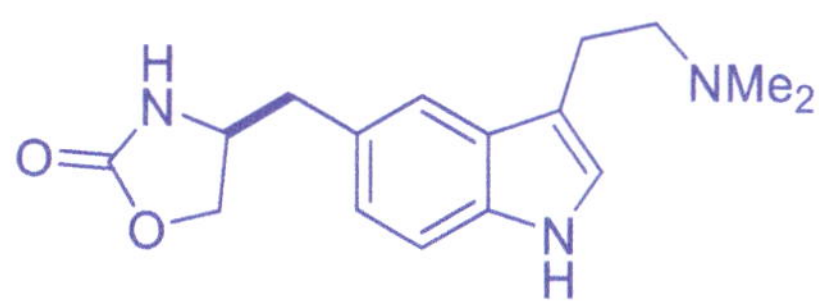

Zolmitriptan (Zomig, **5**)
Wellcome/AstraZeneca, 1997

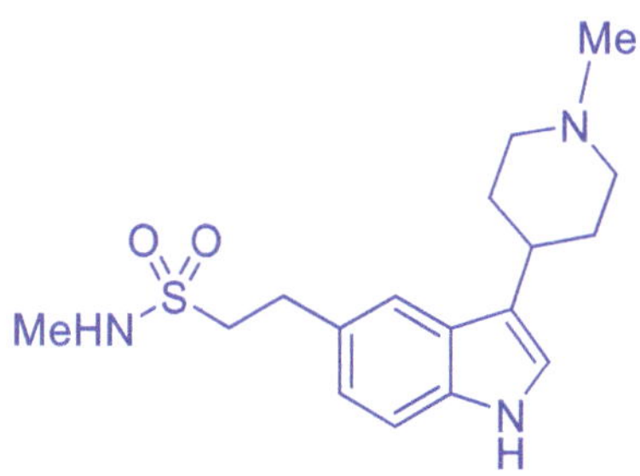

Naratriptan (Amerge, **6**)
GlaxoSmithKline, 1998

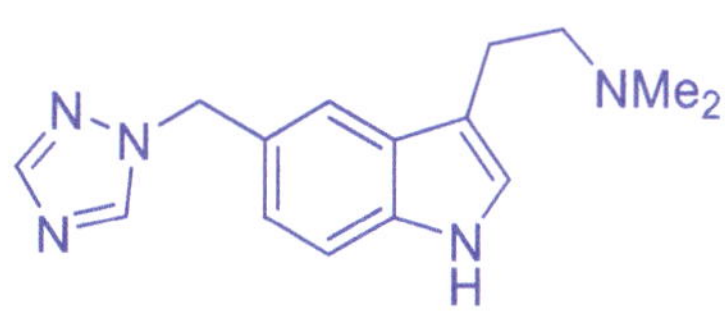

Rizatriptan (Maxalt, **7**)
Merck, 1998

Almotriptan (Axert, **8**)
Almirall/Janssen, 2001

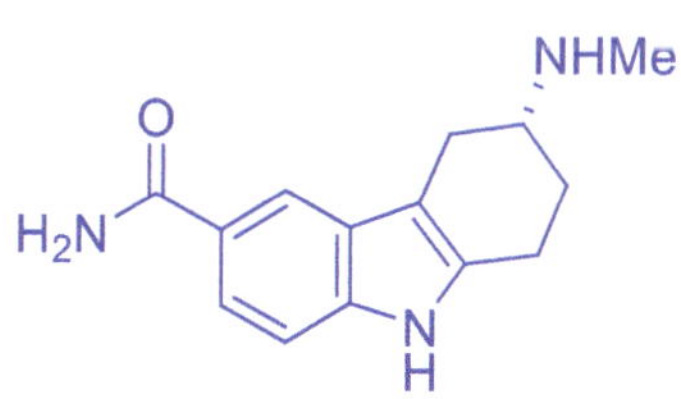

Frovatriptan (Frova, **9**)
SmithKline/Elan, 2001

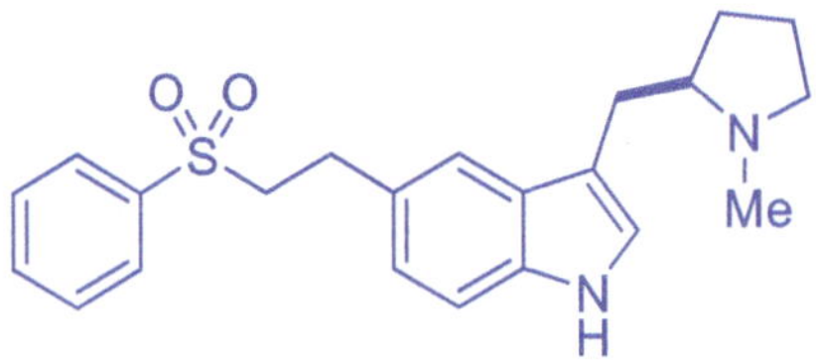

Eletriptan (Relpax, **10**)
Pfizer, 2002

Edvinsson et al. in 2018 put forward a neuroanatomically focused model of migraine initiation that includes three stages: (1) CGPR released from the trigeminal ganglion by C-fibers activates CGRP receptors activates nearby CGRP receptors on Aδ-fiber sensory neurons which are involved in nociperception;[10] (2) CGRP also activates CGRP receptors on satellite glial cells, which release the activating agent nitric oxide and modulate pain sensitivity and transmission;[11–13] (3) elevated levels of CGRP lead to distortion of nociperception in migraineurs and sensitization of neuronal circuits such that normally harmless stimuli (i.e., sounds, odors, and light) could trigger nocuous migraine symptoms.[14] The key difference between the two hypotheses is whether vasodilation is a cause of migraine attack or just a "side effect" of CGRP receptor activation on cranial arteries. This debate leads to a more clinically relevant question that whether the vasoconstriction effect of triptans is desirable to treat migraine or completely an on-target side effect that leads to cardiovascular adverse events and contradictions in patients with cardiovascular disorders. However, no matter which hypothesis is more correct (maybe both are correct), CGRP has been proven to be the absolute protagonist on the stage of migraine, which has led to the development of small-molecule CGRP receptor antagonists (gepants) and antibody treatments targeting CGRP or CGRP receptors.

2. Pharmacology

CGRP is a 37-amino acid neuropeptide that plays a critical role in neurotransmission and vasodilation. Its receptor is a class B G-protein-coupled receptor that consists of a heterodimer of calcitonin receptor-like receptor (CLR), a receptor activity modifying protein 1 (RAMP1), and coupled to a GS protein.

Table 1. Antagonists for CGRP and its receptors approved by the FDA

Generic name (Trade name)	Type of treatment	Target	Company	FDA approval
Small molecules (Gepants)				
Ubrogepant (Ubrelvy)	Acute	CGRP receptor	AbbVie	2019
Rimegepant (Nurtec)	Acute and Prophylactic	CGRP receptor	Pfizer	2020
Atogepant (Qulipta)	Prophylactic	CGRP receptor	AbbVie	2021
Zavegepant (Zavzpret)	Acute	CGRP receptor	Pfizer	2023
Monoclonal antibodies				
Galcanezumab (Emgality)	Prophylactic	CGRP receptor	Eli Lilly	2018
Erenumab (Aimovig)	Prophylactic	CGRP	Amgen	2018
Eptinezumab (Vyepti)	Prophylactic	CGRP	Lundbeck	2020
Fremanezumab (Ajovy)	Prophylactic	CGRP	Teva	2021

The importance of CGRP release in migraine was long discovered in 1990 by Goadsby et al. that elevated levels of CGRP was discovered in blood samples from migraine patients during the headache phase.[14] Intravenous infusion of CGRP could trigger migraine-like headaches in patients prone to migraine.[15] Moreover, treatment with ergotamine or sumatriptan could reduce CGRP level in blood concurrent with mitigation of migraine pain.[14] There are currently eight treatments targeting CGRP and its receptor approved by the FAD, including four gepants, atogepant, and ubrogepant of AbbVie, rimegepant and zavegepant of Pfizer, one antibody treatment targeting CGRP receptor, erenumab of Amgen and Novartis, and three antibody treatment targeting CGRP, galcanezumab of Eli Lilly, eptinezumab of Lundbeck, and fremanezumab of Teva (Table 1).

It took 30 years of extensive research for the first gepant to be approved by the FDA, because development of a small-molecule CGRP receptor antagonist is very challenging for the following two reasons: (1) the affinity of CGRP to its receptor is extremely high (K_i = 27 pM), making displacement of this large endogenous agonist very difficult; (2) very high circulating free plasma levels are required to provide near maximal receptor antagonist coverage to yield substantial efficacy.[16] Moreover, many early CGRP receptor antagonists failed due to liver toxicity, including BI's olcegepant (BIBN-4096 BS, **11**), BI-44370 (**12**), Merck's telcagepant (**13**), and MK-3207 (**14**).

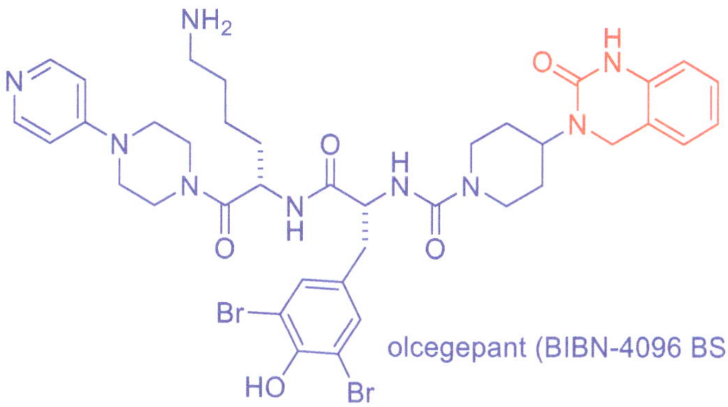

olcegepant (BIBN-4096 BS, **11**)

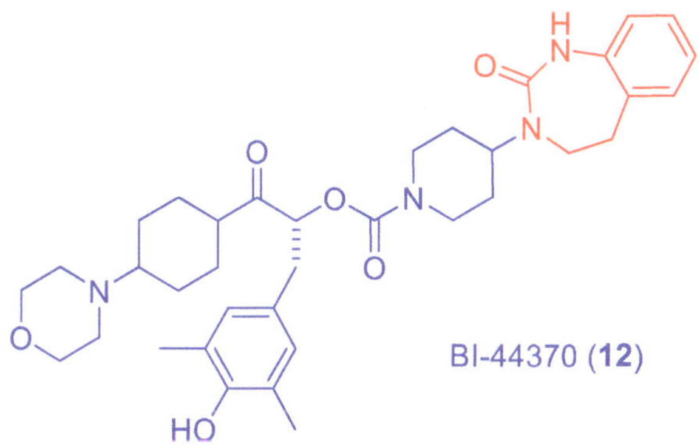

BI-44370 (**12**)

Upon binding with the agonist, the ligand bound CGRP receptor complex is internalized to endosomes with three distinct fates depending on the duration of CGRP ligand exposure.[16] Brief exposure will lead to ligand degradation and recycling of the receptor to the cell surface; while continuous exposure results in proteolytic degradation of the entire complex.[17] Alternatively, CGRP receptor may also serve as an intracellular receptor located within endosomes.[18] If the endosomal signaling also plays a crucial role in migraine, it might favor small-molecule CGRP receptor antagonists that could penetrate the cell membrane and might explain the moderate efficacy with monoclonal antibodies. Preliminary evidence was seen in two case studies where rimegepant (**1**) was consistently effective for the acute treatment of breakthrough migraine attacks in patients receiving erenumab as the preventive treatment.[19]

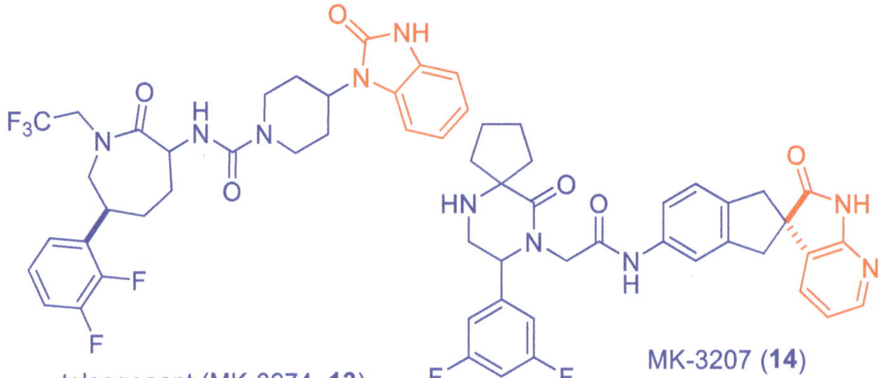

telcagepant (MK-0974, **13**) MK-3207 (**14**)

CGRP receptor is universally expressed in many regions of the CNS and PNS and plays a role in a variety of physiological processes and homeostatic responses in pathological conditions.[20] Besides its role in migraine, in the cardiovascular system, CGRP is found in nerve fibers that innervate blood arteries and the heart. As a powerful vasodilator, CGRP has cardioprotective effect against hypertension, heart failure, and cardiac or cerebral ischemia. However, unlike triptans, no significant cardiovascular AEs were observed in clinical trials of gepants because these gepants have higher functional potency in meningeal arteries than in coronary arteries.[21–23] The differences in tissue response might arise from CGRP receptor heterogeneity,[24] variations in receptor internalization and endosomal signaling,[18] or different influence on intracellular pathways.[1] On the other hand, monoclonal antibodies targeting CGRP or its receptor do raise some safety concerns, especially because they are approved for long-term prophylactic use. Although none of the four antibodies showed any cardiovascular TRAEs in the clinical trials, it should be noted that cardio and cerebrovascular events did take place during the clinical trials, including one fatal event of atherosclerosis in a phase II trial for erenumab.[25] In order to provide convincing evidence for cardiovascular safety of these monoclonal antibodies, more studies should be performed in cardiovascularly compromised patients.[5]

In the gastrointestinal system, CGRP is found in the nerve fiber of the enteric nervous system and is responsible for maintaining mucosal integrity and modulating motility, which partially explains the relatively high occurrence rate of the gastrointestinal adverse events (i.e. nausea and constipation) seen in clinical trials of both atogepant and rimegepant.[26] Clinical trials of atogepant and rimegepant also discovered infections as the most common adverse events, including upper respiratory tract infections and urinary tract infections.[6] However, the mechanisms behind infections and other adverse events still remain unknown; whether CGRP and its receptor might play a role in immune response have not yet been elucidated.

Although the battle against migraine has been fought for centuries, with all these latest weapons in hand, there are still more complications to be solved. First, no matter for triptans, ditans, gepants or monoclonal antibodies, the response rate is about 30–40%—there are always more than half of the patients who do not respond to the current treatments. More research is needed to better understand the mechanism behind migraine and why some patients do not respond to some treatments (Figure 2), with the hope to ultimately identify clinically meaningful biomarkers that could predict the most suitable treatment for each patient. Second, since CGRP signaling is known to be influenced by hormonal fluctuation,[27] which might partially explain the sex difference in migraine, it should be determined whether these CGRP antagonizing treatments have different response in the two sexes. It is hypothesized that women are at an increased risk of cardiovascular adverse events after blocking CGRP.[28]

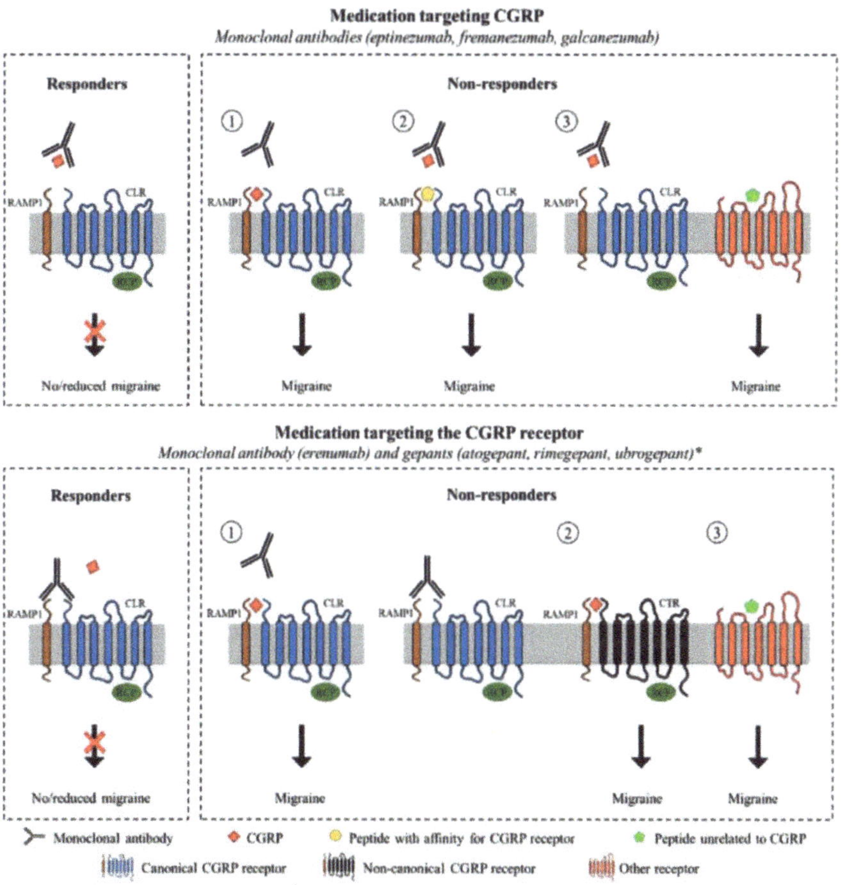

Figure 2. Potential drug resistance mechanisms to medications targeting CGRP and CGRP receptor. Source: de Vries et al.[5]/with permission of Elsevier

3. Structure–Activity Relationship (SAR)

Research work on CGRP receptor antagonist at Bristol-Myers Squibb, Wallingford CT site started in 2001 with a major focus on then newly published BI's clinical compound olcegepant (**11**) and other disclosed chemical matter BI-44370 (**12**). A series of D-amino acid derived ureidomamides as exemplified by the benzothiophene **15**, were discovered to have low single digit nM inhibition potency.[29] The major issue associated with this series was moderately potent CYP3A4 inhibition with IC_{50} of 220 nM for **15** as an example.[30] Further work indicated that compound **16** with the presence of a hydrogen bond donor had greatly attenuated CYP3A4 inhibition with $IC_{50} > 40\ \mu M$, albeit with lowered CGRP antagonist potency.

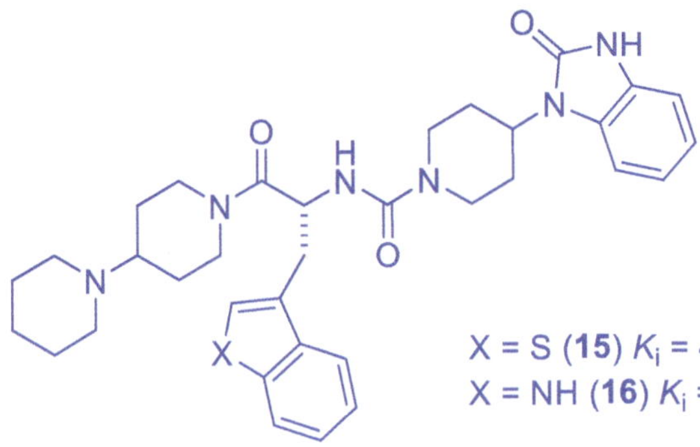

X = S (**15**) K_i = 4.3 nM
X = NH (**16**) K_i = 84 nM

Further SAR optimization led the tyrosine mimetic D-amino acid derivatives such as indazole **17** with further improved activity.[31] As a close mimetic of 2,6-dimethyl phenol moiety in BI-44370 (**12**), incorporation of the 7-methyl group to the indazole analog (**18**) greatly increased the binding activity by 23-fold. All compounds in this series that were tested for oral PK showed poor exposures. This was not surprising considering the high MW, three hydrogen bond donors, and high number of rotating bonds of these analogs. However, the group started to consider an intranasal (IN) delivery route due to high aqueous solubility (e.g. 15 mg/mL at pH 5.0 from amorphous solid of **18**), substantial unbound plasma fractions (f_u = 10–80%), and exceptional potency (low double digit pM) of these analogs. Among these analogs, BMS-694153 (**19**) was nominated as a clinical candidate suitable for IN delivery.[32] Compared to **18**, addition of one fluorine substituent in BMS-694153 (**19**) increased aqueous solubility to over 500 mg/mL (pH < 6.8) from crystalline free base while maintaining desired potency and mitigated benzylic oxidation as hoped. BMS-694153 was robustly active in the marmoset model at 0.03 mg/kg (SC), and had good IN bioavailability of 26% in rabbits with T_{max} < 10 minutes. Unfortunately, it showed nasal mucosal injury findings in rats. Subsequently zavegepant (BMS-742413, **20**) was discovered to overcome the nasal mucosal injury issue with all other properties comparable to BMS-694153 (**19**).[33] BMS-742413 (**20**) was parked at BMS for many years before it was successfully carried through all phase clinical studies by Biohaven Pharmaceuticals, where it demonstrated superiority to placebo at both 10 and 20 mg IN doses in acute treatment of migraine. Zavegepant was approved by FDA as a nasal spray for acute treatment of migraine with or without aura in adults on March 10, 2023, at Pfizer after two pivotal double-blind, placebo-controlled phase III studies.

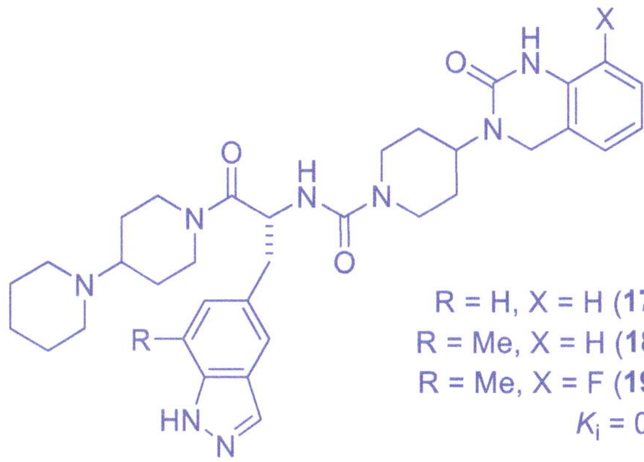

R = H, X = H (**17**) K_i = 0.23 nM
R = Me, X = H (**18**) K_i = 0.010 nM
R = Me, X = F (**19**, BMS-694153)
K_i = 0.013 nM

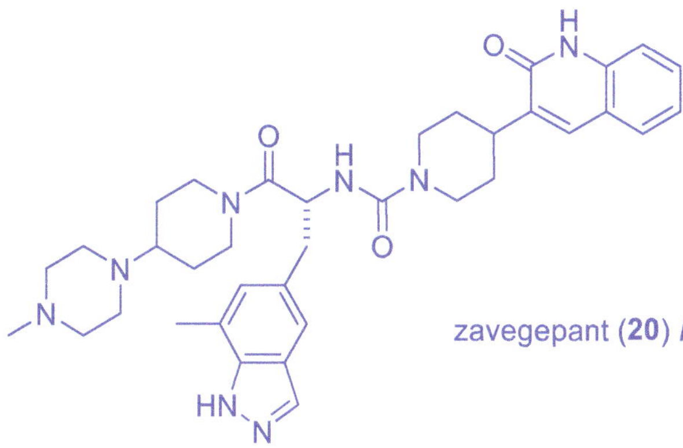

zavegepant (**20**) K_i = 0.023 nM

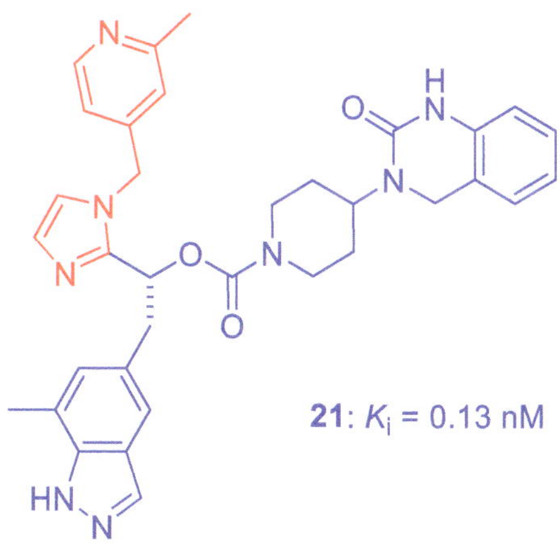

21: K_i = 0.13 nM

In parallel to the IN approach, an oral CGRP receptor antagonist was rigorously pursued. To improve the oral bioavailability of the D-amino acid uredo amide series, the polar amide portion was replaced by nitrogen-containing heterocycles. A representative imidazole derivative **21** showed reasonable potency, but with no measurable oral bioavailability.[34] The pyridine series as represented by **22** and **23**, showed higher level of potency.[35] Among these analogs, *in vitro* predictors of oral exposure such as Caco-2/PAMPA and microsomal stabilities tended not to track with oral exposures, and few compounds depicted low single digit percentage oral bioavailability. Compound **23** showed moderate activity in the marmoset facial blood flow assay with oral dosing at 10 mg/kg and the carbamate connector was consistently favored over the urea in these compounds regarding permeability. Additionally, aspartates (urea NH was replaced by carbon and the benzylic carbon was replaced by NH) and succinate (urea NH was replaced by carbon) derivatives could maintain the binding potency but offered no improvement for oral exposure.[36]

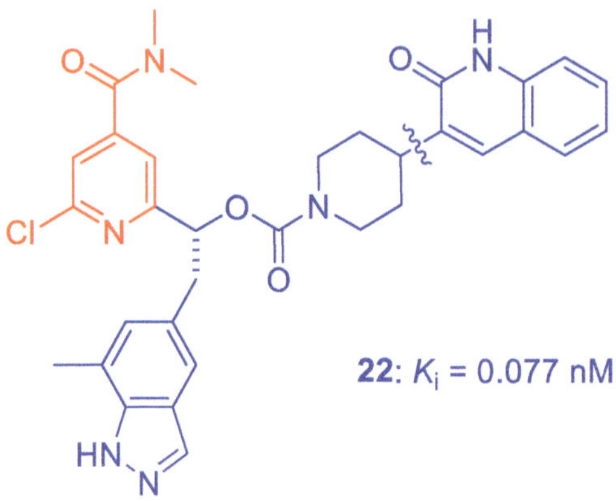

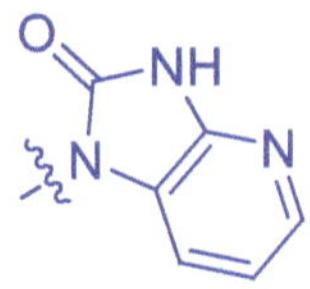

22: K_i = 0.077 nM **23**: K_i = 0.071 nM

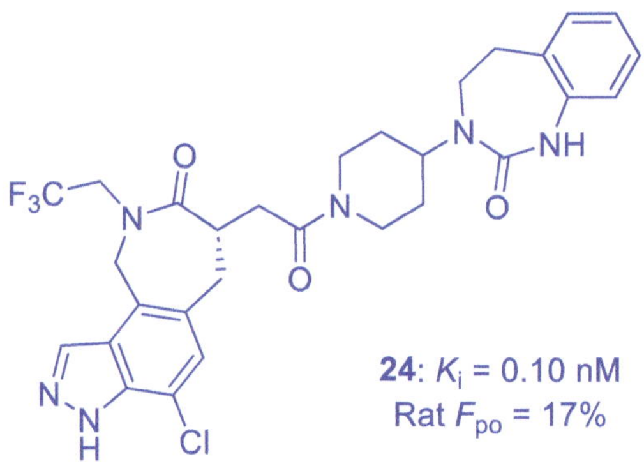

24: K_i = 0.10 nM
Rat F_{po} = 17%

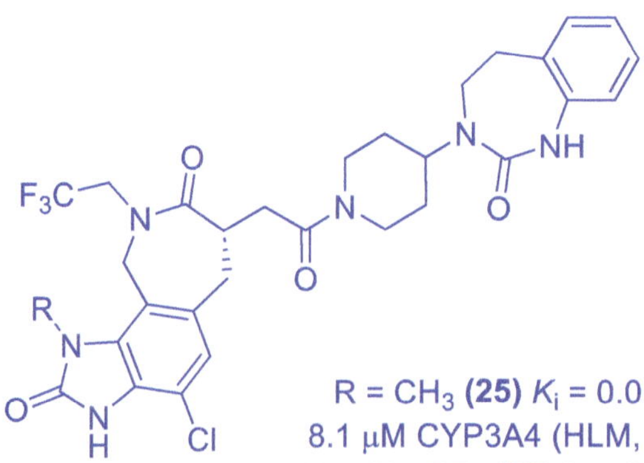

R = CH_3 (**25**) K_i = 0.015 nM
8.1 μM CYP3A4 (HLM, 10 min)
R = CH_2CF_3 (**26**) K_i = 0.020 nM
8.4 μM CYP3A4 (HLM, 10 min)

Alternatively, azepinone-based beta-turn mimetics were used to restrict appropriate conformation of the critical indazole group to maintain potency while removing the large polar amide group for better oral exposure. The efforts resulted in a new series of analogs as represented by compound **24** with good hCGRP receptor binding potency and 17% oral bioavailability in rat.[37] Compound **24** demonstrated *in vivo* efficacy in the marmoset facial blood flow assay. However, compound **24** showed poor amorphous aqueous solubility of only 1 μg/mL. It was also demonstrated to be a potent time dependent CYP3A4 inhibitor with IC_{50} of 0.1 μM at 10 min in human liver microsome.

SAR around compound **24** pointed to indazole moiety as a potential cause of time dependent CYP3A4 inhibition. Efforts to replace the indazole piece with other heterocycles were undertaken to address the CYP issue. Since the indazole NH in **24** had been shown to be critical for high-affinity CGRP receptor binding affinity, heterocycles derivatives with NH at the same position were prepared including imidazolones, oxazolidinone, indoles, imidazole, triazole, and indanone.[38] Among all these analogs, mono-substituted imidazolone compounds **25** and **26** showed much improved potency ($K_i \leq 20$ pM) and time dependent CYP3A4 issue (>8 μM at 10 min in HLM). Both compounds showed very good efficacy in the marmoset facial blood flow assay at 7 mg/kg SC up to 85% and 84% reduction of facial blood flow, respectively. However, oral exposure remained low, emphasizing the challenges everyone in the field had encountered in discovering development candidates for this difficult Class B GPCR target.

Over 6 years had passed and the pursuing of an oral CGRP receptor antagonist at BMS reached an impasse with closure of the program imminent. Inspired by the structure of then newly disclosed Merck phase II compound telcagepant (**13**), coupled with the experience of pyridine replacement of amide group in **22**, a target compound **27** was proposed to capture the better properties of the clinical compound.[39]

The first route of **27** synthesis was extremely tedious with formation of pyridine from a ketone (potential key intermediate to form other heterocycles) as the key step. However, the result was extremely gratifying that BMS-846372 (**27**) offered 11-fold boost of potency vs telcagepant (hCGRP K_i = 0.77 nM) in the same binding assay. BMS-846372 (**27**) has good oral bioavailability in rat, dog, and cynomolgus monkey and overall attractive preclinical properties including >5.0% exposure-dependent *in vivo* efficacy in the marmoset facial blood flow assay. BMS-846372 (**27**) was nominated as a developmental candidate and a robust process route was developed as reported.[40] During the course of human liver microsomal studies, the major metabolite was proposed to have a benzylic hydroxylation of **27** as compound **28**. A stereospecific synthesis of **28** was developed to unequivocally prove the metabolite structure.[41] Compound **28** has substantially lowered potency (20-fold) compared to **27** as a none-active metabolite at the target human dose.

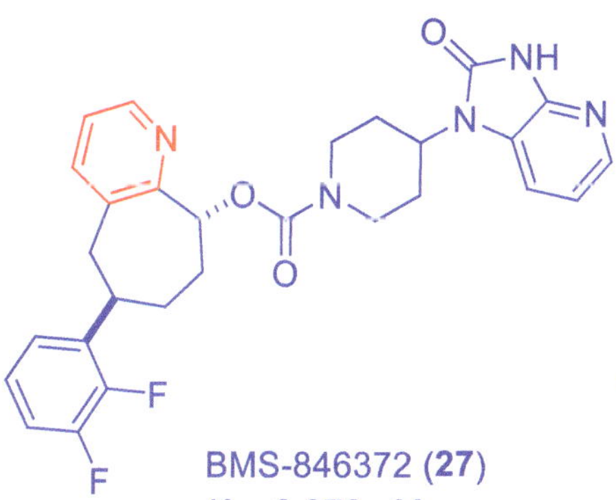

BMS-846372 (**27**)
K_i = 0.070 nM

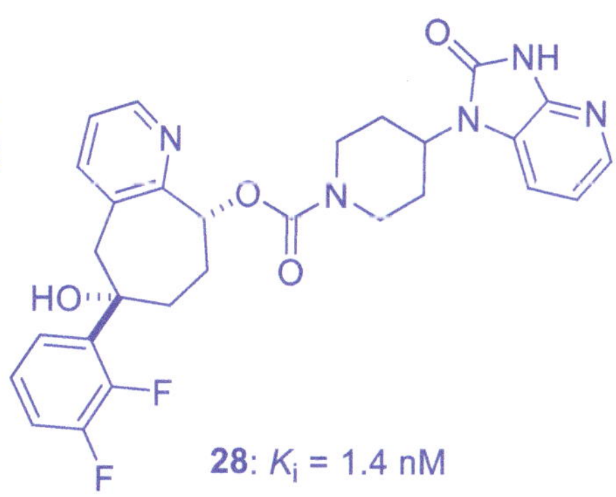

28: K_i = 1.4 nM

With the crystalline free base aqueous solubility of **27** less than 2 μg/mL, formulation of **27** in clinical development would be a potential liability for BMS-846372 (**27**). Further modifications around the phenyl group, the pyridinoimidazolone group, central pyridine substitutions, and replacement of the carbamate group with urea or amide uniformly sacrificed various good properties of BMS-846372 (**27**).[42] Salt formation of central pyridine facilitated the hydrolysis of the carbamate group, presumably through the neighboring group participation effect. There was a strong desire to find a better molecule with improved aqueous solubility. Compound **28** displayed a much better aqueous solubility of 90 μg/mL. Encouraged by this result, small and polar group such as OH or NH_2, was proposed to be added to the cycloheptane core of the molecule to improve the aqueous solubility. Fortunately, a key ketone intermediate from the process synthesis of BMS-846372 (**27**) served the purpose. Analogs **28**, **29**, and finally BMS-927711 (Rimegepant, **1**) were prepared.[43] Compound **27** with *R*-OH was much less active than compound **28** with *S*-OH, which showed aqueous solubility of 66 μg/mL. Rimegepant (**1**) was the most potent with a crystalline free base aqueous solubility of 50 μg/mL. Further preclinical characterization of BMS-927711 (**1**) proved that it surpassed BMS-846372 (**27**) in every aspect of drug properties.

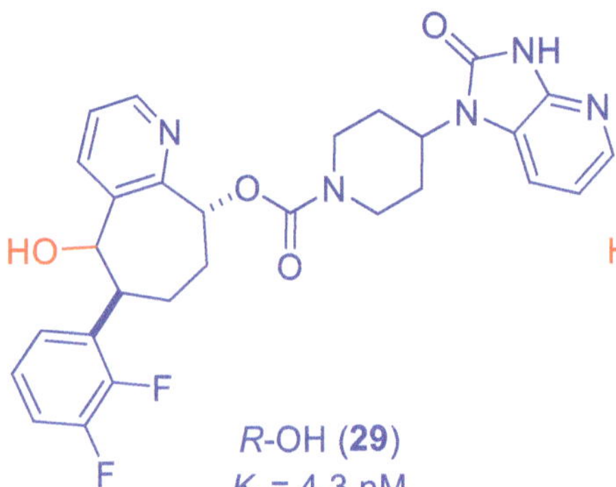

R-OH (**29**)
K_i = 4.3 nM
S-OH (**30**)
K_i = 0.081 nM

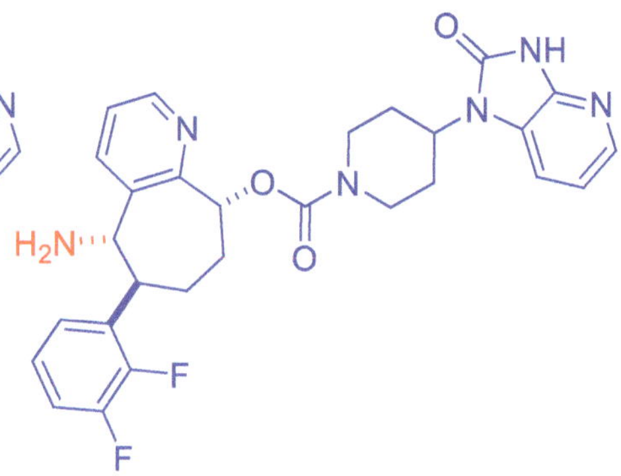

rimegepant (BMS-927711, **1**)
K_i = 0.027 nM

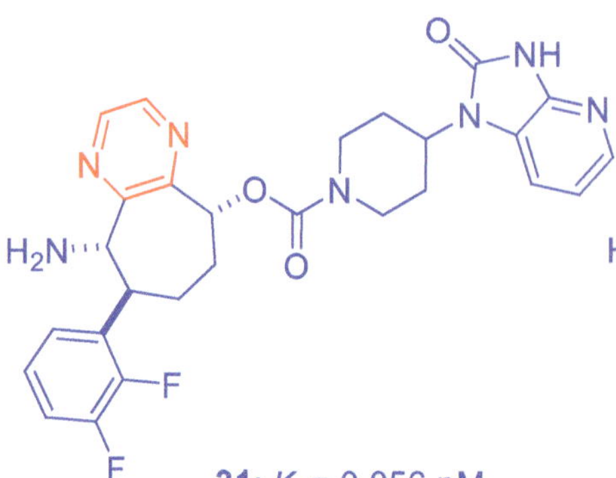

31: K_i = 0.056 nM

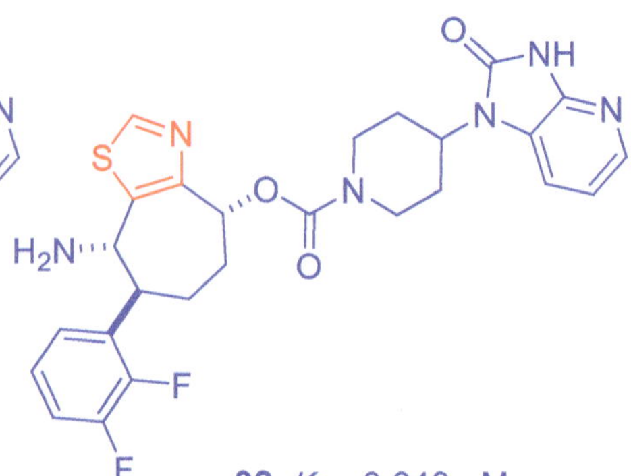

32: K_i = 0.048 nM

Further SAR at the central pyridine targeted synthesis of the pyrazine and thiazole derivatives **31** and **32**, with the critical hydrogen bond acceptor nitrogen preserved.[44] The asymmetric synthesis of **31** and **32** was achieved through a different route as the starting ring closure reaction for rimegepant (**1**) was not applicable to either pyrazine or thiazole. Both **31** and **32** displayed great potency, but with lowered metabolic stability unsuitable for further advancement.

4. Pharmacokinetics and Drug Metabolism

Rimegepant is available in the form of a quick-dissolve orally disintegrating tablet (ODT) for oral administration. The 75 mg free-base equivalent of a suspension of rimegepant hemisulfate in a proprietary matrix is freeze-dried, and the resulting tablet disperses almost instantly in the mouth on contact with saliva without the need of water. This technology allows accelerated absorption of rimegepant (**1**), possibly starting in the oral cavity.[45] For the treatment of acute migraine attacks, the recommended dosage is 75 mg (one tablet) of rimegepant (**1**), while for prophylaxis, the dosage is 75 mg every 2 days. The maximum daily dose is 75 mg.[46]

A Phase I study by BMS confirmed that the 75 mg dose of free base rimegepant led to oral exposures commensurate with robust activity in the marmoset PD assay (≥700 nM, $T_{1/2}$ 10–12 h).[47] A food effect was found that had the undesired effect of delaying T_{max} from 1 to 4 h, post-dose. Co-administration with famotidine significantly reduced C_{max} by ~26% and AUC by ~42%. Alternate salt forms were explored. The hemisulfate sesquihydrate crystalline salt form was found to significantly mitigate the effects. Based on C_{max} and AUC, the ODT showed bioequivalence with a traditional oral tablet form and an earlier T_{max} of about 1.5 h (~30 min faster than the oral tablet form).[48] This salt form became the dominant clinical form beyond Phase II, in addition to free base form.

Rimegepant is a substrate for P-gp and BCRP efflux transporters. Ingestion of the drug with P-gp and BCRP inhibitors may result in higher plasma concentrations (a similar effect may occur with inducers). Majority of the compound (77%) is excreted without biotransformation at $t_{1/2} = 11$ h. Elimination is mainly fecal (78%), while a smaller percentage is excreted in the urine (24%). The drug is metabolized mainly by CYP3A4 and to a lesser extent by CYP2C9. Because rimegepant is a CYP3A4 substrate, increased accumulation with grapefruit may occur. Although most of the drug is excreted unmetabolized, it is advisable to avoid inducers and/or inhibitors of P-gp, BCRP, CYP3A4, and CYP2C9. Gender, race, age, body weight, and CYP2C9 genotype do not affect the pharmacokinetics of rimegepant.[49]

5. Efficacy and Safety

A large double-blind, randomized, placebo-controlled Phase IIb dose-ranging study (885 patients) by BMS tested six doses of rimegepant free base (10, 25, 75, 150, 300, and 600 mg) along with 100 mg oral sumatriptan as an active comparator using an adaptive trial design.[50] The primary endpoint was pain freedom at 2 h, along with secondary, exploratory, and safety endpoints. As predicted preclinically, pain freedom at the 75 mg dose (31.4%), along with the 150 mg (32.9%) and 300 mg (29.7%) doses, and sumatriptan (35%), was statistically superior to placebo (15.3%). The placebo level was notably higher than the 10% seen in historical triptan trials conducted more than a decade ago.[51] The overall safety profile for rimegepant (**1**) resembled placebo with no serious adverse events. This dose-ranging study demonstrated full efficacy at 2 h with 75 mg rimegepant (**1**), and no additional efficacy benefit in pain, nausea, photophobia, or phonophobia was observed when dosing higher. In addition, the 2–24 and 2–48 h sustained pain measures showed that 75 mg efficacy was not increased by dosing higher. Thus 75 mg dose of rimegepant (**1**) was selected for Phase III trials.

In a clinical study of 1186 patients by Biohaven Pharma., 594 and 592 patients were randomly assigned to receive rimegepant 75 mg free base dose and placebo, respectively.[52] Among these, 537 patients in the rimegepant group and 535 patients in the placebo group could be evaluated for efficacy. The overall mean age of the patients evaluated for efficacy was 40.6 years, and 88.7% were women. In a modified intention-to-treat analysis, the percentage of patients who were pain-free 2 h after receiving the dose was 19.6% in the rimegepant group and 12.0% in the placebo group (absolute difference, 7.6 percentage points; 95% confidence interval [CI], 3.3–11.9; $P < 0.001$). The percentage of patients who were free from their most bothersome symptom (MBS) 2 h after the dose was 37.6% in the rimegepant group and 25.2% in the placebo group (absolute difference, 12.4 percentage points; 95% CI, 6.9–17.9; $P < 0.001$). The most common adverse events were nausea (1.8% of users vs 1.1% in the placebo group) and urinary tract infection (1.5% of users vs 1.1% in the placebo group). Serious adverse events (back pain) were reported in one patient receiving the drug and two patients receiving the placebo. 2.4% of rimegepant users had increased AST or ALT activity, as did 2.2% of the placebo group. Although none of the above adverse effects were statistically significant, any patient may develop hypersensitivity reactions even several days after administration of the drug.

With rimegepant ODT 75 mg dose in a separate Phase III trial of 1351 patients who were evaluated for efficacy, 21.2% of 669 patients were pain-free at 2 h (vs 10.9% of 682 patients on placebo; $p < 0.0001$), and 35.1% were free of their MBS (vs 26.8% on placebo; $p = 0.0009$).[53] One dose of rimegepant ODT 75 mg showed rapid onset efficacy with numerical separation from placebo for pain relief as early as 15 min. A single dose of rimegepant ODT showed statistical superiority to placebo in 21 consecutive, prespecified, hierarchically tested efficacy endpoints including six early-onset (60–90 min) measures, six 2 h endpoints, and nine durable (24–48 h) efficacy outcomes, in addition to the two co-primary endpoints. These measures were as follows: at 60 min on pain relief and ability to function normally; at 90 min on pain freedom, freedom from MBS, ability to function normally and pain relief; at 2 h on pain relief, ability to function normally, freedom from photophobia and freedom from phonophobia; no need for rescue medication use within 24 h; and both 2–24 and 2–48 h sustained freedom from pain, sustained ability to function normally, sustained pain relief, and sustained freedom from MBS. The most common adverse events were nausea (rimegepant: 11 (2%); placebo: 3 (<1%)) and urinary tract infection (rimegepant: 10 (1%); placebo: 4 (1%)). One participant in each treatment group had a transaminase concentration of more than 3× the upper limit of normal; neither was related to study medication, and no elevations in bilirubin greater than 2× the upper limit of normal were reported. Treated participants reported no serious adverse events.

A cohort of patients were given rimegepant 75 mg every other day (QOD) over 3 months in the long-term safety study to support acute treatment, where a substantial reduction in the frequency of monthly migraine days (MMDs) was noted.[54] Patients in the long-term safety study, who were given free drug for a year and allowed up to once daily dosing as needed, showed low rates of monthly pill utilization, and subjects with a history of 2–8 and 9–14 moderate to severe migraine attacks per month used only six and eight doses, respectively, on average per month over the year. To further support the safety of acute treatment and achieve higher frequency use, a fixed dosing group was added by amendment that specified treatment with rimegepant 75 mg every other day for 3 months regardless of migraine status. In the subsequent Phase II/III preventive treatment trial, rimegepant 75 mg was dosed every other day during the 12-week treatment phase.[55] The primary endpoint was change in mean MMDs in the last 4 weeks of the treatment phase (weeks 9–12) compared with a pre-treatment 4-week observation period. Participants receiving rimegepant (348) showed a reduction in MMDs (−4.3 days, 95% CI −4.8 to −3.9) that was significantly ($p < 0.01$) greater than placebo (347, −3.5 days, 95% CI −4.0 to −3.0). Participants with rimegepant ODT 75 mg also demonstrated significant superiority vs placebo on secondary efficacy endpoints that include ≥50% reduction in moderate or severe migraine days per month in weeks 9–12 (49% [95% CI 44–54] vs 41% [36–47]), and reduction in total migraine days per month in weeks 1–12 (-3.6 [95% CI −4.0 to −3.2] days vs −2.7 [−3.1 to −2.3] days). The onset of action was rapid, with a 30% reduction observed in weekly migraine days in the first week of dosing. For preventive treatment of migraine, the most common side effects were nausea (2.7% with rimegepant vs 0.8% with placebo) and stomach pain/indigestion (2.4% with rimegepant vs 0.8% with placebo). Hypersensitivity, including dyspnea and rash, occurred in less than 1% of patients treated with rimegepant. With repeated QOD dosing for 3 months in the prevention trial, the safety and tolerability were comparable to placebo and the well-tolerated profile observed across all the acute treatment trials. Discontinuation rates due to adverse events were low at ≤2.7% across all four of the pivotal acute and prevention trials. Most significantly, no sign of medication overuse headache was seen even with frequent rimegepant dosing. On the contrary, proven reduction in migraine burden offers an important advantage over older therapies with highly undesirable profile of drug-induced increases in migraine days with frequent use.

6. Synthesis

The original synthesis of rimegepant (**1**) supporting all the preclinical studies started from the key intermediate **33** available from the process synthesis of BMS-846372 (**27**).[40, 43]

Starting from commercially available dimethyl 2,3-pyridinedicarboxylate (**33**), its Dieckmann cyclization with diethyl succinate (**34**) was facilitated using potassium *t*-butoxide as the base to give adduct **35**. Decarboxylation of bis-ester **35** was readily accomplished using boiling aqueous hydrobromic acid to afford the cyclohepta-pyridinedione (**36**). Screening of asymmetric reduction conditions led to hydrogenation employing 0.02 mol% of Rh(*R*-binapine)(COD)BF_4 to hydroxy-ketone **37** exclusively. Protecting **37** with more reactive TIPSOTf gave **38**, which was arylated with 1-bromo-2,3-difluorobenzene (**39**) using palladium-catalyzed cross-coupling conditions. Optimization of the α-arylation conditions provided adduct **40** in 57% yield after epimerization of the desired diastereomer under the influence of organic base DBU.

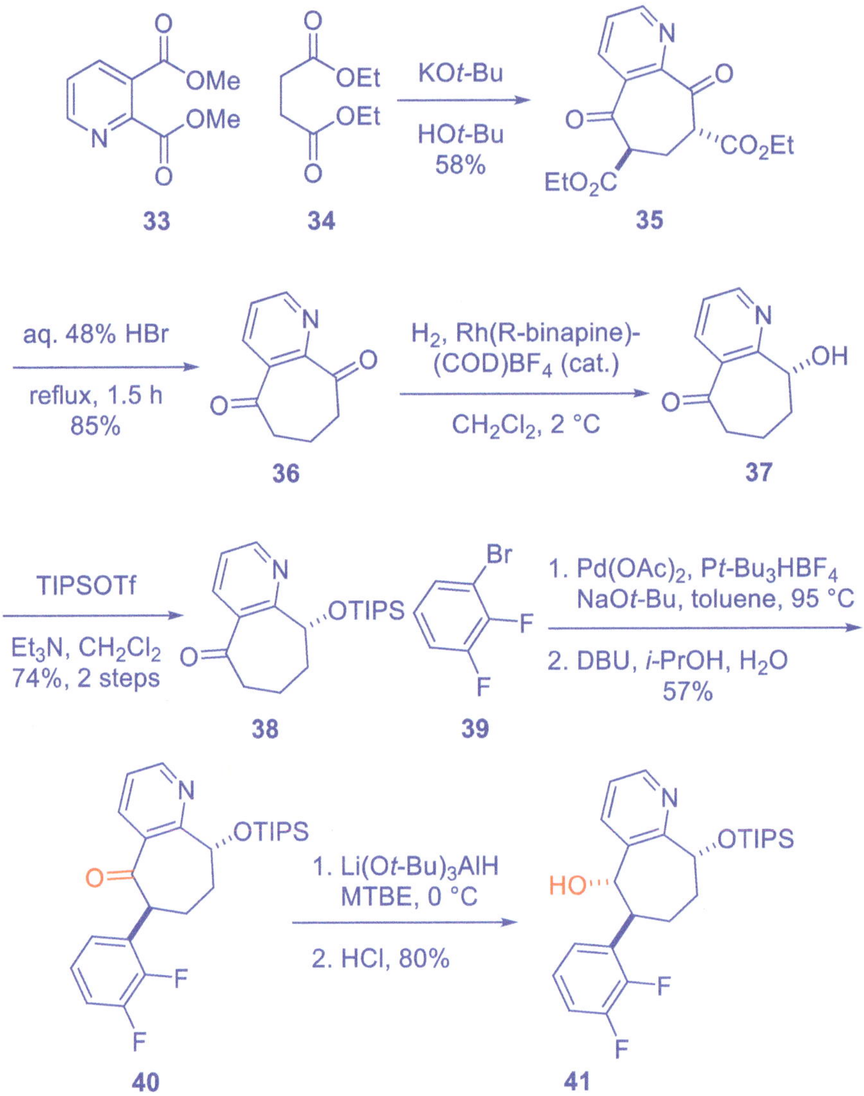

Intermediate **41** was obtained by a diastereoselective reduction of **40**. Initially reduction of **40** by sodium borohydride resulted in a diastereomeric ratio (*dr*) of 3:1 that was in favor of **41** but not sufficient. Subsequent reduction with a bulkier agent, lithium tri-(*tert*-butoxy) aluminum hydride, resulted in a *dr* of 45:1. Extraction of the alcohol **41** as salt with HCl gave a *dr* of 99:1 with a yield of 80%. The diastereoselectivity of the reaction is largely due to steric hindrance of the Si side of the ketone and high torsional forces in the ring system. TIPSO group hinders almost the entire surface of the Si side of the cycloheptane ring, making the larger reducing agent attack unlikely.[40]

Through stereochemical inversion of **41**, the stereospecific chloride intermediate **42** was obtained, which was converted to the azide intermediate **43** with second stereochemical inversion. Further deprotection of TIPS to give alcohol **44**, formation of carbamate **46** by coupling with fragment **45**, and reduction of azide **46** led to final product rimegepant (**1**).[43]

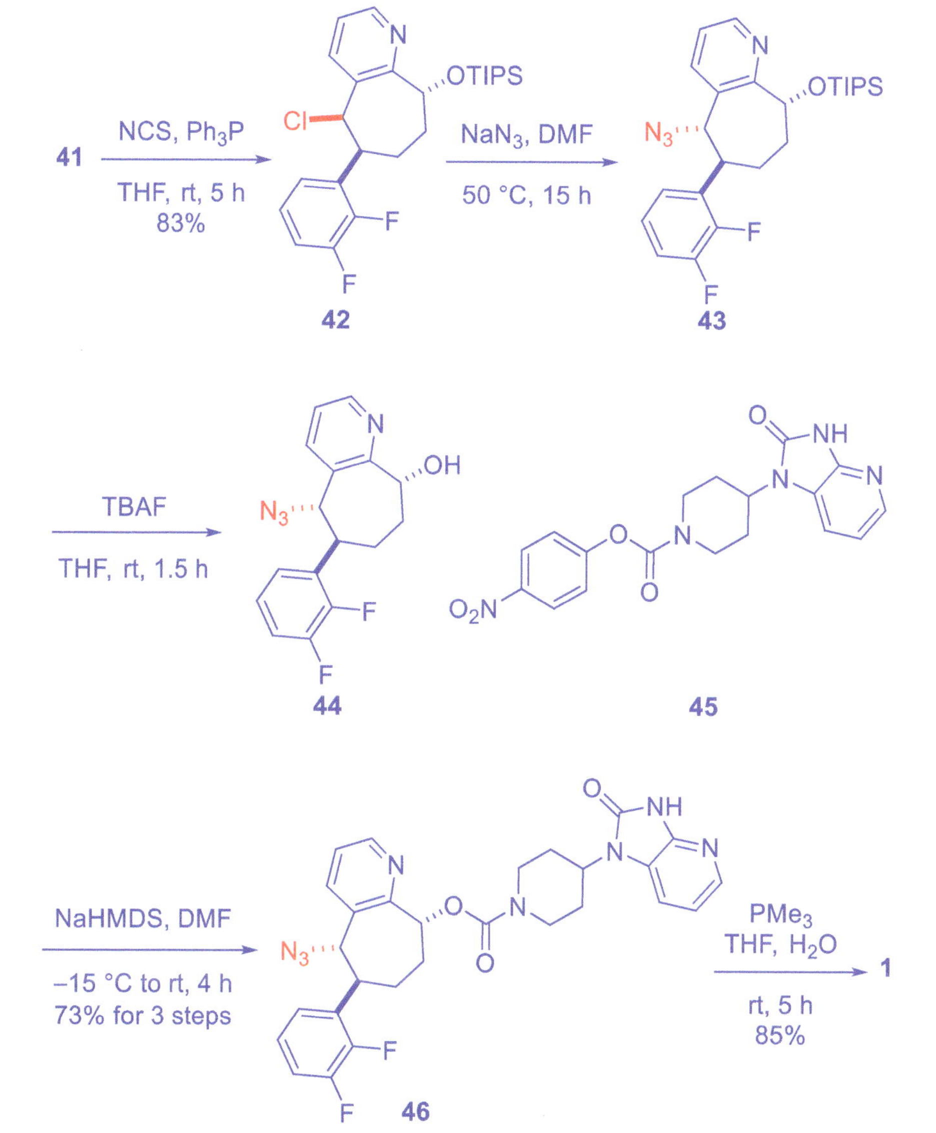

An improved process synthesis of rimegepant (**1**) was developed based on a biocatalysis using transaminases to achieve the direct conversion of the ketone **47** to chiral amine **348**, which allows milder reaction conditions with higher stereoselectivity.[56] A transaminase from *Chromobacterium violaceum* was used as a starting point for the modification to successfully synthesize intermediate **48** from the hydroxyl ketone **47**. Natural transaminases all contain one large and one small binding pocket in the active sites where fragments of the substrate are bound. The small pocket cannot bind fragments larger than a methyl group, making natural transaminases unsuitable for binding of desired substrate **47**. The enzyme needed to be mutated to accept substrate **47**. An alanine substitution was used to enlarge the binding site of enzyme S6, which showed the strongest activity when screened with des-aryl version of ketone **47**. However, there still was no activity with substrate **47** due to the unchanged small pocket.[56] Combination of two mutations that by themselves reversed stereoselectivity demonstrated slow activity for **47**. L59 proved to be locked in the binding site, while F88 was blocking the entry.

Replacement of these groups greatly facilitated substrate entry into the binding site. Further mutations were performed on the flexible loop at the entrance of the substrate tunnel (residues 83–91) and error-prone polymerase chain reaction was used to affect random mutations on the whole enzyme. Several iterations finally afforded a mutant enzyme with 19 amino acids altered, allowing 99.9% conversion of **47** to **48**, under optimal conditions with a *de* of >99.5% and a yield of 80.5% after crystallization at optimal conditions. The results were confirmed up to kg scales with nearly identical results. The reaction was carried out in a buffer with a pH of 9.0 and 15% DMSO, as well as *i*-PrNH$_2$ in 20 equiv excess. A slight nitrogen gas flow through the solution was used to remove acetone as it could slow down the reaction.[56]

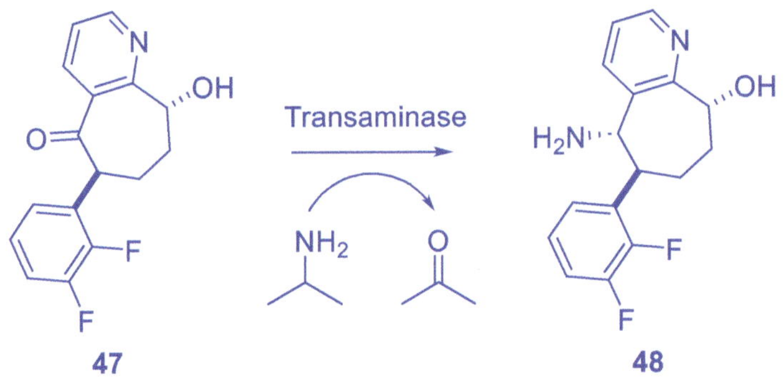

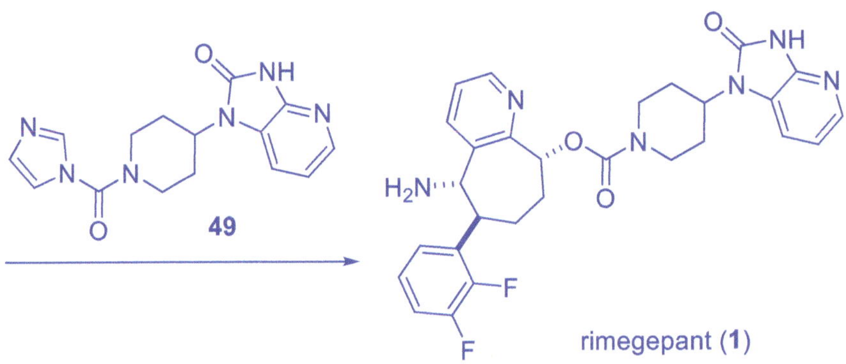

Reaction of **48** with intermediate **49** which was prepared first, led to rimegepant in the following experimental procedures as reported.[57] To a round bottom flask was added 1,1′-carbonyldiimidazole (8.59 g, 51.4 mmol), diisopropylethylamine (12.6 mL, 72.2 mmol), and tetrahydrofuran (100 mL). This mixture was warmed to 40 °C, and aged for 10 min, after which 1-(piperidin-4-yl)-1*H*-imidazo[4,5-*b*] pyridin-2(3*H*)-one dihydrochloride (10 g, 34.3 mmol) was added. The slurry was aged at 40 °C for 3 hours, and then upon reaction completion, the solvent was swapped to acetonitrile which afforded **49** as an off white solid (9.19 g, 85.9%). To a round bottom flask was charged **48** as a dihydrochloride salt (1.00 g, 2.73 mmol) and dichloromethane (15 mL). A solution of sodium carbonate (0.58 g, 5.47 mmol), 20 wt% aqueous sodium chloride (5 mL), and water (10 mL) were added, and the biphasic mixture was aged for 30 min. The phases were allowed to separate, and the organic stream was retained. The dichloromethane solvent was then switched with azeotropic drying to tetrahydrofuran, with a final volume of 15 mL. At 20 °C, **49** (0.95 g, 3.01 mmol) was added, followed by a 20 wt% potassium *tert*-butoxide solution in THF (4 mL, 6.20 mmol). The thin slurry was aged for 1 hours, and then the reaction was quenched with the addition of 20 wt% aqueous sodium chloride (5 mL) and 20 wt% aqueous citric acid (2.5 mL). The layers were allowed to separate, and the organic rich layer was retained. The organic layer was washed with 20 wt% aqueous sodium chloride (15 mL). The organic tetrahydrofuran stream was then concentrated in vacuo to afford an oil that was resuspended in dichloromethane (20 mL) and dried with $MgSO_4$. The dichloromethane stream was concentrated in vacuo to afford an oil, which was crystallized from a mixture of EtOH and heptane to afford rimegepant (**1**) as a white solid (1.14 g 78.3%).[57]

7. Summary

The quest for effective migraine treatment with oral CGRP receptor antagonists took almost two decades after olcegepant (**11**) first showed clinical efficacy with iv dosing. This was largely due to the significant challenges to achieve free plasma exposures higher than 100-fold over the protein-binding adjusted affinity with the need to bind in an extended conformation at the interface of a heteromeric receptor to prevent a large peptide ligand binding with picomolar affinity.[58] It is now well accepted that $\geq EC_{90}$-level CGRP receptor occupancy by unbound drug is needed for efficacy in migraine. Even compounds with sub-nanomolar binding potency require relatively high plasma levels of exposure, which made the search for orally bioavailable antagonists extremely difficult. This challenge has finally led to success with three FDA approved gepants including rimegepant (**1**) as Nurtec ODT. Nurtec ODT as the only gepant approved for both acute (February 27, 2020) and preventive (May 27, 2021) uses, represents a new paradigm in migraine treatment. To have both acute and preventive indications in a single medicine gives patients a new level of personal control over their treatment regimen and provides treatment benefits without the risk of medication overuse headache. Migraine patients now have an opportunity to address their treatment needs with a single medication of Nurtec, available as an orally dissolving tablet, requiring no water. This flexibility has enabled personalized treatment patterns with improved control over the significant and debilitating migraine burden.

References

1. Moreno-Ajona, D.; Pérez-Rodríguez, A.; Goadsby, P. J. Small-molecule CGRP receptor antagonists: a new approach to the acute and preventive treatment of migraine. *Med. Drug Discovery* **2020**, *7*, 100053.
2. Charles, A. The pathophysiology of migraine: implications for clinical management. *Lancet Neurol.* **2018**, *17*, 174–182.
3. Vos, T.; Lim, S. S.; Abbafati, C.; Abbas, K. M.; Abbasi, M.; Abbasifard, M.; Abbasi-Kangevari, M.; Abbastabar, H.; Abd-Allah, F.; Abdelalim, A.; Abdollahi, M. Global burden of 369 diseases and injuries in 204 countries and territories, 1990–2019: a systematic analysis for the global burden of disease study 2019. *The Lancet* **2020**, *396*, 1204–1222.
4. Steiner, T. J.; Stovner, L. J.; Jensen, R.; Uluduz, D.; Katsarava, Z. Migraine remains second among the world's causes of disability, and first among young women: findings from GBD2019. *J. Headache Pain* **2020**, *21*, 137.
5. de Vries, T.; Villalón, C. M.; Maassen Van Den Brink, A. Pharmacological treatment of migraine: CGRP and 5-HT beyond the triptans. *Pharmacol. Ther.* **2020**, *211*, 107528.

6. dos Santos, J. B. R.; da Silva, M. R. R. Small molecule CGRP receptor antagonists for the preventive treatment of migraine: a review. *Eur. J. Pharmacol.* **2022**, *922*, 174902.
7. Tepper, S. J.; Rapoport, A. M.; Sheftell, F. D. Mechanisms of action of the 5-HT1B/1D receptor agonists. *Arch. Neurol.* **2002**, *59*, 1084–1088.
8. Li, J. J.; Johnson, D. S.; Sliskovic, D. R.; Roth, B. D. Chapter 12. Triptans for migraine. In *Contemporary Drug Synthesis*; Wiley: Hoboken, NJ, **2004**. pp. 161–187.
9. Gomez-Mancilla, B.; Cutler, N. R.; Leibowitz, M. T.; Spierings, E. L.; Klapper, J. A.; Diamond, S.; Goldstein, J.; Smith, T.; Couch, J. R.; Fleishaker, J.; Azie, N.; Blunt, D. E. Safety and efficacy of PNU-142633, a selective 5-HT1D agonist, in patients with acute migraine. *Cephalalgia* **2001**, *21*, 727–732.
10. Edvinsson, J. C. A.; Warfvinge, K.; Krause, D. N.; Blixt, F. W.; Sheykhzade, M.; Edvinsson, L.; Haanes, K. A. C-fibers may modulate adjacent Aδ-fibers through axon-axon CGRP signaling at nodes of Ranvier in the trigeminal system. *J. Headache Pain* **2019**, *20*, 105.
11. Haanes, K. A.; Edvinsson, L. Pathophysiological mechanisms in migraine and the identification of new therapeutic targets. *CNS Drugs* **2019**, *33*, 525–537.
12. Afroz, S.; Arakaki, R.; Iwasa, T.; Oshima, M.; Hosoki, M.; Inoue, M.; Baba, O.; Okayama, Y.; Matsuka, Y. CGRP induces differential regulation of cytokines from satellite glial cells in trigeminal ganglia and orofacial nociception. *Int. J. Mol. Sci.* **2019**, *20*, 711.
13. Edvinsson, L.; Haanes, K. A.; Warfvinge, K. Does inflammation have a role in migraine? *Nat. Rev. Neurol.* **2019**, *15*, 483–490.
14. Edvinsson, L.; Haanes, K. A.; Warfvinge, K.; Krause, D. N. CGRP as the target of new migraine therapies—successful translation from bench to clinic. *Nat. Rev. Neurol.* **2018**, *14*, 338–350.
15. Lassen, L. H.; Haderslev, P. A.; Jacobsen, V. B.; Iversen, H. K.; Sperling, B.; Olesen, J. CGRP may play a causative role in migraine. *Cephalalgia Int. J. Headache* **2002**, *22*, 54–61.
16. Dubowchik, G. M.; Conway, C. M.; Xin, A. W. Blocking the CGRP pathway for acute and preventive treatment of migraine: the evolution of success. *J. Med. Chem.* **2020**, *63*, 6600–6623.
17. Cottrell, G. S.; Padilla, B.; Pikios, S.; Roosterman, D.; Steinhoff, M.; Grady, E. F.; Bunnett, N. W. Post-endocytic sorting of calcitonin receptor-like receptor and receptor activity-modifying protein 1. *J. Biol. Chem.* **2007**, *282*, 12260–12271.
18. Gingell, J. J.; Hendrikse, E. R.; Hay, D. L. New insights into the regulation of CGRP-family receptors. *Trends Pharmacol. Sci.* **2019**, *40*, 71–83.
19. Mullin, K.; Kudrow, D.; Croop, R.; Lovegren, M.; Conway, C. M.; Coric, V.; Lipton, R. B. Potential for treatment benefit of small molecule CGRP receptor antagonist plus monoclonal antibody in migraine therapy. *Neurology* **2020**, *94*, e2121–e2125.
20. Deen, M.; Correnti, E.; Kamm, K.; Kelderman, T.; Papetti, L.; Rubio-Beltrán, E.; Vigneri, S.; Edvinsson, L.; Maassen Van Den Brink, A. Blocking CGRP in migraine patients – a review of pros and cons. *J. Headache Pain* **2017**, *18*, 96.

21. Gupta, S.; Mehrotra, S.; Villalón, C. M.; Garrelds, I. M.; de Vries, R.; van Kats, J. P.; Sharma, H. S.; Saxena, P. R.; Maassenvandenbrink, A. Characterisation of CGRP receptors in human and porcine isolated coronary arteries: evidence for CGRP receptor heterogeneity. *Eur. J. Pharmacol.* **2006**, *530*, 107–116.

22. Gupta, S.; Mehrotra, S.; Avezaat, C. J. J.; Villalón, C. M.; Saxena, P. R.; Maassenvandenbrink, A. Characterisation of CGRP receptors in the human isolated middle meningeal artery. *Life Sci.* **2006**, *79*, 265–271.

23. Rubio-Beltran, E.; Chan, K. Y.; Danser, A. J.; Maassen Van Den Brink, A.; Edvinsson, L. Characterisation of the calcitonin gene-related peptide receptor antagonists ubrogepant and atogepant in human isolated coronary, cerebral and middle meningeal arteries. *Cephalalgia Int. J. Headache* **2020**, *40*, 357–366.

24. Haanes, K. A.; Chan, K. Y.; Maassen Van Den Brink, A. Comment on "A second trigeminal CGRP receptor: function and expression of the AMY1 receptor." *Ann. Clin. Transl. Neurol.* **2016**, *3*, 307–308.

25. Ashina, M.; Dodick, D.; Goadsby, P. J.; Reuter, U.; Silberstein, S.; Zhang, F.; Gage, J. R.; Cheng, S.; Mikol, D. D.; Lenz, R. A. Erenumab (AMG 334) in episodic migraine: interim analysis of an ongoing open-label study. *Neurology* **2017**, *89*, 1237–1243.

26. L'Heureux, M.-C.; St-Pierre, S.; Trudel, L.; Plourde, V.; Lepage, R.; Poitras, P. Digestive motor effects and vascular actions of CGRP in dog are expressed by different receptor subtypes. *Peptides* **2000**, *21*, 425–430.

27. Labastida-Ramírez, A.; Rubio-Beltrán, E.; Haanes, K. A.; Chan, K. Y.; Garrelds, I. M.; Johnson, K. W.; Danser, A. H. J.; Villalón, C. M.; MaassenVanDenBrink, A. Lasmiditan inhibits calcitonin gene-related peptide release in the rodent trigeminovascular system. *Pain* **2020**, *161* (5), 1092–1099.

28. Maassen Van Den Brink, A.; Meijer, J.; Villalón, C. M.; Ferrari, M. D. Wiping out CGRP: potential cardiovascular risks. *Trends Pharmacol. Sci.* **2016**, *37*, 779–788.

29. Chaturvedula, P. V.; Pin, S.; Tholady, G.; Conway, C. M.; Macor, J. E.; Dubowchik, G. M. Design and synthesis of potent antagonists containing rigid spirocyclic privileged structures for the CGRP receptor. *Bioorg. Med. Chem. Lett.* **2012**, *22*, 4719–4722.

30. Han, X.; Civiello, R. L.; Conway, C. M.; Cook, D. A.; Davis, C. D.; Macci, R.; Pin, S. S.; Ren, S. X.; Schartman, R.; Signor, L. J.; Thalody, G.; Widmann, K. A.; Xu, C.; Chaturvedula, P. V.; Macor, J. E.; Dubowchik, G. M. The synthesis and SAR of calcitonin gene-related peptide (CGRP) receptor antagonists derived from tyrosine surrogates. Part 1. *Bioorg. Med. Chem. Lett.* **2012**, *22*, 4723–4727.

31. Han, X.; Civiello, R. L.; Conway, C. M.; Cook, D. A.; Davis, C. D.; Degnan, A. P.; Jiang, X.-J.; Macci, R.; Mathias, N. R.; Moench, P.; Pin, S. S.; Schartman, R.; Signor, L. J.; Thalody, G.; Tora, G.; Whiterock, V.; Xu, C.; Macor, J. E.; Dubowchik, G. M. The synthesis and SAR of calcitonin gene-related peptide (CGRP) receptor antagonists derived from tyrosine surrogates. Part 2. *Bioorg. Med. Chem. Lett.* **2013**, *23*, 1870–1873.

32. Degnan, A. P.; Chaturvedula, P. V.; Conway, C. M.; Cook, D. A.; Davis, C. D.; Denton, R.; Han, X.; Macci, R.; Mathias, N. R.; Moench, P.; Pin, S. S.; Ren, S. X.; Schartman, R.; Signor, L. J.; Thalody, G.; Widmann, K. A.; Xu, C.; Macor, J. E.; Dubowchik, G. M.

Discovery of (*R*)-4-(8-fluoro-2-oxo-1,2-dihydroquinazolin-3(4*H*)-yl)-*N*-(3-(7-methyl-1*H*-Indazol-5-yl)-1-oxo-1-(4-(piperidin-1-yl)piperidin-1-yl)propan-2-yl)piperidine-1-carboxamide (BMS-694153): a potent antagonist of the human calcitonin gene-related peptide receptor for migraine with rapid and efficient intranasal exposure. *J. Med. Chem.* **2008**, *51*, 4858–4861.

33. Chaturvedula, P. V.; Mercer, S. E.; Pin, S. S.; Thalody, G.; Xu, C.; Conway, C. M.; Keavy, D.; Signor, L.; Cantor, G. H.; Mathias, N.; Moench, P.; Denton, R.; Macci, R.; Schartman, R.; Whiterock, V.; Davis, C.; Macor, J. E.; Dubowchik, G. M. Discovery of (*R*)-*N*-(3-(7-methyl-1*H*-Indazol-5-yl)-1-(4-(1-methylpiperidin-4-yl)-1-oxopropan-2-yl)-4-(2-oxo-1,2-dihydroquinolin-3-yl)piperidine-1-carboxamide (BMS-742413): a potent human CGRP antagonist with superior safety profile for the treatment of migraine through intranasal delivery. *Bioorg. Med. Chem. Lett.* **2013**, *23*, 3157–3161.

34. Tora, G.; Degnan, A. P.; Conway, C. M.; Kostich, W. A.; Davis, C. D.; Pin, S. S.; Schartman, R.; Xu, C.; Widmann, K. A.; Macor, J. E.; Dubowchik, G. M. Preparation of imidazoles as potent calcitonin gene-related peptide (CGRP) antagonists. *Bioorg. Med. Chem. Lett.* **2013**, *23*, 5684–5688.

35. Luo, G.; Chen, L.; Civiello, R.; Pin, S. S.; Xu, C.; Kostich, W.; Kelley, M.; Conway, C. M.; Macor, J. E.; Dubowchik, G. M. Calcitonin gene-related peptide (CGRP) receptor antagonists: pyridine as a replacement for a core amide group. *Bioorg. Med. Chem. Lett.* **2012**, *22*, 2917–2921.

36. Luo, G.; Chen, L.; Pin, S. S.; Xu, C.; Conway, C. M.; Macor, J. E.; Dubowchik, G. M. Calcitonin gene-related peptide (CGRP) receptor antagonists: novel aspartates and succinates. *Bioorg. Med. Chem. Lett.* **2012**, *22*, 2912–2916.

37. Mercer, S. E.; Chaturvedula, P. V.; Conway, C. M.; Cook, D. A.; Davis, C. D.; Pin, S. S.; Macci, R.; Schartman, R.; Signor, L. J.; Widmann, K. A.; Whiterock, V. J.; Chen, P.; Xu, C.; Herbst, J. J.; Kostich, W. A.; Thalody, G.; Macor, J. E.; Dubowchik, G. M. Azepino-indazoles as calcitonin gene-related peptide (CGRP) receptor antagonists. *Bioorg. Med. Chem. Lett.* **2021**, *31*, 127624.

38. Luo, G.; Jiang, X.-J.; Chen, L.; Conway, C. M.; Gulianello, M.; Kostich, W.; Keavy, D.; Signor, L. J.; Chen, P.; Davis, C.; Whiterock, V. J.; Schartman, R.; Widmann, K. A.; Macor, J. E.; Dubowchik, G. M. Calcitonin gene-related peptide (CGRP) receptor antagonists: heterocyclic modification of a novel azepinone lead. *Bioorg. Med. Chem. Lett.* **2021**, *43*, 128077.

39. Luo, G.; Chen, L.; Conway, C. M.; Denton, R.; Keavy, D.; Gulianello, M.; Huang, Y.; Kostich, W.; Lentz, K. A.; Mercer, S. E.; Schartman, R.; Signor, L.; Browning, M.; Macor, J. E.; Dubowchik, G. M. Discovery of BMS-846372, a potent and orally active human CGRP receptor antagonist for the treatment of migraine. *ACS Med. Chem. Lett.* **2012**, *3*, 337–341.

40. Leahy, D. K.; Fan, Y.; Desai, L. V.; Chan, C.; Zhu, J.; Luo, G.; Chen, L.; Hanson, R. L.; Sugiyama, M.; Rosner, T.; Cuniere, N.; Guo, Z.; Gao, Q. Efficient and scalable enantioselective synthesis of a CGRP antagonist. *Org. Lett.* **2012**, *14*, 4938–4941.

41. Luo, G.; Chen, L.; Conway, C. M.; Johnson, B. M.; Ng, A.; Macor, J. E.; Dubowchik, G. M. Discovery of BMS-846372, a potent and orally active human CGRP receptor antagonist for the treatment of migraine. *J. Org. Chem.* **2017**, *82*, 3710–3720.

42. Luo, G. *et al.* Unpublished results.

43. Luo, G.; Chen, L.; Conway, C. M.; Denton, R.; Keavy, D.; Signor, L.; Kostich, W.; Lentz, K. A.; Santone, K. S.; Schartman, R.; Browning, M.; Tong, G.; Houston, J. G.; Dubowchik, G. M.; Macor, J. E. Discovery of (5*S*,6*S*,9*R*)-5-amino-6-(2,3-difluorophenyl)-6,7,8,9-tetrahydro-5*H*-cyclohepta[*B*]pyridin-9-yl 4-(2-oxo-2,3-dihydro-1*H*-Imidazo[4,5-*B*]pyridin-1-yl)piperidine-1-carboxy-late (BMS-927711): an oral calcitonin gene-related peptide (CGRP) antagonist in clinical trials for treating migraine. *J. Med. Chem.* **2012**, *55*, 10644–10651.

44. Luo, G.; Chen, L.; Conway, C. M.; Kostich, W.; Macor, J. E.; Dubowchik, G. M. Asymmetric synthesis of heterocyclic analogues of a CGRP receptor antagonist for treating migraine. *Org. Lett.* **2015**, *17*, 5982–5985.

45. Parkash, V.; Maan, S.; Deepika; Yadav, S. K.; Hemlata; Jogpal, V. Fast disintegrating tablets: opportunity in drug delivery system. *J. Adv. Pharm. Technol. Res.* **2011**, *2*, 223–235.

46. Drug label. https://dailymed.nlm.nih.gov/dailymed/drugInfo.cfm?setid=9ef08e09-1098-35cc-e053-2a95a90a3e1d (accessed 4 July 2023).

47. Tong, G.; Savant, I.; Jariwala, N.; Burt, D.; Zheng, N.; Buzescu, A.; Bertz, R.; Keswani, S.; Marcus, R. Phase I single and multiple dose study to evaluate the safety, tolerability, and pharmacokinetics of BMS-927711 in healthy subjects. *J. Headache Pain* **2013**, *14*, P118.

48. Croop, R.; Ivans, A.; Stock, D.; Hould, J.; Morris, B. A.; Stringfellow, J.; Moulin, J.; Larouche, R.; Tanguay, M.; Coric, V.; Lipton, R. B. A phase 1 study to evaluate the bioequivalence of oral tablet and orally dissolving tablet formulations of rimegepant in healthy adult subjects under fasting conditions. *Headache* **2018**, *58*, 1303–1304.

49. Szkutnik-Fiedler, D. Pharmacokinetics, pharmacodynamics and drug–drug interactions of new anti-migraine drugs—lasmiditan, gepants, and calcitonin-gene-related peptide (CGRP) receptor monoclonal antibodies. *Pharmaceutics* **2020**, *12*, 1180.

50. Marcus, R.; Goadsby, P. J.; Dodick, D.; Stock, D.; Manos, G.; Fischer, T. Z. BMS-927711 for the acute treatment of migraine: a double-blind, randomized, placebo controlled, dose-ranging trial. *Cephalalgia* **2014**, *34*, 114–125.

51. Ferrari, M. D.; Goadsby, P. J.; Roon, K. I.; Lipton, R. B. Triptans (serotonin, 5-HT1B/1D agonists) in migraine: detailed results and methods of a meta-analysis of 53 trials. *Cephalalgia* **2002**, *22*, 633–658.

52. Lipton, R. B.; Croop, R.; Stock, E. G.; Stock, D. A.; Morris, B. A.; Frost, M.; Dubowchik, G. M.; Conway, C. M.; Coric, V.; Goadsby, P. J. Rimegepant, an oral calcitonin gene-related peptide receptor antagonist, for migraine. *N. Engl. J. Med.* **2019**, *381*, 142–149.

53. Croop, R.; Goadsby, P. J.; Stock, D. A.; Conway, C. M.; Forshaw, M.; Stock, E. G.; Coric, V.; Lipton, R. B. Efficacy, safety, and tolerability of rimegepant orally disintegrating tablet for the acute treatment of migraine: a randomised, phase 3, double-blind, placebo-controlled trial. *Lancet* **2019**, *394*, 737–745.

54. L'Italien, G.; Popoff, E.; Johnston, K.; McGrath, D.; Conway, C. M.; Powell, L.; Harris, L.; Kowalczyk, N.; Croop, R.; Coric, V. Rimegepant 75 mg for acute treatment of migraine is associated with significant reduction in monthly migraine days: results from a long-term, open-label study. *Cephalalgia Rep.* **2022**, *5*, 1–11.

55. Croop, R.; Lipton, R. B.; Kudrow, D.; Stock, D. A.; Kamen, L.; Conway, C. M.; Stock, E. G.; Coric, V.; Goadsby, P. J. Oral rimegepant for preventive treatment of migraine: a phase 2/3, randomised, double-blind, placebo-controlled trial. *Lancet* **2021**, *397*, 51–60.
56. Ma, Y.; Jiao, X.; Wang, Z.; Mu, H.; Sun, K.; Li, X.; Zhao, T.; Liu, X.; Zhang, N. Engineering a transaminase for the efficient synthesis of a key intermediate for rimegepant. *Org. Process Res. Dev.* **2022**, *26*, 1971–1977.
57. Leahy, D. K.; Fan, Y.; Chan, C.; Desai, L. V.; Patel, S. S.; Sugiyama, M. Process for the preparation of Cycloheptapyridine Cgrp receptor antagonists. US 2015*/0099887A1*, April 9, **2015**.
58. Dubowchik, G. M.; Croop, R.; Conway, C. M. Case history: Nurtec® ODT (rimegepant) for the acute and preventive treatment of migraine. *Med. Chem. Rev.* **2022**, *57*, 517–542.

14

Daridorexant (Quviviq): An Antagonist of Orexin Receptors for Treating Insomnia

Dexi Yang

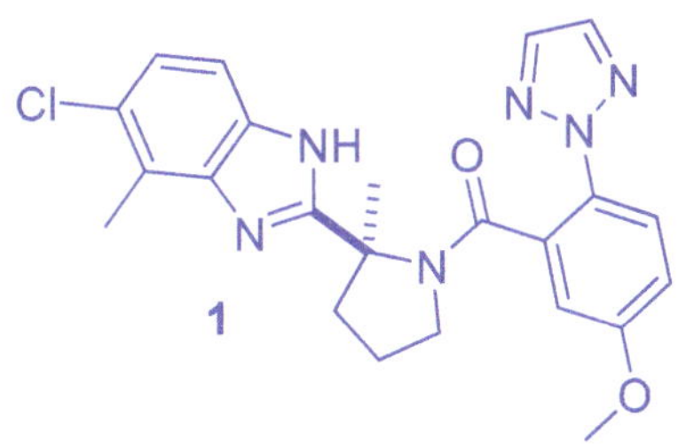

USAN: Daridorexant
Trade name: Quviviq
Idorsia, 2022
Antagonist of
Orexin Receptors

1. Background

As human beings, we spend one-third of our lives sleeping. Sleep plays a vital role in our health and well-being throughout our lives. Inadequate sleep can hurt our body and mind, affect our abilities to focus, and in a long term cause serious health problems such as diabetes, metabolic disorder and depression.[1] Symptoms of insomnia include difficulty in falling asleep, frequent awakenings, short sleep duration, early waking up, etc.[2, 3] It is estimated 10–20% of people suffer from sleeping disorder, but very few of them are under therapy. Although insomnia is not fatal, it directly hurts quality of life, causes daytime fatigue, inattention, bad mood, low energy levels, and increases risks of traffic and work-related accidents.[4]

Ironically, insomnia costs the US economy more than $100 billion per year, but very few people get effect therapy. Although 70–80% of insomnia can be improved via changing sleep habits, many people still need medicines.[5, 6] Traditional pharmacotherapies have many limitations, and we will explain them in the following

Chemistry and Pharmacology of Drug Discovery, First Edition. Edited by Jie Jack Li.

paragraph. It is still an unmet medical need. But since insomnia affects so many people, it is also a significant commercial opportunity for the pharmaceutical industry.

The pharmacotherapy of insomnia mainly includes four types of medications approved by the FDA: the "Z" sedative hypnotics, benzodiazepines, melatonin receptor agonists and the most recent orexin receptors antagonists.[7]

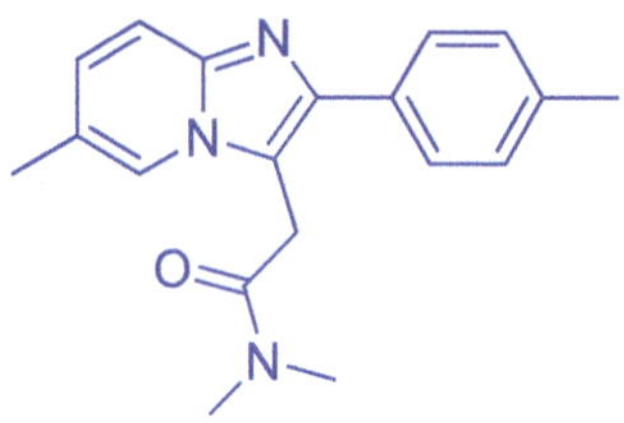

zolpidem (**2**, Ambien)

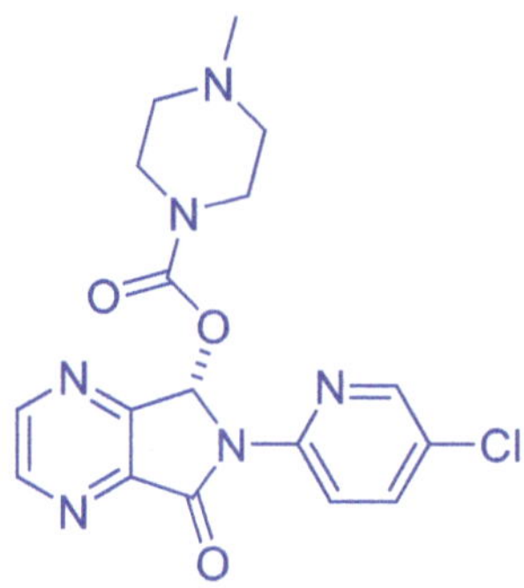

eszopiclone (**3**, Lunesta)

The "Z-drugs," such as zolpidem (**2**), eszopiclone (**3**) and zaleplon are allosteric GABA-A subunit alpha-1 receptor modulators. They are generally selected as the first-line medications for insomnia, with each drug providing a different range of suppression of CNS (central nervous system) activity as well as different onset and duration for patients. The common side effects of Z-drugs are mainly the next-day hangover effect of residual drowsiness, dizziness, and ataxia. The FDA has specifically warned that zolpidem (**2**) and eszopiclone (**3**) can impair next-day operation of machinery and driving.[8]

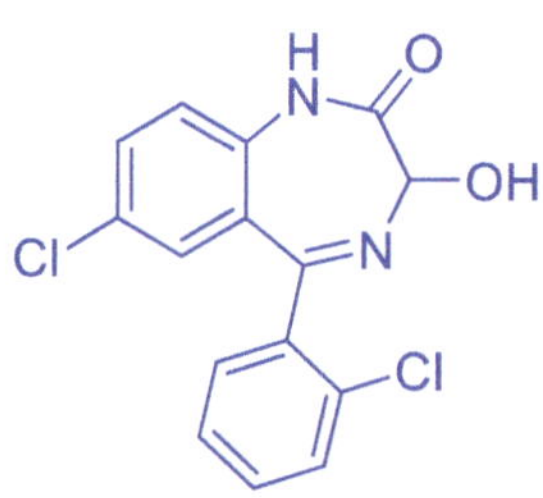

lorazepam (**4**, Ativan/Orfidel)

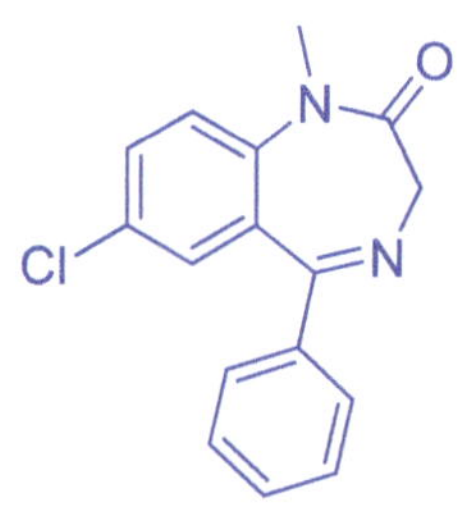

diazepam (**5**, Valium)

The other class of drugs for insomnia are benzodiazepines. There are several benzodiazepines on the market, such as lorazepam (**4**) and diazepam (**5**). They generally have short half-life, and less next-day hangover effect in comparing with the Z-drugs. However, they disrupt normal sleep patterns, reduce rapid eye movement (REM) sleep, and consequently decrease coordination the next morning after administration.[4] For these reasons, they are still not ideal treatment for insomnia, especially for the older population.

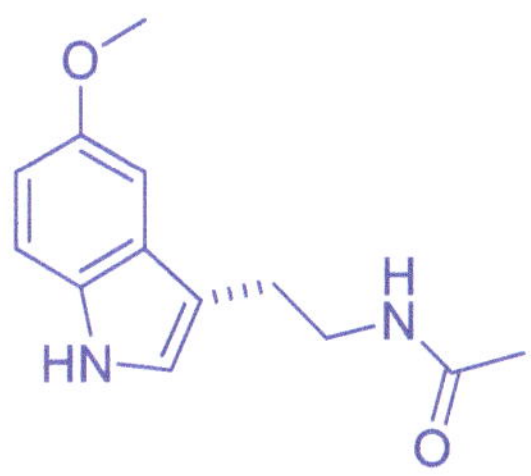

melatonin (**6**)

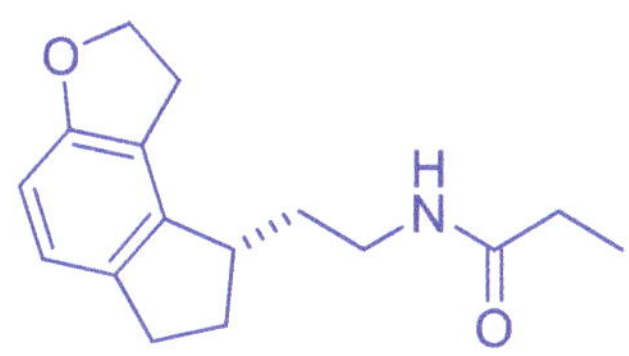

ramelteon (**7**, Rozerem)

Melatonin (**6**) is a hormone produced in the brain in response to darkness. People believe melatonin can help with sleep for a long time, so it is sold as a dietary supplement in drug stores. There are more than 30 million people using it to alleviate insomnia. In fact, it is not even enlisted in clinical guidelines due to lack of efficacy and safety studies.[9] Interestingly, based on this natural product, Takeda further optimized it and discovered ramelteon (Rozerem, **7**) approved for treating insomnia by the FDA in 2005.[10] It is an agonist of melatonin receptor in the suprachiasmatic nucleus (SCN), not targeting the GABA-A receptors as Z-type of drugs we discussed earlier. However, ramelteon was approximately 33% less effective at improving latency to persistent sleep (LPS) in adults when compared with placebo.[11] So far, the pursuit of more effective pharmacological treatment is still in progress. New medications must consider more factors, especially sleeping patterns, drug safety, efficacy and no next-day hangover effect.

In the last two decades, orexin receptors OX1R and OX2R attracted a lot of attention. They are two G protein-coupled receptors (GPCRs) modulated by two neuropeptides orexin A (33 amino acids) and orexin B (28 amino acids) generated in hypothalamus (Figure 1). Two research groups, one is the Yanagisawa group at the Howard Hughes Medical Institute in Dallas, and the other is the Sutcliffe group in The Scripps Research Institute in La Jolla identified the two neuropeptides independently.[13, 14] Orexins were named after Greek word "orexis" because initially they were considered as peptides relating to appetite. It was soon found that orexin signaling played a key role in the sleep/wake circadian cycle based on several solid facts. One is the correlation that narcolepsy or cataplexy in dogs was linked to a loss-of-function mutation in the OX_2R GPCR in dogs. The other is that low level of the orexin peptides was observed in the CSF of narcolepsy patients. Low concentration of orexin A or orexin B in the brain causes low energy level or sleep, while high concentration of the neuropeptides brings more energy or less sleep. Based on these observations, as well as gene knocked off rat experiments, it was hypothesized that small molecule orexin agonists could keep people awake to cure patients with narcolepsy, while orexin antagonists could help people sleep. Guided by this hypothesis, small molecule orexin antagonists were invented to cure patients with insomnia.

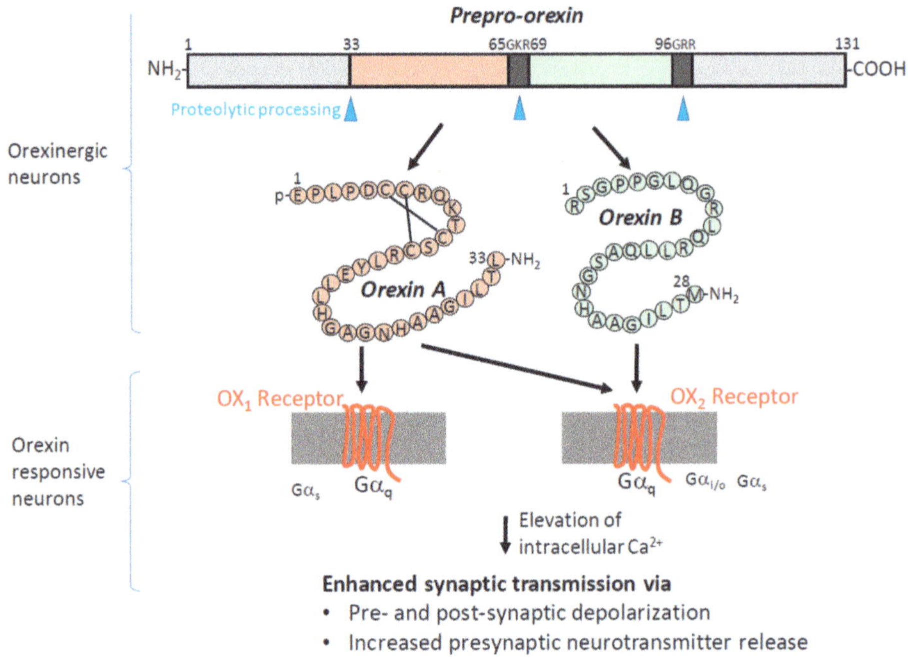

Figure 1. Diagram of the generation of Orexin A and Orexin B from prepro-orexin and their activation of OX1 and OX2 receptors. Subsequent signaling cascades involve phospholipase C (PLC), leading to elevated Ca^{2+} concentration to enhance synaptic transmission. Source: Boss et al. [12]/with permission of John Wiley & Sons

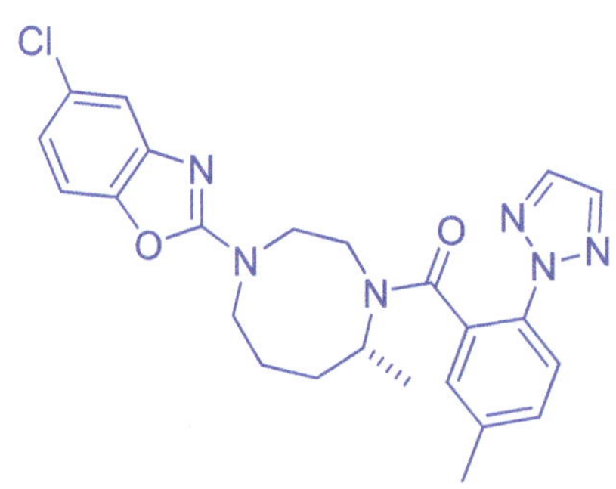

8 suvorexant

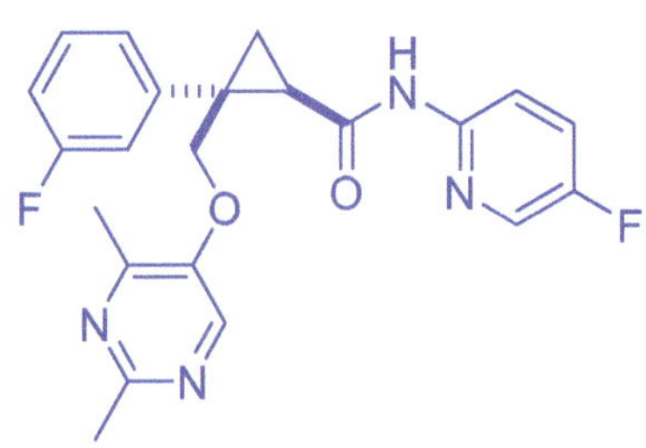

9 lemborexant

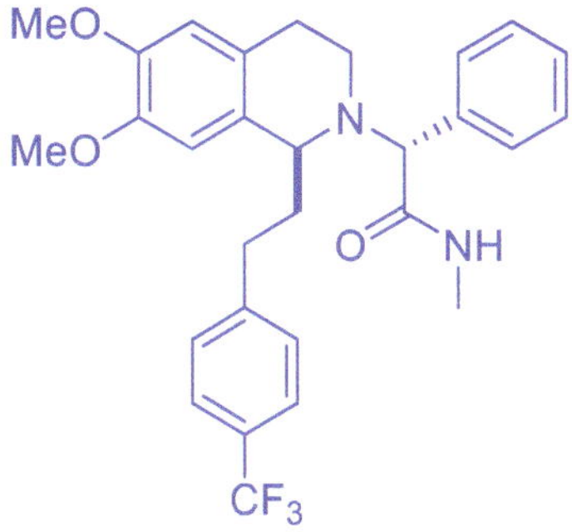

10 almorexant

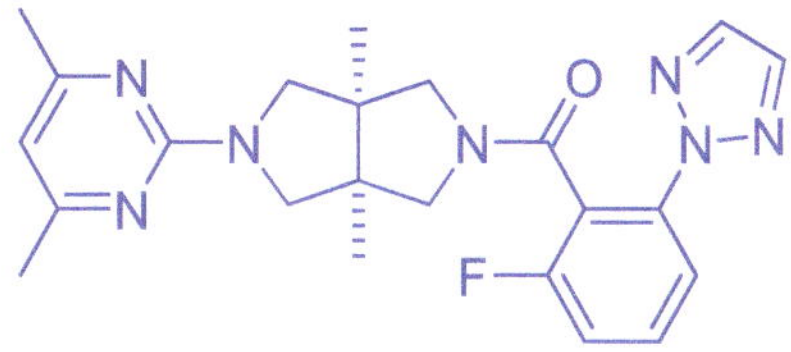

11 seltorexant

In the past decade, several structurally diverse orexin receptor antagonists have been clinically investigated. Of all the drug candidates, those inhibit both OX1R and OX2R are called dual orexin agonists (DOAR), while those inhibit only one orexin are called SORA. The FDA approved Merck's suvorexant (**8**) as the first-in-class (FIC) in 2014. Later in 2019, Eisai's lemborexant (**9**) was approved.[15, 16] Soon after, the third compound daridorexant (**1**) from Idorsia was approved in 2022.[17] They are all DOARs and share many features. Other than these three approved drugs, there are several clinical compounds either failed or still under development. One is almorexant (**10**), a DOAR, but failed due to tolerability issues.[18] The other is seltorexant (**11**) which is still under development by Minerva Neurosciences and Janssen Pharmaceuticals. Different from other DOAR antagonists, it is a selective antagonist of the orexin OX2 receptor (2-SORA), with 100-fold greater binding affinity for the OX2 receptor over the OX1 receptor.[19] One limitation of all orexin receptor antagonist is that they exclude narcolepsy patients or people suffering from low concentrations of orexin neuropeptides.[20]

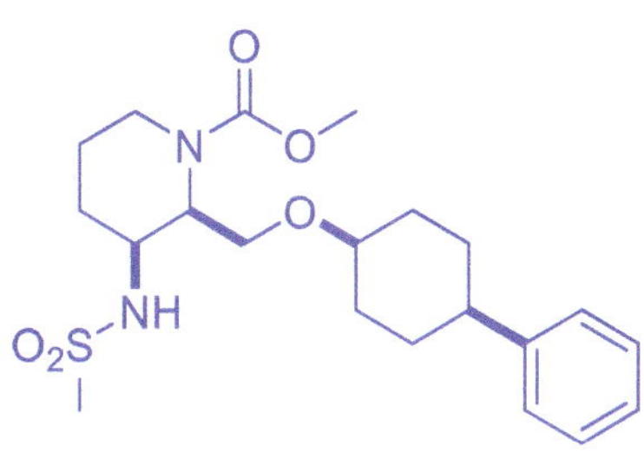

TAK-925

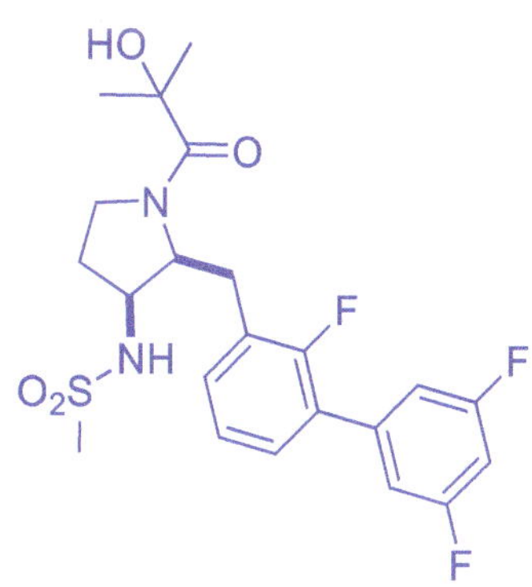

TAK-994

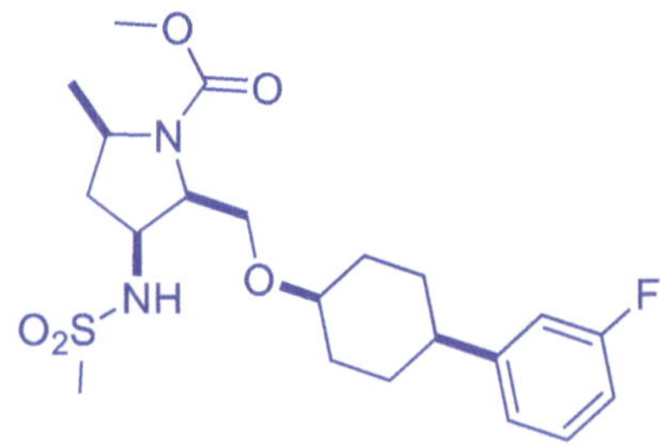

OX2R agonist (Merck)

It is worth noting that on the contrary to orexin antagonists, small molecule orexin agonists have also been validated to show efficacy on narcolepsy patients. In this arena, Taketa was the earliest runner and had two clinical compounds TAK-925 and TAK-944.[21, 22] TAK-925 has been used to validate MOA via IV administration. Due to high metabolism, it was stopped in Phase 1b. The other is an oral candidate TAK-944, which was also stopped in Phase 2 due to hepatotoxicity. Now the only small molecule orexin agonist in clinical trial is MK-6552.[23] Limited publication revealed that Merck's orexin agonists have more complicated cores and substituents. All examples in their patents feature multiple congested chiral centers, and hopefully they can fix high metabolism and hepatotoxicity issues observed in TAK candidates, while maintain good brain penetration and low dose required by CNS drugs for everyday administration.[24, 25]

2. Pharmacology

Daridorexant (**1**) is a highly potent small-molecule antagonist on both GPCR receptors OX1R and OX2R. As a CNS drug to promote sleep, it is a good brain penetrating antagonist with fast onset, and a duration of action just long enough to cover nighttime sleep, while avoiding residual morning effects.[26] It promotes sleeping via a new mechanism, that is to selectively inhibit only OX1R and OX2R to allow sleep to occur, bypassing other receptors such as GABA-A, thereby to avoid adverse effects associated with inhibitors of those receptors. Different from previous medications, daridorexant (**1**) promotes sleep without altering the sleep architecture.[12, 27] It increases REM sleep but shortens non-REM sleep in physiological proportions.[28] As a precaution, narcolepsy patients or people suffering from low concentrations of orexin neuropeptides, should avoid taking daridorexant (**1**) due to its MOA of deactivating orexin receptors which aggravates narcolepsy.[20]

3. Structure–Activity Relationship (SAR)

The structure of daridorexant (**1**) features a proline core, an aryl amide and a benzimidazole pendant attached at α-position. The proline hit was identified at Actelion via high-throughput screening using Ca^{2+} release assays in the search for dual orexin receptor antagonists. The proline sulfonamide (**12**) in Figure 2 has a low molecular weight of 461.42 and shows significant activity in the calcium release assay. It also showed good properties necessary for CNS drugs, especially good brain penetration, no liabilities of being P-glycoprotein (Pgp) substrate, low *in vitro* human liver microsomal metabolic stability and no obvious cytochrome inhibitory activity.[12, 29]

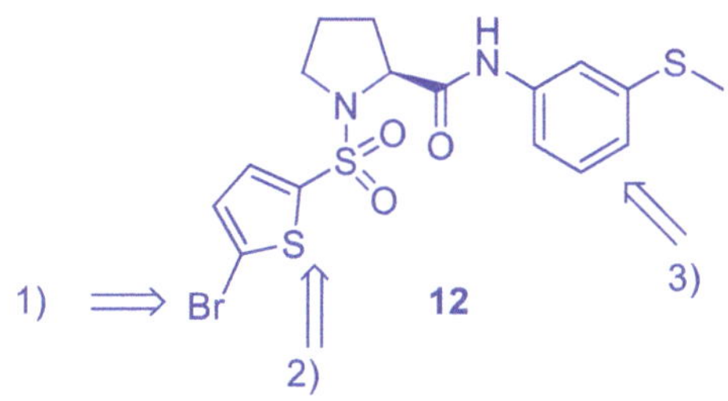

	in vivo BBB	MDR1-MDRK assay
$IC_{50}(OX_1)$: 21 nM	@ 100 mg•kg^{-1} in rat, at 3h	E ratio: 1.3
$IC_{50}(OX_2)$: 9.9 nM	[P]: 524 ng•mL^{-1}	P_{app}AB: 34.5 x 10^{-6} cm/s
IC_{50}(3A4): 6.1 uM	[B]: 403 ng•g^{-1}	P_{app}AB: 46 x 10^{-6} cm/s
HLM: 541 uL•min^{-1}•mg^{-1}	B/P: 77%	
RLM: >1250 uL•min^{-1}•mg^{-1}		

Figure 2. Structure and properties of OX1/OX2 HTS hit

Starting from hit **12**, earlier optimization mainly focused on these three aspects to improve both potency, brain penetration and general ADME properties: (1) replacement of the bromothiophene with more drug-like aryl or alkyl groups; (2) optimization of the sulfonamide moiety with better bioisostere; (3) screening of better heterocycles other than aniline which may have hERG liability, as well as identifying more robust groups other than alkyl sulfide that could be easily oxidized.[29]

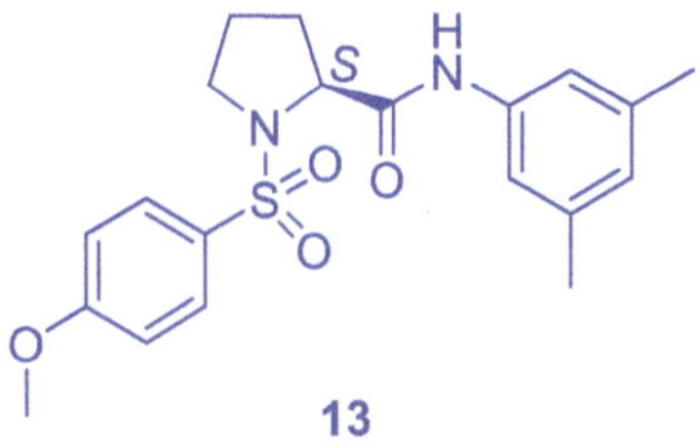

13

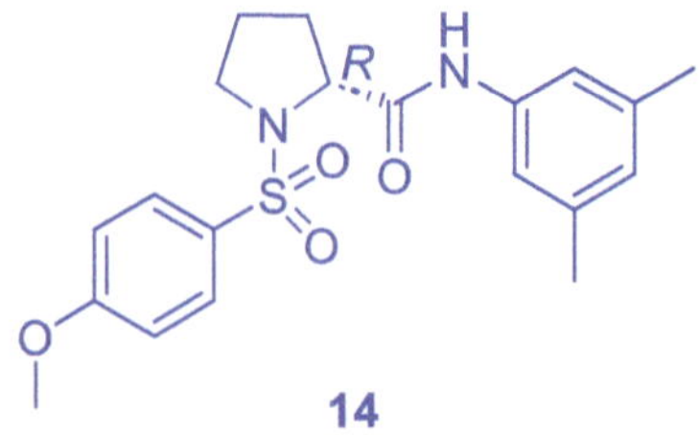

14

	13	14
$IC_{50}(OX_1)$ [nM]	60	6215
$IC_{50}(OX_2)$ [nM]	11	6044
[P] (ng/mL)	2667	1309
[B] (ng/g)	1219	1041
[B]/[P] (%)	46	79

Figure 3. Influence of proline core chirality

Lead optimization was conducted by (1) screening aryl groups on sulfonamide and (2) replacing aniline on amide. It has been observed that derivatives from this practice generally maintained potencies toward OX1 and OX2, but inhibition toward CYP3A4 varied a lot. After thoroughly screening the two vectors, **13** (Figure 3) was discovered with high potencies on both OX1 and OX2, while with the least liability toward CYP3A4 (15 μM). When **13** was further profiled in an *in vivo* blood-brain barrier penetration experiment in male Wistar rats, it showed excellent brain penetration with brain concentration of $[B] = 1219$ ng/g comparing to plasma concentration of $[P] = 2667$ ng/mL, and $[B]/[P]$ ratio is 46%. Its enantiomer **14** has also been prepared and the chirality has been confirmed to be important to potency. Different to the potent *S* enantiomer **13**, the *R* enantiomer **14** tested to have no antagonist activity for either orexin receptor.

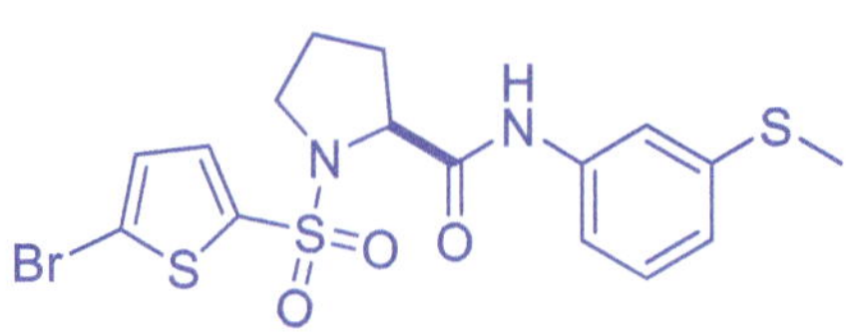

12

IC_{50}(OX1): 20 nM
IC_{50}(OX2): 10 nM

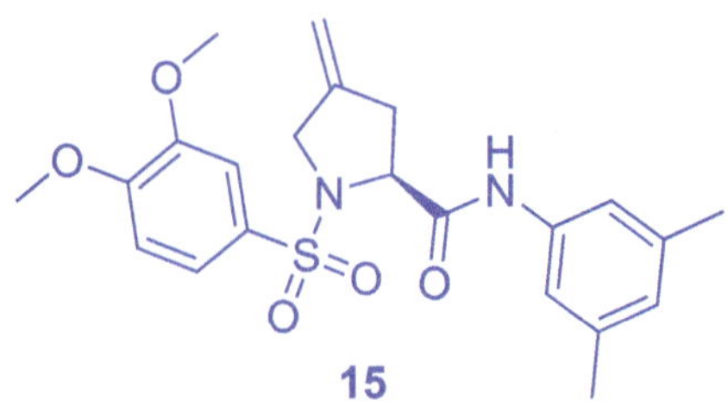

15

IC_{50}(OX1): 1,410 nM
IC_{50}(OX2): 79 nM

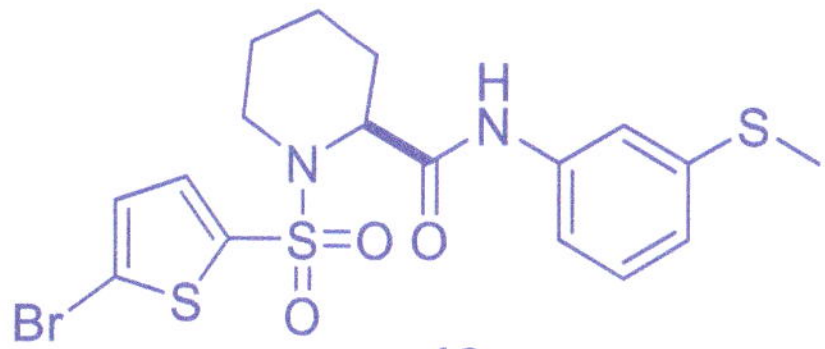

16

IC_{50}(OX1): 351 nM

IC_{50}(OX2): 1,170 nM

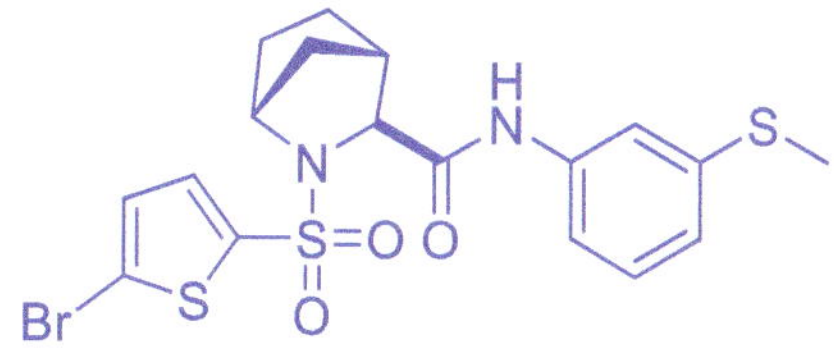

17

IC_{50}(OX1): 23 nM

IC_{50}(OX2): 30 nM

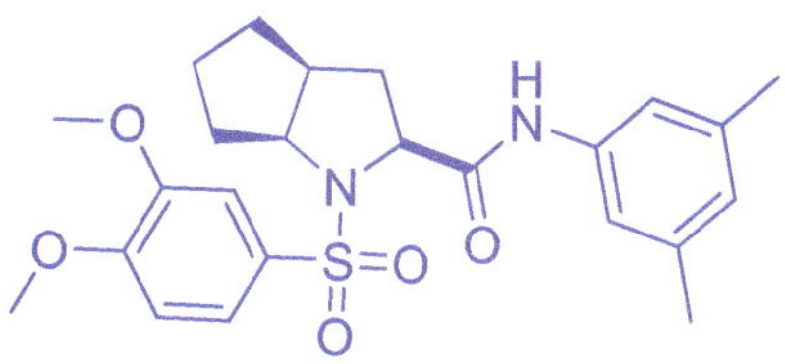

18

IC_{50}(OX1): 184 nM

IC_{50}(OX2): 78 nM

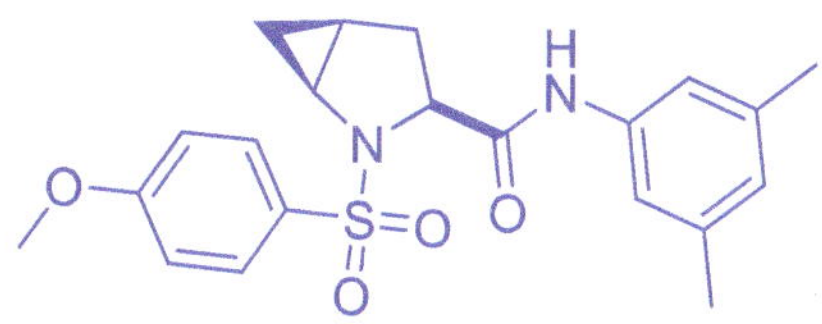

19

IC_{50}(OX1): 791 nM

IC_{50}(OX2): 245 nM

Figure 4. Selected OX1R/OX2R agonists via scaffold hopping

In addition to changing both aryl groups, scaffold hopping has also been conducted and some examples were collected in Figure 4. SAR is very narrow in the scaffold optimization. A simple addition of 3-methylene gave compound **15**, which retained some potency, but inferior to the parent compound **12**. Further investigation of this type of compounds showed very low brain concentration. Another strategy was to replace pyrrolidine with piperidine to give compound **16**. Unfortunately, although only added one carbon, **16** substantially lost potency. Interestingly, analog **17** with a bridged bicyclic template containing both piperidine and pyrrolidine regained potency. Further exploration into bicyclic templates showed the bicyclic [3.3.0] as in **18** and [3.1.0] ring as in **19** did not have advantages over the simple prolidine core, but added a lot of synthetic complexities for the scaffold synthesis.

After all these seminal studies, **13** was chosen as a tool compound. It was then fully profiled. In addition to exhibiting good potency on both GPCR receptors, it was also proved to be a non Pgp substrate (efflux ratio = 1.1) based on a human multidrug resistant protein transporter assay (MDR-1). It also showed human plasma protein binding (hPPB) as well as rat plasma protein binding (rPPB) of 99.3%, and metabolic stability in human liver microsomes (HLM) of 140 μL/min//mg. Further more, the compound showed no significant cytochrome P450 inhibition. Unfortunately, its PK data showed short half lives of only 1.9 h in rat and 1.7 h in dog respectively. In general, although **13** showed certain limitation, its overall profile still good and **13** is a starting point for further optimization.

Further medicinal chemistry practice started from replacing the aniline amide moiety in **13** with more variations. Both non-aromatic amides and heteroaromatic were screened. Non aromatic amides studied examplified with **20** totally lost orexin receptor antagonistic activity. Delightly, heteroaromatic replacements of the aniline with oxadiazole **21** maintained potency. Further exploration of fused ring heterocycles showed benzimidazole group as in **22** also tolerated. Di-methyl substituted benzimidazole groups as in **23** and **24** improved OX1 and OX2 potencies by more than 10–20 times in comparing with **22**. Interestingly, replacement of both methyl groups in **24** with CF_3 to give **25** and it completely lost potency. It was later discovered that halogen and methoxy substitutions were also tolerated. Both substitution groups and their patterns have significant influence on the orexin receptor antagonist potency.[29]

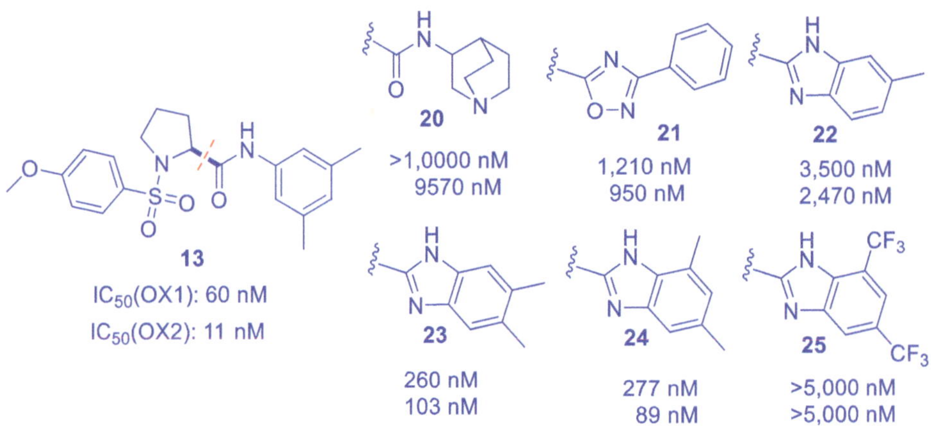

After identifying a pool of substituted benzimidazoles and 1,2,4-oxadiazoles, the next step was to replace the aryl sulfonamide moiety. In general, smaller molecular weight compounds have better permeability for CNS candidates. When sulfonamides were changed to much smaller amides, the bis-aryl-amide moieties attached to the pyrrolidine core showed similar potency with combination with either 1,2,4-oxadiazole (**26**) or benzimidazole (**27**). Many bi-aryl moieties were also tolerated, with triazole-substituted toluene the best. The piperidine core (**28**) is still inferior to pyrrolidine (**27**). The [2.2.1]-fused core (**29**) showed high potency, but same as compound **17** discussed before, it was confirmed to be a Pgp substrate in human MDR-1 assay where the efflux ratio was 53 and $Papp_{AB}$ was 1.1×10^{-6} cm/s. With all this data, pyrrolidine core was chosen for further optimization.

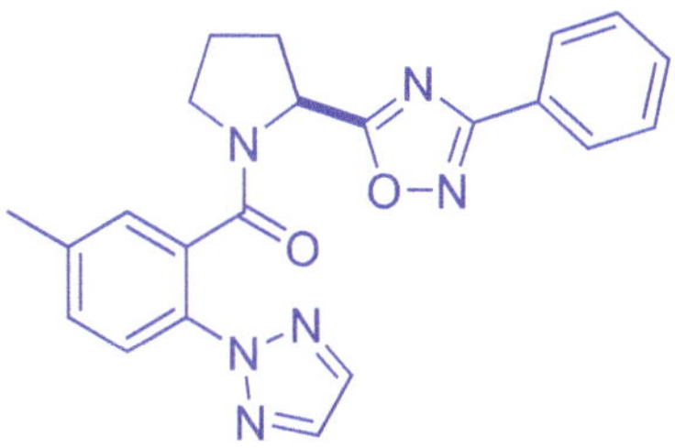

26

$IC_{50}(OX_1)$: 486 nM
$IC_{50}(OX2)$: 76 nM

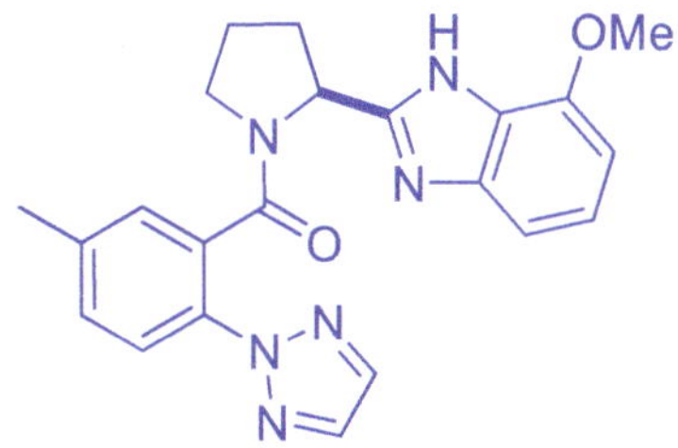

27

$IC_{50}(OX1)$: 241 nM
$IC_{50}(OX2)$: 30 nM

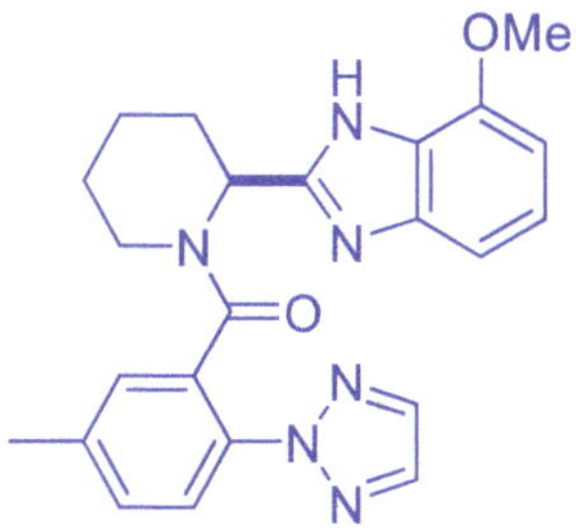

28

$IC_{50}(OX1)$: 296 nM
$IC_{50}(OX2)$: 369 nM

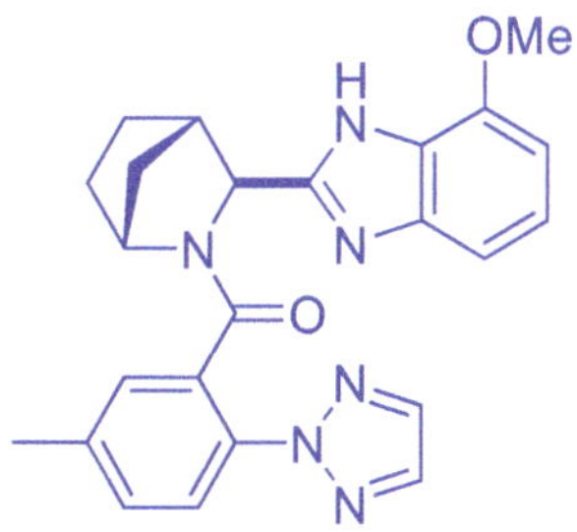

29

$IC_{50}(OX1)$: 13 nM
$IC_{50}(OX2)$: 12 nM

Extensive core modification showed that alkyl or aryl substituents were tolerated, but no benefit observed. Analog with phenyl group (**31**) is more potent in OX1 than the one with methyl group (**30**). Other substituents such as OMe and F did not show obvious change. Chirality of F at C4 on pyrrolidine has very limited influence on potency. It is delighted to see when a chiral methyl group was introduced at C2 on pyrrolidine as in **35**, its OX1R antagonist potency increased by 10-fold, although the OX2R potency decreased by 2-fold. Considering the methyl group can lock the chiral center to prevent racemization, this core with a quaternary carbon was chosen for further modification.

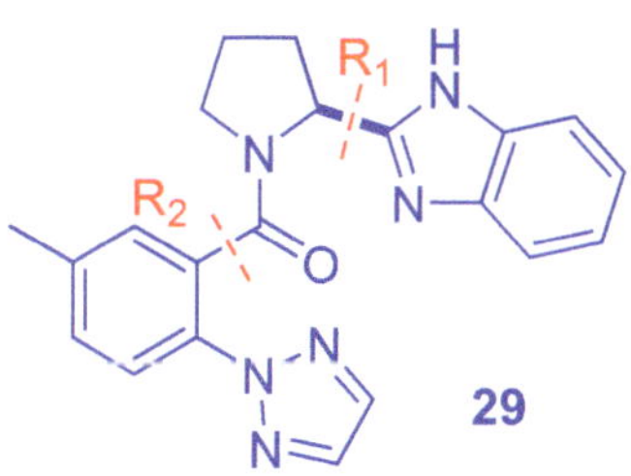

29

$IC_{50}(OX1)$: 2,020 nM
$IC_{50}(OX2)$: 397 nM

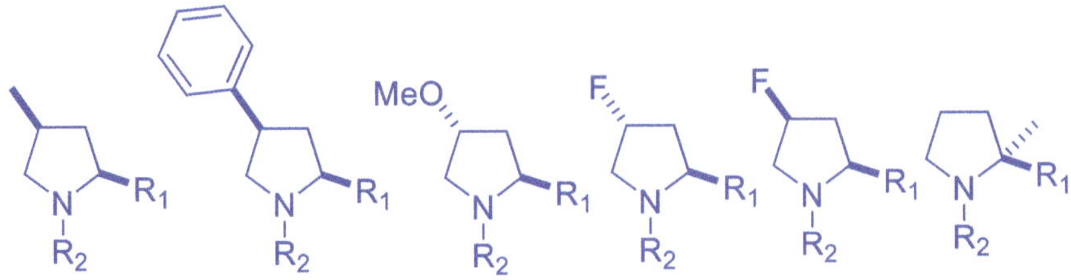

30	31	32	33	34	35
2,860	1,540	1,610	1,980	5,430	280
237	619	329	1,080	1,840	1,010

With the optimal di-substituted pyrrolidine core as in **35** chosen, the team then systematically screened substituents on both benzimidazole and bi-aryl moiety for the optimal mix-match. Table 1 collected some examples from their 2-vector library. Compounds **36**, **40**, **43,** and **47** containing the 4-methyl-5-chlorobenzimidazole are clearly the most potent group, and all of them were further profiled with PK/PD study. Compounds **37**, **44,** and **48** showed preferential activities toward only OX1R. For compounds featuring di-methoxy substituted benzimidazole, they are Pgp substrates, and were also abandoned. The analogs featured OCF_3 substituted benzimidazole **39**, **42**, **46**, and **50** were mostly Pgp substrates, and were all discarded. Please read the original publication to see detailed discussion of other similar analogs.[12] After PK/PD study, compound **40** showed excellent properties and therefore was selected for preclinical studies. It was finally developed as daridorexant (**1**).

Table 1. Final 2-vector library with pyrrolidine core

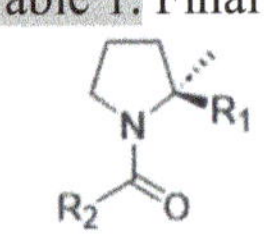	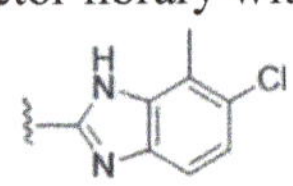	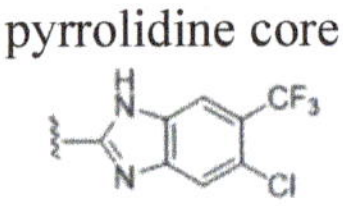	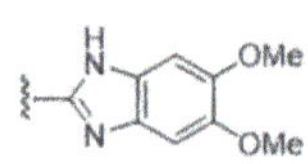	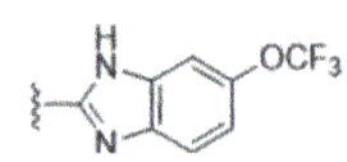
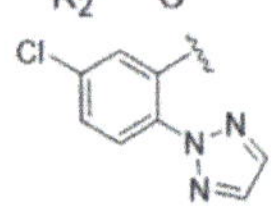	3 8 **36**	2.3 21 **37**	10 23 **38**	99 119 **39**
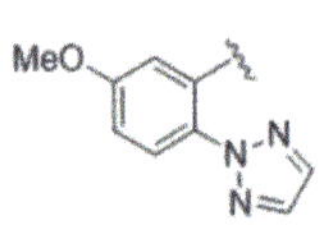	1.4 3.2 **40**	N.D.	25 26 **41**	46 199 **42**
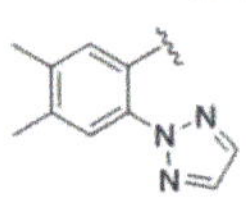	1.9 3.1 **43**	1.5 4.5 **44**	15 25 **45**	5 16.5 **46**
	10 13 **47**	8.9 36 **48**	414 1200 **49**	197 1870 **50**

4. Pharmacokinetics and Drug Metabolism

The pharmacokinetics of daridorexant (**1**) are similar following a single dose and multiple doses, with no clinically relevant accumulation. The plasma exposure of daridorexant (**1**) is dose-proportional between the therapeutic doses of 25–50 mg. The data of pharmacokinetics of daridorexant (**1**) are collected in Table 2. It has a T_{max} of 1–2 h at the therapeutic dose range. Its absolute bioavailability is 62%. The max of daridorexant was delayed by ≈2 h and the peak plasma concentration decreased by 24% after a high-fat, high-calorie meal in healthy subjects, but AUC was not affected.[20] The volume of distribution of daridorexant (**1**) is 31 L. Its plasma protein-bound is 99.7% with a blood/plasma ratio of 0.64.[30]

Table 2. Pharmacokinetic features of daridorexant (**1**)

Pharmacokinetic feature	Daridorexant (**1**)
Time to reach peak plasma concentration (T_{max})	1–2 h
Bioavailability	62%
Volume of distribution (V_d)	31 L
Plasma protein binding	99.7%
Half-life ($t_{1/2}$)	8 h
Metabolism	Extensively by CYP3A4 (89%)
Elimination	57% feces, 28% urine, rest in bile

Source: Adapted from [1]

It is worth noting that daridorexant (**1**) is extensively metabolized, primarily by CYP3A4 (89%), and other CYP enzymes individually contribute to <3% of the metabolic clearance of daridorexant (**1**). It is mostly excreted via feces (57%) and urine (28%) as metabolites.[31] There are a total of 22 daridorexant metabolites detected and major metabolites are M3, M4, and M5. Their chemical structures based on mass spectrometric grounds are depicted in Figure 5. These chemical references of metabolites were also synthesized and exhibited identical retention times and fragmentation patterns.[31] Metabolite M3 was the result of aliphatic hydroxylation of the methyl benzimidazole moiety, which further oxidized to M11. M4 was formed by O-demethylation on the phenol. M5 was formed in two stages, firstly, hydroxylation at the α-carbon of the pyrrolidine ring, which was further underwent ring-opening to give an amino aldehyde and followed by cyclized with the benzimidazole moiety to yield a piperidinol.[31]

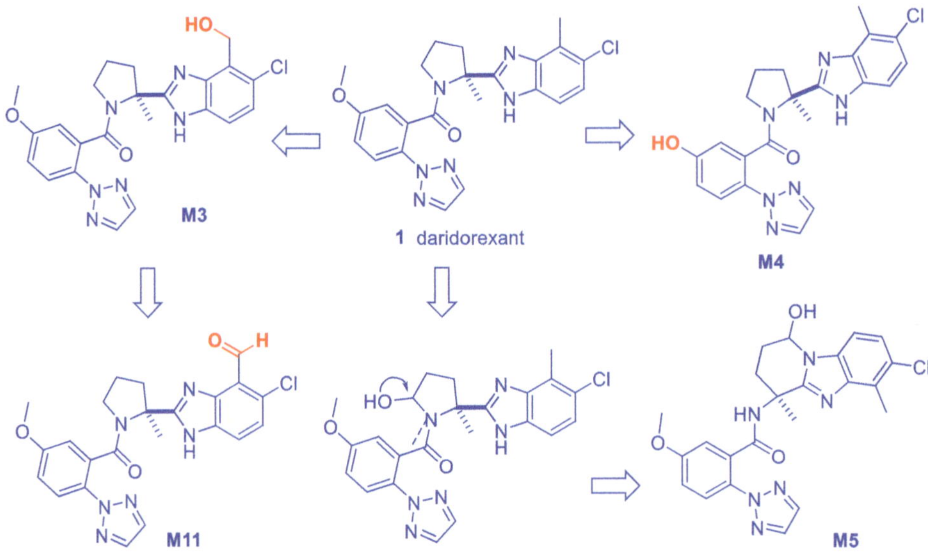

Figure 5. Major metabolites of daridorexant (**1**). Source: Adapted from [1]

Due to high metabolism by CYP3A4, DDI needs to be evaluated. Coadministration of 25 mg daridorexant with the mild CYP3A4 inhibitor diltiazem increased daridorexant's AUC by 240%, and coadministration with the strong CYP3A4 inhibitor itraconazole was expected to increase AUC by more than 400%, according to physiologically based pharmacokinetic modeling.

Coadministration with the moderate CYP3A4 inducer efavirenz, on the other hand, reduced daridorexant AUC by 35%, and coadministration with the strong CYP3A4 inducer rifampin is expected to reduce AUC by more than 50%. On this basis, the maximum recommended dose of daridorexant is 25 mg when used in conjunction with a moderate CYP3A4 inhibitor but use in conjunction with a powerful CYP3A4 inhibitor or a modest or strong CYP3A4 inducer is not advised.[32]

Co-administration with alcohol increased the daridorexant T_{max} and was related to the additive effects on motor coordination performance. As a result, patients should avoid drinking alcohol while taking daridorexant.[30] When given with other CNS depressants, it should be used with caution, and dosage adjustments for either or both drugs should be considered.[32, 33]

5. Efficacy and Safety

The evaluations of daridorexant were conducted in multiple studies. Muehlan et al. conducted a random, double-blind, placebo-controlled study of a single oral dose on 40 healthy males. They were divided into groups to receive a dose of 5, 25, 100, 100, or 200 mg of daridorexant (**1**) in the morning under fasting conditions. The effects on the group with 5 mg dose were hardly detected. But for all other groups with higher dose, the subjects were observed to have reduced vigilance, attention, vasomotor symptoms, coordination, and postural stability as indicated by a decreased saccadic peak velocity (SPV). In addition, adaptive tracking performance and increased body sway were also observed. All these effects were observed 1 h after administration, with maximum effects occurring at 1.5–2 h for different doses. The effects of daridorexant (**1**) at 25 mg of dose returned to baseline within 3–6 h, and 50 mg dose took 6–8 h to return to baseline. For groups with 100 and 200 mg doses, it took 8–10 h to return to baseline.[34]

It is also important to evaluate dose-response of daridorexant (**1**) on waking after sleep onset. One study was conducted by Zammit et al. on a group of 58 participants diagnosed with insomnia. They were divided into five groups to receive 5, 10, 25, 50 mg of daridorexant or a placebo. Every five treatment periods consisted of two treatment nights, followed by a 5- to 12-day period of washout. The main efficiency endpoints were the absolute change from baseline in wakefulness after sleep onset (WASO) (primary endpoint) and latency to persistent non-REM sleep (LPS) (secondary endpoint) for days 1 and 2 in each period. Both WASO and LPS were dose-dependently reduced from baseline to days 1 and 2 after administration of daridorexant (**1**). The most statistically meaningful reduction was for doses higher or equal to 10 mg compared to placebo, but not detectible for 5 mg.[35]

Mignot et al. reported a phase 3 trial study of daridorexant (**1**) in *Lancet* in 2022. The group conducted two multicenter, randomized, double-blind, placebo-controlled phase 3 trials on 930 participants in 156 cities in 17 countries.[36] The subjects were randomly assigned to receive for 3 months 50 or 25 mg daridorexant or placebo (study 1), or 25 or 10 mg or placebo (study 2) every evening. The results showed both WASO and LPS were significantly reduced in a group who received daridorexant at the 25 and 50 mg doses compared to placebo at months 1 and 3. What is important is that there was no significant difference between the placebo group and the 10 mg daridorexant group. This finding corresponded with the research conducted by Muehlan et al.

Based on all these clinical studies, daridorexant is available in the form of 25 and 50 mg oral tablets. It is provided as daridorexant hydrochloride, with each tablet containing 27 or 54 mg of this substance (equivalent to 25 or 50 mg daridorexant).[37]

The adverse events of patients treated with daridorexant were revealed in one trial (NCT03545191). Among those participants, these were nasopharyngitis (6% among 310 study participants who received daridorexant 50 mg, 7% among 310 trial participants treated with daridorexant 25 mg and 6% with placebo received by 310 trial participants), headache (6%, 5%, and 4%), drug overdose (3%, 1%, and 2%), fatigue (2%, 2%, and 1%), dizziness (2%, 2%, 2%, and 1%), nausea (2%, 1%, and 1%), and somnolence (2%, 4%, and 2% among 930 total study participants. Adverse events leading to treatment discontinuation occurred in 1% of daridorexant 50 mg recipients, 2% of daridorexant

25 mg recipients, and 3% of placebo recipients. Serious adverse events occurred in 1% of daridorexant 50 mg recipients, 1% of daridorexant 25 mg recipients, and 2% of placebo recipients.[38]

6. Synthesis

6.1. Discovery Synthesis

Daridorexant (**1**) is a relatively simple small-molecule compound. Its chemical name is (*S*)-(2-(5-chloro-4-methyl-1*H*-benzo[*d*]imidazol-2-yl)-2-methylpyrrolidin-1-yl)(5-methoxy-2-(2*H*-1,2,3-triazol-2-yl)phenyl)methanone, with molecular weight of 450.93, or 487.38 g/mol for its hydrochloride salt as the drug approved by the FDA.[39] Daridorexant hydrochloride is a white to light yellowish powder only slightly soluble in water.[40]

The synthesis of daridorexant (**1**) was first reported in 2013.[41] As a simple compound with only one chiral center on proline, its medicinal chemistry synthesis route is straight forward. Although minor modifications were observed in different patents of this series of compounds, the key steps did not change too much. A typical medicinal chemistry synthesis is shown in Scheme 1.

The synthesis starts with commercially available *N*-Boc protected (*S*)-2-methylproline (**51**), which forms an amide bond with 3-methly-4-chloro phenylenediamine (**52**) under the standard conditions with HATU as the coupling reagent in dichloromethane using DIPEA as the base. After stirring the mixture at room temperature for overnight, two regio isomers **53** and **54** were formed in quantitative yield, which were further cyclized under glacial acetic acid at 100 °C in 1 h to give identical benzimidazole derivative **55** in 49% yield. Deprotection of Boc can be achieved by HCl in dioxane (1.6 M) at room temperature to give the HCl-salt of intermediate **56**.

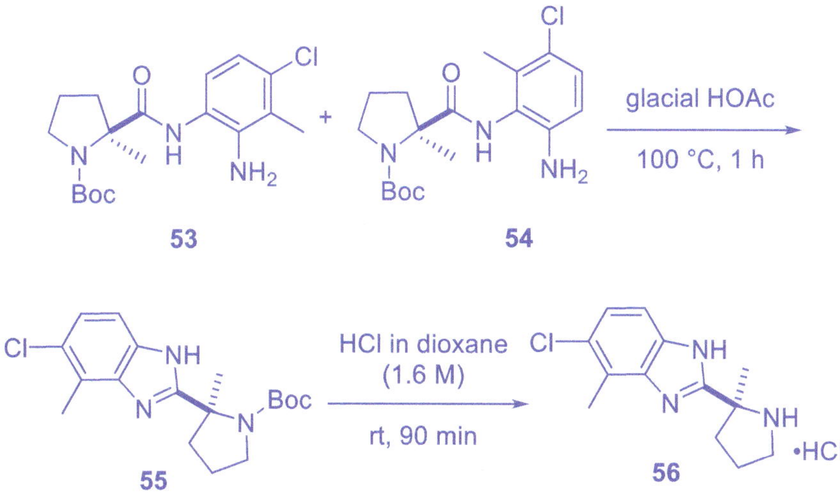

Scheme 1. Medicinal chemistry synthesis route of daridorexant (**1**).

In parallel with the synthesis of the key intermediate **56**, another intermediate 2-triazolo-5-methoxy-benzoic acid (**58**) was prepared by reacting the commercially available 2-iodo-5-methoxy-benzoic acid (**57**) in a copper catalyzed coupling with 1,2,3-triazole in the presence of cesium carbonate in DMF at 80 °C for 3 h to give **58** in 76%.

With both intermediates **56** and **58** ready, the final step was to couple 2-triazolo-5-methoxy-benzoic acid (**58**) with substituted chiral proline **56** by HATU in dichloromethane with DIPEA as the base, yielding **1** in 63%. After purification, pure **1** was converted to its HCl salt by adding 1.0 equiv of 1.6 M HCl in dioxane solution to a slurry of **1** in dioxane, followed by evaporation of the solvent and drying of the final material under high vacuum conditions to give the HCl salt of **1** as a white solid.

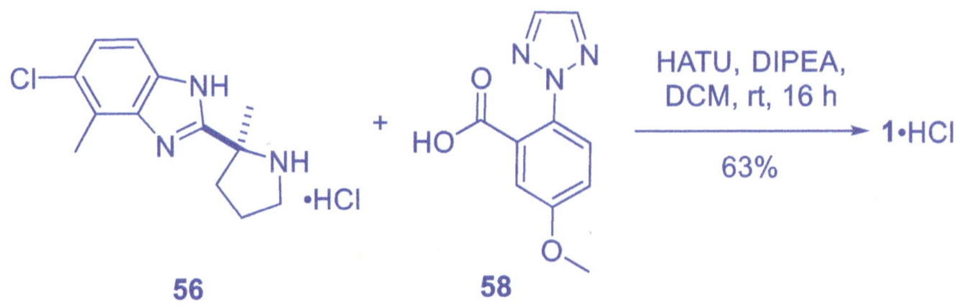

The efficient synthesis route furnished the final product **1** in only five steps. The chemistry is robust enough to prepare from multiple grams to kilograms. Based on the first generation of medicinal chemistry route, several routes with minor modifications were published. One example was to use HBTU to replace HATU to achieve a more efficient synthetic route. However, both HATU and HBTU are expensive coupling reagents. They are not suitable for big scale synthesis. A more economic route has also been reported by the team to avoid expensive coupling reagents or column chromatography separation.

6.2. Alternative Synthesis (Multi-kilogram Scale)

In drug discovery, medicinal chemistry routes are generally efficient, but are also expensive due to expensive reagents, building blocks and purification methods. For multiple-kilogram synthesis, process chemists always avoid using protecting groups, expensive amide coupling reagents, and column chromatography for purification. The new route we will discuss is a good example to save the aforementioned cost and finished most of the synthesis in two efficient one-pot reactions.[12] The first one-pot reaction started from 5-methoxy-2-(2*H*-1,2,3-triazol-2-yl)-benzoic acid (**58**) which also used in medicinal chemistry synthesis. To avoid using HATU as coupling reagent, carboxylic acid was converted to more reactive acid chloride **60**. Due to the short synthesis route, we also share detailed procedure of this synthesis. The reaction was run in a 30 L Büchi reactor, to which was added sequentially **58** (1000 g, 4.56 mol), toluene (5 L), and DMF (18 mL, 0.05 equiv). Thionyl chloride (600 g, 5.04 mol) was then added at 40–45 °C, and the reaction mixture was stirred at 45 °C for 35 min, then was cooled to 0 °C. With good stirring, the acid chloride (**59**) solution was added into a mixture of methyl (*S*)-2-methylpyrrolidine-2-carboxylate hydrochloride (**60**) (861 g, 4.79 mol), toluene (5 L), water (10 L), and K_2CO_3 (3151 g, 22.8 mol) at 5 °C. The resulting reaction mixture was heated to 56 °C and the aqueous layer was separated. The organic layer was washed with 1N HCl (2 L) and water (2 L). The solvent was swapped from toluene to MeOH via distillation. The practice gave a solution of the intermediate ester (**61**) in c. 13 L MeOH. Finally, a 16% aqueous NaOH solution (3.75 L) was added at 64 °C to hydrolyze the ester. After acidifying the reaction mixture, a suspension was obtained, while was filtered, and the product cake was dried under reduced pressure to yield **62** (1431 g, 95%) as a white solid.

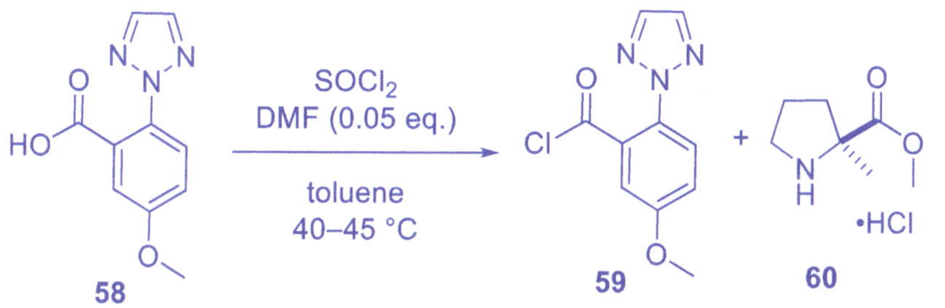

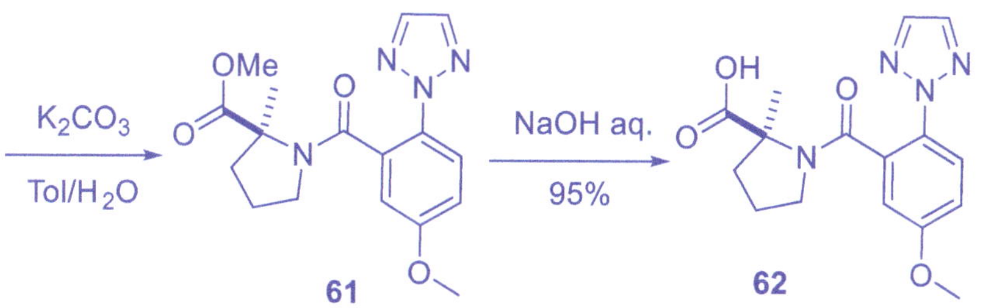

With kilograms of carboxylic acid **62** in hand, the next step was to prepare benzimidazole via condensation with diamine **52**. To avoid using expensive coupling reagent such as HATU, carboxylic acid was directly converted into more reactive acid chloride **63** with thionyl chloride in a mixed solvent of toluene (5 L) and 11 g of DMF. The mixture was stirred for 30 min at 47 °C to give amide **63**. Meanwhile, a suspension of 3-methyl-4-chloro-phenylendiamine HCl (**52**, 643 g, 3.32 mol) in THF (12 L) was neutralized by stirring with triethylamine (1530 g, 15.1 mol) at 60 °C for 1 h, then cooled to 10 °C. The prepared acid chloride **(63)** solution was then added at 5–15 °C over a period of 15 min. The reaction mixture was extracted with water (2 L) at 23 °C. The solvent was swapped from THF/toluene to *i*PrOH via distillation. The product **(64)** recrystallized in *i*-PrOH to yield **64** (1,152 g, 81%) as a white solid.

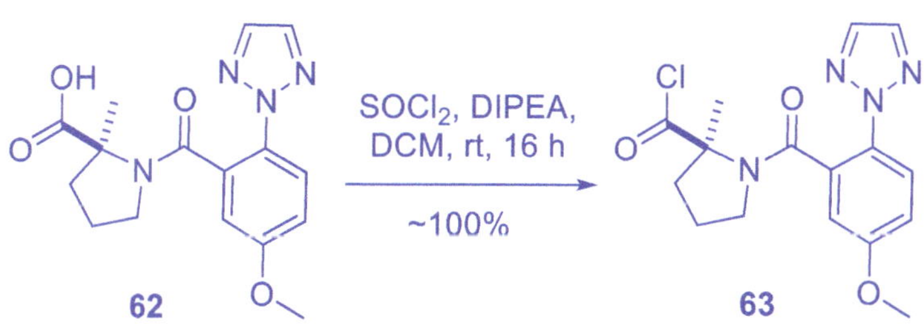

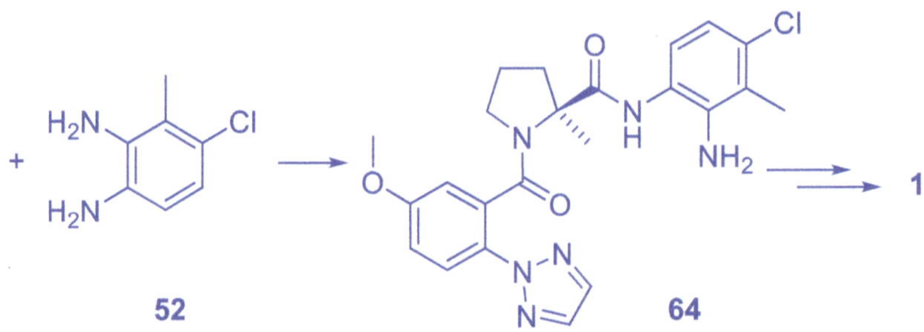

The amide **64** (1000 g, 2.13 mmol) was further condensed to give the final product **1** by dissolving first in EtOH (5 L), followed by addition of 32% aq. HCl (500 mL) at RT to be converted to HCl salt. After filtration, the filtrate was heated at 75 °C for 4 h. The resulting suspension was cooled to 0 °C, then filtered to give the final product **1** as a white solid (922 g, 89%). From starting material, the product was prepared as a white crystal with 69% total yield without using expensive protecting groups, amide coupling reagents or column chromatography. Only cheap and environmentally friendly solvents such as toluene, THF, i-PrOH and EtOH were used. The robust and convenient route provides a good solution for further scaling up of the drug molecule.[12, 41]

7. Summary

In summary, daridorexant, a new selective DORA agent, acts as a competitive orthosteric antagonist for GPCRs OX1R and OX2R, exhibits good brain penetration, promotes sleep with preserved sleep-architecture, and leaves no next-morning side effects. It is a better substitution of traditional pharmacotherapies of hypnotic agents such as benzodiazepines. To medicinal chemists' interests, PK/PD study, metabolism, SAR study and synthesis of the drug were all discussed. Other orexin drug candidates in the clinical trials, both antagonists and agonists, have been briefly overviewed.

References

1. See: https://www.nhlbi.nih.gov/health-topics/education-and-awareness/sleep-health
2. Scammel, T. E.; Winrow, C. J. *Annu. Rev. Pharmacol. Toxicol.* **2011**, *51*, 243–266.
3. Daley, M.; Morin, C. M.; LeBlanc, M.; Gregoire, J. P.; Savard, J.; Baillargeon, L. *Sleep Med.* **2009**, *10*, 427–438.

4. Colten, H. R.; Altevogt, B. M. *Sleep Disorders and Sleep Deprivation: An Unmet Public Health Problem*, National Academies Press (US): Washington, DC **2006**.

5. Sullivan, S. S.; Guilleminault, C. *Expert Opin. Emerging Drugs* **2009**, *14*, 411–422.

6. Renger, J. J. *Curr. Top. Med. Chem*. **2008**, *8*, 937–953.

7. Sateia, M. J.; Buysse, D. J.; Krystal, A. D.; Neubauer, D. N.; Heald, J. L. *J. Clin. Sleep Med.* **2017**, *13* (2), 307–349.

8. http://www.fda.gov/drugs/drugsafety/ucm352085.htm (accessed 19 Jan 2024).

9. Schutte-Rodin, S.; Broch, L.; Buysse, D.; Corsey, C.; Sateia, M. *J. Clin. Sleep Med.* **2008**, *4* (5), 487–504.

10. Kato, K; Hirai, K; Nishiyama, K; Uchikawa, O; Fukatsu, K; Ohkawa, S; Kawamata, Y; Hinuma, S; Miyamoto, M. *Neuropharmacology* **2005**, *48,* 301–310.

11. Mayer, G.; Wang-Weigand, S.; Roth-Schechter, B.; Lehmann, R.; Staner, C.; Partinen, M. *Sleep* **2009**, *32*, 351–360.

12. Boss, C.; Gatfield, J.; Brotschi, C.; Heidmann, B.; Sifferlen, T.; von Raumer, M.; Schmidt, G.; Williams, J. T.; Treiber, A.; Roch, C. *ChemMedChem* **2020**, *15*, 2286.

13. Sakurai, T.; Amemiya, A.; Ishii, M.; Matsuzaki, I.; Chemelli, R.; Tanaka, H.; Williams, S. C.; Richardson, J. A.; Kozlowski, G. P.; Wilson, S.; et al. *Cell* **1998**, *92*, 573–585.

14. De Lecea, L.; Kilduff, T. S.; Peyron, C.; Gao, X.-B.; Foye, P. E.; Danielson, P. E.; Fukuhara, C.; Battenberg, E. L. F.; Gautvik, V. T.; Bartlett, F. S., II; Frankel, W. N.; Van Den Pol, A. N.; Bloom, F. E.; Gautvik, K. M.; Sutcliffe, J. G. *PNAS* **1998**, *95*, 322–327.

15. Jacobson, L. H.; Callander, G. E.; Hoyer, D. *Expert Rev. Clin. Pharmacol.*, **2014**, *7* (6), 711–730.

16. Yoshida Y.; Naoe, Y.; Terauchi, T.; Ozaki, F.; Doko, T.; Takemura, A.; Tanaka, T.; Sorimachi, K.; Beuckmann, C. T.; Suzuki, M.; Ueno, T.; Ozaki, S.; Yonaga, M. *J. Med. Chem.* **2015**, *58*, 4648–4664.

17. Ziemichód, W.; Grabowska, K.; Kurowska A.; Biała G. *Molecules* **2022**, *27*, 6041.

18. Brisbare-Roch, C.; Dingemanse, J.; Koberstein, R.; Hoever, P.; Aissaoui, H.; Flores, S.; Mueller, C.; Nayler, O.; van Gerven, J.; de Haas, S. L.; Hess, P.; Qiu, C.; Buchmann, S.; Scherz, M.; Weller, T.; Fischli, W.; Clozel, M.; Jenck, F. *Nat. Med.* **2007**, *13*, 150–155.

19. Letavic, M. A.; Bonaventure, P.; Carruthers, N. I.; Dugovic, C.; Koudriakova, T.; Lord, B.; Lovenberg, T. W.; Ly, K. S.; Mani, N. S.;

Nepomuceno, D.; Pippel, D. J.; Rizzolio, M.; Shelton, J. E.; Shah, C. R.; Shireman, B. T.; Young, L. K.; Yun, S. *J. Med. Chem.* **2015**, *58*, 5620–5636.

20. Sutton, E. L. *Drug Des. Dev. Ther.* **2015**, 9, 6035–6042.
21. Evans, R.; Kimura, H.; Nakashima, M.; Ishikawa, T.; Yukitake, H.; Suzuki, M.; Hazel, J.; Faessel, H.; Wu, J.; Hang, Y.; Alexander, R.; Rosen, L.; Hartman, D. S.; Ratti, E. *J. Sleep Res.* **2023**, *32* (5), e13878.
22. Dauvilliers, Y.; Mignot, E.; del Río Villegas, R.; Du, Y.; Hanson, E.; Inoue, Y.; Kadali, H.; Koundourakis, E.; Meyer, S.; Rogers, R.; et al. *N. Engl. J. Med.* **2023**, *389*, 309–321.
23. Study of MK-6552 in Participants with Narcolepsy Type 1 NCT06179407. https://clinicaltrials.gov/study/NCT06179407
24. Merck patents: Bogen, S. L.; Clausen, D. J.; Guiadeen, D. G.; Rudd, M. T.; Yang, D. Bicycloheptane Pyrrolidine Orexin Receptor Agonists, Int. Patent Appl. WO2022/040070 A1, Feb 24, 2022.
25. a) Sabnis, R. W. *ACS Med. Chem. Lett.* **2020**, *11* (11), 2085–2086. b) Sabnis, R. B. *ACS Med. Chem. Lett.* **2022**, *13* (7), 1010–1011.
26. Treiber, A.; de Kanter R.; Roch, C.; Gatfield, J.; Boss, C.; von Raumer, M.; Schindelholz, B.; Muehlan, C.; van Gerven, J.; Jenck, F. *J. Pharmacol. Exp. Ther.* **2017**, *362* (3), 489–503.
27. Markham, A. *Drugs* **2022,** *82* (05), 601–607.
28. Grandjean, C.; Kiry, M.; Vaillant, C.; Nayler, O.; Gatfield, J.. *Sleep (Basel)* **2021**, *44* (Supplement_2), A25.
29. Boss, C.; Roch-Brisbare, C.; Steiner, M. A.; Treiber, A.; Dietrich, H.; Jenck, F.; von Raumer, M.; Sifferlen, T.; Brotschi, C.; Heidmann, B.; Williams, J. T.; Aissaoui, H.; Siegrist, R.; Gatfield, J. *ChemMedChem* **2014**, *9*, 2486–2496.
30. a) Muehlan, C.; Heuberger, J.; Juif, P. E. et al. *Clin. Pharmacol. Ther.* **2018**, *104* (5), 1022–9. b) Nie, T.; Blair, H. A. *CNS Drugs* **2023**, *37*, 267–274. c) Muehlan, C.; Brooks, S.; Zuiker, R. *Eur. Neuropsychopharmacol.* **2019**, *29* (7), 847–57.
31. a) Muehlan, C.; Fischer, H.; Zimmer, D.; Aissaoui, H.; Grimont, J.; Boss, C.; Croft, M.; van Gerven, J.; Krähenbühl, S.; Dingemanse, J. *Curr. Drug Metab.* **2019**, 20, 254–265. b) Treiber A.; Delahaye, S.; Weigel, A.; Aeänismaa, P.; Gatfield, J.; Seeland, S. *Xenobiotica* **2023**, *53* (3), 173–183.
32. Muehlan, C.; Boehler, M.; Brooks, S.; Zuiker, R.; van Gerven, J.; Dingemanse, J. *J. Psychopharmacol.* **2020**, *34* (3), 326–35.
33. Boof, M. L.; Alatrach, A.; Ufer, M. *Eur. J. Clin. Pharmacol.* **2019**, *75* (2), 195–205.

34. Muehlan, C.; Heuberger, J.; Juif, P.E.; Croft, M.; van Gerven, J.; Dingemanse, J. *Clin. Pharmacol. Ther.* **2018**, *104*, 1022–1029.

35. Zammit, G.; Dauvilliers, Y.; Pain, S.; Sebök Kinter, D.; Mansour, Y.; Kunz, D. *Neurology* **2020**, *94*, e2222–e2232.

36. Mignot, E.; Mayleben, D.; Fietze, I.; Leger, D.; Zammit, G.; Bassetti, C.L.A.; Pain, S.; Kinter, D.S.; Roth, T. *Lancet Neurol.* **2022**, *21*, 125–139.

37. QUVIVIQ™ (Daridorexant). https://www.idorsia.us/documents/us/label/Quviviq_PI.pdf (accessed on Jan 21, 2024).

38. https://clinicaltrials.gov/study/NCT03545191

39. Park, J.; Render, K. P.; Cates, D. W. *Ann. Pharmacother.* **2023**, *57* (9), 1076–1087.

40. https://pubchem.ncbi.nlm.nih.gov/compound/91801202 (Retrieved Jan 17, 2024)

41. Boss, C.; Brotschi, S.; Gde, M.; Heidmann, B.; Sifferlen, T.; Williams, J. Use of Benzimidazole-Proline Derivatives, Int. Patent Appl. WO2015/083094 A1, June 11, 2015.

Section IV. ANTI-INFLAMMATORY DRUGS

15

Deucravacitinib (Sotyktu): A First-in-Class Deuterated TYK2 Inhibitor for the Treatment of Plaque Psoriasis

Daljit Matharu

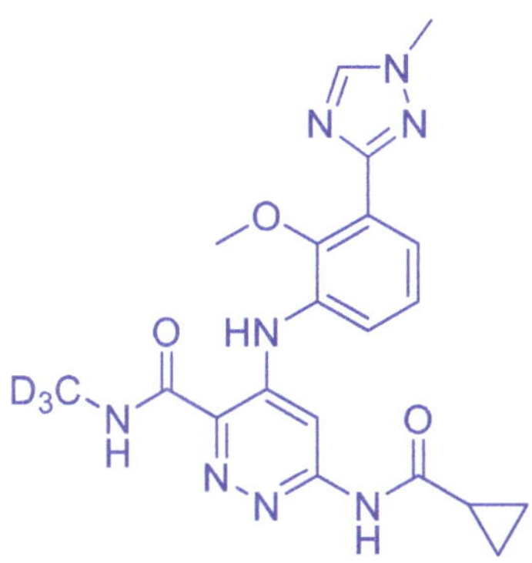

Deucravacitinib (**1**, Sotyktu)
Bristol Myers Squibb, 2022
Deuterated TYK2 inhibitor

- FDA approved in 2022 for patients with plaque psoriasis
- Deuteromethylamide supresses N-demethylation via deuterium kinetic isotope effect
- High selectivity for TYK2 over JAK1, JAK2, and JAK3

1. Background

In recent years, immune-mediated and autoimmune chronic inflammatory diseases have emerged as a leading cause of morbidity, with conditions such as psoriasis, psoriatic arthritis, axial and peripheral spondyloarthritis, as well as inflammatory bowel diseases

Chemistry and Pharmacology of Drug Discovery, First Edition. Edited by Jie Jack Li.

(IBDs) having an estimated prevalence of 5–7% in the Western world.[1] Research into the pathogenesis of immune-mediated inflammatory diseases has led to significant advances in understanding human autoimmunity, which in turn has enabled improvements in diagnosis and, most importantly, armed researchers with key information needed to develop effective therapies.[1]

Amongst the numerous autoimmune chronic inflammatory diseases, psoriasis is a debilitating skin condition that causes a rash with itchy, scaly patches most commonly occurring on the knees, elbows, trunk, and scalp, of which plaque psoriasis is the most widespread. According to current studies, 125 million people worldwide (2–3% of the total population) have psoriasis.[2] The disorder is caused by an over-reactive immune system that causes skin cells to multiply up to ten times faster than normal, leading to the accumulation of bumpy patches on the skin. Common triggers for psoriasis include emotional stress, infection, skin injury, certain medications, such as beta-blockers, and changes in body temperature due to the weather.[3]

Widely used treatment options include corticosteroid creams such as hydrocortisone, vitamin D3 ointment, retinoid creams, medication to slow skin cell production (anthralin), and moisturizers for dry skin and coal tar. While these treatments provide sub-optimal relief of symptoms, effective treatments that address the underlying causes of inflammatory disease have remained elusive.

The Janus kinases (JAKs) are a family of intracellular tyrosine kinases (JAK1, 2, 3, and TYK2) that bind to distinct cell surface cytokine receptors, playing an essential role in the signaling of numerous cytokines that have been implicated in the pathogenesis of inflammation. As a result, small-molecule inhibitors of the JAK kinases have offered a promising approach to treat a variety of serious inflammatory and autoimmune diseases.[4]

The JAKs contain seven distinct homologous regions consisting of four structural domains (Figure 1).[5] One of their characteristic features is the two structurally related domains JH1 and JH2. JH1 is the active kinase catalytic domain, while JH2 is referred to as the pseudokinase domain that closely resembles a catalytic domain but possesses a series of individual residue and conformational differences that prevents it from being catalytically active.

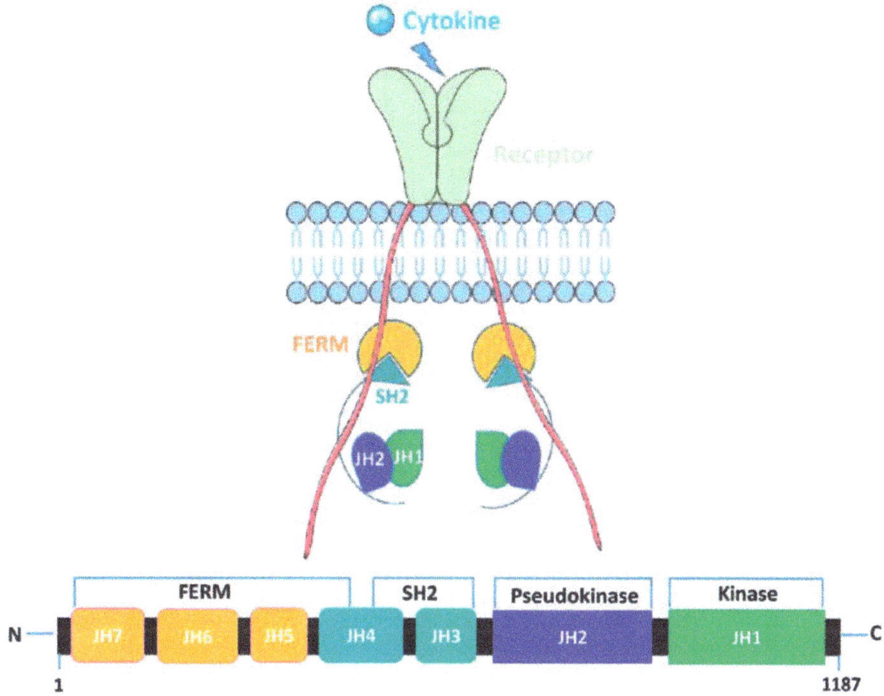

Figure 1. Schematic representation of functional domains of the JAK family of kinases

To date, a number of small-molecule pan-JAK inhibitors or modestly selective JAK inhibitors have been approved by regulatory agencies or are in different stages of clinical development. These first- and second-generation inhibitors are active site-directed inhibitors that bind to the ATP site of the catalytic domain (JH1 domain) of the JAK protein (selected examples shown in Figure 2). Because of the high structural homology of the ATP-pockets across the kinome and especially within the ATP binding site of the JAKs, the development of highly selective JAK inhibitors has presented researchers with a momentous challenge. As a result of cross-inhibition, dose-limiting side effects have been documented, including risk of infection associated with over-immunosuppression, anemia or leukopenia and neutropenia, and other rare yet serious adverse effects that have led the FDA to issue a "black box" warning for some JAK inhibitors.[6] Hence, the narrow therapeutic index of current JAK inhibitors represents an unmet medical need to be addressed to discover safer alternative treatments for cytokine-driven inflammatory conditions.

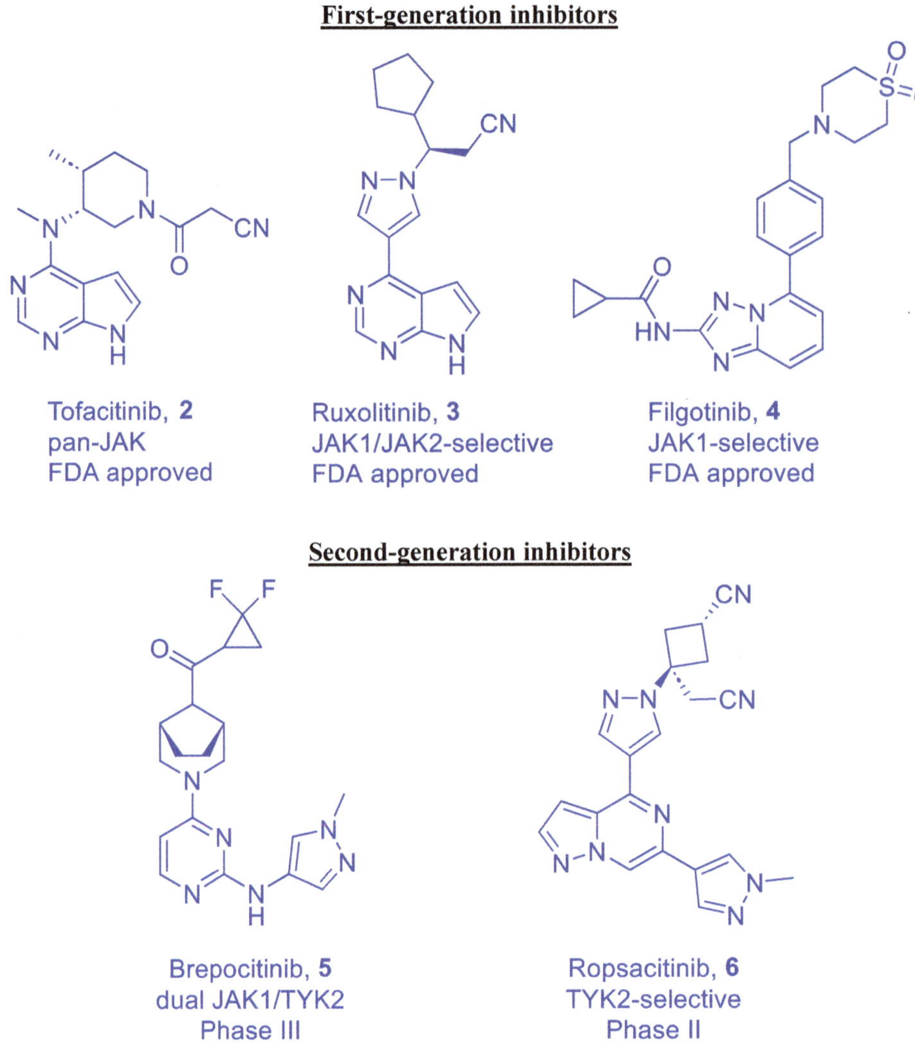

Figure 2. Structures of select examples of first-generation clinically approved JAK inhibitors **2–4** and second-generation experimental JAK family inhibitors **5–6**

1.1. TYK2 Pseudokinase: The Opportunity for Selectivity

The non-receptor tyrosine kinase 2 (TYK2) has garnered considerable interest recently due to the pivotal role it plays in mediating signal transduction pathways downstream of the pro-inflammatory receptors for IL-23, IL-12, and Type 1 interferons (IFNα and IFNβ), all important drivers of autoimmune and inflammatory diseases.[7] TYK2 represents an ideal therapeutic target due to its association with a small number of cytokine receptors relative to the other JAKs. Genetic studies have revealed that TYK2-deficient mice are protected from various models of experimental autoimmunity, including collagen-induced arthritis and experimental autoimmune encephalomyelitis.[8, 9] The TYK-2 dependent pathways have also been validated in treating human disease with antibody therapeutics. The IL-12/IL-23 antibody ustekinumab (Stelara) is currently marketed for the treatment of psoriasis, with clinical development underway for treatment of Crohn's disease.[10, 11] The anti-type 1 interferon receptor antibody anifrolumab has been reported to provide benefit for the treatment of systemic lupus erythematosus (SLE).[12]

Despite the potential of TYK2 inhibition, only modest progress has been reported in identifying potent and highly selective small molecules that target the catalytically active JH1 domain.[13, 14] The pseudokinase (JH2) domains of JAK family kinases have previously been implicated to play an autoinhibitory role in regulating activation of the adjacent catalytic domains.[15, 16] Using a chemogenic approach, the elegant work of researchers at Bristol-Myers Squibb (BMS) led to the identification of molecules that act on TYK2 by preventing the receptor-mediated activation of the TYK2 JH1 domain as a consequence of binding and stabilization of the adjacent catalytically inactive JH2 domain, ultimately blocking downstream signal transduction (Figure 3).[17] It was also found that there is a greater residue differentiation in the TYK2 pseudokinase domain from the other JAK family members pseudokinase domains, offering the greatest opportunity to design inhibitors that would yield both kinase and family selectivity. This breakthrough provided the first example of TYK2 protein function inhibition through an allosteric mechanism and set the stage for capitalizing on this novel approach to optimize potent and selective molecules for the treatment of autoimmune and inflammatory disorders.

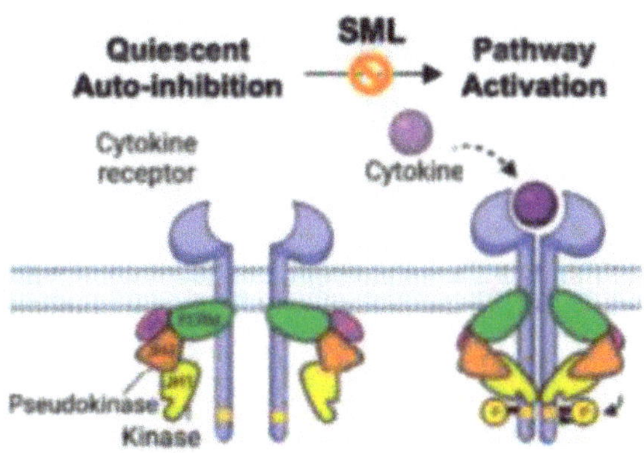

Figure 3. Schematic depicting prevention of receptor-mediated activation. Source: Adapted from [17, 18]

The combined limitations of existing therapies for psoriasis and the novel opportunity seen through TYK2 inhibition, led BMS to undertake a research program that culminated in the successful launch of deucravacitinib, a first-in-class highly selective allosteric TYK2 inhibitor small molecule approved by the FDA in September 2022 for treating moderate-to-severe plaque psoriasis. The launch of the only approved TYK2 inhibitor worldwide and the first innovation in oral treatment for this disease in nearly 10 years was a defining moment. The path to approval will be discussed in the remainder of the chapter.

2. Pharmacology

Deucravacitinib, marketed as Sotyktu, works by stabilizing an auto-inhibitory interaction between the regulatory (JH2) and catalytic (JH1) domains of the enzyme, thereby trapping the kinase in its inactive state and preventing receptor mediated activation and its downstream functions in cells.[19]

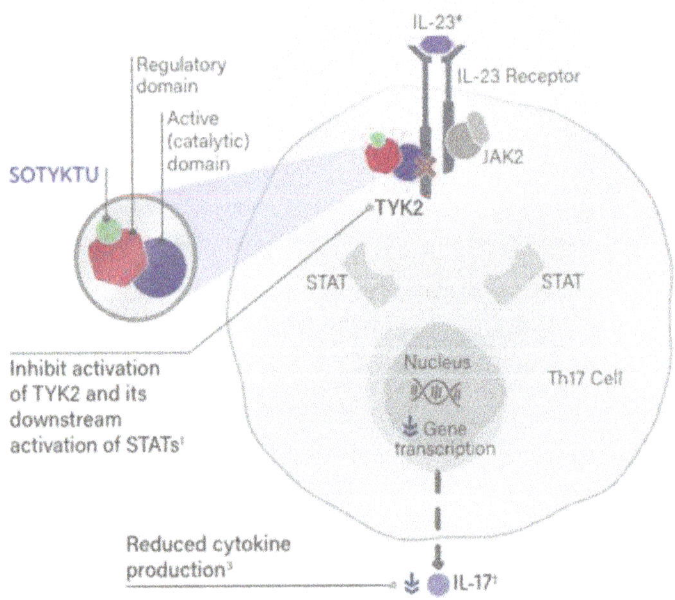

Figure 4. Proposed mechanism of action of deucravacitinib. Source: [20]/with permission of Bristol-Myers Squibb Company

Inhibition of signal transduction downstream of the IL-23 receptor is being actively pursued as an intriguing approach to the treatment of autoimmunity and reports strongly linking IL-23 to the pathogenesis of psoriasis have recently emerged.[19, 21] While the intricate mechanistic details linking TYK2 inhibition to its efficacy are not currently known, a proposed mechanism is outlined in Figure 4.[20, 22] IL-23 is secreted by skin cells and induces production of the proinflammatory mediator IL-17 by T helper 17 (Th17) cells. By blocking this IL-23/Th17 axis through inhibition of TYK2 activation and its downstream activation of STATs, gene transcription is halted, leading to a reduction in IL-17 production and subsequent therapeutic effectiveness for autoimmune and chronic inflammatory diseases.

3. Structure–Activity Relationship (SAR)

3.1. Imidazopyridazine Series

In their quest to develop an orally dosed small-molecule TYK2 inhibitor, BMS focused their efforts on an allosteric approach to target the JH2 domain. Herein, the key aspects of the long and arduous SAR campaign that culminated in the discovery of deucravacitinib will be highlighted.

A phenotypic screen of a large collection of kinase inhibitors using an IL-23/IFNα stimulated reporter assay led to the identification of imidazopyridazine hit compound **7**, which displayed relatively weak affinity for the TYK2 pseudokinase but exhibited excellent selectivity (with only 0.5% of kinases in the 386-kinase panel being inhibited by greater than 67% at 1 μM compound concentration).[23] Modifications at the C6 and C8 positions resulted in compound **8**, that exhibited improved binding affinity and cellular potency, was fully selective over the Janus catalytic domains and showed excellent ligand and ligand lipophilic efficiency (LE = 0.35, LLE = 6.2) (Figure 5).

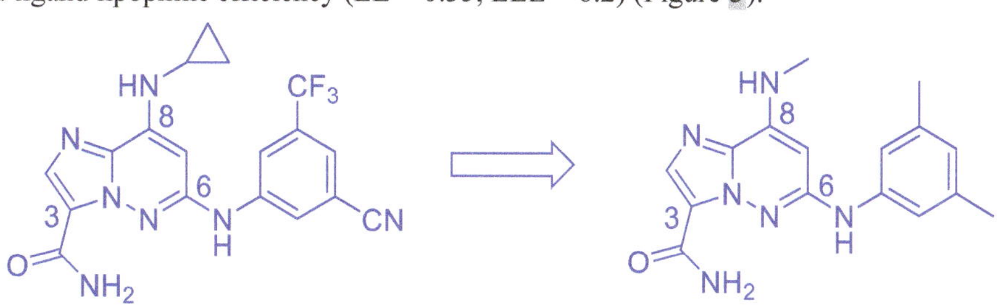

7

TYK2 JH2 IC_{50} = 480 nM
TYK2 JH1 IC_{50} = >50 μM
IL-23/IFNα IC_{50} = 8.2/12.6 μM

8

TYK2 JH2 IC_{50} = 7 nM
TYK2 JH1 IC_{50} = >50 μM
IL-23/IFNα IC_{50} = 0.1/0.05 μM

Figure 5. Left, high-throughput screen hit **7** for IL-23 inhibition; Right, structure and associated data of **8**

The pocket proximal to the C3 amide in the TYK2 JH2 domain contains a combination of residues that are largely unique relative to the kinome, such as the small Ala671 residue under the "gatekeeper" (Thr687). Of note is the single residue change preceding the activation loop (Ser 758), that alters the positions of the conserved catalytic Lys642 and Asp759 (Figure 6). It was postulated that interactions made by the C3 amide in this pocket, accounted for the kinome selectivity of compound **8**. The co-crystal structure revealed the two key hydrogen bond interactions that **8** makes with TYK2 JH2. One takes place at the hinge region between the C8 methylamino NH and the carbonyl of Val690 and between the N of the imidazopyradazine and the NH of Val690. The other occurs near the gatekeeper involving the C3 amide carbonyl to the NH of Lys642 and to the carbonyl of Glu688 mediated by a bridging water molecule.[23]

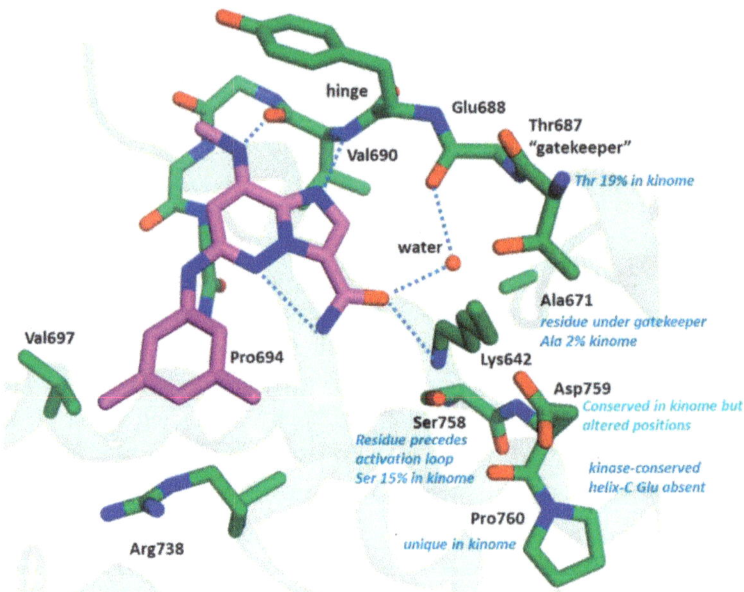

Figure 6. Close-up of the binding site highlighting key interactions made by **8**. Source: Image design by Dr. Yi Li

Seeking to improve permeability, further SAR exploration, guided by these critical structure-based insights led to 2-pyridyl substituted pyridone **9**. The intramolecular hydrogen bonds effectively masked the molecule's polarity and consequently increased the permeability (Figure 7).[24]

Modifications of the C3 amide side chain to improve metabolic stability resulted in the enantiomeric (1R,2S)-2-fluorocyclopropyl group as the optimal substituent, leading to a four-fold enhancement in Tyk2 JH2 affinity, cellular and human whole blood (hWB) activities. Of note, **10** remarkably displayed >10,000-fold selectivity for Tyk2 JH2 and demonstrated *in vivo* proof of concept, giving confidence that a small-molecule TYK2 JH2 ligand could be developed into a therapeutic for autoimmune and inflammatory disorders.[24]

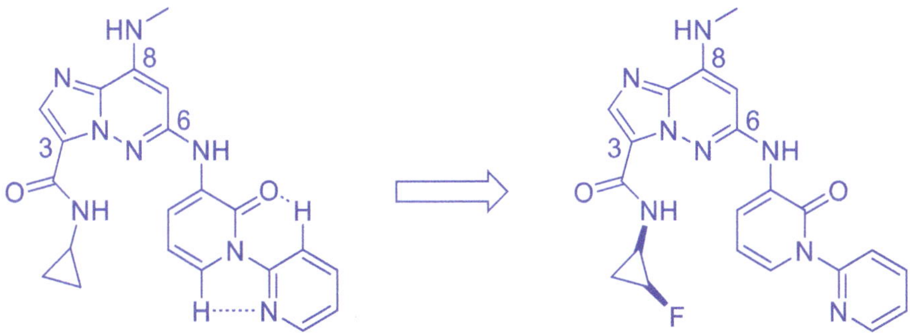

9

TYK2 JH2 IC_{50} = 0.34 nM
IFNα IC_{50} = 94 nM
hWB IC_{50} = 398 nM
Caco-2 AB (nm/s) = 169

10

TYK2 JH2 IC_{50} = 0.086 nM
IL-23/IFNα IC_{50} = 25 nM
hWB IC_{50} = 81 nM
Caco-2 AB (nm/s) = 110

Figure 7. Left, structure and associated *in vitro* data for **9**; Right, structure and associated *in vitro* data for **10**

3.2. Nicotinamide Series

In parallel, BMS simultaneously conducted a second high-throughput screening (HTS) campaign of the larger corporate compound database utilizing a scintillation proximity assay (SPA) and identified a novel nicotinamide TYK2 JH2-binding scaffold **11** (Figure 8).[25] The SPA assay was replaced with a more sensitive homogenous time-resolved fluorescence (HTRF) assay due to the notably higher potencies observed in this series. The good functional potency displayed by **11** was offset by its poor kinome and family selectivity (Figure 8). However, its good metabolic stability and high ligand efficiency (LE = 0.30, LLE = 5.8) warranted further multiparameter optimization of potency, selectivity, and drug-like properties and this series was prioritized over the imidazopyridazines.

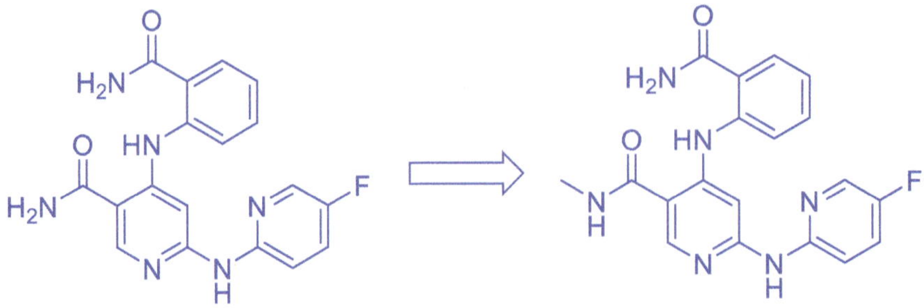

11
IC$_{50}$ < 1 μM: 84/261 kinases

12
IC$_{50}$ < 1 μM: 4/265 kinases
Caco-2 AB (nm/s) [er] = <15 [>35]

Figure 8. Structures and kinase selectivity of **11** and **12**

Applying structural lessons from the imidazopyridazine series, methyl amide **12** was made (Figure 8). Binding affinity SAR was rationalized on the basis of co-crystal structures, with the key feature being projection of the C3 amide methyl toward the distinctive "alanine pocket" containing Ala671, an uncommon residue found in the TYK2 pseudokinase domain, but only found in 2% of the entire kinome. The methyl amide **12** retained the TYK2 JH2 affinity while imparting a remarkable improvement in both kinome and family selectivity relative to **11**. This improved family selectivity was reflected in the cellular assays, with **12** showing no measurable activity in IL-2 (JAK 1 and JAK 3 dependent) and GM-CSF (JAK 2 dependent) control assays (Table 1).

	TYK2 JH2 (IC$_{50}$, nM)	JH1 TYK2/JAK1/ JAK2 (IC$_{50}$, nM)	TYK2 dependent IL-23/IFNα (IC$_{50}$, nM)	TYK2 independent IL-2/GM-CSF (IC$_{50}$, nM)
11	0.5	15/26/24	89/37	400/210
12	1.5	>2000	790/280	>12500/>12500

Table 1. Associated *in vitro* data comparing **11** and **12**

The less than desirable mouse pharmacokinetics of **12** (C_{max} = 36 nM; AUC = 150 nM h, following a po dose of 10 mg/kg) were improved by lowering the polar surface area (PSA) through replacement of the amide with a methyl sulfonyl group. Sulfone **13** maintained the binding affinity of **12**, while being slightly more potent in the reporter assay and a hWB assay of IFNα-induced STAT5 phosphorylation (Figure 9). The boost in permeability accompanied by a decrease in efflux ratio (er) led to **13** exhibiting a much more acceptable PK profile in a 10 mg/kg mouse oral PK study (C_{max} = 1620 nM; AUC = 9830 nM h).

An X-ray co-crystal structure of **11** bound to TYK2 JH2 confirmed the two critical hydrogen bond interactions made by one of the sulfone oxygens and the methyl amide carbonyl oxygen to the Lys642 residue, stabilizing the active conformation of **13**.

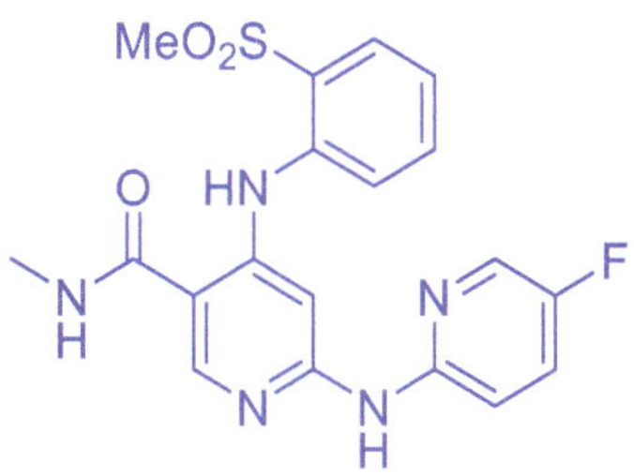

13

TYK2 JH2 IC_{50} = 0.92 nM
hWB IFNα pSTAT5 IC_{50} = 180 nM
kinases IC_{50} <1 μM= 2/248

LM % rem (h,r,m) = 90/85/87
cLogP/PSA = 3.0/113
Caco-2 AB (nm/s) [er] = 48 [10]

C_{max} (10 mpk, p.o.) = 1620 nM
AUC =(10 mpk, p.o.) = 9830 nM·hr

Figure 9. Structure and associated *in vitro*/*in vivo* data for Compound **13**

Expecting *in vivo* demethylation of the methyl amide to occur and the lack of selectivity displayed by primary amide **11**, there was concern that the anticipated primary amide metabolite of **13** could falsify *in vivo* efficacy data and lead to off-target effects. The putative primary amide metabolite of **13** was prepared and its potency and poor selectivity profile was confirmed (**14**, Figure 10). An analysis of plasma samples from a mouse PK study of **14** dosed orally and intravenously, revealed circulating primary amide metabolite levels between 5% and 30%.

A prior report of metabolic demethylation of tertiary amides being significantly diminished by the kinetic isotope effect,[26] prompted the preparation of trideuteromethyl secondary amide **15** in an attempt to curtail formation of the primary amide. A mouse PK study of **15** dosed orally and intravenously revealed circulating primary amide metabolite levels below the lower limit of quantification (LLQ) at all time points tested (Figure 10).

While deuterium incorporation circumvented the formation of a non-selective primary amide metabolite, **13** presented with a hERG liability (2.5 μM IC_{50} measured in a whole cell patch clamp assay). To avoid a narrow therapeutic window toward undesirable cardiovascular effects, achieving selectivity over hERG while maintaining all the necessary interactions, required some clever medicinal chemistry. Reducing hERG affinity can often be accomplished by removing peripheral rings to disrupt potential π-stacking within the aromatic residue-rich hERG channel,[27] and as such C6 amide exploration was carried out. It turned out that replacement of the 2-aminopyridine with cyclopropylamide **16** greatly reduced hERG channel inhibition as confirmed in the patch clamp assay (IC_{50} > 30 μM). A cocrystal structure of **13** and **16** overlaid with TYK2 JH2 revealed that both bind in an almost identical fashion concluding that the cyclopropyl amide in essence, serves as an isostere for 2-aminopyridine, lending credence to the similar binding potencies observed between **13** and **16** (Figure 11).

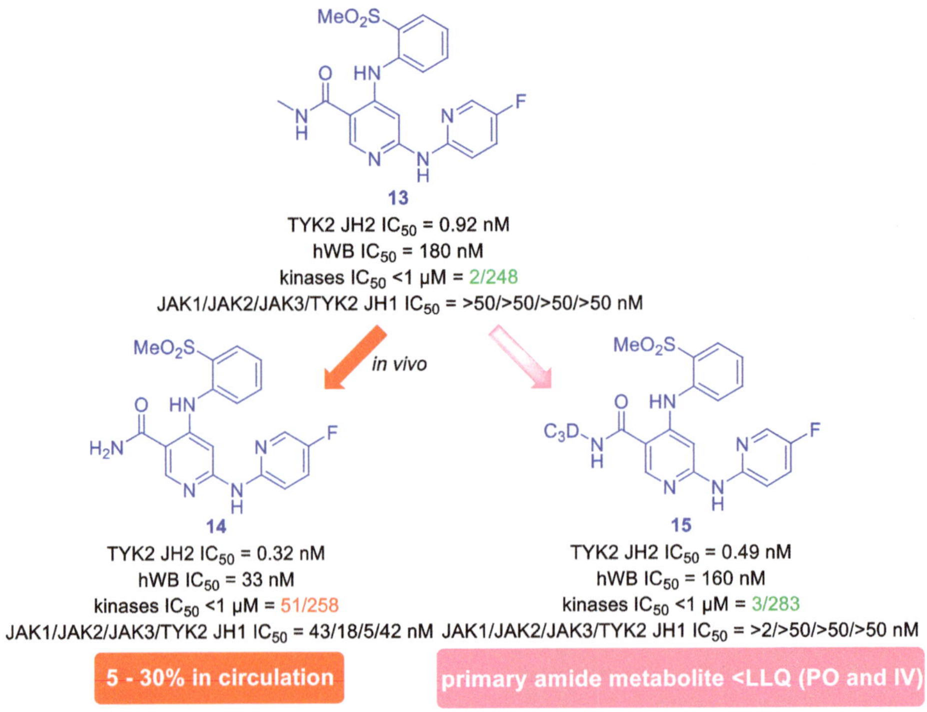

Figure 10. Profiles of **13**, primary metabolite **14** and trideuteromethyl analog **15**

Unfortunately, the good hERG selectivity of **16** was accompanied with poor mouse oral PK presumably due to its poor permeability (Figure 11). This limitation was overcome by replacing the central pyridine ring with the less basic and more lipophilic pyridazine core, which is also a strong hydrogen bond acceptor. Several pyridine- and corresponding pyridazine-matched molecular pairs were synthesized and evaluated. The pyridazine analogues consistently provided improved exposures without sacrificing potency. Pyridazines **17** and **18** emerged as the most promising based on their composite *in vitro*/*in vivo* properties (Figure 12).[25]

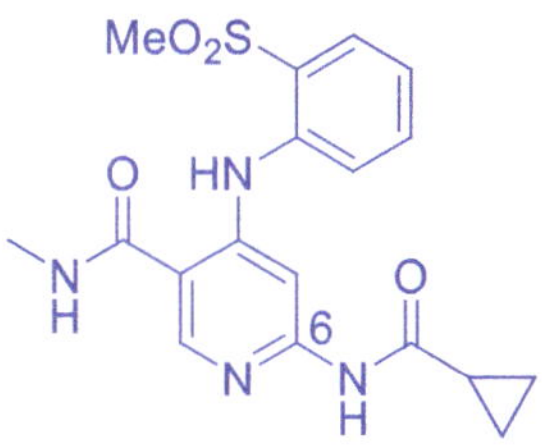

16

TYK2 JH2 IC_{50} = 1.6 nM
hWB IC_{50} = 309 nM
hERG IC_{50} = > 30 μM
Caco-2 AB (nm/s) [er] = 17 [29]
C_{max} (10 mpk, p.o.) = 186 nM
AUC =(10 mpk, p.o.) = 550 nM·hr
hERG selective, potency maintained,
poor exposure

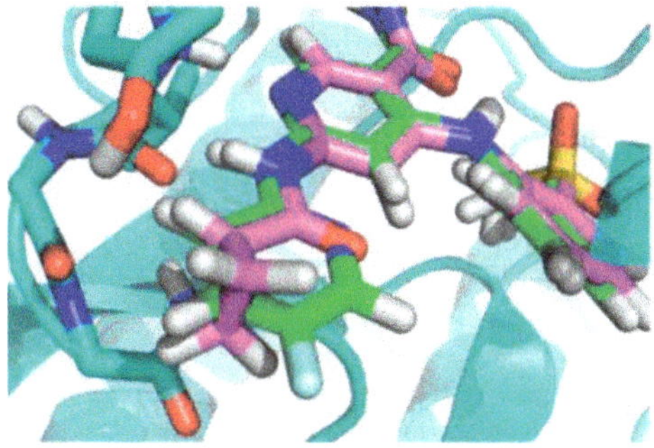

Figure 11. Left, structure and associated *in vitro*/*in vivo* data for compound **16**; Right, overlay of **13** and **16** with TYK2 JH2 showing cyclopropylamide as isostere for 2-aminopyridine (PDB ID: 6NZH)

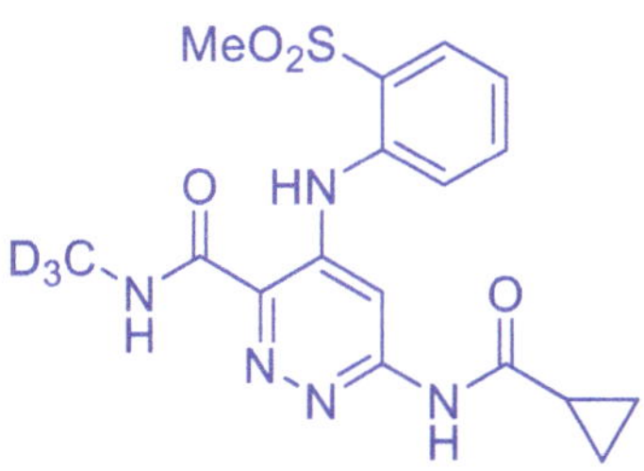

17

TYK2 JH2 IC_{50} = 1.3 nM
hWB IC_{50} = 230 nM
hERG IC_{50} = > 30 μM
Caco-2 AB (nm/s) [er] = 69 [7.7]
C_{max} (10 mpk, p.o.) = 3420 nM
AUC =(10 mpk, p.o.) = 6610 nM·hr
potent inhibitor, hERG selective,
good exposure

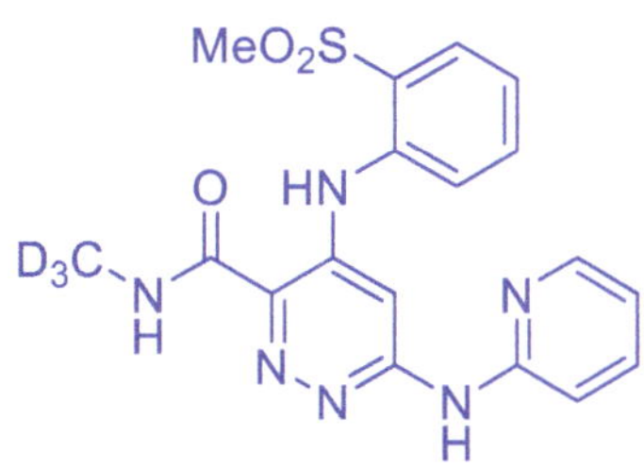

18

TYK2 JH2 IC_{50} = 0.53 nM
hWB IC_{50} = 130 nM
hERG IC_{50} = 10.9 μM
Caco-2 AB (nm/s) [er] = 280 [2.1]
C_{max} (10 mpk, p.o.) = 10500 nM
AUC =(10 mpk, p.o.) = 24300 nM·hr
potent inhibitor, poor hERG selectivity,
very good exposure

Figure 12. Structure and associated *in vitro*/*in vivo* data for compounds **17** and **18**

Compound **18** displayed a dose-dependent response in a mouse model of IBD, albeit with effective inhibition achieved at a high 250 mg/kg dose.[25] Entering into the lead optimization campaign, the objectives of the program were to retain the key discoveries made during hit-to-lead efforts and focus on fine-tuning these leads to further improve whole blood potency, mitigate a potential cardiovascular liability while allowing for a lower projected human dose.

As the strategic replacement of the 2-aminopyridine side chain with the cyclopropyl carboxamide successfully reduced hERG affinity and the deuteromethyl amide group binding to the atypical "alanine pocket" of TYK2 JH2 furnished high selectivity while also blocking an N-demethylation metabolic pathway, modifications centered around the aryl methyl sulfone group. X-ray crystal structures of sulfone analogues such as **18** bound to TYK2 JH2 proved very revealing.[28] In addition to key interactions already described, of note was the presence of a structural water molecule aiding indirect hydrogen bonds between the second sulfone oxygen of **18** with Arg738 and Gln597 of the P-loop region (Figure 13). To further improve potency, a targeted investigation to replace an energetically disfavored water molecule and form an additional direct hydrogen bond interaction with the protein, was pursued.

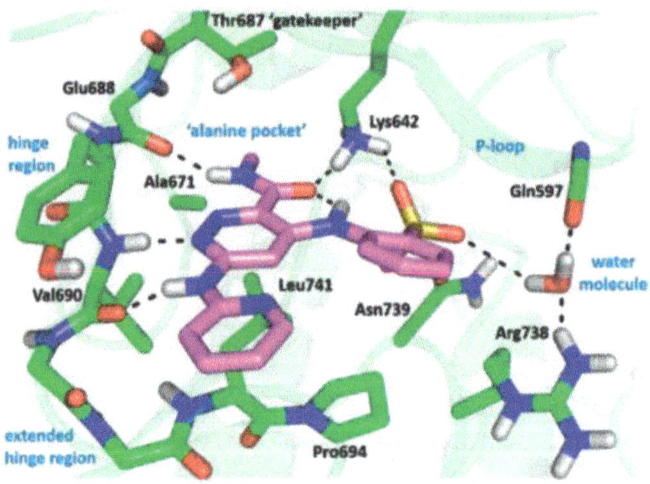

Figure 13. X-ray crystal structure of **18** bound with TYK2 JH2, highlighting the key hydrogen bond interactions within the ligand binding site and the observed structural water (PDB ID: 6NZR)

Investigating moieties that could replace the water molecule while also lowering PSA led to initial SAR studies being pursued on a prior compound within the pyridine-3-carboxamide series. Replacement of the methyl sulfone group with a methoxy substituent yielded a compound that had a lower PSA and although ~five-fold potency was lost, this change formed the basis of further SAR efforts. A review of modelling results revealed that introduction of hydrogen bond accepting groups ortho to the C2′ methoxy (C3′ position) could replace the structural water and form a direct hydrogen bond with the

Arg738 side chain. This led to a comprehensive SAR investigation of C3′ amides that provided some key findings: the C2′ methoxy group was critical, as its removal led to a dramatic loss in potency and the carbonyl O of the C3′ amides formed a direct hydrogen bond with Arg738. Despite good potency and validation of the water replacement strategy, the amides suffered from metabolic instability and permeability issues.

To address these, preparation of a variety of five-membered heterocycles that contained a heteroatom at the ortho position to the ring connection were found to successfully mimic the hydrogen bond interaction with Arg738 made by the carbonyl O of the C3′ amides. Triazole **19** exhibited remarkable potency and exemplified >1000-fold selectivity for TYK2 JH2 in an in-house 260 kinase panel. However, it was found to be a modest hERG inhibitor (31 μM IC_{50}) (Figure 14).

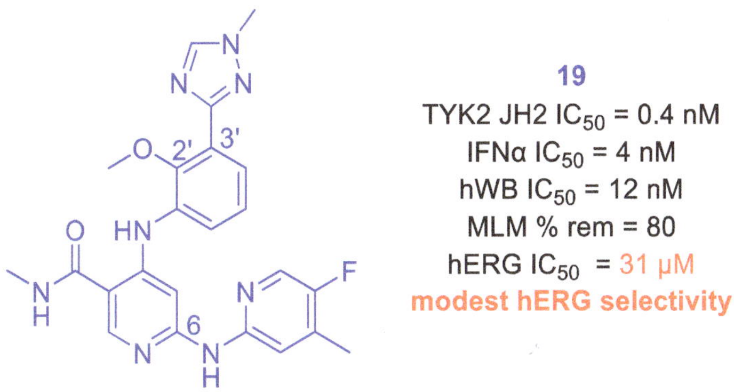

Figure 14. Structure and associated *in vitro* data for compound **19**

In an attempt to mitigate this potential cardiac liability, the C6 cyclopropyl amide side chain was reintroduced and as anticipated, compound **20** gave reduced hERG inhibition and excellent hWB potency (Table 2). The deuterated methyl amide **21** was made to prevent *in vivo* formation of the primary amide but unfortunately it exhibited poor exposure which was attributed to the reduced permeability (Table 2). This issue was overcome by making the pyridazine variant of **21**. This led to **1**, which unknowingly at the time, would eventually become deucravacitinib!

Compound **1** maintained excellent potency in hWB and exhibited functional selectivity in hWB of ~47-fold against JAK1/JAK2 (IL-6 IC_{50} = 609 nM), ~150-fold against JAK1/JAK3 (IL-2 IC_{50} = 1900 nM), and high kinome selectivity against 249 kinases. *In vitro* ADME and *in vivo* pharmacokinetics in rodent and higher species revealed high stability in liver microsomes ($T_{1/2}$ > 120 min), significantly improved permeability leading to much improved exposure (C_{max} = 7500 nM for **1** vs 310 nM for **21**), low overall drug–drug interaction (DDI) risk and no significant hERG inhibition in the flux assay (IC_{50} > 80 μM) (Table 2). With it also demonstrating robust efficacy in murine models of psoriasis, lupus nephritis, and IBD, **1** was selected as a potential candidate for advancement.[28]

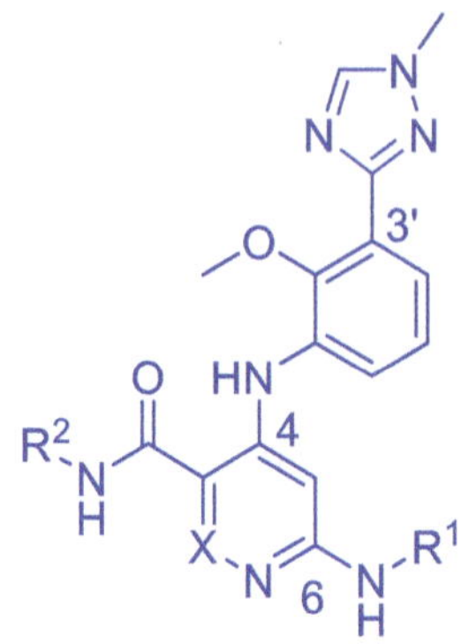

	R^1	X	R^2	TYK2 JH2 (IC_{50}, nM)	IFNα (IC_{50}, nM)	hWB (IC_{50}, nM)	MLM (% rem)	Caco-2 (nm/sec)	hERG flux (IC_{50}, μM)
20	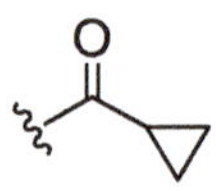	C	CH_3	0.7	14	16	93	23	>80
21	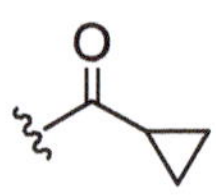	C	CD_3	0.3	8	13	89	<15	>80
1	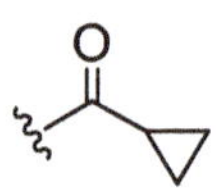	N	CD_3	0.2	5	13	89	70	>80

Table 2. *In vitro* data for select analogues from C3′ *N*-methyl triazole series

The proposed binding mode was confirmed from a solved X-ray structure of **1** bound to TYK2 JH2, reinforcing key interactions that have already been described (Figure 15). The observed increase in potency is most probably due to the C2′ methoxy forming a direct hydrogen bond with the conserved Lys642. Remarkably, the des-methoxy variant of **1** was determined to be ~100-fold less potent in TYK2 JH2 binding affinity. Replacement of the structural water molecule occurred through a direct hydrogen bond engagement with Arg738 of the N-2 triazole nitrogen, as thoughtfully designed.[28]

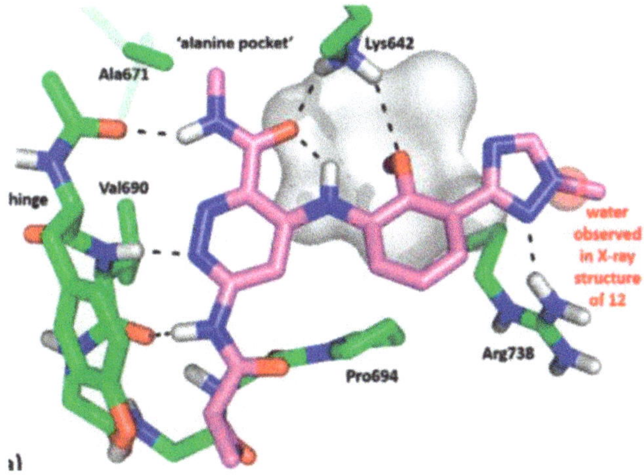

Figure 15. X-ray crystal structure of **1** bound with TYK2 JH2 showing key interactions (PDB ID: 6NZP)

4. Pharmacokinetics and Drug Metabolism

In preclinical pharmacokinetic studies of intravenously (iv) and orally administered (po) deucravacitinib (**1**) across multiple species, the low rate of metabolism observed in the microsomal assays ($T_{1/2} > 120$ min) translated into low to modest *in vivo* clearance rates, excellent exposures, and high bioavailability ($\%F > 85$) in mouse, dog, and monkey.[28] Circulating primary amide metabolite formation from N-dealkylation of the deuteromethyl amide of **1** was near the lower limits of detection (<2 nM) in these studies, consistent with deuteration serving as an effective strategy for blocking of this metabolic pathway.

Following oral administration in humans, deucravacitinib's plasma Cmax and AUC increased dose proportionally over a 3–36 mg dose range (0.5- to 6-fold the approved recommended dosage) in healthy subjects.[29] Following once-daily dosing in healthy subjects, the accumulation of deucravacitinib was <1.4-fold. The PK of deucravacitinib and its active demethylated metabolite, BMT-153261, were comparable between healthy subjects and subjects with psoriasis. The steady state C_{max} and AUC_{24} of deucravacitinib following administration of 6 mg once daily were 45 ng/mL and 473 ng h/mL, respectively. The steady state C_{max} and AUC_{24} of the active deucravacitinib metabolite, BMT-153261, following administration of 6 mg once daily were 5 ng/mL and 95 ng h/mL, respectively.

The absolute oral bioavailability of deucravacitinib was 99% and the median T_{max} ranged from 2 to 3 h in healthy subjects. No clinically meaningful differences in the pharmacokinetics were observed after administration of a high-fat, high-calorie meal. C_{max} and AUC of deucravacitinib when administered with food were decreased by approximately 24% and 11%, respectively, and T_{max} was prolonged by 1 h. C_{max} and AUC of BMT-153261 when administered with food were decreased by approximately 23% and 10%, respectively, and T_{max} was prolonged by 2 h.[29]

The volume of distribution of deucravacitinib at steady state was 140 L. Plasma protein binding of deucravacitinib was 82–90% and its blood-to-plasma concentration ratio was 1.26. The terminal half-life of deucravacitinib was 10 h. The renal clearance of deucravacitinib ranged from 27 to 54 mL/min.[29]

Deucravacitinib is metabolized by cytochrome P-450 (CYP) 1A2 to form major metabolite BMT-153261. Deucravacitinib is also metabolized by CYP2B6, CYP2D6, carboxylesterase (CES) 2, and uridine glucuronyl transferase (UGT) 1A9. Although the active deucravacitinib metabolite, BMT-153261, has comparable potency to the parent drug, its circulating exposure accounts for approximately 20% of the systemic exposure of all drug-related components.[29]

5. Efficacy and Safety

In preclinical studies in mice, deucravacitinib (**1**) was determined to be efficacious against both type I IFN, IL-12, and IL-23-dependent pathobiology. The *in vivo* efficacy from orally administered **1** was demonstrated in multiple murine models of psoriasis, colitis, and lupus and correlated well with the whole blood IC_{50} value coverage over the dosing intervals and was at least as effective as the blocking antibody controls used in these studies.[28]

In humans, the efficacy and safety of deucravacitinib (6 mg once daily) was evaluated in two multinational, multicenter, randomized, double-blind, placebo- and active comparator-controlled 52-week Phase 3 clinical trials, POETYK PSO-1 (NCT03624127) and POETYK PSO-2 (NCT03611751).[30] All enrolled subjects were 18 years of age and older with moderate-to-severe plaque psoriasis and were candidates for phototherapy or systemic therapy. Participants had a body surface area involvement of $\geq$10%, a Psoriasis Area and Severity Index (PASI) score $\geq$ 12, and a static Physician's Global Assessment (sPGA) $\geq$ 3 (moderate or severe).[30]

Efficacy was assessed in POETYK PSO-1 (664 enrolled patients) and POETYK PSO-2 (1,020 enrolled patients) randomized to either *Sotyktu* (6 mg once daily), placebo, or Otezla® (apremilast) (30 mg twice daily). The two co-primary endpoints of both trials, assessed at week 16 vs placebo, were the proportion of patients who achieved at least a 75% improvement in PASI scores from baseline (PASI 75) and the proportion of patients who achieved a sPGA score of 0 (clear) or 1 (almost clear) with at least a 2-grade improvement from baseline (Table 3).[30]

Key secondary endpoints assessed at week 16 and week 24, that compared Sotyktu and placebo, included the percentage of subjects who achieved PASI 75, PASI 90 and sPGA 0/1 (Table 3).[30] As shown, response rates across all primary and secondary endpoints at both 16 and 24 weeks were highest for deucravacitinib, demonstrating its superiority over both placebo and twice-daily Otezla for the treatment of patients with moderate-to-severe plaque psoriasis.

POETYK PSO-1 (*n* = 664) Results at Week 16 and Week 24

Endpoint	Time	*Sotyktu* 6 mg (*n* = 330)	Placebo (*n* = 166)	Otezla 30 mg (*n* = 168)
PASI 75	Week 16	58%[a]	13%[a]	35%
	Week 24	69%	—	38%
PASI 90	Week 16	36%	4%	20%
	Week 24	42%	—	22%
sPGA 0/1	Week 16	54%[a]	7%[a]	32%
	Week 24	59%	—	31%

POETYK PSO-2 (*n* = 1020) Results at Week 16 and Week 24

Endpoint	Time	*Sotyktu* 6 mg (*n* = 511)	Placebo (*n* = 255)	Otezla 30 mg (*n* = 254)
PASI 75	Week 16	53%[a]	9%[a]	40%
	Week 24	58%	—	38%
PASI 90	Week 16	27%	3%	18%
	Week 24	32%	—	20%
sPGA 0/1	Week 16	50%[a]	9%[a]	34%
	Week 24	49%	—	30%

[a]Co-primary endpoints for POETYK PSO-1 and POETYK PSO-2 at Week 16.

Table 3. Efficacy results in adults with moderate-to-severe plaque psoriasis from POETYK PSO-1 and POETYK PSO-2 clinical trials at Week 16 and Week 24[30]

Responses persisted through Week 52, as 78% (151/194) of patients who achieved an sPGA 0/1 response at week 24 maintained their response at week 52. Amongst the patients who achieved PASI 75 and PASI 90 with *Sotyktu* at Week 24, 82% (187/228) and 74% (103/140) respectively, maintained their response at Week 52, in POETYK PSO-1. In POETYK PSO-2, 80% (119/148) of patients who continued on *Sotyktu* maintained PASI 75 response compared to 31% (47/150) of patients who were withdrawn from *Sotyktu*.[31] Another noteworthy finding was in patients that transitioned over from placebo to *Sotyktu* treatment at week 16 had clinical responses at week 52 that were comparable to patients who received continuous *Sotyktu* treatment from day 1.[31]

At week 16, In the POETYK PSO trials, the most common adverse reactions (≥1% and higher than placebo) in patients on *Sotyktu* were upper respiratory infections (19.2%), blood creatine phosphokinase increase (2.7%), herpes simplex (2.0%), mouth ulcers (1.9%), folliculitis (1.7%), and acne (1.4 %). The frequency of serious adverse events during weeks 0-16 was lowest in the *Sotyktu* group (2.1% vs 5.5% with placebo and 2.4% with Otezla). In addition, discontinuation due to adverse reactions was encountered for 2.4% of patients on *Sotyktu*, 3.8% of patients on placebo, and 5.2% of patients on Otezla.[30]

6. Synthesis

6.1. Discovery Synthesis Route

The general route to make the *N*-methyl nicotinamides and *N*-methyl pyridazine-3-carboxamides (not shown) employed established chemistry and proved to be highly versatile, lending itself well for both SAR exploration and for larger scale preparation of key compounds that were needed for advanced preclinical testing.

The synthesis of core pyridazinone intermediate **27**[25] initiated with diazo transfer to diethyl malonate **22** to provide **24**. Subsequent chemistry involved a diaza-Wittig reaction followed by hydrolysis and saponification to provide crude acid diol **25**. Chlorination with $POCl_3$ furnished the trichloro-intermediate **26** that was subsequently trapped *in situ* with the methyl-d_3-amine hydrochloride in the presence of Hunig's base to provide amide **27** (Scheme 1).

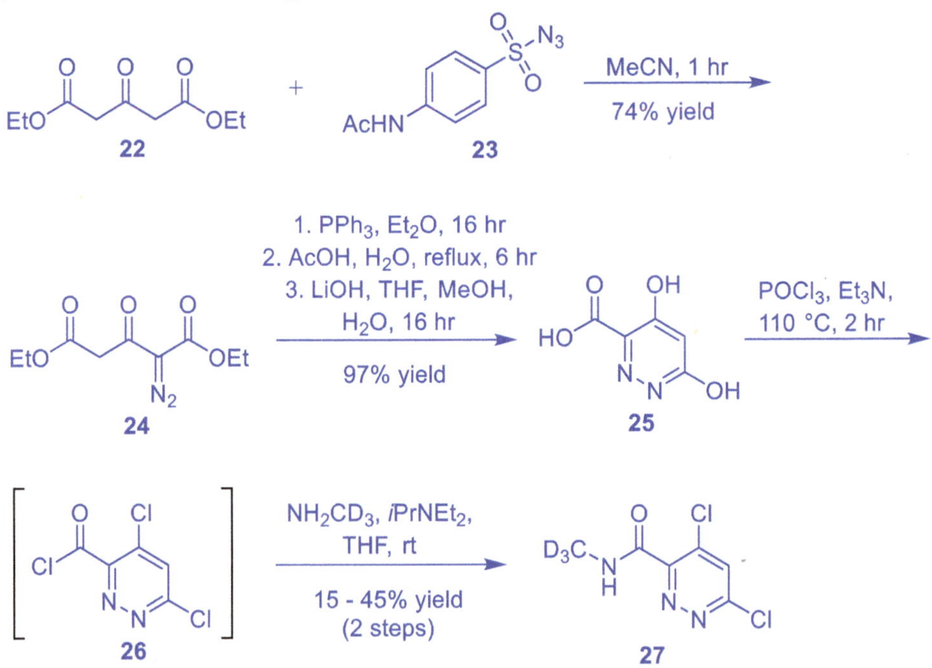

Scheme 1. Synthesis of intermediate **27**

The synthesis of **1** commenced with methylation of commercially available methyl-2-hydroxy-3-nitrobenzoate **28** (Scheme 2).[28] Subjecting the intermediate methyl ester to ammonia afforded amide **29**. Incorporation of the triazole motif was achieved through reaction of **29** with DMF-DMA followed by condensation with hydrazine hydrate to afford **30**. Methylation of the triazole with methyl iodide in the presence of potassium carbonate afforded a 2:1 mixture of the desired regioisomer **31** and undesired regioisomer **32**, respectively. Separation of the regioisomers was attained by supercritical fluid chromatography (SFC) to afford the desired isomer **31** in 67% overall yield. It was found that a significant improvement in the regioselectivity (~8:1) in favor of the desired isomer **31** could be accomplished by replacing potassium carbonate with potassium hexamethyldisilazide as the base and THF instead of DMF, as the solvent. The crude product enriched in the desired isomer could be crystallized to afford isomerically pure **31**, eliminating the need to separate the isomers by SFC on a larger scale.

Pure **31** was subjected to standard palladium-catalyzed hydrogenation conditions for the nitro group reduction to yield aniline **33** in 92% yield. Subsequent coupling to pyridazine core intermediate **27** using lithium hexamethyldisilazide as the base afforded the penultimate intermediate **34** in 66% yield. Initial Buchwald–Hartwig coupling of **34** with cyclopropyl amide **35** using XantPhos as the palladium ligand afforded **1** in a modest 46% yield. This was significantly improved by replacing the XantPhos ligand with 1,1′-bis(dicyclohexylphosphino)ferrocene (dppf) and switching out the cesium carbonate base with aqueous potassium triphosphate. These adjusted conditions allowed a lower reaction temperature and afforded an improved 76% yield enabling preparation of multigram quantities of **1** for advanced preclinical studies (Scheme 2).[28]

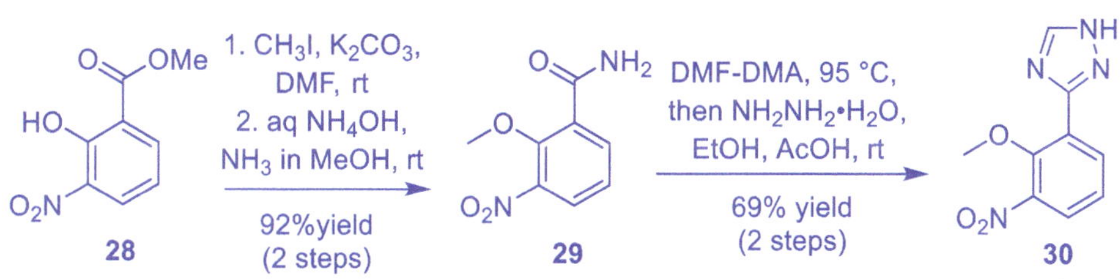

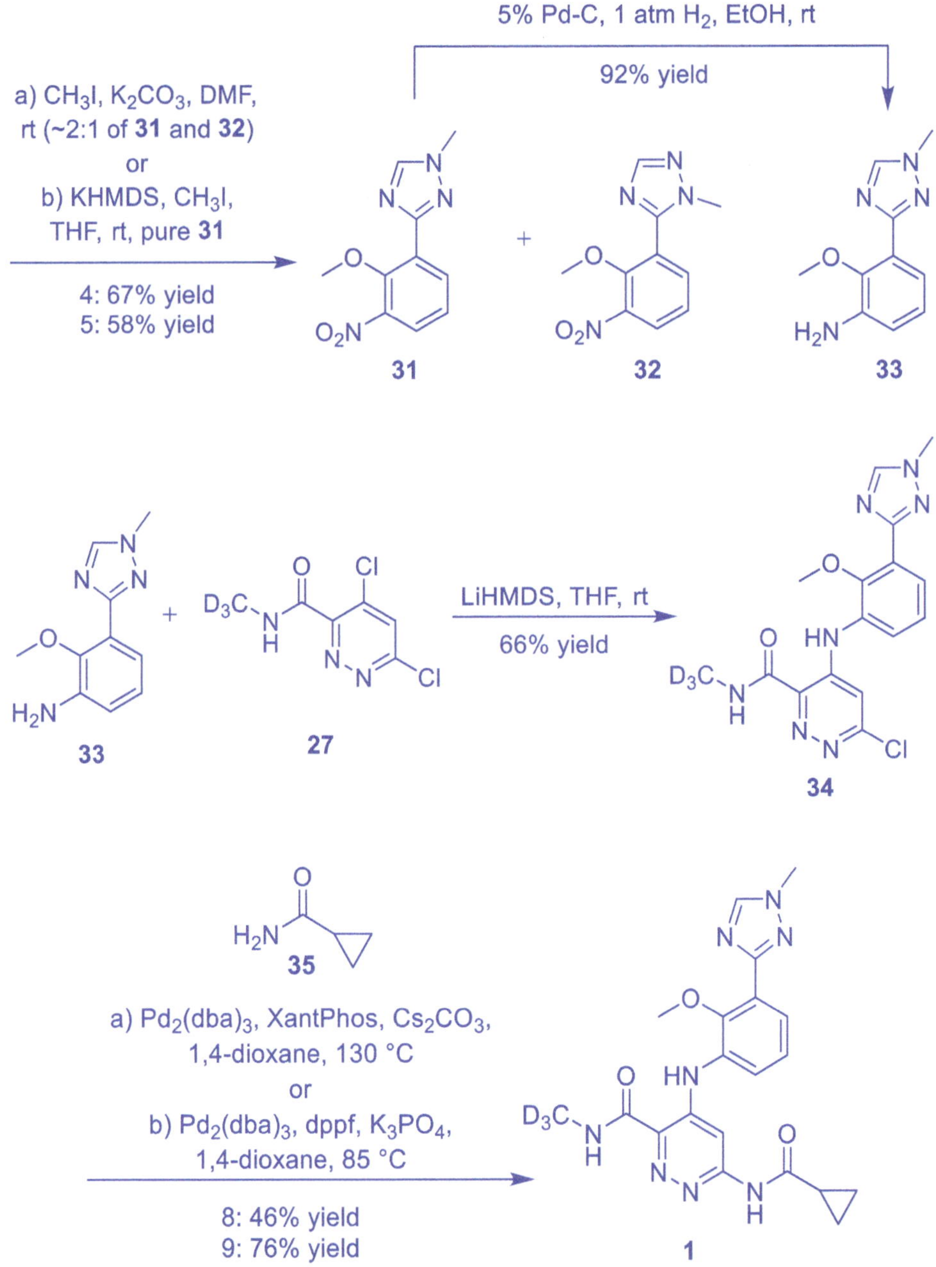

Scheme 2. Discovery synthesis of deucravacitinib (**1**)

6.2. Process Development Route

Anticipating several metric tons of commercial API to be needed over the lifetime of a successfully marketed drug, a thorough evaluation of the discovery chemistry route was undertaken. Although the synthesis was modular and enabled preparation of sufficient material to support SAR studies, several liabilities in this first-generation route were identified[32]:

(a) **Expensive, low-yielding synthesis of 33**. This was ascribed to the poor inherent regioselectivity in the triazole methylation.

(b) **Poor yield in the deuteromethyl amide formation step**. This was mainly due to the inherent instability of the trichloro-intermediate **26**, as significant decomposition was encountered during the work-up of the chlorination reaction. This low yield necessitated a large excess of expensive trideuteromethyl amine (11 equivalents per equivalent of **1**).

(c) **Poor Yield in the API step**. With the use of a palladium catalyst in the last step, remediation efforts accounted for the poor yield. Due to Nitrogen-rich **1** being an excellent metal chelator, multiple aqueous washes, a resin treatment, and a recrystallization with methanol were required to afford **1** with acceptable levels of residual palladium.

Efforts to devise new routes to **1** were initiated with cost and sustainability at the forefront of route planning. Due to the simple connectivity of **1**, the number of potential routes was limited. It was envisioned that the central bis-aminoacylpyridazine core could be derived from three sequential nucleophilic displacements to form three new C–N bonds (Figure 16). Initial route design efforts focused on trying to identify the potential order of formation for these three C–N bonds, considering the expected cost of the three nitrogen nucleophiles. With the cost of trideuteromethylamine plaguing the discovery synthesis, it was reasoned that the ideal route would incorporate amidation with trideuteromethylamine in the final step. This would avoid the use of a Pd catalyst in the API step, helping to solve the residual metal control issue. With respect to formation of the pyridazine C–N bonds, earlier studies revealed that much faster substitution took place at the C4 position over the C6 position. Armed with this information, the planned general synthetic strategy would incorporate the aniline fragment first, the cyclopropanecarboxamide fragment second, followed by the trideuteromethyl amide last (Figure 16).

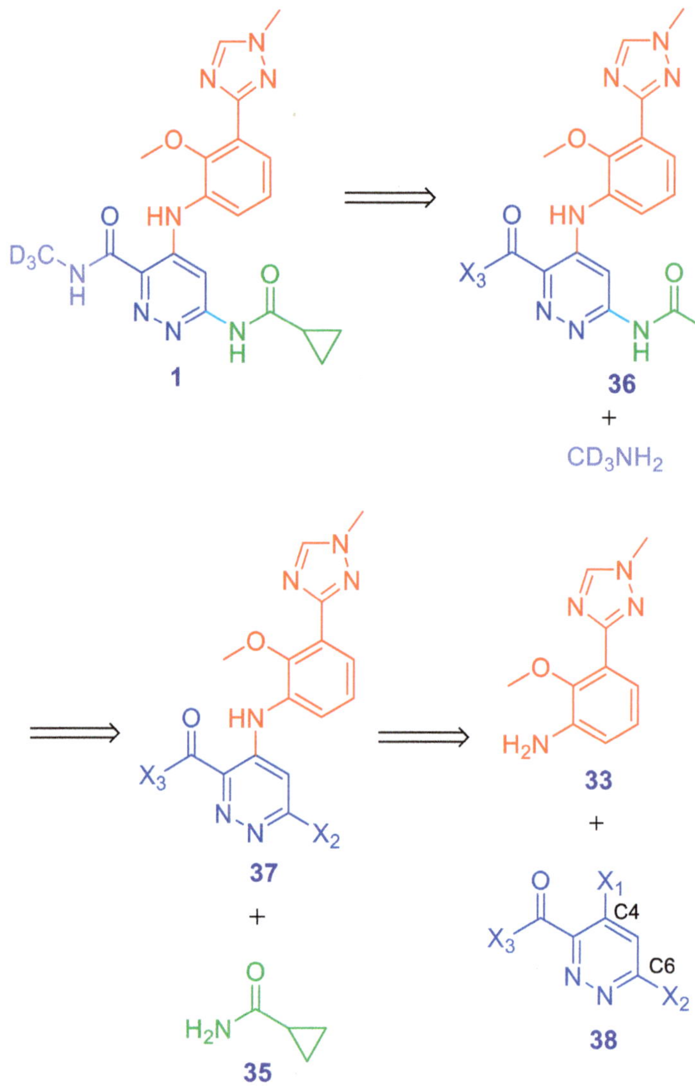

Figure 16. Retrosynthetic analysis of **1**

The first task was to determine the appropriate functional groups on **38** that would enable a robust, high-yielding synthesis of this intermediate. After experimentation, the so-called "acid route" was developed commencing with bis-deoxychlorination of commercially available dihydroxy pyridazine carboxylic ester **39** and subsequent saponification to yield carboxylic acid **40**, which was able to be carried through the S_NAr and Pd-catalyzed amidation steps.

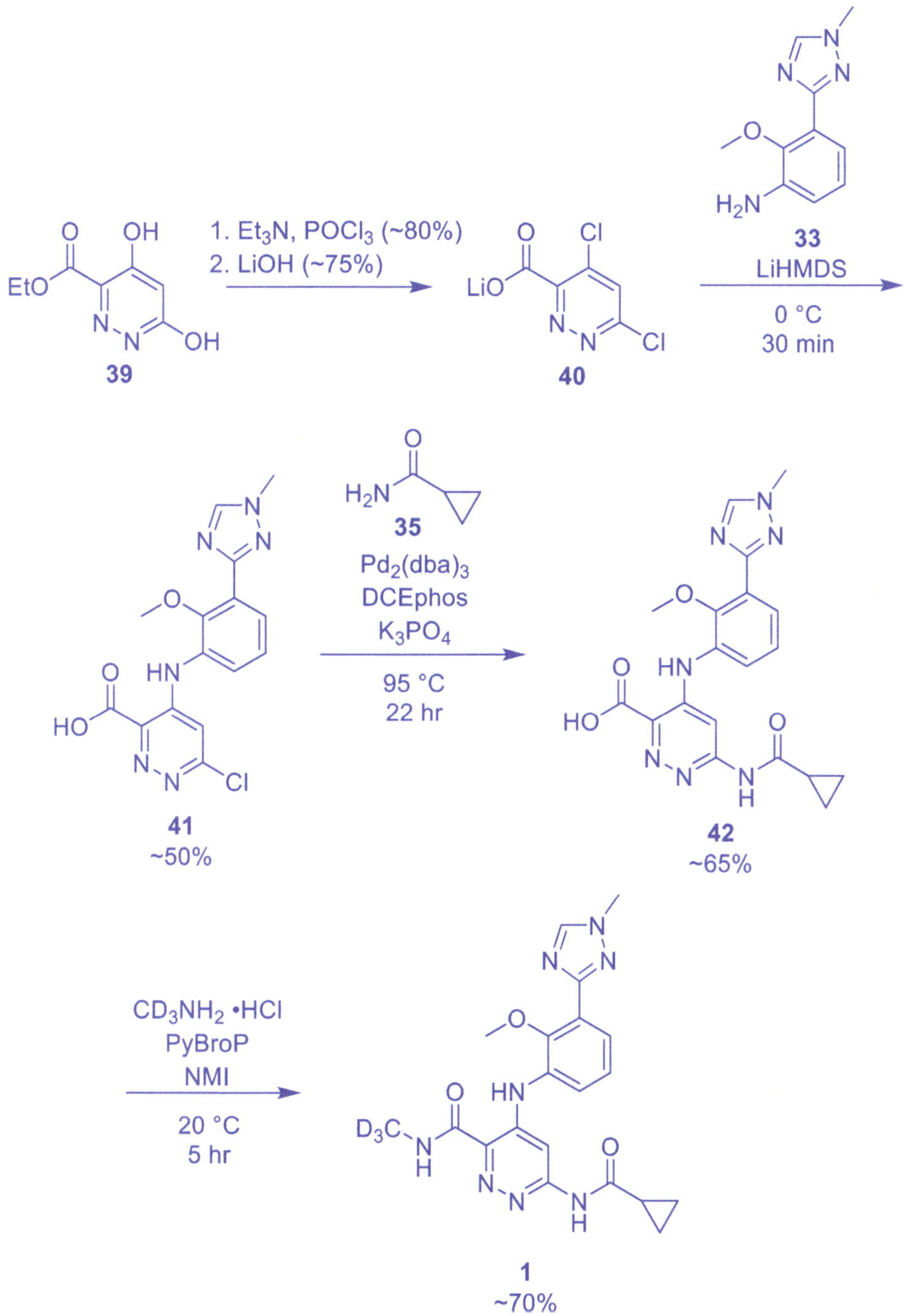

Scheme 3. First pass proof of concept for "acid route"

Proof of concept for this route was successfully achieved on the first pass and ultimately evolved into the commercial process (Scheme 3).[32]

After extensive development and multiple iterations of design and testing, details hereon in highlight optimization of this first pass process including key reactions and conditions that were ultimately settled on and employed in the final commercial process.

6.2.1. Synthesis of Aniline Fragment **33**

Devising an efficient and cost-effective synthesis to aniline **33** was of paramount importance due to the modest yielding and poorly regioselective triazole methylation step. The optimized synthesis to aniline **33** commenced with a cyclocondensation reaction between cheap commercially available 5-chloro-2-methoxybenzonitrile **43** and *N*-methylformohydrazide **44**, affording the free base which was transformed to the bisulfate salt **45** (Scheme 4). The nitration reaction to form **46** was extremely clean (>99% in-process purity). In the final step, a one-pot hydrogenation/hydrogenolysis to reduce both the nitro group and the aryl chloride moiety was readily achieved using gaseous hydrogen and palladium on carbon at modest pressure. Selective choices of hydrogen pressure, catalyst loading, reaction temperature, and reaction concentration (guided by design of experiment studies) minimized impurity formation to <0.1% (Scheme 4).

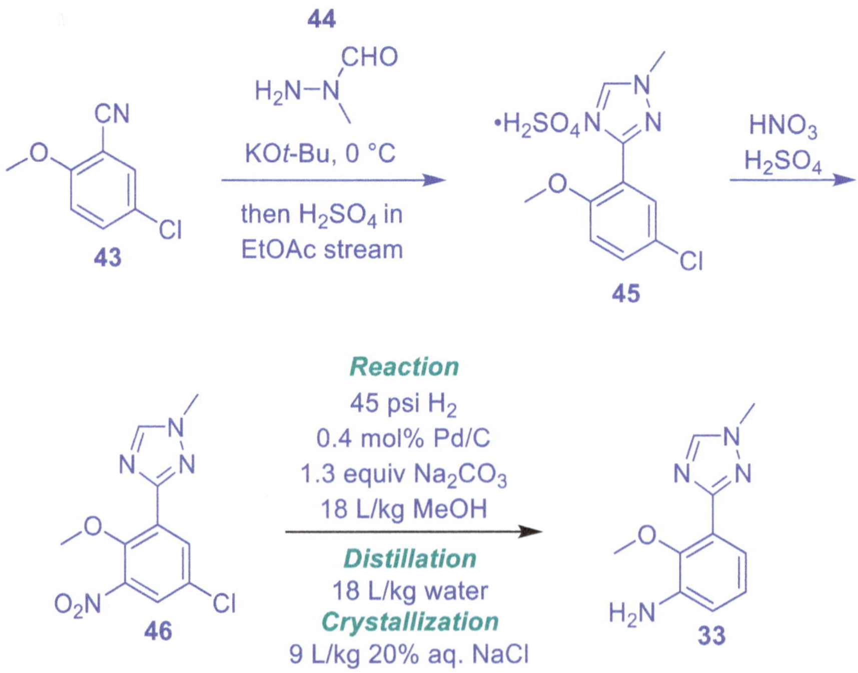

Scheme 4. Synthesis of aniline **33**

6.2.2. Synthesis of Pyridazine Core **40-01**

The process chemistry synthesis to dihydroxypyridazine **39** was not too dissimilar to the discovery route and will not be discussed further here (shown in Scheme 1). Double deoxychlorination of **39** with $POCl_3$ to the corresponding dichloropyridazine carboxylic ester **47** was straightforward but required careful quenching with water to ensure homogeneity of the reaction mixture (Scheme 5). This proved crucial for minimizing the formation of hydrolytic impurities. Ester **47** was telescoped into the subsequent hydrolysis step but the resulting free pyridazine carboxylic acid underwent spontaneous decarboxylation. It was found that addition of lithium bromide and water followed by slow addition of DIPEA effected the hydrolysis resulting in the stable crystalline lithium monohydrate salt **40-01**. Overall, the chlorination-hydrolysis telescope reproducibly afforded **40-01** in good yield and high purity on a >200 kg scale (Scheme 5).

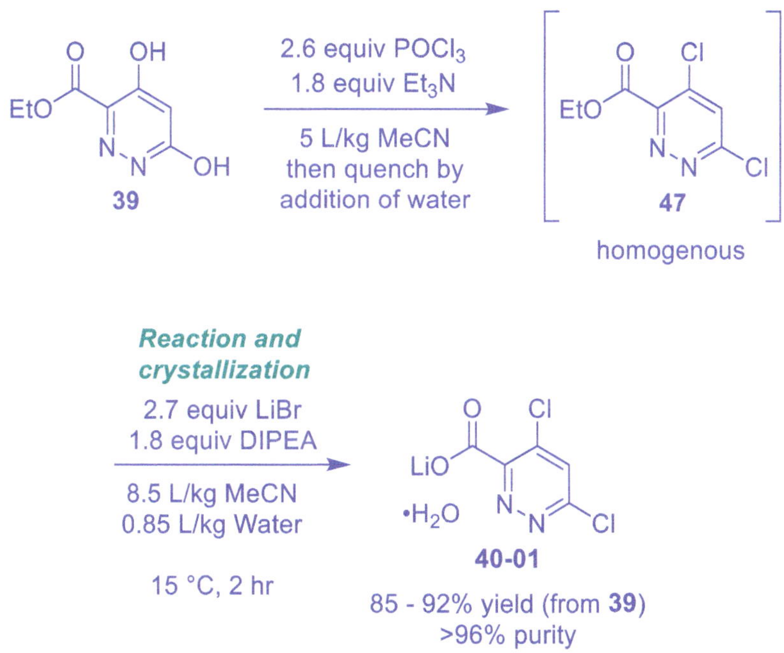

Scheme 5. Synthesis of dichloropyridazine carboxylic acid salt **40-01**

6.2.3. Nucleophilic Aromatic Substitution Between **33** and **40-01** to Afford **48**

Initially, the desired product in its free acid form (presumably a zwitterion) was found to be both hygroscopic and susceptible to decarboxylation at elevated temperatures. To circumvent these issues and improve selectivity for the desired product over the undesired regioisomer and the disubstituted adduct (structures not shown), a screen of Lewis acids revealed that stoichiometric $Zn(OAc)_2$ could mediate the reaction with near-perfect selectivity. Conveniently, the stable product salt precipitated spontaneously from the reaction solution as the zinc salt **48**. Adding small amounts of isopropanol as co-solvent, expedited product filtering (due to improved crystal morphology). The zinc salt was reproducibly isolated in excellent yield and purity on >100 kg scale (Scheme 6).

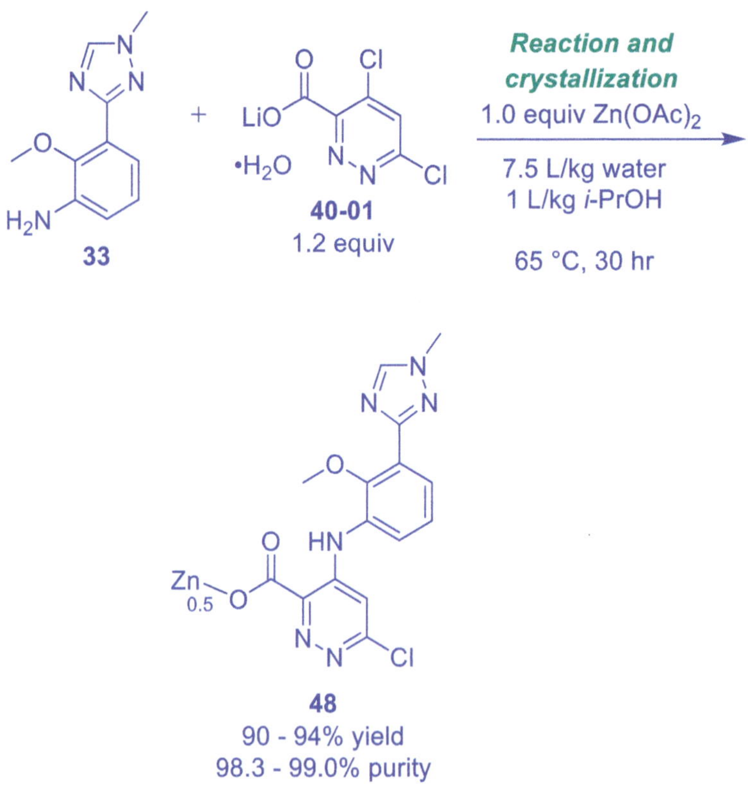

Scheme 6. Synthesis of carboxylic acid zinc salt **48**

6.2.4. Penultimate Step: Palladium-Catalyzed C–N Coupling of **48** and Cyclopropyl Amide **35** to Afford **49**

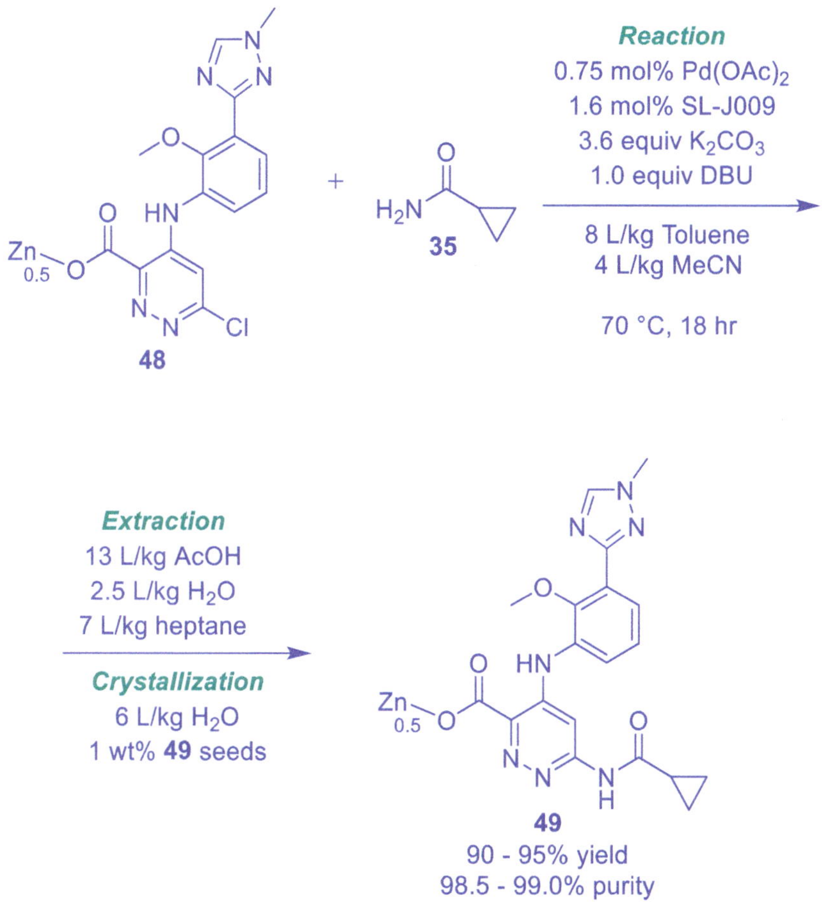

Scheme 7. Penultimate Step: Synthesis of carboxylic acid zinc salt **49**

This was the last step in the discovery synthesis but was switched to a penultimate step in the process route to minimize cost and avoid the use of a Pd catalyst in the final step. High-throughput experimentation (HTE) was employed to identify the most suitable catalyst/ligand/base replacements. From a cost, availability, and stability standpoint, $Pd(OAc)_2$ was chosen as the catalyst for the commercial process (Scheme 7). The base had a significant impact on reaction kinetics and a comprehensive investigation led to the finding that a combination of DBU and K_2CO_3 was optimal for complete dissolution of the zinc carboxylate as well as enhancing the rate of the reaction. Ligand screening with the dual-base combination revealed SL-J009 as the preferred ligand due to lower palladium catalyst loading requirements. With respect to solvent, an iterative process determined that an acetonitrile/toluene combination ensured the correct balance between reactivity and solubility of the zinc carboxylate **48**. The product was isolated as zinc salt **49** in excellent yield and purity, after a seeded crystallization. The robust process was successfully and reproducibly scaled multiple times to >80 kg/batch (Scheme 7).

6.2.5. API Step: Amidation with CD_3NH_2 to Afford **1**

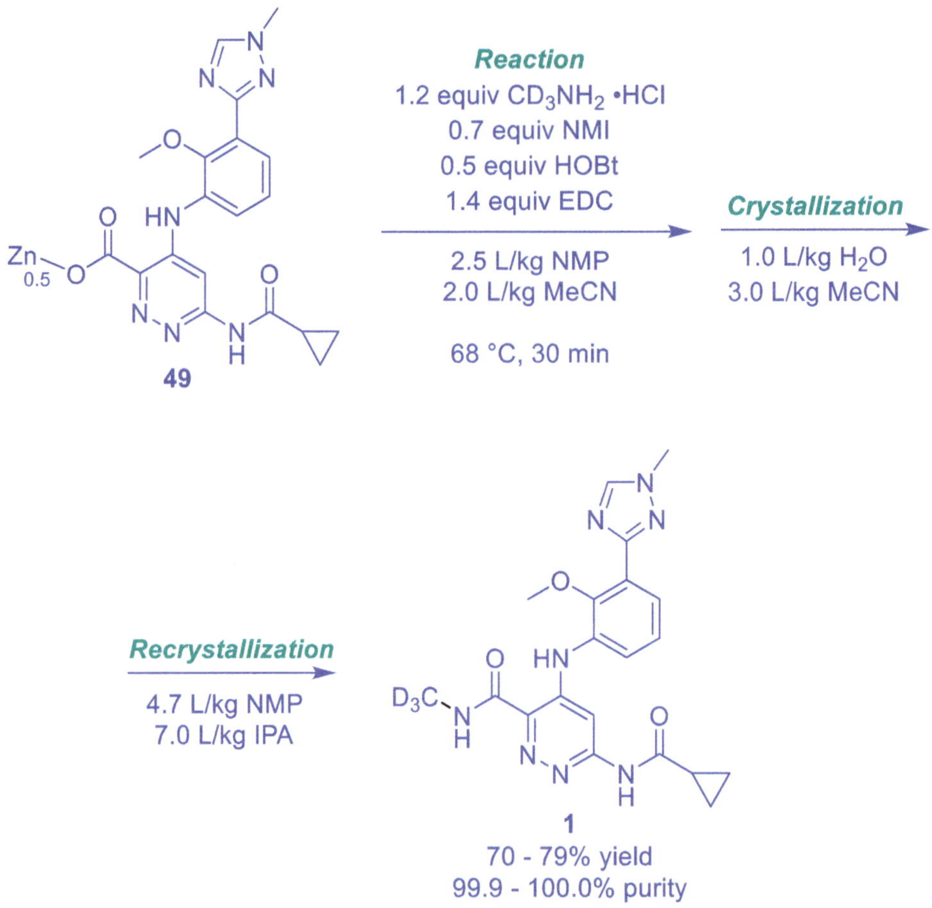

Scheme 8. Commercial API step: synthesis of **1**

The final step in the commercial process to deucravacitinib was amidation between zinc carboxylate **49** and trideuteromethylamine. Due to handling and commercial availability concerns, trideuteromethylamine (a gas) was procured as its crystalline hydrochloride salt.

Conducting laboratory scale optimization based on HTS data revealed the most effective coupling reagents to be EDC/HOBt. Polar aprotic solvents worked best (presumably due to solvation of poorly soluble salt **49**), and the nucleophilic sp^2-hybridized amine base, *N*-methylimidazole (NMI), proved superior. The optimal reaction and crystallization/recrystallization parameters gave suitable impurity purge, excellent yield and consistently pure deucravacitinib **1** (>99.5% purity) at scales of >100 kg per batch (Scheme 8).

In summary, an efficient, high-yielding synthesis of **1** in eight steps from inexpensive commercial materials was developed, enabling production of more than a metric ton of deucravacitinib for clinical and commercial use (Scheme 9). The convergence of the synthesis was maximized through development of novel, mild conditions for the regioselective synthesis of a 1-methyl-3-aryl triazole. Subsequent optimization afforded protocols that produced intermediates and drug substance of very high purity, while in tandem achieving sustainability goals and a greener synthesis.[32]

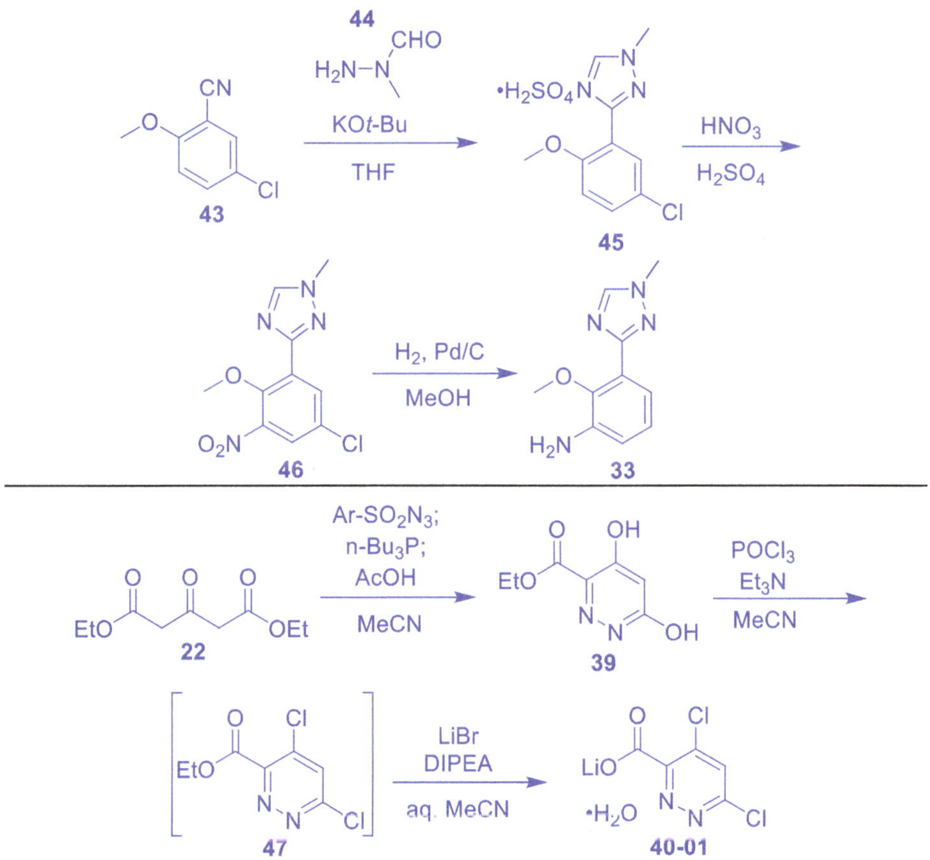

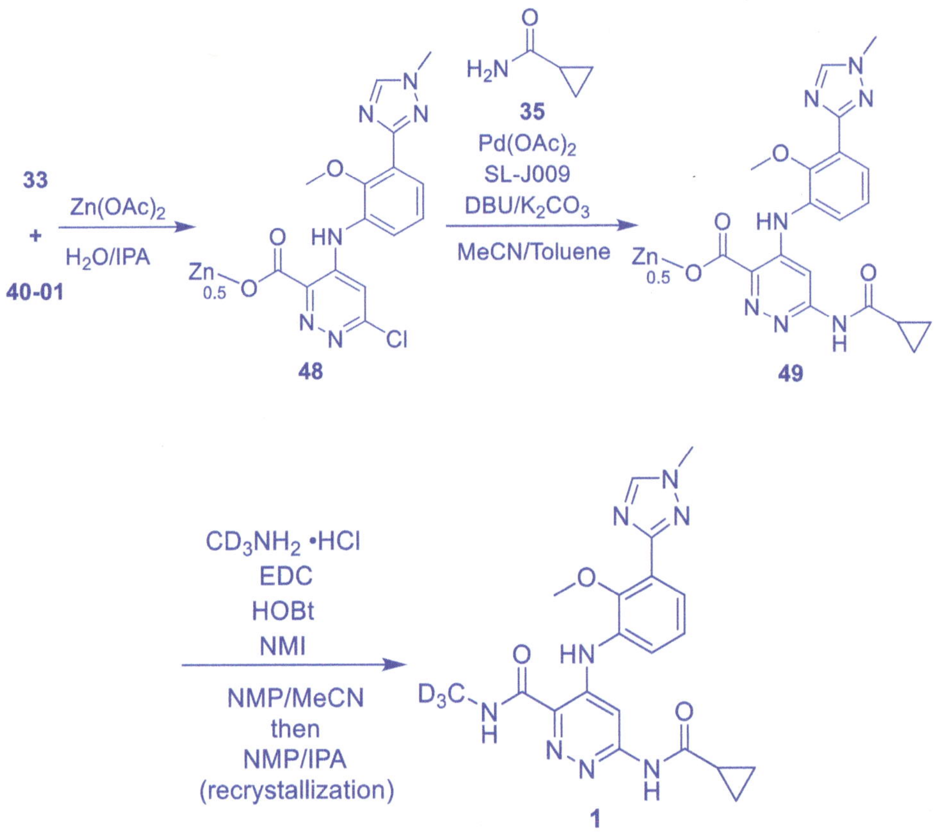

Scheme 9. Commercial process to deucravacitinib (**1**)

7. Summary

In summary, deucravacitinib (*Soyktu*) is a first-in-class, oral, selective, allosteric tyrosine kinase 2 (TYK2) inhibitor that was granted approval by the FDA in September 2022 for moderate-to-severe plaque psoriasis. The approval was based on results from two pivotal Phase 3 POETYK PSO-1 and POETYK PSO-2 clinical trials, which demonstrated superior efficacy of once-daily *Sotyktu* compared to placebo and twice-daily Otezla in 1684 patients aged 18 years and older with moderate-to-severe plaque psoriasis.

The approval of deucravacitinib represented a watershed moment for manufacturer and originator, BMS, with it being the first of its kind in more ways than one: it is the first approved drug targeting a pseudokinase domain. This unique allosteric mechanism differentiates deucravacitinib from all other reported JAK/TYK2 inhibitors due to its exquisite selectivity for TYK2 and its impressive benefit-safety profile that is not attainable through JAK1/2/3 inhibition. It is also the first approved de novo deuterated drug. A "magic methyl" on the amide moiety was found to be responsible for the remarkable selectivity through binding within a rare "alanine pocket" in the ligand binding domain of TYK2 JH2. This was protected from N-demethylation via deuteration, thus preventing formation of the less selective primary amide.

From identification of hits from an HTS campaign to extensive lead optimization, the discovery of deucravacitinib ran into many issues that are routinely encountered in a typical drug discovery program, including potency, selectivity, metabolism, permeability, bioavailability, hERG/cardiotoxicity etc. All of these issues were solved through innovative structure-based drug design strategies and very clever medicinal chemistry. Of note was the heterocyclic core modifications (pyridine to pyridazine) to provide optimal permeability and exposure, replacement of the C6 amino heterocycle with a cyclopropyl carboxamide to mitigate hERG activity, deuteration to mitigate levels of an undesired metabolite, and intelligent use of modelling to replace a structural water molecule within the binding site, leading to significant potency enhancement. These optimizations culminated in a molecule that had excellent PK properties across species with minimal profiling liabilities.

Key highlights from the process chemistry route included the development of a novel cyclocondensation reaction to afford a methylated 1,2,4-triazole with excellent regioselectivity, and implementation of the amidation with trideuteromethylamine as the final step, serving to simultaneously reduce costs and control residual metal levels. This exceptional work led to a highly cost-effective, efficient and sustainable commercial route that enabled the production of more than a metric ton of deucravacitinib.

The fascinating drug discovery and development story of deucravacitinib provides the reader with a cornucopia of knowledge in drug design, medicinal chemistry, and process development. The tour de force of many scientists across several disciplines, involved in bringing this innovative oral treatment to patients with moderate-to-severe plaque psoriasis, constitutes a truly heroic effort.

References

1. El-Gabalawy, H.; Guenther, L. C.; Bernstein, C.N. Epidemiology of immune-mediated inflammatory diseases: incidence, prevalence, natural history, and comorbidities. *J. Rheumatol.* **2010**, *37* (Suppl 85), 2–10.
2. https://www.psoriasis.org/psoriasis-statistics/
3. See: https://my.clevelandclinic.org/health/diseases/6866-psoriasis
4. Clark, J. D.; Flanagan, M. E.; Telliez, J.-B. Discovery and development of Janus kinase (JAK) inhibitors for inflammatory diseases. *J. Med. Chem.* **2014**, *57*, 5023–5038.

5. Zhang, K.; Ye, K.; Tang, H.; Qi, Z.; Wang, T.; Mao, J.; Zhang, X.; Jiang, S. Development and therapeutic implications of tyrosine kinase 2 inhibitors. *J. Med. Chem.* **2023**, *66*, 4378–4416.
6. JAK Inhibitors for Atopic Dermatitis (webmd.com)
7. Bronson, J. J. *Medicinal Chemistry Reviews*. **2018**, *53*, Chapter 10, 177–200.
8. Oyamada, A.; Ikebe, H.; Itsumi, M.; Saiwai, H.; Okada, S.; Shimoda, K.; Iwakura, Y.; Nakayama, K. I.; Iwamoto, Y.; Yoshikai, Y.; Yamada, H. Tyrosine kinase 2 plays critical roles in the pathogenic CD4 T cell responses for the development of experimental autoimmune encephalomyelitis. *J. Immunol.* **2009**, *183*, 7539–7546.
9. Ortmann, R.; Smeltz, R.; Yap, G.; Sher, A.; Shevach, E. M. A heritable defect in IL-12 signaling in B10.Q/J mice. I. In vitro analysis. *J. Immunol.* **2001**, *166*, 5712–5719.
10. J. D. Croxtall. Ustekinumab: a review of its use in the management of moderate to severe plaque psoriasis. *Drugs* **2011**, *71*, 1733–1753.
11. Tuskey, A.; Behm, B. W. Profile of ustekinumab and its potential in patients with moderate-to-severe Crohn's disease. *Clin. Exp. Gastroenterol.* **2014**, *7*, 173–179.
12. Furie, R.; Merrill, J. T.; Werth, V. P.; Kalunian, K.; Brohawn, P.; Illei, G. G.; Drappa J, Wang L, Yoo S.; CD1013 Study Investigators Anifrolumab, an anti-interferon alpha receptor monoclonal antibody, in moderate to severe systemic lupus erythematosus (SLE) [abstract #3223]. *Ann. Rheum. Dis.* **2016**, *75*, 168.
13. Norman, P. Selective JAK1 inhibitor and selective Tyk2 inhibitor patents. *Expert Opin. Ther.* **2012**, *22*, 1233–1249.
14. Liang, J.; van Abbema, A.; Balazs, M.; Barrett, K.; Berezhkovsky, L.; Blair, W.; Chang, C.; Delarosa, D.; De oss, J.; Driscoll, J.; Eigenbrot, C. Lead optimization of a 4-aminopyridine benzamide scaffold to identify potent, selective, and orally bioavailable TYK2 inhibitors. *J. Med. Chem.* **2013**, *56*, 4521–4536.
15. Saharinen, P.; Silvennoinen, O. The pseudokinase domain is required for suppression of basal activity of Jak2 and Jak3 tyrosine kinases and for cytokine-inducible activation of signal transduction. *J. Biolumin. Chemilumin.* **2002**, *277*, 47954–47963.
16. Saharinen, P.; Vihinen, M.; Silvennoinen, O. Autoinhibition of Jak2 tyrosine kinase is dependent on specific regions in its pseudokinase domain. *Mol. Biol. Cell* **2003**, *14*, 1448–1459.
17. Tokarksi, J. S.; Zupa-Fernandez, A.; Tredup, J. A.; Pike, K.; Chang, C.; Xie, D.; Cheng, L.; Pedicord, D.; Muckelbauer, J.; Johnson, S. R.; Wu,

S. Tyrosine kinase 2-mediated signal transduction in T lymphocytes is blocked by pharmacological stabilization of its pseudokinase domain. *J. Biolumin. Chemilumin.* **2015**, *290*, 11061–11074.

18. Zhou, Y.; Li, X.; Shen, R.; Wang, X.; Zhang, F.; Liu, S.; Li, D.; Liu, J.; Li, P.; Yan, Y.; Dong, P. Novel small molecule tyrosine kinase 2 pseudokinase ligands block cytokine-induced TYK2-mediated signaling pathways. *Front. Immunol.* **2022**, *13*, 1–17.
19. Burke, J. R.; Cheng, L.; Gillooly, K. M.; Strnad, J.; Zupa-Fernandez, A.; Catlett, I. M.; Zhang, Y.; Heimrich, E. M.; McIntyre, K. W.; Autoimmune pathways in mice and humans are blocked by pharmacological stabilization of the TYK2 pseudokinase domain. *Sci. Transl. Med.* **2019**, *11*, 1-16.
20. https://www.sotyktuhcp.com/mechanism-of-action
21. Nogueira, M.; Puig, L.; Torres, T. JAK inhibitors for treatment of psoriasis: focus on selective TYK2 inhibitors. *Drugs*, **2020**, *80*, 341–352.
22. Cesare, A. D.; Meglio, P. D.; Nestle, F. O. The IL-23/Th17 axis in the immunopathogenesis of psoriasis. *J. Invest. Dermatol.* **2009**, *129*, 1339–1350.
23. Moslin, R.; Gardner, D.; Santella, J.; Zhang, Y.; Duncia, J. V.; Liu, C.; Lin, J.; Tokarski, J. S.; Strnad, J.; Pedicord, D.; Chen, J. Identification of imidazo[1,2-*b*]pyridazine TYK2 pseudokinase ligands as potent and selective allosteric inhibitors of TYK2 signalling. *Med. Chem. Commun.* **2017**, *8*, 700–712.
24. Liu, C.; Lin, J.; Moslin, R.; Tokarski, J. S.; Muckelbauer, J.; Chang, C.; Tredup, J.; Xie, D.; Park, H.; Li, P.; Wu, D. R. Identification of imidazo[1,2-*b*]pyridazine derivatives as potent, selective, and orally active TYK2 JH2 inhibitors. *ACS Med. Chem. Lett.* **2019**, *10*, 383–388.
25. Moslin, R.; Zhang, Y.; Wrobleski, S. T.; Lin, S.; Mertzman, M.; Spergel, S.; Tokarski, J. S.; Strnad, J.; Gillooly, K.; McIntyre, K. W.; Zupa-Fernandez, A. Identification of N-methyl nicotinamide and N-methyl pyridazine-3-carboxamide pseudokinase domain ligands as highly selective allosteric inhibitors of tyrosine kinase 2 (TYK2). *J. Med. Chem.* **2019**, *62*, 8953–8972.
26. Hall, L. R.; Hanzlik, R. P. Kinetic deuterium isotope effects on the N-demethylation of tertiary amides by cytochrome P-450. *J. Biolumin. Chemilumin.* **1990**, *265*, 12349–12355.
27. Jamieson, C.; Moir, E. M.; Rankovic, Z.; Wishart, G. Medicinal chemistry of hERG optimizations: highlights and hang-ups. *J. Med. Chem.* **2006**, *49*, 5029–5046.

28. Wrobleski, S. T.; Moslin, R.; Lin, S.; Zhang, Y.; Spergel, S.; Kempson, J.; Tokarski, J. S.; Strnad, J.; Zupa-Fernandez, A.; Cheng, L.; Shuster, D. Highly selective inhibition of tyrosine kinase 2 (TYK2) for the treatment of autoimmune diseases: discovery of the allosteric inhibitor BMS-986165. *J. Med. Chem.* **2019**, *62*, 8973–8995.

29. See: https://www.accessdata.fda.gov/drugsatfda_docs/label/2022/214958s000lbl.pdf

30. Armstrong, A. W.; Gooderham, M.; Warren, R. B.; Papp, K. A.; Strober, B.; Thaçi, D.; Morita, A.; Szepietowski, J. C.; Imafuku, S.; Colston, E.; Throup, J. Deucravacitinib versus placebo and apremilast in moderate to severe plaque psoriasis: Efficacy and safety results from the 52-week, randomized, double-blinded, placebo-controlled phase 3 POETYK PSO-1 trial. *J. Am. Acad. Dermatol.* **2022**, *88*, 29–39.

31. Bristol Myers Squibb - U.S. Food and Drug Administration Approves Sotyktu™ (deucravacitinib), Oral Treatment for Adults with Moderate-to-Severe Plaque Psoriasis (bms.com)

32. Treitler, D. S.; Soumeillant, M. C.; Simmons, E. M.; Lin, D.; Inankur, B.; Rogers, A. J.; Dummeldinger, M.; Kolotuchin, S.; Chan, C.; Li, J.; Freitag, A. Development of a commercial process for deucravacitinib, a deuterated API for TYK2 inhibition. *Org. Process Res. Dev.* **2022**, *26*, 1202–1222.

Section V. MISCELLANEOUS DRUGS

16

Bremelanotide (Vyleesi): A Melanocortin Receptor Agonist for Treating Female Hypoactive Sexual Desire Disorder

Yan Wang

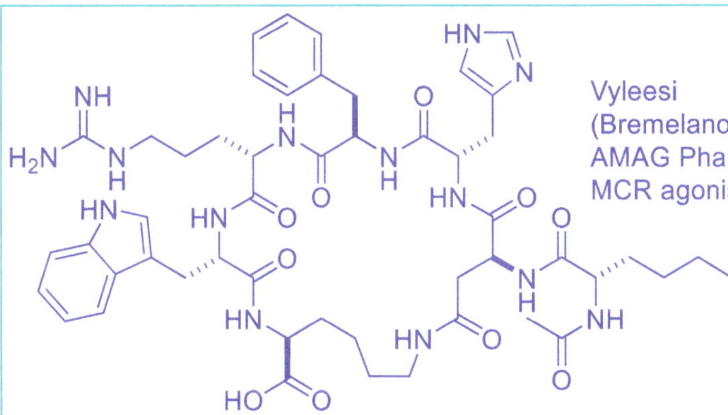

Vyleesi
(Bremelanotide, **1**)
AMAG Pharm., 2019
MCR agonist

1. Background

Hypoactive sexual desire disorder (HSDD)[1, 2] refers to a person's chronic or ongoing lack of interest in sex to the point of personal or relationship distress. It affects both men and women but is most common in women. This sexual dysfunction affects about 10% of adult women. Women with HSDD lack or loss of motivation to participate in sexual activity. Sexual dysfunction adversely impacts the quality of life in women and general well-being.

Bremelanotide (BMT, Vyleesi™, **1**) was approved by FDA to treat acquired, generalized HSDD in June 2019.[3] Vyleesi is administered by patients as a 1.75 mg subcutaneous injection in the abdomen or thigh using a single-use auto-injection pen,

Chemistry and Pharmacology of Drug Discovery, First Edition. Edited by Jie Jack Li.

45 minutes before anticipated sexual activity. Before BMT approval, Flibanserin (Addyi™, **2**)[4] was the only FDA-approved medication for female HSDD. Bupropion (**3**) and buspirone (**4**) may be considered off-label treatments for female HSDD despite limited safety and efficacy data.

Sexual desire is believed to be regulated by neuromodulators (neurotransmitters and hormones) of excitatory pathways [e.g. dopamine (**5**), norepinephrine (**6**), melanocortins (see Section 2.2), oxytocin (**7**)] and inhibitory pathways (e.g. serotonin (**8**), opioids, endocannabinoids (**9, 10**)).[5] See Figure 1 on excitatory and inhibitory effects of neurotransmitters and hormones on sexual desire.

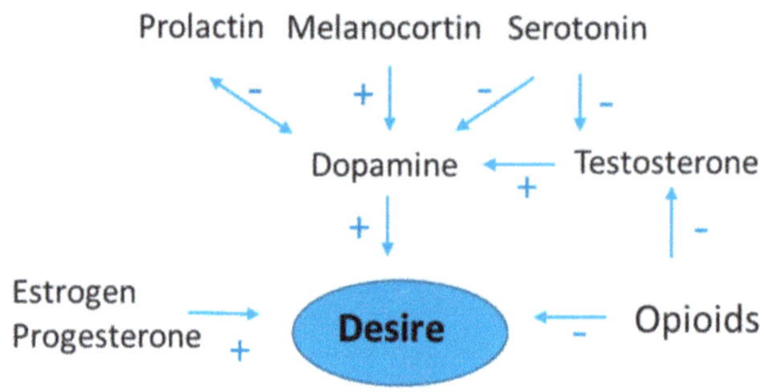

Figure 1. Excitatory (+) and inhibitory (–) effects of neurotransmitters and hormones on sexual desire

Flibanserin's mechanism of action is exerted through serotonin 1A (5HT-1A) receptor agonism and serotonin 2A (5HT-2A) receptor antagonism. This process reduces serotonin inhibition of excitatory neurotransmitters and thus indirectly increases the release of dopamine (**5**) and norepinephrine (**6**).[6, 7] BMT (**1**) is an agonist of melanocortin receptors (MCRs), and it nonselectively activates several of the receptor subtypes, of which subtype 4 is at relevant therapeutic doses.[8] Melanocortin-4 receptor (MC4R) is predominantly expressed in the brain's medial preoptic area (mPOA) of the hypothalamus and is important for female sexual function.

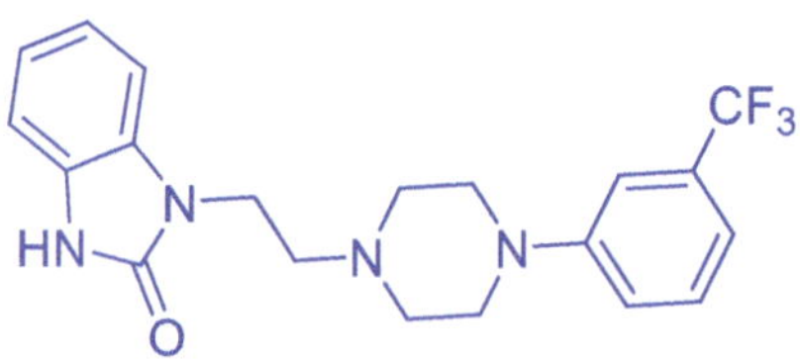

flibanserin (Addyi, **2**)
Sprout Pharmaceutical, 2015
Serotonin agonist and antagonist

Addyi (**2**) is an oral tablet taken once daily. It must be taken at bedtime to reduce the risk of low blood pressure, etc. There is concern about consuming alcoholic drinks before taking Addyi (**2**) at bedtime. Vyleesi (**1**) is a subcutaneously administered, on-demand therapy. It is used as needed, at least 45 minutes before anticipated sexual activity. It can be safely co-administered with ethanol.[9, 10]

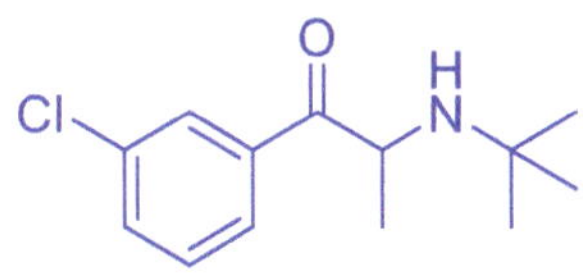

bupropion (**3**)

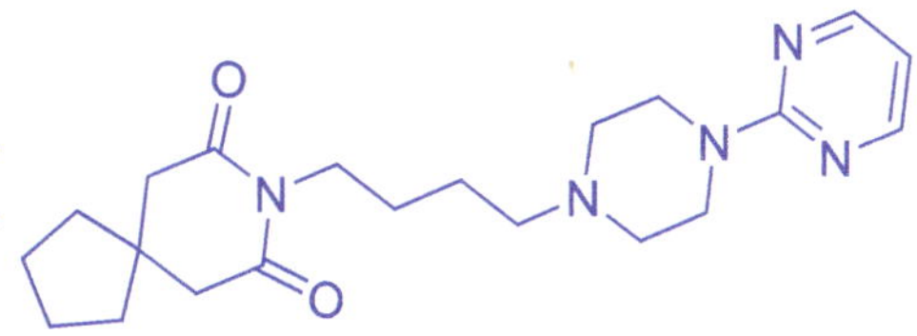

buspirone (**4**)

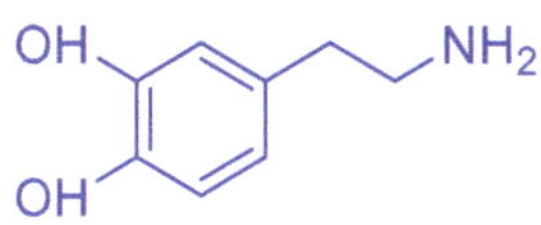

dopamine (**5**)

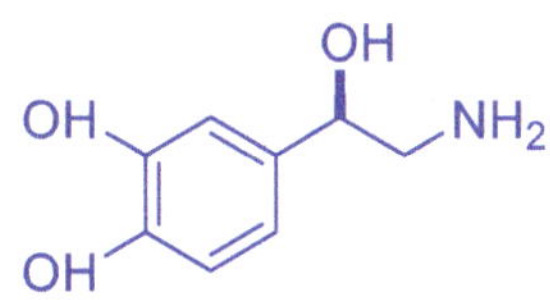

norepinephrine (**6**)

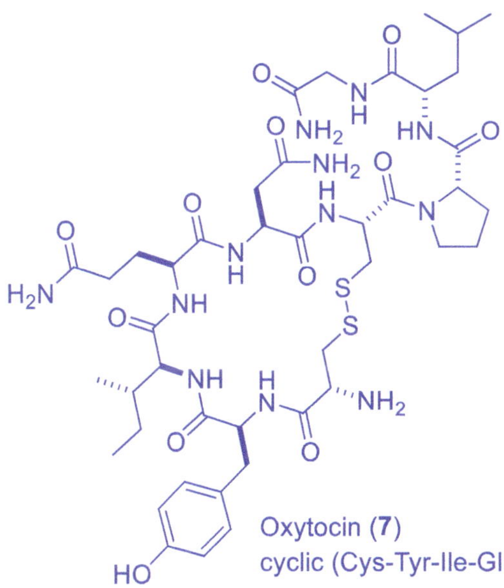

Oxytocin (**7**)
cyclic (Cys-Tyr-Ile-Gln-Asn-Cys)-Pro-Leu-Gly-NH_2

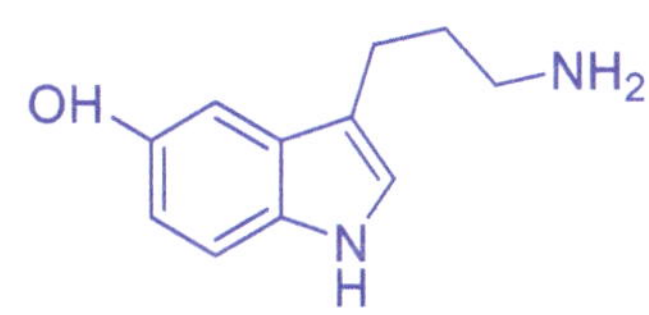

serotonin (**8**)

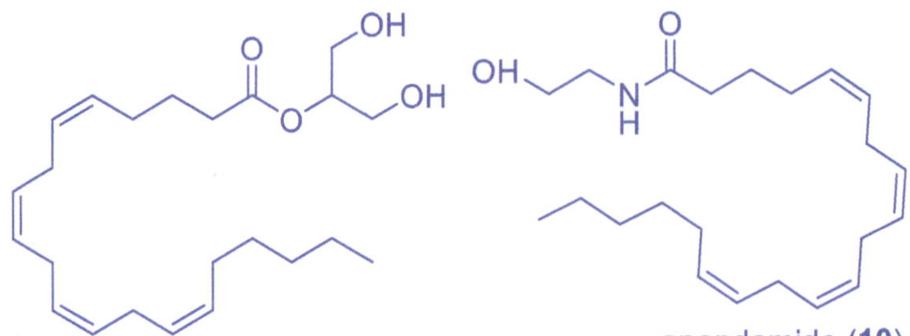

2. Pharmacology

2.1. MCRs Location and Functions

The MCR system consists of five 7-transmembrane G-protein-coupled receptors (GPCRs), generally called MC1R, MC2R, MC3R, MC4R, and MC5R. They signaled mainly through intracellular cyclic adenosine monophosphate.[11–13]

MC1R is found primarily in the periphery, especially in the skin; the MC2R in the adrenal gland; the MC3R in the brain and periphery; the MC4R in the brain and periphery; and the MC5R throughout the body.

The MCRs are involved in a diverse number of physiological functions. MC1R governs the mammalian skin and hair color by regulating melanin production. It plays an important role in the pigmentation process. MC2R, also known as the ACTH receptor, is located in the adrenal cortex and controls glucocorticoids (stress hormones). MC3R is responsible for body weight and appetite control. Recent studies also revealed that MC3R has crucial functions such as regulating hunger, appetite, and body weight. MC5R plays a key role in governing immune reactions and inflammatory responses.

MC4R is predominantly expressed in the mPOA of the hypothalamus in the brain. It is a critical regulator of energy homeostasis, including food intake and expenditure. It is also important for female sexual function. Animal studies suggest that BMT (**1**) may affect female sexual desire by activating presynaptic MC4Rs on neurons in the mPOA of the hypothalamus, leading to increased release of dopamine (**5**). This excitatory neurotransmitter increases sexual desire (see Figure 1).

2.2. Melanocortin Peptides

The natural agonist ligands (Table 1) for MCRs are derived by processing a primordial animal gene product, proopiomelanocortin (POMC). The ligand for the MC2R is adrenocorticotropic hormone (ACTH, **11**), a larger processed peptide from POMC. The natural ligands for the other four MCRs are smaller peptides, including α-melanocyte stimulating hormone (α-MSH, **12**) and related peptides from POMC (β-MSH **13** and γ^1-MSH, **14**, and γ^3-MSH, **15**). They all contain the sequence His-Phe-Arg-Trp that is conserved throughout evolution.[14]

Table 1. Sequences of human melanocortins

Peptide name	Sequences
ACTH (**11**)	Ser-Tyr-Ser-Met-Glu-**His-Phe-Arg-Trp**- Gly-Lys-Pro-Val-Gly-Lys-Lys-Arg-Arg-Pro-Val-Lys-Val-Tyr-Pro-Asn-Gly-Ala-Glu-Asp-Glu-Ser-Ala-Glu-Ala-Phe- Pro-Leu-Glu-Phe
α-MSH (**12**)	Ac-Ser-Tyr-Ser-Met-Glu-**His-Phe-Arg-Trp**-Gly-Lys-Pro-Val-NH_2
β-MSH (**13**)	Asp-Glu-Gly-Pro-Tyr-Arg-Met-Glu-**His-Phe-Arg-Trp**-Gly-Ser-Pro-Pro-Lys-Asp
γ^1-MSH (**14**)	Lys-Tyr-Val-Met-Gly-**His-Phe-Arg-Trp**-Asp-Arg-Phe-NH_2
γ^3-MSH (**15**)	Lys-Tyr-Val-Met-Gly-**His-Phe-Arg-Trp**-Asp-Arg-Phe-Gly-Arg-Arg-Asn-Ser-Ser-Ser-Ser-Gly-Ser-Ser-Gly-Ala-Gly-Gln

Agouti signaling protein (ASIP) and AgRP are endogenous MCRs antagonists (Table 2). ASIP is selective for MC1R and MC4R, but AgRP is a selective antagonist for MC3R and MC4R. See Table 2 on pharmacological properties of MCR subtypes.

Table 2. Pharmacological properties of MCR subtypes

MCRs	Agonist potency/endogenous antagonists
MC1R	α-MSH = ACTH > β-MSH >> γ-MSH/Agouti
MC2R	ACTH
MC3R	α-MSH = β-MSH = γ-MSH = ACTH/AgRP
MC4R	α-MSH = ACTH > β-MSH >> γ-MSH/Agouti, AgRP
MC5R	α-MSH > ACTH > β-MSH > γ-MSH

2.3. Bremelanotide

Palatin Technologies developed BMT (also known as PT-141, **1**). After the Phase III trials, Palatin Technologies partnered with AMAG Pharmaceuticals to complete the development and commercialization further.[15]

BMT (**1**) was originally developed as a potential sunless tanning agent and then was discovered to cause sexual arousal. Therefore, it was investigated for the treatment of male and female sexual dysfunction and turned into the treatment of premenopausal women with HSDD.[3, 16, 17] The BMT (**1**) mechanism of action for treating male sexual dysfunction is better understood. BMT (**1**) nonselectively activates several MCR subtypes. In males, it primarily acts on MC3R and MC4R to help treat erectile dysfunction. The stimulation of the MCRs, in general, causes a local increase of nitric oxide in the penis and leads to vasodilation and penile erection (Figure 2).

In females who suffer from HSDD, it is considered that an imbalance of various neurotransmitters causes abnormal sexual responses. Amongst the neurotransmitters, dopamine (**5**) and melanocortin stimulate attention and desire, while norepinephrine (**6**) and oxytocin (**7**) stimulate sexual arousal.[16] BMT (**1**) administration acts primarily on the presynaptic MC4R. It stimulates the release of dopamine (**5**) to portions of the nucleus accumbens, mPOA, verbal tegmental area, arcuate nucleus, and the medial and basolateral amygdala. These brain areas regulate the motivational, arousal, and appetitive aspects of sexual behavior (Figure 1).[18, 19]

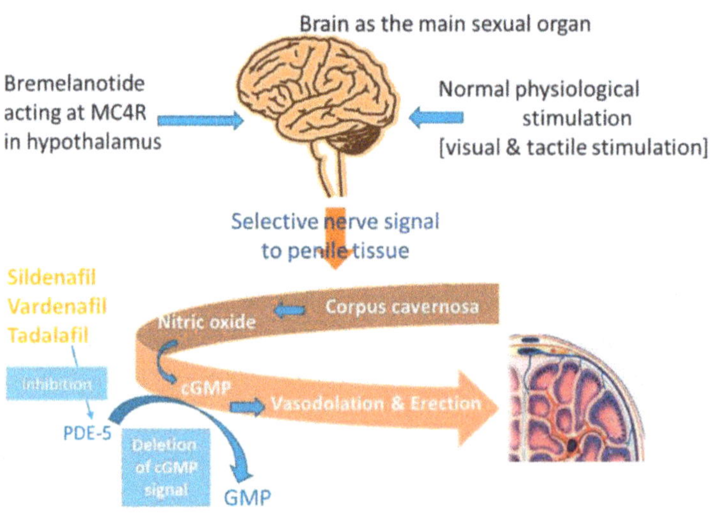

Figure 2. BMT (**1**), mechanism of action. Source: Adapted from Clayton et a l.[17]

3. Structure–Activity Relationship (SAR)

3.1. Structure–Activity Relationship of Bremelanotide

BMT (**1**) is a synthetic, cyclic heptapeptide with a free acid at the carboxyl terminus and an acetylated amino group at the amino terminus. The cyclization is between the side chains of Asp2 and Lys7 to form the lactam analog of α-MSH (**12**). BMT (**1**) sequence is Ac-Nle-cyclo-(Asp-His-DPhe-Arg-Trp-Lys)-OH.[16]

The des-α-MSH (**16**) peptide is derived from the N-terminal 13 residues of ACTH (**11**).[14] Both termini of α-MSH (**12**) are modified, with the N-terminal acetylation and the C-terminal carboxyamidation (Figure 3). The N-terminal acetylation increases the stability of α-MSH (**12**) compared to des-α-MSH (**16**). The plasma half-life of α-MSH (**12**) is 20–25 minutes in humans.[20] The full-length peptide possesses nonselective sub-nanomolar to nanomolar potencies to MCRs.

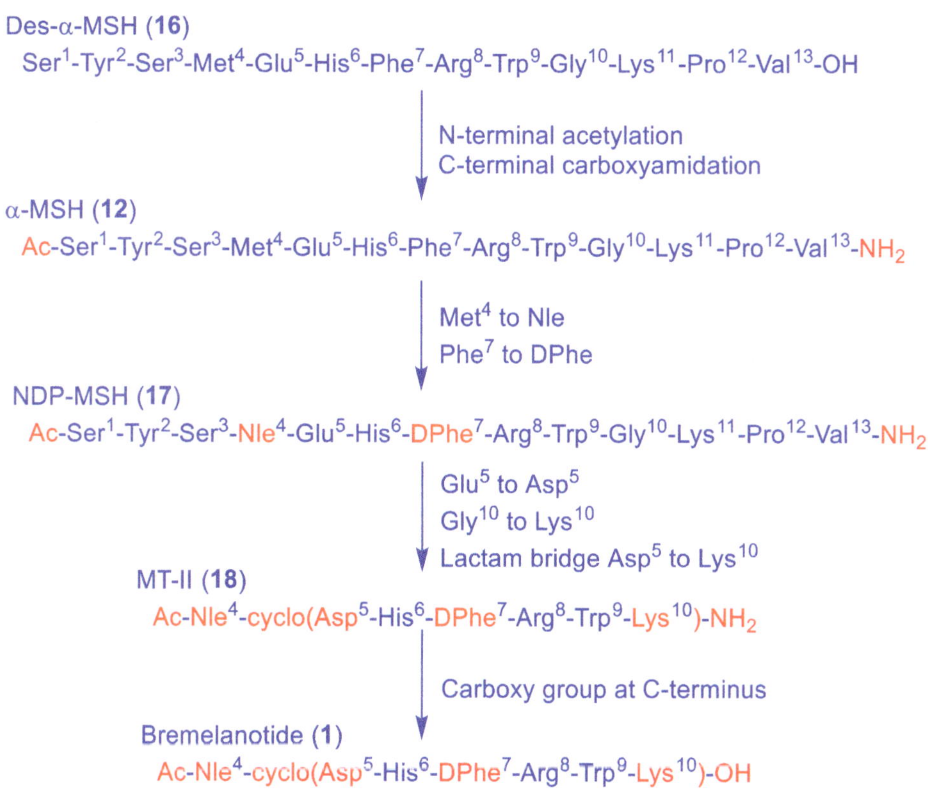

Figure 3. SAR of MCRs agonists

Since the sequence of α-MSH (**12**) was reported in 1957, this peptide has been subjected to numerous SAR studies, including the sequence's amino acid truncation and alanine scan.[21] As indicated in Section 2.2, the natural agonist ligands all have the invariant sequence His-Phe-Arg-Trp. It is the pharmacophore of the endogenous melanocortin ligands. Except that, Met4 and Glu5 also affect functional activity.

NDP-MSH (MT-I, **17**) is a synthetic α-MSH analog.[22] Two amino acids, Met4 and Phe7, were substituted in α-MSH (**12**) to yield NDP-MSH (**17**). Methionine was reported to be prone to oxidation when attempting to radiolabel α- or β-MSH. It was replaced with norleucine and increased the potency relative to α-MSH. The Phe7 to DPhe7 substitution was explored due to the observation that heat-alkali treatment of α-MSH (**12**) enhanced activity, and the Phe7 position was a major site of racemization. The DPhe-containing diastereoisomers (Figure 3) were at least an order of magnitude more potent than the L-Phe analogs. In 2024, NDP-MSH (**17**) was approved in the European Union as a treatment for adult erythropoietic protoporphyria.[23]

MT-II (**18**) is also a synthetic analog of α-MSH and MCR agonist.[24, 25] MT-II (**18**) is shorter than NDP-MSH (**17**)/α-MSH (**12**) after the truncation of three residues from both the N- and C-termini. It was cyclized through a lactam bridge between Asp5 and Lys10, maintaining the His-DPhe-Arg-Trp active tetrapeptide sequence of NDP-MSH (Figure 3). The rationale design of the cyclization site was based on the hypothesis that there is salt bridge between the Glu5 and Lys11 of α-MSH (**12**)/NDP-MSH (**17**) based upon NMR and computer modeling. In addition to the truncation and cyclization of MT-II (**18**), Glu5 and Gly10 were substituted with Asp5 and Lys10. MT-II (**18**) is still a potent, nonselective melanocortin ligand with agonist activity; K_i values are 0.67, 6.6, 34, and 46 nM for MC1, MC4, MC3, and MC5 receptors.

Competitive Technologies, a technology transfer company operating on behalf of the University of Arizona, licensed MT-II to Palatin Technologies as a sexual dysfunction agent.[26] Palatin ceased MT-II development (**18**) in 2000 and synthesized, patented, and began to develop BMT (**1**). It is a likely metabolite of MT-II (**18**) that differs in the C-terminus carboxy group where MT-II (**18**) has an amide.[27]

4. Pharmacokinetics and Drug Metabolism

After subcutaneous administration of BMT (**1**), its maximal levels occur after about 1 h, with a range of 0.5 to 1.0 h. And BMT (**1**) has 100% bioavailability with a subcutaneous injection. Its maximum plasma concentration reaches 72.8 ng/mL, and AUC is 276 h*ng/mL. Renal and hepatic impairment causes an increase in BMT's AUC. BMT's C_{max} level reaches its plateau after a 7.5 mg dose administration. Its plasma protein binding is 21%. BMT (**1**) is metabolized via hydrolysis of its peptide bonds. The elimination half-life of BMT is 2.7 h, with a range of 1.9 to 4.0 h. BMT (**1**) has a mean clearance of 6.5 ± 1.0 L/h. It is excreted 64.8% in urine and 22.8% in feces. BMT (**1**) decreases gastric emptying and has been shown to reduce the rate and extent of absorption of other orally administered drugs, particularly indomethacin (**19**) and naltrexone (**20**).[3, 28]

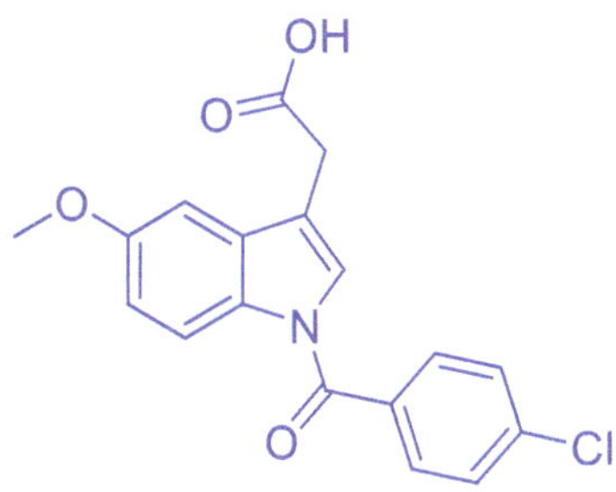

indomethacin (**19**)

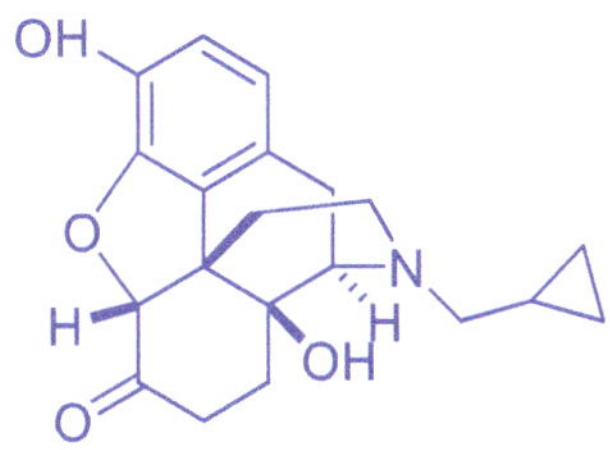

naltrexone (**20**)

5. Efficacy and Safety

In preclinical studies, PT-141 (**1**) effect on cAMP accumulation in cells expressing the human MC4R was measured.[16] PT-141 (**1**) inhibits ^{125}I-NDP-α-MSH binding to human MC4R and MC3R. PT-141 (**1**) shows a higher affinity for MC4R than it does for MC3R. PT-141 (**1**), as an MC4R agonist for the treatment of sexual dysfunction, was administrated to rats with different routes. It results in a significant increase in the number of erections. Direct injection into the lateral verntricle was eretogenic at doses 100- to 1000-fold lower than those required when given systemically.

In a placebo-controlled feasibility study, eighteen premenopausal women with a primary diagnosis of female sexual arousal disorder were randomly assigned to receive a single intranasal dose of 20 mg BMT (**1**).[29] More women reported moderate or high sexual desire following BMT (**1**) treatment, and a trend toward more positive responses regarding feelings of genital arousal occurred. The initial clinical evidence of potential BMT (**1**) benefit in female patients with impaired sexual arousal.

To evaluate the safety and tolerability of BMT (**1**), a phase I randomized double-blind study assessed the administration of BMT (**1**) in conjunction with ethanol to analyze pharmacokinetic interactions.[30] There were also no clinically significant changes in blood pressure and no noticeable pharmacokinetic interactions with BMT (**1**) administration alongside ethanol.

The phase IIB, placebo-controlled trial (NCT01382719) of BMT (**1**) was conducted in a well-characterized population of premenopausal women with HSDD, FSAD, or both to explore its safety and efficacy.[31] Subcutaneous doses of 0.75, 1.25, and 1.75 mg were selected. Self-administered subcutaneous BMT (**1**), taken as needed for up to 12 weeks, and indicated dose-responsive improvements in desire, arousal, and associated distress, as well as increases in the number of satisfying sexual events compared with placebo. Effective dosing with 1.25 and 1.75 mg was established.

In two identical phase III, placebo-controlled trials (NCT02333071 and NCT02338960) evaluated the safety and efficacy of BMT 1.75 mg administered subcutaneously as needed in premenopausal women with HSDD.[32] Both studies indicated that BMT significantly improved sexual desire and related distress in premenopausal women with HSDD.

The phase I study results indicated that single doses of up to 10 mg (healthy male subjects) and 6 mg (erectile dysfunction patients) of BMT (**1**) were safely administered and well tolerated.[33] No dosing adjustments are recommended for patients with mild to moderate renal and hepatic impairment.[3] The most commonly reported side effects are flushing, headaches, nausea, and skin irritation. Most events were reported to be mild (31%) to moderate (40%) in intensity and transient.[31, 32, 34]

BMT (**1**) can significantly decrease the absorption of oral naltrexone (**20**), and its concomitant use is not recommended. Intramuscular naltrexone (**20**) can be an alternative for this subset of patients.[3]

6. Synthesis

BMT (**1**) is synthesized using the standard solid-phase peptide synthesis (SPPS) method and supplied in acetate salt.[35–38]

In the patent of Palatin Technologies, they reported the linear peptide synthesis using SPPS with Fmoc chemistry. The patent indicated that the lysine side chain protecting group can be Adpoc, Aloc, and Mtt. The aspartic acid side chain was protected with an allyl group.

As an example, Fmoc-Lys(Aloc)-OH was coupled to the resin (**21**) to yield the first amino acid loaded resin (**22**). The resin (**21**) was not specified in the patent. After linear peptide assembly (**23**), the Lys and Asp side chain deprotected peptidyl resin (**24**) was suspended in a suitable solvent (DMF, DCM, or NMP), an appropriate coupling reagent (TBTU, TATU, TPTU, or DCC/HOBt) was added, and coupling reaction was initiated using a suitable base (DIPEA or NMM) for cyclization. Then, the cyclized peptidyl resin (**25**) was washed, and the peptide was cleaved from the resin using trifluoroacetic acid (TFA) in the presence of water and ethanedithiol (EDT) (Scheme 1).

The crude of the final product (**1**) was precipitated with cold ether and collected by filtration. Final purification was done using reversed-phase HPLC with a C18 column. The purified peptide was converted to acetate salt with an ion-exchange column.

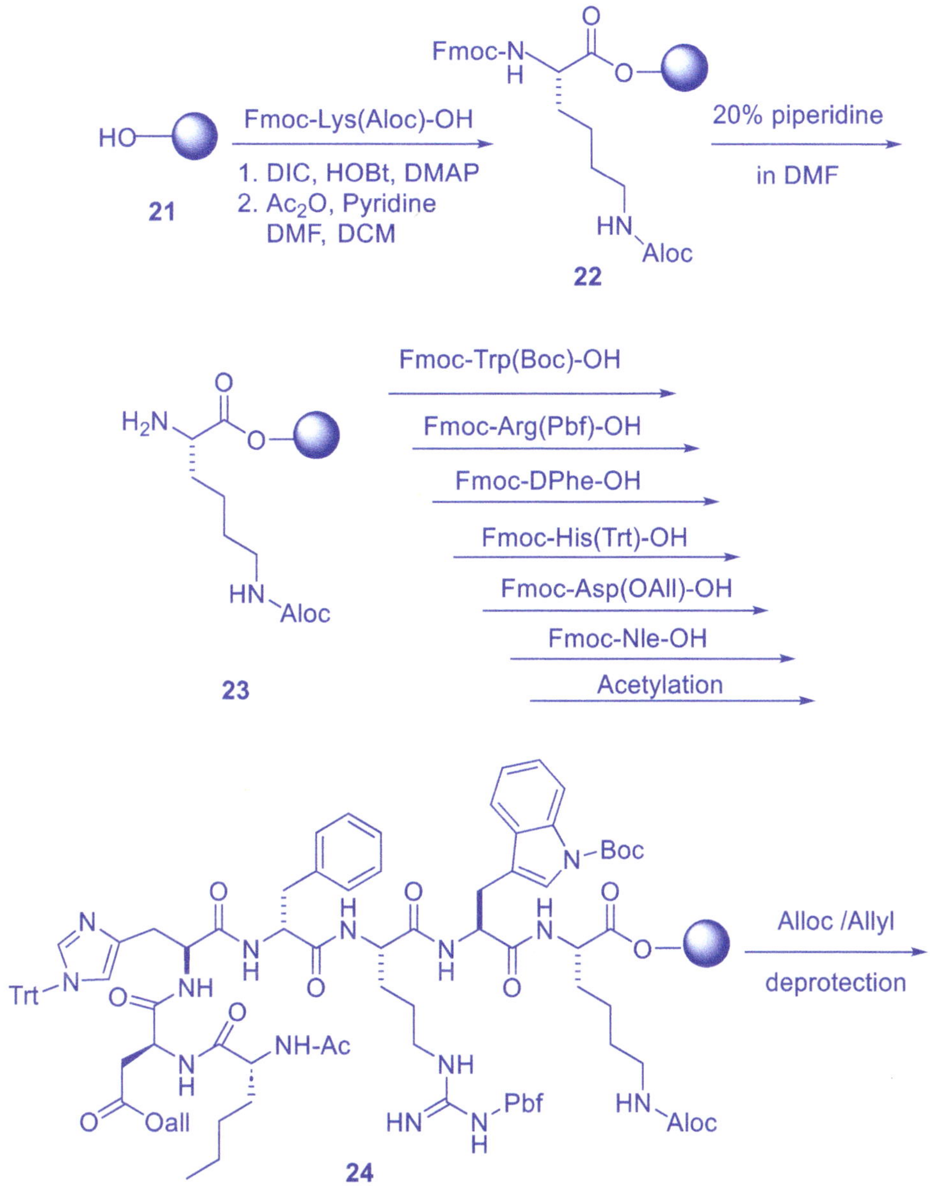
HO
21
Fmoc-Lys(Aloc)-OH
1. DIC, HOBt, DMAP
2. Ac2O, Pyridine
DMF, DCM
Fmoc-N
H
O
O
HN
Aloc
22
20% piperidine
in DMF
H2N
O
O
HN
Aloc
23
Fmoc-Trp(Boc)-OH
Fmoc-Arg(Pbf)-OH
Fmoc-DPhe-OH
Fmoc-His(Trt)-OH
Fmoc-Asp(OAll)-OH
Fmoc-Nle-OH
Acetylation
N-Boc
Trt-N
N
NH
NH-Ac
Oall
NH
HN
Pbf
HN
Aloc
24
Alloc /Allyl
deprotection

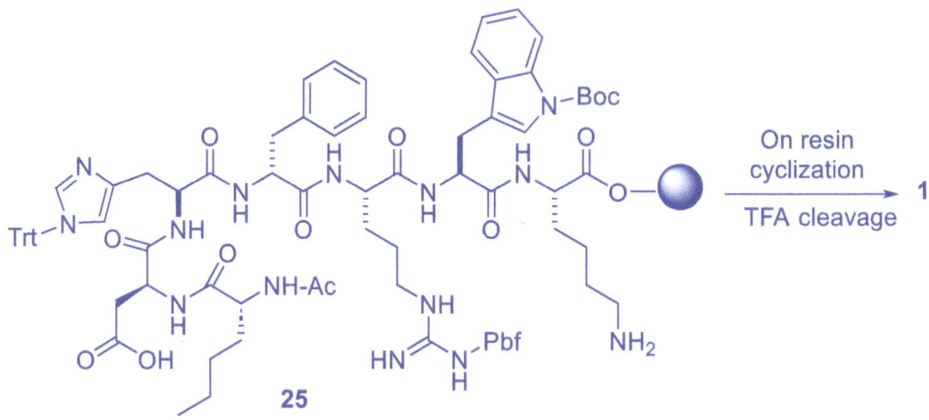

Scheme 1. Solid phase peptide synthesis of BMT (**1**)

In another patent, due to the concern of aspartimide formation during aspartic acid coupling step, a different synthesis strategy is adopted.[35, 38]

The carrier resin is Fmoc-Linker-AM resin (Scheme 2). The linker, 4-hydroxybenzoic acid (HMBA), was coupled to the AM-resin (**26**) using conventional condensation conditions with DIC and HOBt, furnishing modified HMBA-Rink-Amide-AM-resin (**27**, Scheme 2).

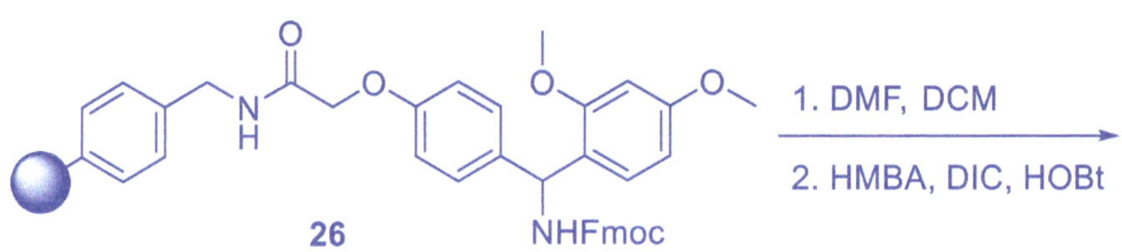

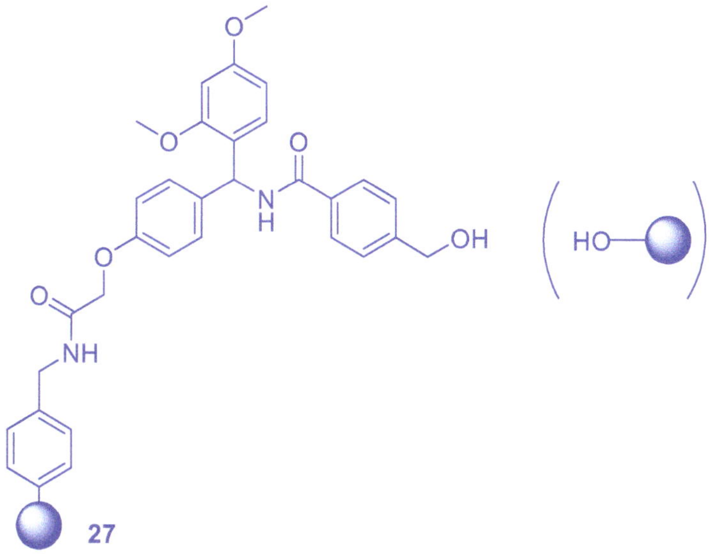

27

Scheme 2. HMBA-Rink-Amide-AM-resin preparation

The first amino acid, Fmoc-Lys(ivDde)-OH, is loaded to the HMBA-Rink-Amide-AM-resin (**27**) through esterification to give **28** (Scheme 3). Then *N*-Fmoc protecting group was deprotected with 20% piperidine/DMF solution to give **29**. The coupling of amino acids and Fmoc deprotection were continuously repeated to give **30**. The SPPS does not continue with aspartic acid coupling in the next step due to the concern of aspartimide formation. Therefore, the ivDde protection group was removed on resin using hydrazine monohydrate to yield **31**. Coupling of Ac-Nle-Asp-OtBu (**33**) to **32** generates final linear peptidyl resin **34**. The peptide was cleavage from the resin using TFA cocktail (TFA:PhSMe:Phenol:EDT = 90:5:2:3). The crude peptide was cyclized in DMF:DCM = 2.5:1 using PyBop as condensation reagent. The protecting group of C-termini COOH was hydrolyzed to get the final crude product of BMT (**1**). Final purification was done as described above.

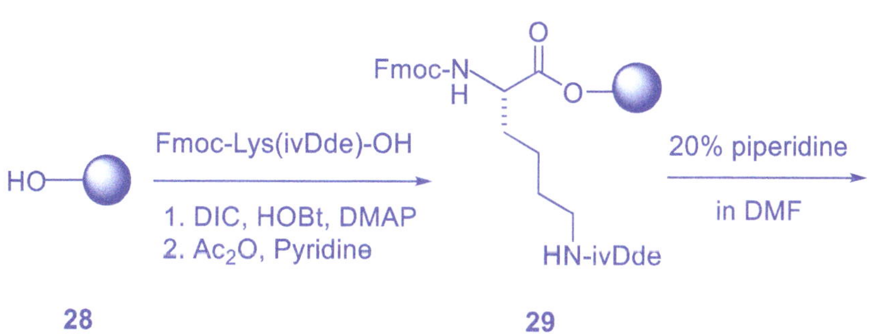

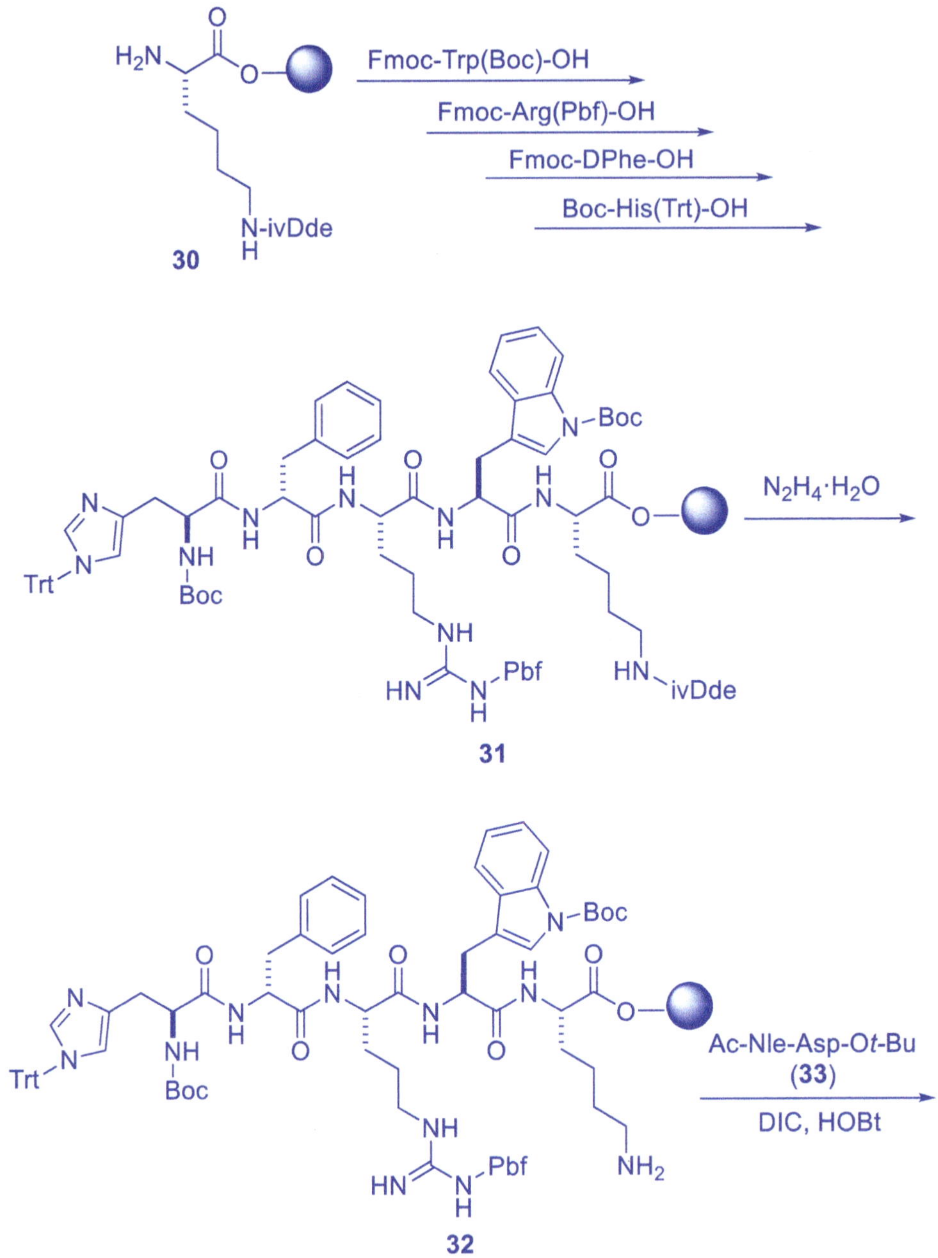
Fmoc-Trp(Boc)-OH
Fmoc-Arg(Pbf)-OH
Fmoc-DPhe-OH
Boc-His(Trt)-OH
30
31
$N_2H_4 \cdot H_2O$
32
Ac-Nle-Asp-Ot-Bu
(33)
DIC, HOBt

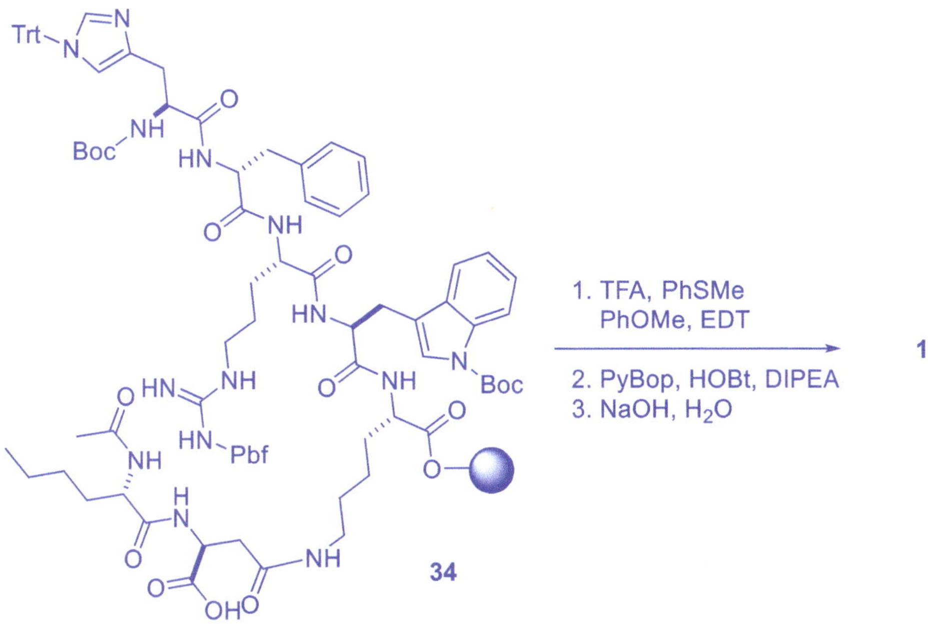

Scheme 3. Synthesis of BMT (**1**)

7. Summary

HSDD is a multifaceted disorder involving biological, psychological, and pharmacological influences. It often goes underdiagnosed and undertreated due to the private nature of the condition, making it difficult for patients to discuss with physicians. Female sexual dysfunction has not been studied as extensively as male sexual dysfunction. FDA approval of Vyleesi (**1**) provides women with another treatment option for HSDD. However, Spielmans reanalyzed phase III BMT (**1**) trials and suggested that the data reporting and measurement practices were incomplete and lacked transparency. Therefore, the drug is generally not useful. The debate on BMT (**1**) benefit to the patient and the unknown exact mechanism of action in improving female HSDD indicate that female HSDD treatment is still an unmet medical need.

References

1. (a) Clayton, A.; Kingsberg, S.; Goldstein, I. Evaluation and management of hypoactive sexual desire disorder. *J. Sex. Med.* **2017**, *6*, 59–74. (b) Goldstein, I.; Noel, K.; Clayton, A. Hypoactive sexual desire disorder: International Society for the Study of Women's Sexual Health

(ISSWSH) expert consensus panel review. *Mayo Clin Proc.* **2017**, *92*(1), 114–128.

2. FDA approves new treatment for hypoactive sexual desire disorder in premenopausal women, U.S. Food & Drug Administration. https://www.fda.gov/news-events/press-announcements/fda-approves-new-treatment-hypoactive-sexual-desire-disorder-premenopausal-women. Released June, **2019**.
3. Dooley, E.; Melanie, M.; Clayton, A. Flibanserin: from bench to bedside. *Sex. Med. Rev.* **2017**, *5*, 461–469.
4. Clayton, A. The pathophysiology of hypoactive sexual desire disorder in women. *Int. J. Gynaecol. Obstet.,* **2010**, *10* (1), 7–11.
5. Stahl, S.; Sommer, B., Allers, K. Multifunctional pharmacology of flibanserin: possible mechanism of therapeutic action in hypoactive sexual desire disorder. *J. Sex. Med.* **2011**, *8*, 15–27.
6. Katz, M.; DeRogatis L.; Ackerman, R.; Hedges, P.; Lesko, L.; Garcia, M.; Sand, M. Efficacy of flibanserin in women with hypoactive sexual desire disorder: results from the BEGONIA trial. *J. Sex. Med.* **2013**, *10*, 1807–1815.
7. Yuan, X.; Tao, Y. Ligands for melanocortin receptors: beyond melanocyte-stimulating hormones and adrenocorticotropin. *Biomolecules* **2022**, *12*, 461–469.
8. Shou, Y.; Yu, B.; Liu, H. New drug approvals for 2019: synthesis and clinical applications. *Eur. J. Med. Chem.* **2019**, *205*, 1–5.
9. Vyleesi vs Addyi: What is the difference? https://www.drugs.com/medical-answers/vyleesi-addyi-difference-3501189/. Last updated Feb. **2023**.
10. Cone, R. Studies on the physiological functions of the melanocortin system. *Endocr. Rev.* **2006**, *27*(7), 736–749.
11. Yanik, T.; Durban, S. Specific functions of melanocortin 3 receptor (MC3R). *J. Clin. Res. Pediatr. Endocrinol.* **2023**, *15* (1), 1–6.
12. Wolf Horrell, E.; Boulanger, M.; D'Orazio, J. Melanocortin 1 receptor: structure, function, and regulation. *Front. Genet.* **2016**, *7*, 1–16.
13. Jégou, S.; Cone, R.; Eberlé, A.; Vaudry, H. Chapter-111. Melanocortins. *Handbook of Biologically Active Peptides*, Elsevier Inc., **2013**, 838–844.
14. Palatin Technologies. https://palatin.com/press_releases/palatin-technologies-announces-mutual-termination-of-license-agreement-with-amag-pharmaceuticals-for-vyleesi-2/. Released July, **2020**.

15. Molinoff, P.; Shadiack, A.; Earle, D.; Dimond, L. PT-141: a melanocortin agonist for the treatment of sexual dysfunction. *Ann. N. Y. Acad. Sci.* **2003**, *994*, 96–102.

16. Shadiack, A.; Sharma, S.; Earle, D.; Spana, C.; Hallam, T. Melanocortins in the treatment of male and female sexual dysfunction. *Curr. Top. Med. Chem.* **2007**, *7*, 1137–1144.

17. Clayton A.; Lucas, J.; Jordan, R.; Spana, C.; Pfaus, J. The neurobiology and efficacy of bremelanotide in HSDD. *J. Sex. Med.* **2017**, *14*, E95.

18. Pfaus, J.; Kippin, T.; Coria-Avila, G. What can animal models tell us about human sexual response? *Annu. Rev. Sex Res.* **2003**, *14*, 1–63.

19. Berna, I.; Oktar, K. The actions of the α-melanocyte stimulating hormone (α-MSH) in inflammatory conditions. *Marmara Med. J.* **2001**, *14* (2), 113–118.

20. Hadley, M.; Dorr, R. Melanocortin peptide therapeutics: historical milestones, clinical studies and commercialization. *Peptides* **2006**, *27*, 921–930.

21. Sawyer, T.; Sanfilippo, P.; Hruby, V.; Engel, M.; Heward, C.; Burnett, J.; Hadley, M. 4-Norleucine, 7-D-phenylalanine-alpha-melanocyte-stimulating hormone: a highly potent alpha-melanotropin with biological activity. *PNAS* **1980**, *77* (10), 5754–5758.

22. Luger, T.; Bohn, M. An α-MSH analog in erythropoietic protoporphyria. *J. Infect. Dis.* **2015**, *135*, 929–931.

23. Al-Obeidi, F.; Castrucci, A.; Hadley, M.; Hruby, V. Potent and prolonged acting cyclic lactam analogues of α-melanotropin: design based on molecular dynamics. *J. Med. Chem.* **1989**, *32*, 2555–2561.

24. Al-Obeidi, F.; Hadley, M.; Pettitt, B.; Hruby, V. Design of a new class of superpotent cyclic α-melanotropins based on quenched dynamic simulations. *JACS* **1989**, 111, 3413–3416.

25. Hadley, M. Discovery that a melanocortin regulates sexual functions in male and female humans. *Peptides,* **2005**, *26* (10), 1687–1689.

26. Hedlund, P. PT-141 palatin. *Curr. Opin. Invest. Drugs,* **2021**, *26* (10), 456–462.

27. Bremelanotide. https://go.drugbank.com/drugs/DB11653. Accessed December, **2021**.

28. Diamond, L.; Earle, D.; Heiman, J.; Rosen, R.; Perelman, M.; Harning, R. An effect on the subjective sexual response in premenopausal women with sexual arousal disorder by bremelanotide (PT-141), a melanocortin receptor agonist. *J. Sex. Med.* **2006**, *3* (4), 628–638.

29. Clayton, A. H.; Lucas, J.; DeRogatis, L. R.; Jordan, R. Phase I randomized placebo-controlled, double-blind study of the safety and

tolerability of BMT coadministered with ethanol in healthy male and female participants. *Clin. Ther.* **2017**, *39*, 514–526.

30. Clayton, A.; Altof, S.; Kingsbery, S.; DeRogatis, L.; Kroll, R.; Goldstein, R.; Kaminetsky, J.; Spana, C.; Lucas, J.; Jordan, R.; Portman, D.; Bremelanotide for female sexual dysfunctions in premenopausal women: a randomized, placebo-controlled dose-finding trial. *Womens Health* **2016**, *12* (3), 325–337.

31. Kingsberg, S.; Clayton, A.; Portman, D.; Willams, L.; Krop, J.; Jordan, R.; Lucas, J.;Simon, J. Bremelanotide for the treatment of hypoactive sexual desire disorder. *OBGYN* **2019**, *134* (5), 899–908.

32. Rosen, R.; Diamond, L.; Earle, D.; Shadiack, A.; Molinoff, P. Evaluation of the safety, pharmacokinetics and pharmacodynamic effects of subcutaneously administered PT-141, a melanocortin receptor agonist, in healthy male subjects and in patients with an inadequate response to Viagras. *Int. J. Impot. Res.* **2004**, *16*, 135–142.

33. https://www.accessdata.fda.gov/drugsatfda_docs/ label/2019/ 210557s000lbl.pdf. Revised June **2019**.

34. Flora, D.; Mo, H.; Mayer, J.; Khan, M.; Yan, L. Detection and control of aspartimide formation in the synthesis of cyclic peptides. *Bioorg. Med. Chem. Lett.* **2005**, *15*, 1065–1068.

35. Blood, C.; Shadiack, A.; Bernstein, J.; Herbert, G. Compositions and methods for sexual dysfunction. Patent US6794489B2, **2004**.

36. Liao, Z.; Yang, Y. Solid phase synthesis of bremelanotide. Patent CN101857629, **2010**.

37. Yan, L.; Wang, B.; Li, J.; Jin, Y.; Wang, J.;Yang, Z. A synthetic method for bremelanotide. Patent CN106589111A, **2017**.

38. Spielmans, G. Reanalyzing phase III bremelanotide trials for "hypoactive sexual desire disorder" in women. *JSR* **2021**, *58*, 1–20.

17

Odevixibat (Bylvay): A Selective Inhibitor of the Ileal Bile Acid Transporter

Andrew Outlaw and
Timothy A. Cernak

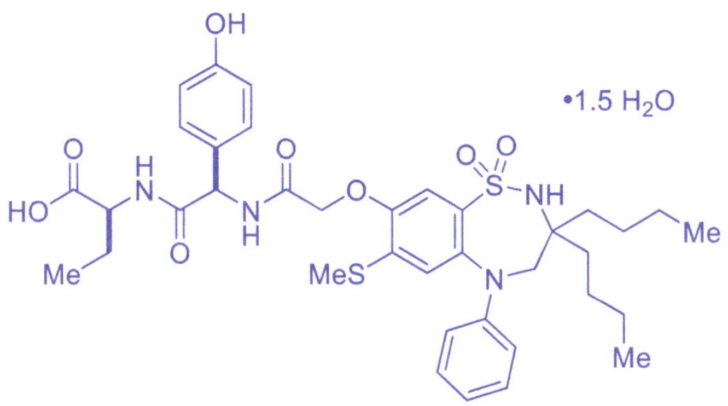

odevixibat (Bylvay, **1**)
Albireo Pharma/Ipsen, 2021/2023
Ileal Bile Acid Transport Inhibitor

Much work has been done in recent years for developing novel treatment options for orphan diseases. One recent case is that of Albireo Pharma's Ileal bile acid transporter (IBAT) inhibitor odevixibat (**1**). As of 2021, odevixibat (**1**) has been approved by the FDA as a first-in-class treatment option for pruritus in patients 3 months of age and older with progressive familial intrahepatic cholestasis (PFIC). Following the 2023 acquisition of Albireo by Ipsen, odevixibat (**1**) has also been approved to treat pruritus in patients 12 months and up with Alagille syndrome (ALGS). Odevixibat (**1**) is currently marketed

Chemistry and Pharmacology of Drug Discovery, First Edition. Edited by Jie Jack Li.

under the commercial name Bylvay and sold as daily oral pellets or capsules for patient populations affected by cholestatic liver disease.

The road toward safe and effective treatment options for cholestasis-inducing disorders such as PFIC and ALGS is one lined with many hurdles. Cholestasis treatments that properly balance efficacy while still managing adverse effects have largely remained elusive. While a host of treatment options exist for managing cholestasis, if the adverse effects of a treatment match the severity of the symptoms it aims to alleviate, the very point of the treatment is lost for many. The approval of odevixibat (**1**) marks a significant hallmark in treating cholestatic liver disease for its efficacy and general tolerability, as well as its validation of targeting inhibition of the ileal bile acid transporter mechanism.

1. Background

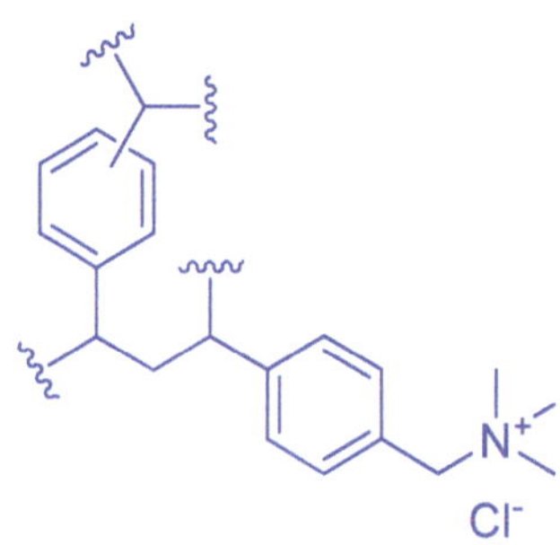

cholestyramine (**2**)
Bile Acid Sequestrant

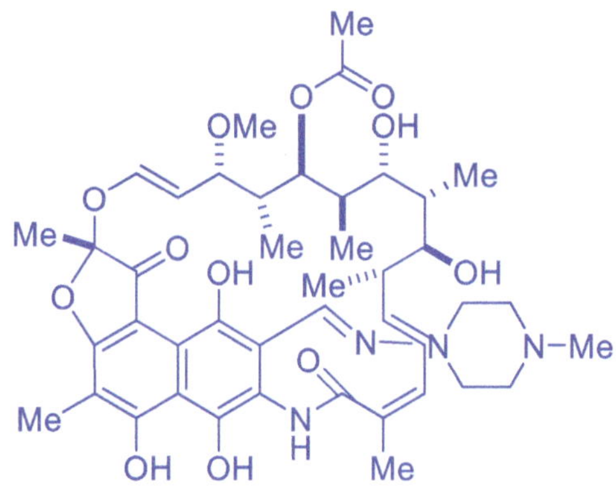

rifampicin (**3**)
Antimycobacterial

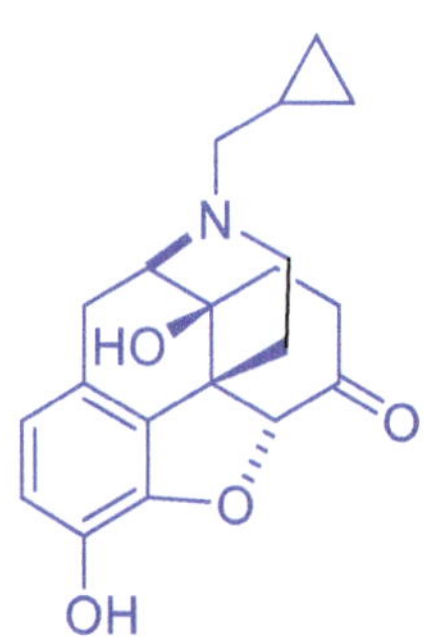

naltrexone (**4**)
μ-Opioid Receptor Antagonist

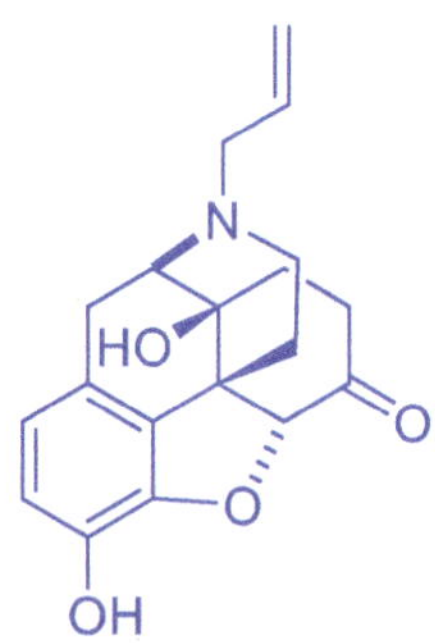

naloxone (**5**)
μ-Opioid Receptor Antagonist

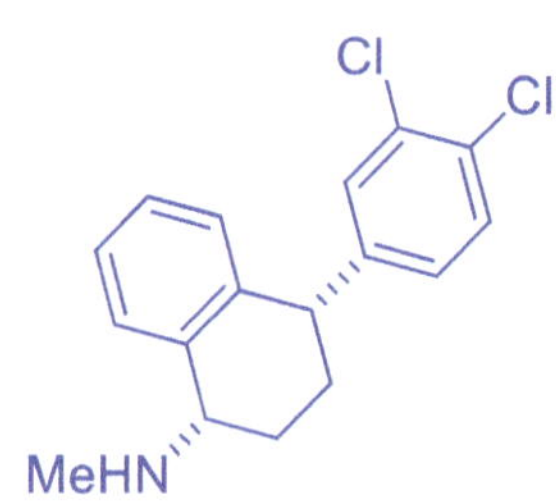

sertraline (**6**)
SSRI

Historically, medications for cholestatic liver disease include first-line treatment with the bile acid sequestrant cholestyramine (**2**), the antimycobacterial second-line treatment rifampicin (**3**), and alternative options like the μ-opioid receptor antagonists naltrexone (**4**) or naloxone (**5**) or the selective serotonin re-uptake inhibitor (SSRI) sertraline (**6**).[1] The glaring issue with many of these options lies in poor palatability and patient compliance due to adverse effects like gallstone formation, hepatitis, vomiting, or loss of control of pain.[2–5]

Ileal bile acid transport inhibitors were largely considered experimental treatments prior to the recent development of IBAT inhibitors. Albireo's odevixibat (**1**) and elobixibat (**7**) along with Mirum Pharmaceutical's maralixibat (**8**) present compelling cases for treating cholestasis via modulation of bile acid reuptake.[6–8] IBAT inhibitors work to lower the bile acid pool in the liver by preventing reuptake of bile acids in the intestinal ileum via competitive inhibition of IBAT.[9] Thus bile acid load in the colon increases, preventing bile acids recirculating back into the liver.

Odevixibat (**1**) is a "Me-Too" drug based on modification of the IBAT inhibitor elobixibat (**7**) earlier developed by Albireo.[10] While elobixibat (**7**) showed a promising early example of the benzothiazepine scaffold for IBAT inhibition, it never completed clinical trials in the United States.[11,12] Instead, Albireo came to a licensing agreement with Ajinomoto Pharmaceuticals to market the drug in Japanese and Asian markets in 2012.

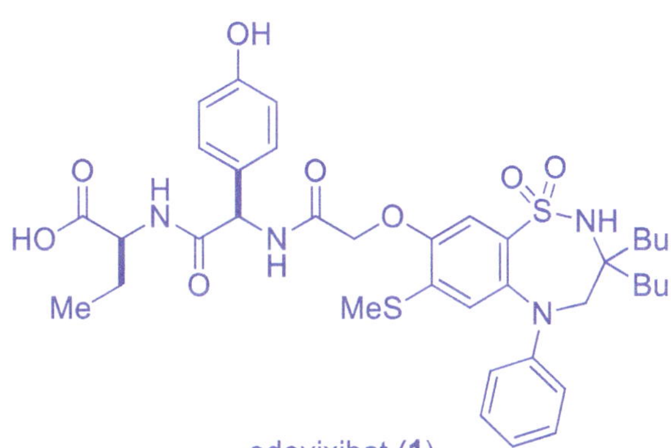

odevixibat (**1**)
IBAT IC_{50} = 0.16 nM

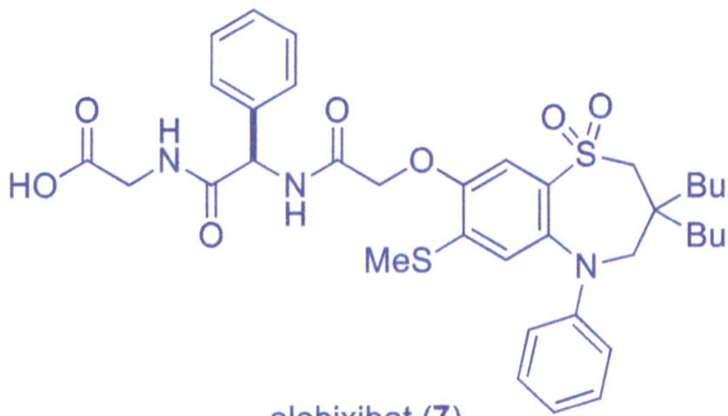

elobixibat (**7**)
IBAT IC_{50} = 1.2 nM

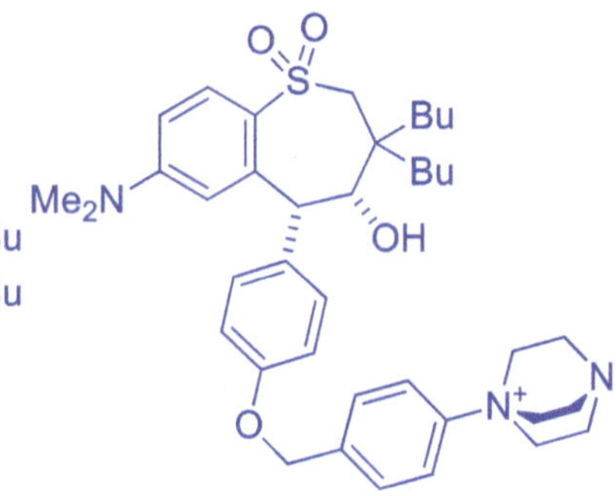

maralixibat (**8**)
IBAT IC_{50} = 0.28 nM

Picking back up on the benzothiazepine structure of elobixibat in the late 2010s, Albireo made further modification of the peripheral structure of elobixibat (**7**) to include alteration of the core sulfone to a sulfonamide along with a chiral ethyl chain and phenolic hydroxyl group. These alterations give the final odevixibat (**1**) structure with improved potency for IBAT in humans (IC_{50} = 0.16 nM vs 1.2 nM) while maintaining its tolerability and safety profile.[13] Completing clinical trials in the United States and receiving FDA approvals in 2021 and 2023, odevixibat (**1**) is considered a first-in-class medication for treatment of cholestatic pruritus resulting from PFIC and ALGS.[14] As we are now done outlining the history of drugs for treating cholestatic liver disease, we now dive into the pharmacology of ileal bile acid transport inhibitors.

2. Pharmacology

PFIC and ALGS are both disorders that affect the enterohepatic flow of bile in the body. PFIC is a hereditary disorder that affects the ability of liver cells to properly secrete bile within the liver.[15] Similarly, ALGS is a hereditary disorder that results in the development of too few bile ducts to export bile out of the liver.[16] Both conditions lead to cholestasis, or reduced flow of bile out from the liver, in neonates. Proper diagnosis and treatment within the first several months of life is often necessary to prevent scarring and potential failure of the liver.

2.1. Enterohepatic Circulation of Bile Acids

Bile is a physiological solution released from the liver that helps to separate the nutrients from the waste in the digestive system. Bile is primarily composed of bile acids and bile acid-conjugates derived from cholesterol (**9**) such as cholic acid (**10**) and taurocholic acid

(**11**) along with a range of other components including cholesterol, bilirubin, and phospholipids.[17] Amongst these, bile acids make up roughly 80% of the total bile composition and play a critical role in emulsifying and breaking down fats and fat-soluble vitamins for digestion.[18]

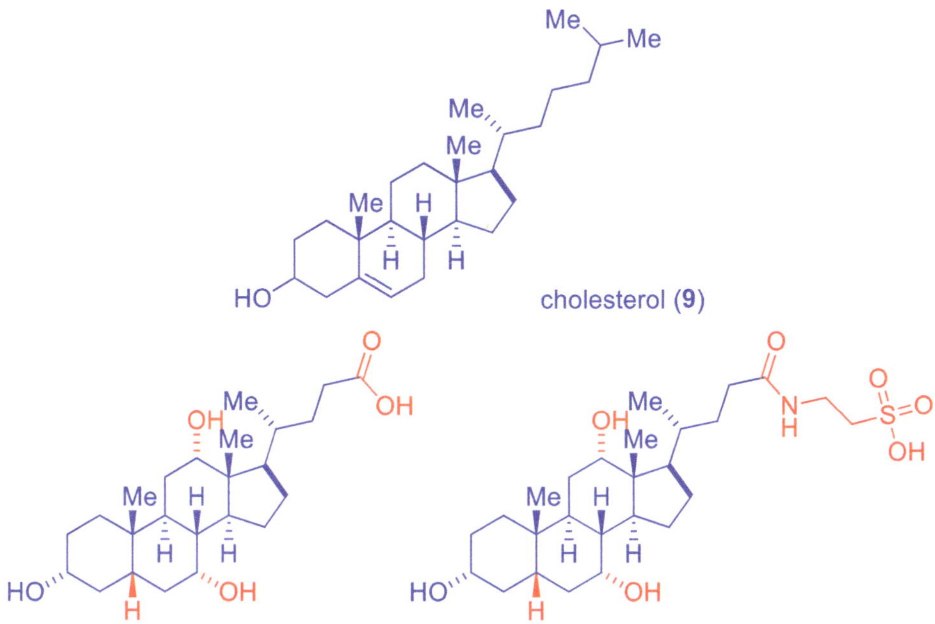

cholesterol (**9**)

cholic acid (**10**)
Bile Acid

taurocholic acid (**11**)
Bile Acid Conjugate

During the digestive process, bile is secreted from liver cells and released from the liver through a series of bile ducts to perform its job as a digestive aid and detergent in the GI tract. Upon reaching the ileum of the small intestine, the digested nutrients are absorbed into the hepatic portal vein and filtered through the liver to enter systemic circulation. Up to 95% of the bile acids initially secreted during digestion re-enter enterohepatic circulation at this point via the IBAT.[19] IBAT, also known as the apical sodium-dependent bile acid transport (ASBT), is a cotransporter glycoprotein. It can recognize cholesterol derived bile-acids for transport into the portal vein to be recycled in the liver for later digestive cycles.[18]

Any dysfunction affecting enterohepatic bile flow can have serious long-term health effects for patients. As both PFIC and ALGS are genetic conditions, they are present at birth and often necessitate rapid diagnosis and treatment of the resulting liver diseases.

2.2. Genetic Disorders Affecting Bile Acid Homeostasis

The prevalence of genetically inherited disorders such as PFIC and ALGS are quite rare. PFIC is estimated to occur only 1 in 50,000 to 100,000 births while ALGS is found in 1 out of 70,000 to 100,000 births. While both disorders are hereditary, PFIC is autosomal recessive, requiring the mutant copy of the appropriate gene to be passed down from both parents simultaneously for a chance the child develops PFIC.[20] ALGS on the other hand is autosomal dominant and only requires a single copy of the mutant gene to be passed from parent to child to have potential affect.[21]

The genetic mutations associated with PFIC occur on the ATP8B1, ABCB11, ABCB4, and Myo5B genes along with loss of function of the TJP2 and FXR proteins.[22] The ATP8B1 gene codes for the protein responsible for helping bile acids cross cell membranes so that they can be released from liver cells. The ABCB11 gene encodes information for the bile salt export pump which also assists in bile acid export out of the intracellular matrix. The ABCB4 gene is responsible for a protein that enables phospholipid binding of bile acids to help carry them in the extracellular matrix. Mutations of these three genes are responsible for PFIC types 1, 2, and 3, respectively. Also, loss of function of the Tight Junction Protein 2 (TJP2), Farnesoid X Receptor (FXR), and genetic mutation of the Myo5B gene are associated with the newer PFIC types 4, 5, and 6.

In the case of ALGS, the culprit is loss of function in the NOTCH signaling pathway due to mutation of the JAG1 gene in most cases, or a small portion of cases resulting from NOTCH2 gene mutation.[21] Mutation of the JAG1 gene either results in the production of an abnormally short Jagged-1 protein that does not properly span the transmembrane region of the cell or interferes with transport of the protein to the cell membrane. In either case, loss of function of the Jagged-1 protein affects intercellular communication via the highly conserved NOTCH signaling pathway responsible for development of functioning bile ducts and proper cardiac function.[23]

2.3. Structure and Function of the Ileal Bile Acid Transporter

The human ileal bile acid transporter is a sodium-dependent 48-kDa glycoprotein consisting of 348 amino acids that is encoded by the SLC10A2 gene. It is primarily expressed on the apical surface of intestinal ileal enterocytes along with lesser expression in renal tubule cells and the lining of biliary cholangiocytes. While the exact structure of the human ileal bile acid transporter has not been fully elucidated, it is believed to express seven transmembrane topologies, similar to other bile acid transporters like NTCP.[24] As mentioned previously, its main purpose is the reabsorption of bile acids and bile acids conjugated to taurine or glycine from the intestines to the liver. IBAT functions as a cotransporter which utilizes a sodium gradient to transport bile acids across the cellular membrane.[25] For every equivalent of bile acid transported, two equivalents of sodium ions transport in the same direction.

Photo affinity labeling and enzymatic digestion studies have helped discern that the substrate binding domain of the human IBAT likely lies in the seventh transmembrane helix and C-terminus 56 through 67th amino acids. While heavily debated, computational models for IBAT inhibitors tend to fall into the regime of suggesting a hydrophobic core with hydrogen-bond acceptor and potentially hydrogen-bond donor characteristics that extend deeper into the substrate binding pocket.[26, 27]

Work on the structures of bacterial homologs of IBAT in *Yersinia frederiksenii* and *Neisseria meningitidis* has further unveiled some of the mechanisms of the transporter.[25] Each of the bacterial homologs host a ten-transmembrane helical structure instead of the seven helices in humans. In bacterial homologs, TM helices 3–5 and 8–10 form the sodium-binding core motif of the transporter which can accept two sodium ions largely coordinated by adjacent carbonyl backbone oxygen atoms. TM helices 1, 2, 6, and 7 form separate V-shaped motifs, which together with the adjacent core transmembrane helices form the hydrophobic bile acid binding pocket (Figure 1).

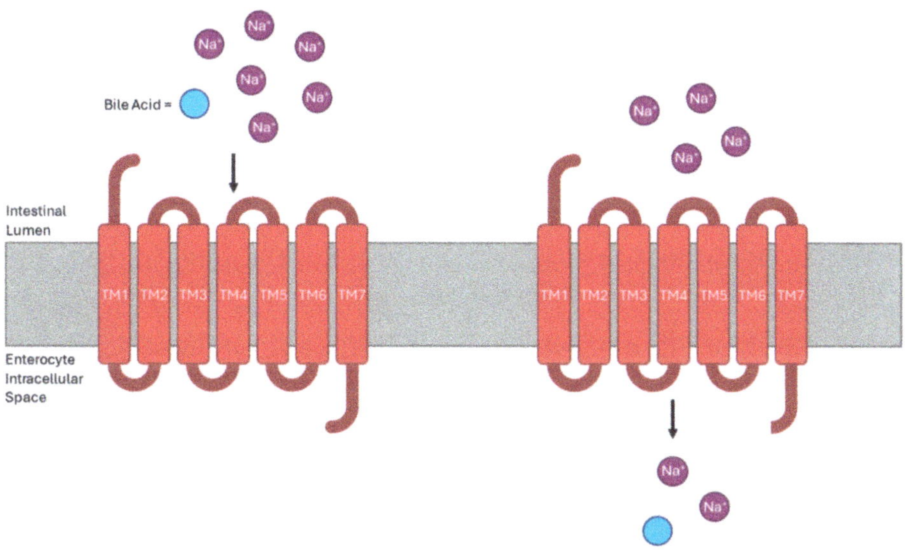

Figure 1. Function of human ileal bile acid transporter (IBAT).

The core motif shuttles two sodium ions from the intestinal lumen through the enterocyte's apical membrane through a sequence of oxygen and nitrogen atom chelation. At the same time, bile acids fit into the substrate-binding cavity between the core and V-shaped motifs to be transported through the apical membrane as well. The bile acid then crosses the basolateral membrane via the organic solute transporter alpha-beta (Ostα-Ostβ) to feed into the portal vein and back to the liver for use in later digestive cycles.

3. Early Inhibitors of the Ileal Bile Acid Transporter

There have been many approaches to modulating bile acid levels including bile acid sequestrants, direct dosing of secondary bile acids like ursodeoxycholic acid, or modulation of receptors involved in bile acid synthesis like FXR[28]. Aside from bile acid sequestrants, most medications in this space have historically mimicked cholesterol structure.

In the pursuit of novel treatment methods for hypercholesterolemia, the Burroughs Wellcome Company published preliminary work on a benzothiazepine structure in 1995. Compound 2164U90 (**12**) was unveiled with an IC_{50} of 2 μM against the transport of taurocholic acid by human ileal brush border membrane vesicles.[29, 30] They also discovered that the 2164U90 structure functions as a reversible competitive inhibitor of the ileal bile acid transporter through a series of experiments with tritium labeled taurocholic acid. In contrast to the practice of using cholesterol derivatives to modulate bile acid levels, the discovery of this potent benzothiazepine scaffold for IBAT inhibition offered a novel perspective in the field.

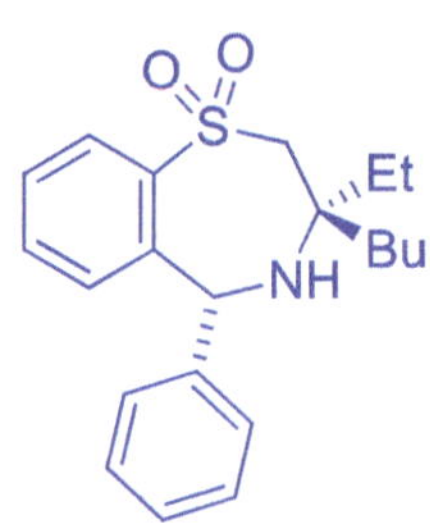

2164U90 (**12**)
human ileal membrane vessicle IC_{50} = 2 μm

4. Structure–Activity Relationship (SAR)

To follow up on compound 2164U90 (**12**), several teams explored the SAR for benzothiazepine based IBAT inhibitors. In 2004, a team at the University of Missouri worked on expanding the SAR for benzothiepine inhibitors related to 2164U90 (**12**).[31, 32] Their early studies focused on increasing potency and identifying the number of chiral centers necessary for activity in the thiepine ring. Through a seven-step synthesis to reach thiepines **13** and **14**, the chirality effects of the 3, 4, and 5 positions of the core ring system were interrogated (Table 1).

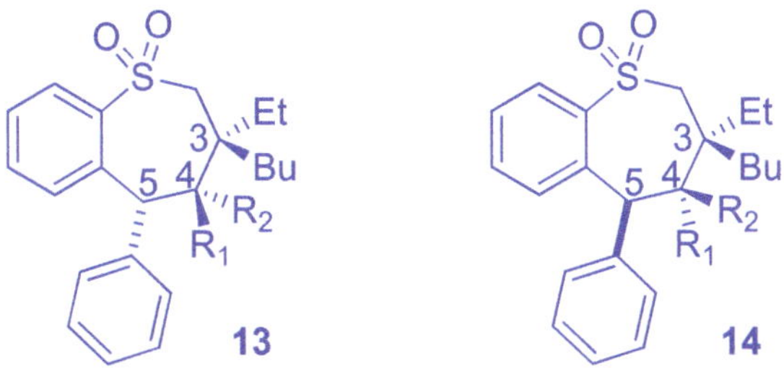

13 **14**

Compound	R1	R2	IBAT IC_{50} (μM)
(±)-**13a**	H	OH	0.28
(±)-**13b**	OH	H	4.8
(±)-**14a**	H	OH	2
(±)-**14b**	OH	H	37% at 10 μM

Table 1. Evaluation of the C3, 4, and 5 chiral centers

Both compounds **13a** and **14a** with cis confirmations between the C4 hydroxyl and C5 phenyl ring outperformed their trans counterparts, **13b** and **14b**. Additionally, between these two, *cis* orientation of the phenyl, hydroxyl, and C3 ethyl group further boost the potency, with compound **13a** giving an *in vitro* IC_{50} of 0.28 μM against human IBAT.

From here, substitution patterns on benzo-fused ring were looked at (Table 2). It was clear that substitution at the 7-position is preferred, compared to similar substitution at the 8-position, with electron-donating amines outperforming electron-withdrawing amide counterparts. There is also some inference as to binding pocket size found in their series of bulkier alkyl amine-containing derivatives. Where the amine **17** and dimethyl amine **18** occupy appropriate space in the binding pocket, hexylamine **19** is over 300-fold less potent than **18**. The same trend follows in amide derivatives **20** and **21**.

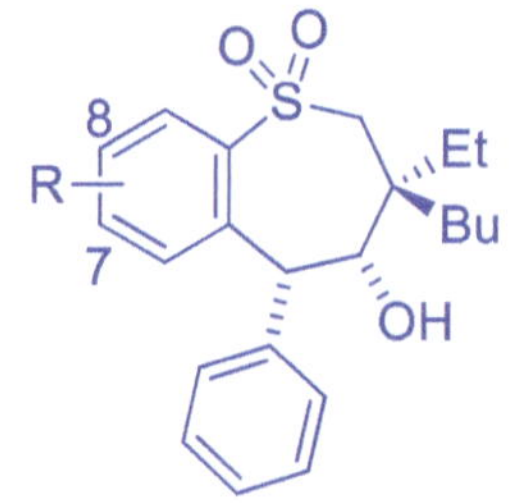

Compound	R	IBAT IC_{50} (μM)
(±)-**13a**	H	0.28
15	7-OMe	0.055
16	8-OH	0.09
17	7-NH_2	0.068
18	7-NMe_2	0.005
19	7-NHC_6H_{13}	1.67
20	7-$NHCO_2CH_2Ph$	0.4
21	7-$NHCOCH_3$	0.09

Table 2. Optimization of fused benzene ring substituents

From this point, the C3 stereocenter of **18** was amended to an achiral di-butyl substitution to simplify the structure down to only two chiral centers with compound **22**. With the 5 nM compound **22** in hand, focus turned toward improving PK properties of their lead.

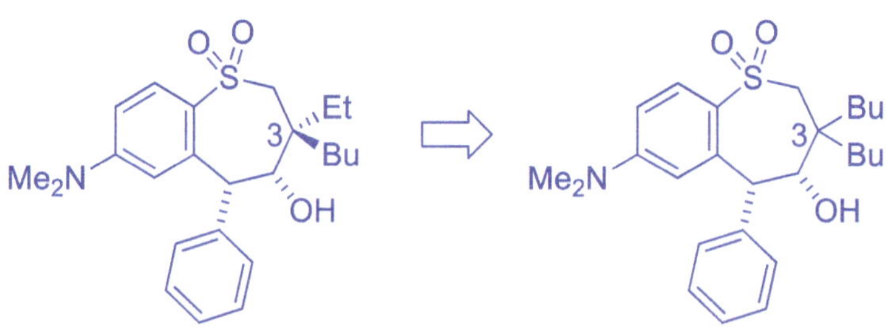

18
IBAT IC_{50} = 5 nM

22
IBAT IC_{50} = 5 nM

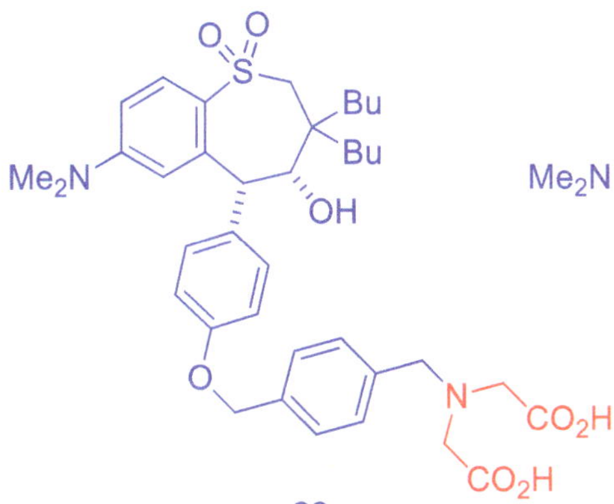

23
IBAT IC_{50} = 2 nM
C_{max} portal = 0 μg/mL
C_{max} artery = 0 μg/mL
% in bile = 0.12

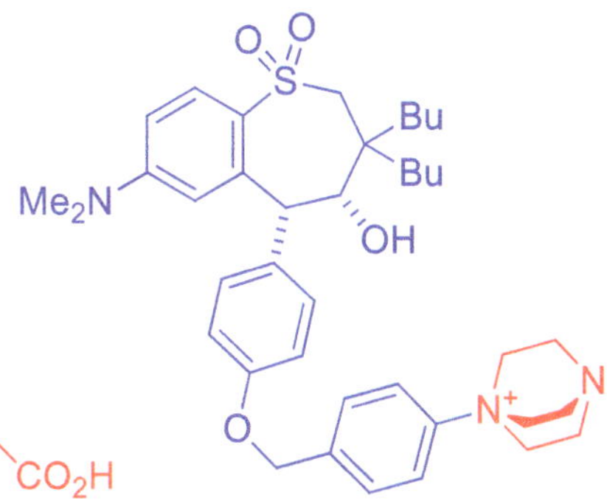

maralixibat (**8**)
IBAT IC_{50} = 0.28 nM
C_{max} portal = 0 μg/mL
C_{max} artery = 0 μg/mL
% in bile = 0.11

To prevent undesirable systemic exposure, they worked on further developing compound **22** by amending highly polar functionalities to the C5-phenyl ring to promote intestinal localization. Dicarboxylic acid **23** and quaternary ammonium salt **8** (which would later come to be known as maralixibat) improved the potency down to 2 and 0.28 nM, respectively, while eliminating systemic exposure to these compounds.

At the same time, AstraZeneca and its spinoff Albireo began pumping out patents in the same space on benzothiazepine and benzothiadiazepine inhibitors bearing peptide chains as the means of modulating PK properties.[33–35] With these patents came the release of elobixibat (**2**), which despite its impressive 1.2 nM IC_{50}, only exhibited a 28% inhibition of human IBAT *in vivo* at 0.156 μmol/kg dosage.[13]

Many derivatives synthesized from elobixibat's (**7**) core included short peptides with 1–3 amino acids in length and all tended toward low nanomolar activity. Trending once again toward improved gut localization with more heteroatoms for decreased intestinal permeability, amending on 2-amino acid peptides became the most popular option to simplify synthesis (Table 3).

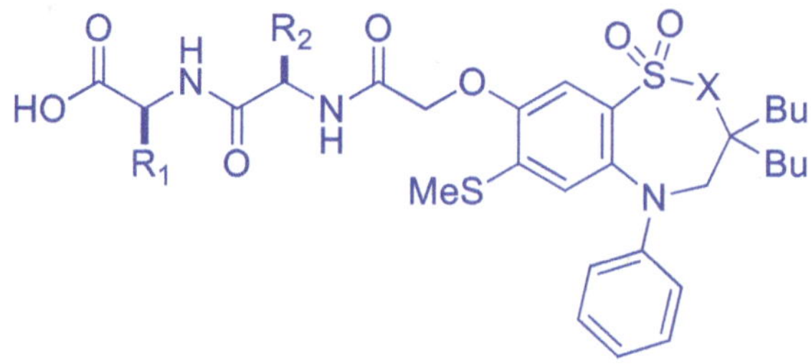

Compound	X	R_1	R_2	IBAT IC_{50} (nM)
Elobixibat (7)	CH_2	H	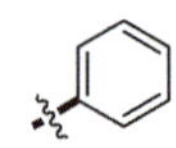	1.2
24	NH	H	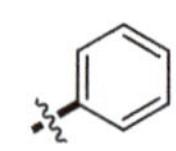	0.45
25	NH	Me	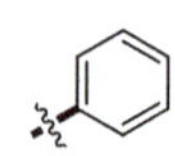	0.2
26	NH	CH_2SMe	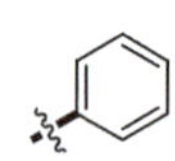	0.35
27	NH	Me	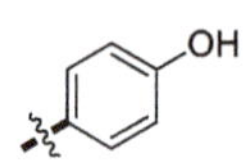	0.3
Odevixibat (1)	NH	Et	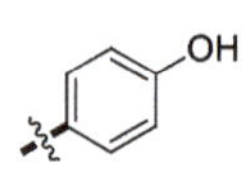	0.16

Table 3. Optimization of peptide chain substituents

Comparing otherwise similar analogues, insertion of the sulfonamide functionality into the benzothiazepine ring resulted in almost a 3-fold increase in potency from elobixibat's structure (**2**) to compound **24**. At the R_1 position, various small alkyl substituents were all well tolerated, with ethyl groups showing the best results. Finally, swapping from an unsubstituted phenyl group at the R_2 position to a phenol leads us to the structure of odevixibat (**1**), showing an excellent 0.16 nM IC_{50} and 74% inhibition of IBAT activity at 0.156 μmol/kg dosage during *in vivo* mouse model studies.[13]

5. Pharmacokinetics and Drug Metabolism

Due to the extensive development efforts to ensure intestinal localization of odevixibat (**1**), the drug has very minimal systemic exposure.[36] Up to 83% of the initial dose is excreted in feces. Doses of anywhere from 40 up to 120 μg/kg have no detectable plasma concentrations in most patients. Following a 7.2 mg dosage, mean C_{max} was determined to be 0.47 ng/mL and mean $AUC_{0-24\,h}$ was 2.19 ng*h/mL. Additionally, odevixibat (**1**) is >99% protein bound and 97% of the excreted drug remains unmetabolized with mono-hydroxylation being the only detectable metabolite during *in vitro* studies. Its mean half-life after a single dose is 2.5 h.

It is also important to note that in patients with ABCB11 gene variants that result in dysfunction or absence of bile salt export pump protein (BSEP), odevixibat (**1**) may not be an effective treatment for PFIC.

6. Efficacy and Safety

During development, odevixibat (**1**), formulated as the sesquihydrate, performed well across several bouts of *in vitro* and *in vivo* mouse model studies. Odebixibat (**1**) was not found to inhibit any CYP isoforms, P-gp, BCRP, OCT2, MATE1, or MATE2K *in vitro*.[37] It also has no notable affinity for other bile acid transporters as IBAT is the only uptake transporter for bile acids in the intestine. While the Ostα-Ostβ, NTCP, and BSEP transporters all recognize bile acids, they occur further downstream in enterohepatic circulation and odevixibat (**1**) first inhibits IBAT prior to traveling further into enterohepatic circulation.

Translating well into clinical trials, the safety and dosage or odevixibat (**1**) was first assessed. Phase 1 studies enrolled 62 patients with PFIC type 1 or 2.[37] Following once daily dosages of 40 μg/kg odevixibat (**1**), a mean 22.2% of patients reported significant reduction of pruritus over the 24-week trial over the placebo group. At dosages of 120 μg/kg, 16.9% of patients reported significant improvement in pruritus relative to placebo. The most reported adverse effects included diarrhea, elevation of transaminase levels, serum bilirubin elevation, and fat-soluble vitamin deficiency. No major adverse effects were reported in phase 1 studies.

An exploratory phase 2 trial included administration of 10–200 μg/kg daily doses to 20 patients.[38] Similar endpoints were reported, as in phase 1 trials. Primary adverse events included liver test abnormalities such as elevated ALT, AST, as well as direct and total bilirubin levels. Four patients re-enrolled in a second-dose level along with seven others. Serum bile acid levels were reportedly reduced in all cohorts. Reductions in pruritus of 43% to 98% for 11 of the 13 enrolled patients were reported, providing further support for odevixibat (**1**)'s efficacy *in vivo*.

Rapid phase 3 trials[39] found a similar reduction in pruritus and greater than 70% reduction in serum bile acid levels in 33% of patients over a 24-week period. Following these results, odevixibat (**1**) was granted its first FDA approval for treating pruritus in PFIC patients ages 3 months and up by the FDA in 2021. Similar clinical evaluations took place to assess efficacy and safety in patients with ALGS in 2023.[40] Leading directly into a second FDA approval for pruritus in patients 12 months and up with ALGS, odevixibat (**1**) became the first approved IBAT inhibitor for PFIC and the second for ALGS, following maralixibat (**8**).

7. Synthesis

The synthesis of odevixibat (**1**) occurs through an 11-step linear synthesis[41, 42] beginning with functionalization and cyclization of the core benzothiadiazepine ring system. Following, peripheral functionalization to carry out methanethiol substitution and sequential amide coupling steps to form the chiral dipeptide chain completes the synthesis to give odevixibat (**1**) in straightforward fashion.

Starting from commercially available thiol **28**, bromination and disulfide formation gives dibromide **29** in 19% yield over two steps. Following oxidative chlorination of **29**, the resulting sulfonyl chloride **30** is prepped for a two-step cyclization to form the core thiadiazepine motif.

Selective addition of the primary amine of **31** into sulfonyl chloride **30** gives the tethered sulfonamide **32**. The difficulty in this step likely originates from attempting addition of a diamine substrate into the electrophilic system of **30** which can potentially react at three different places, through the sulfonyl chloride or S_NAr at either bromide. Despite the complexity of this step, sulfonamide **32** can still be isolated to proceed with the synthesis.

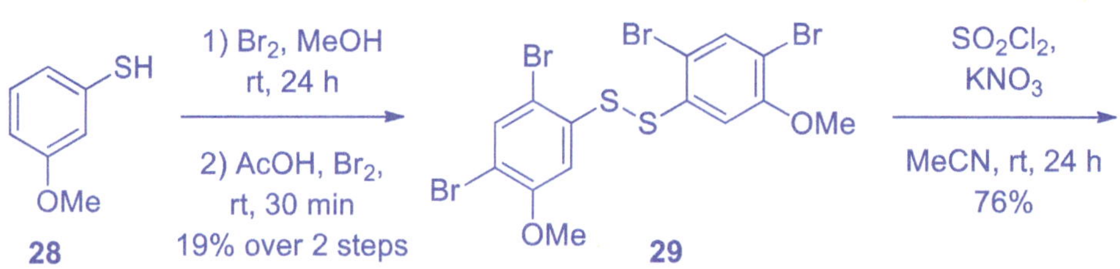

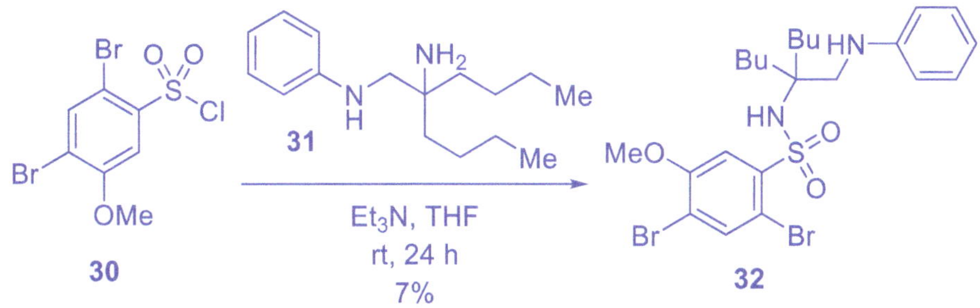

The intramolecular Ullmann-type coupling of **32** proceeds in a moderate 52% yield to give cyclized benzothiadiazepine **33**. With the core of the molecule complete, only peripheral modification and amending of the peptide chain are left. The most difficult steps are out of the way with isolation of the tetrabrominated disulfide **29** from the first step and selective addition with the sterically hindered primary amine **31**.

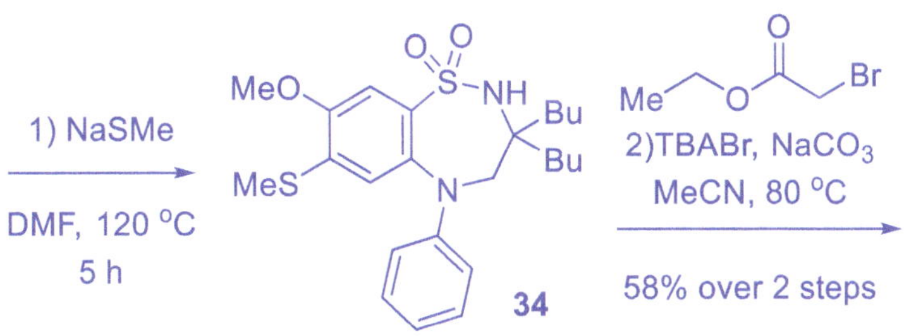

To continue functionalizing the periphery, the bromide of **33** undergoes S_NAr reaction with sodium methanethiolate to give compound **34**. The crude mixture of **34** is then resuspended in acetonitrile to complete the coupling with ethyl bromoacetate to give ethyl ester **35** in 58% yield over two steps. Saponification of **35** gives free acid **36**.

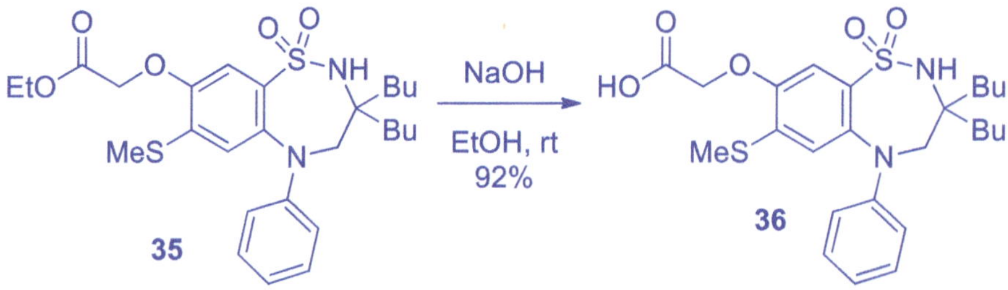

Carboxylic acid **36** can then efficiently couple with amino acid **37** under TBTU-catalyzed conditions to give *t*-butyl ester **38**. A second TBTU-catalyzed coupling of ester **38** with amino acid **39** and a TFA deprotection of the resulting *t*-butyl protected dipeptide gives odevixibat (**1**) in a final yield of 66% over the last two steps.

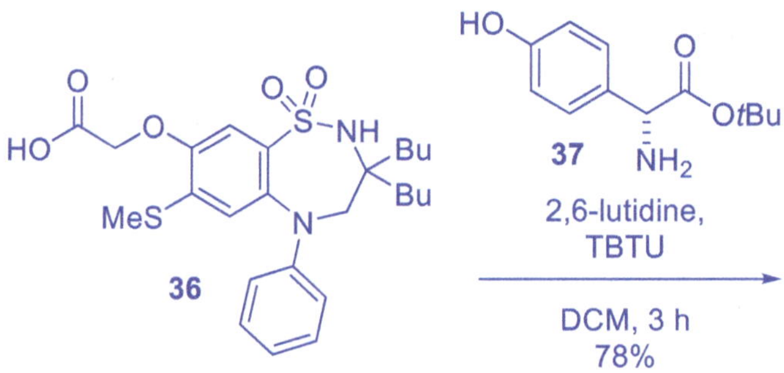

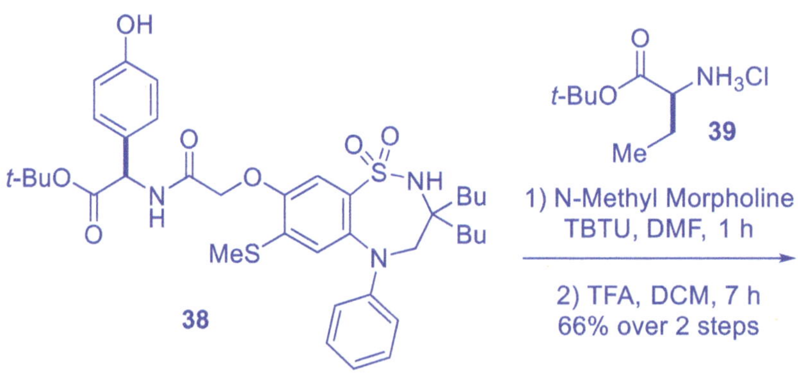

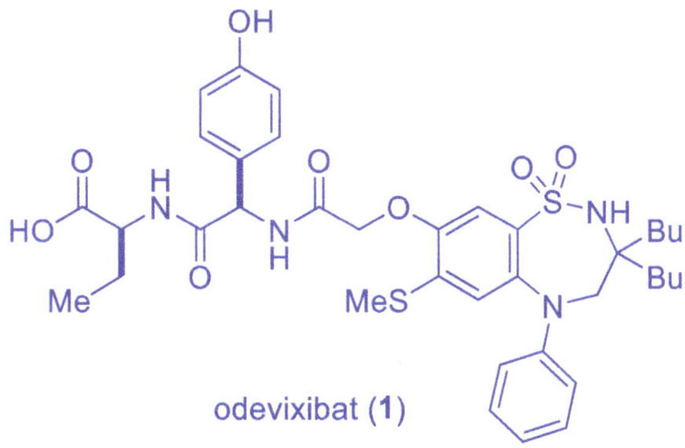

odevixibat (1)

While all the steps in odevixibat's (**1**) synthesis appear straightforward at first glance, the difficulty arises when considering isolation of a single desired product from an array of competing products and isomers in the early steps. Through careful setup and cyclization of the core ring system in five concise steps, the peripheral functionalities can easily be amended to complete the synthesis in a facile manner.

8. Summary

The development process of the ileal bile acid transporter odevixibat (**1**), was an arduous but inspiring one. Through the effort of many groups, the discovery of a simple but effective benzothiazepine scaffold decades ago was able to develop into something much more fruitful. The rigorous efforts made to expand our understanding of bile acid function and transport in the body have made the development of such highly selective and potent inhibitors possible.

Reflecting on the milestones and achievements in odevixibat (**1**)'s development process, it is important to highlight the many improvements and iteration cycles made along the way. Starting from the discovery of the scaffold, derived from compound 2164U90 (**12**), early studies on SAR and chirality made significant progress toward validating benzothiazepine-based drugs for bile acid-related diseases. Later PK studies using ammonium salts, carboxylates, and peptide chains to target intestinal localization helped further pave the road toward the clinical successes witnessed later. Together, these achievements brought forth a first-in-class treatment option for two rare and debilitating genetic disorders, PFIC and ALGS.

Odevixibat (**1**), being the first IBAT inhibitor to reach FDA approval in 2021 along with its second approval in 2023, paved significant ground in the landscape of medicinal chemistry. As such, the success of Albireo's IBAT inhibitor program has done much to spur the development of new projects in cholestasis research, with many more, sure to follow.

References

1. Patel, S. P.; Vasavda, C.; Ho, B.; Meixiong, J.; Dong, X.; Kwatra, S. G. Cholestatic pruritus: emerging mechanisms and therapeutics. *J. Am. Acad. Dermatol.* **2019**, *81* (6), 1371–1378.
2. Kronsten, V.; Fitzpatrick, E.; Baker, A. Management of cholestatic pruritus in paediatric patients with alagille syndrome: The King's College Hospital experience. *J. Pediatr. Gastroenterol. Nutr.* **2013**, *57* (2), 149–154.
3. Tandon, P.; Rowe, B. H.; Vandermeer, B.; Bain, V. G. The efficacy and safety of bile acid binding agents, opioid antagonists, or rifampin in the treatment of cholestasis-associated pruritus. *Am. J. Gastroenterol.* **2007**, *102* (7), 1528–1536.
4. Bachs, L.; Parés, A.; Elena, M.; Piera, C.; Rodés, J. Effects of long-term rifampicin administration in primary biliary cirrhosis. *Gastroenterology* **1992**, *102* (6), 2077–2080.
5. McRae, C. A.; Prince, M.; Hudson, M.; Day, C. P.; James, O.; Jones, D. Pain as a complication of use of opiate antagonists for symptom control in cholestasis. *Gastroenterology* **2003**, *125* (2), 591–596.
6. Mosińska, P.; Fichna, J.; Storr, M. Inhibition of ileal bile acid transporter: an emerging therapeutic strategy for chronic idiopathic constipation. World J. Gastroenterol. **2015**, *21* (24), 7436–7442.
7. Acosta, A.; Camilleri, M. Elobixibat and its potential role in chronic idiopathic constipation. *Therap. Adv. Gastroenterol.* **2014**, *7* (4), 167–175.
8. Hegade, V. S.; Kendrick, S. F.; Jones, D. E. Drug treatment of pruritus in liver diseases. *Clin. Med.* **2015**, *15* (4), 351–357.
9. Al-Dury, S.; Marschall, H.-U. Ileal bile acid transporter inhibition for the treatment of chronic constipation, cholestatic pruritus, and NASH. *Front. Pharmacol.* **2018**, *9*, 387830.
10. Yuan, S.; Wang, D.-S.; Liu, H.; Zhang, S.-N.; Yang, W.-G.; Lv, M.; Zhou, Y.-X.; Zhang, S.-Y.; Song, J.; Liu, H.-M. New drug approvals for 2021: synthesis and clinical applications. *Eur. J. Med. Chem.* **2023**, *245*, 114898.

11. Efficacy and Safety Trial of Elobixibat in Patients with Chronic Idiopathic Constipation, Identifier NCT01833065, 2013.

12. 26 Week Efficacy and Safety Trial for Patients with Chronic Idiopathic Constipation, Identifier NCT01827592, 2015.

13. Gilberg, P.; Graffner, H.; Starke, I. IBAT Inhibitors for the Treatment of Liver Disease, US Patent Appl. US 9,694,018 B1, Jul 4, 2017.

14. US FDA approves Bylvay for patients living with cholestatic pruritus due to Alagille syndrome: https://www.ipsen.com/press-releases/u-s-fda-approves-bylvay-for-patients-living-with-cholestatic-pruritus-due-to-alagille-syndrome/. Accessed Dec. 15, 2023.

15. Harris, M. J.; Le Couteur, D. G.; Arias, I. M. Progressive familial intrahepatic cholestasis: genetic disorders of biliary transporters. *J. Gastroenterol. Hepatol.* **2005**, *20* (6), 807–817.

16. Turnpenny, P. D.; Ellard, S. Alagille syndrome: pathogenesis, diagnosis and management. *Eur. J. Hum. Genet.* **2011**, *20* (3), 251–257.

17. Samant, H.; Manatsathit, W.; Dies, D.; Shokouh-Amiri, H.; Zibari, G.; Boktor, M.; Alexander, J. S. Cholestatic liver diseases: an era of emerging therapies. *World J. Clin. Cases*. **2019**, *7* (13), 1571–1581.

18. Hofmann, A. F. The continuing importance of bile acids in liver and intestinal disease. *Arch. Intern. Med.* **1999**, *159* (22), 2647–2658.

19. Chothe, P. P.; Czuba, L. C.; Moore, R. H.; Swaan, P. W. Human bile acid transporter ASBT (SLC10A2) forms functional non-covalent homodimers and higher order oligomers. *Biochim. Biophys. Acta* **2018**, *1860* (3), 645–653.

20. Davit-Spraul, A.; Gonzales, E.; Baussan, C.; Jacquemin, E. Progressive familial intrahepatic cholestasis. *Orphanet J. Rare Dis.* **2009**, *4* (1), 1.

21. Prussak, E. Medical and dental management of alagille syndrome: a review. *Med. Sci. Monit.* **2014**, *20*, 476–480.

22. Vinayagamoorthy, V.; Srivastava, A.; Sarma, M. S. Newer variants of progressive familial intrahepatic cholestasis. *World J. Hepatol.* **2021**, *13* (12), 2024–2038.

23. Lemaigre, F. P. Notch signaling in bile duct development: new insights raise new questions. *Hepatology* **2008**, *48* (2), 358–360.

24. Alrefai, W. A.; Gill, R. K. Bile acid transporters: structure, function, regulation and pathophysiological implications. *Pharm. Res.* **2007**, *24* (10), 1803–1823.

25. Hu, N.-J.; Iwata, S.; Cameron, A. D.; Drew, D. Crystal structure of a bacterial homologue of the bile acid sodium symporter ASBT. *Nature* **2011**, *478* (7369), 408–411.

26. Li, M.; Wang, Q.; Li, Y.; Cao, S.; Zhang, Y.; Wang, Z.; Liu, G.; Li, J.; Gu, B. Apical sodium-dependent bile acid transporter, drug target for bile acid related diseases and delivery target for prodrugs: current and future challenges. *Pharmacol. Ther.* **2020**, *212*, 107539.

27. Zhang, E. Y.; Phelps, M. A.; Banerjee, A.; Khantwal, C. M.; Chang, C.; Helsper, F.; Swaan, P. W. Topology scanning and putative three-dimensional structure of the extracellular binding domains of the apical sodium-dependent bile acid transporter (SLC10A2). *Biochemistry* **2004**, *43* (36), 11380–11392.

28. Bertolini, A.; Fiorotto, R.; Strazzabosco, M. Bile acids and their receptors: modulators and therapeutic targets in liver inflammation. *Semin. Immunopathol.* **2022**, *44* (4), 547–564.

29. Lewis, M. C.; Brieaddy, L. E.; Root, C. Effects of 2164U90 on ileal bile acid absorption and serum cholesterol in rats and mice. *J. Lipid Res.* **1995**, *36* (5), 1098–1105.

30. Root, C.; Smith, C. D.; Winegar, D. A.; Brieaddy, L. E.; Lewis, M. C. Inhibition of ileal sodium-dependent bile acid transport by 2164U90. *J. Lipid Res.* **1995**, *36* (5), 1106–1115.

31. Tremont, S. J.; Lee, L. F.; Huang, H.-C.; Keller, B. T.; Banerjee, S. C.; Both, S. R.; Carpenter, A. J.; Wang, C.-C.; Garland, D. J.; Huang, W.; et al. Discovery of potent, nonsystemic apical sodium-codependent bile acid transporter inhibitors (Part 1). *J. Med. Chem.* **2005**, *48* (18), 5837–5852.

32. Huang, H.-C.; Tremont, S. J.; Lee, L. F.; Keller, B. T.; Carpenter, A. J.; Wang, C.-C.; Banerjee, S. C.; Both, S. R.; Fletcher, T.; Garland, D. J.; et al. Discovery of potent, nonsystemic apical sodium-codependent bile acid transporter inhibitors (Part 2). *J. Med. Chem.* **2005**, *48* (18), 5853–5868.

33. Starke, I.; Dahlstrom, M.; Blomberg, D. 1,5 Benzothiazepines and their use as antihyperlipidemics, Int. Patent Appl. WO 02/50051 A1, Jun 27, **2002**.

34. Starke, I.; Dahlstrom M.; Alenfalk, S.; Skjaret, T.; Lemurell, M. Peptides derivatives comprising thiazepine group for the treatment of hyperlipidemic conditions, Int. Patent Appl. WO 03/106482 A1, Dec 24, **2003**.

35. Abrahamsson, H. R.; Gillberg, P. G. Use of an IBAT Inhibitor for the Treatment of Constipation, US Patent Appl. US 7,514,421 B2, Apr 7, **2009**.

36. Deeks, E. D. Odevixibat: first approval. *Drugs* **2021**, *81* (15), 1781–1786.

37. FDA Approval: Bylvay (odevixibat) capsules, for oral use: https://www.accessdata.fda.gov/drugsatfda_docs/label/2021/215498s000lbl.pdf Accessed Dec. 15, **2023**.

38. Baumann, U.; Sturm, E.; Lacaille, F.; Gonzalès, E.; Arnell, H.; Fischler, B.; Jørgensen, M. H.; Thompson, R. J.; Mattsson, J. P.; Ekelund, M.; Lindström, E. Effects of odevixibat on pruritus and bile acids in children with cholestatic liver disease: phase 2 study. *Clin. Res. Hepatol. Gastroenterol.* **2021**, *45* (5), 101751.

39. Thompson, R.; Arnell, H., Artan, R.; Baumann, U.; Calvo, P. L.; Czubkowski, P.; Dalgic, B.; D'Antiga, L.; Durmaz, Ö.; Fischler, B.; Gonzalès, E. Odevixibat treatment in progressive familial intrahepatic cholestasis: a randomized, placebo-controlled, phase 3 study. *Lancet Gastroenterol. Hepatol.* **2022**, 7, 830–42.

40. Efficacy and Safety of odevixibat in Patients with Alagille Syndrome (ASSERT), Identifier NCT04674761, 2023.

41. Handlon, A.; Hodgson, G.; Hyman, C. Hypolipidemic bicyclic derivatives, Int. Patent Appl. WO 98/38182, Sep. 3, 1998.

42. Starke, I.; Dahlstrom M.; Blomberg, D.; Alenfalk, S.; Skjaret, T.; Lemurell, M. Benzothiazepine and benzothiadiazepine derivatives with ileal bile acid transport (IBAT) inhibitory activity for the treatment hyperlipidaemia, Int. Patent Appl. WO 03/022286 A1, Mar. 20, 2003.

Index

Chemistry and Pharmacology of Drug Discovery, First Edition. Edited by Jie Jack Li.

Printed in the USA/Agawam, MA
December 9, 2024

878526.031